All About Hops: The Crop, its Cultivation, and its Uses in Brewing and Beyond

Katya Carbone
Editor

All About Hops: The Crop, its Cultivation, and its Uses in Brewing and Beyond

Springer

Editor
Katya Carbone
CREA – Research Centre for Olive, Fruit and Citrus Crops
Rome, Italy

ISBN 978-3-031-96605-7 ISBN 978-3-031-96606-4 (eBook)
https://doi.org/10.1007/978-3-031-96606-4

This Springer imprint is published by the registered company Springer Nature Switzerland AG
The registered company address is: Gewerbestrasse 11, 6330 Cham, Switzerland

Dedicated to my incredible staff

Dear colleagues,
the completion of this book would not have been possible without your hard work, dedication, and passion. Just as hops contribute balance, character, and depth to a brew, each of you adds something invaluable to our team. Your commitment to excellence, your relentless pursuit of quality, and your ability to grow even in tough conditions are an inspiration to me every day.

Preface

Hops (*Humulus lupulus* L.) are considered an industrial crop in numerous countries. However, due to the success of craft beer, which has experienced a period of significant growth from the latter half of the 1990s to the present, it is now regarded as an alternative crop in countries not traditionally dedicated to brewing.

Hops represent a subject of considerable interest and possess significance beyond their role as an ingredient in beer production. It is, in fact, a plant that weaves history, science, and culture into a compelling narrative across the centuries.

Ethnobotanical research has elucidated the therapeutic use of the plant since ancient times, and presently, many of its secondary metabolites are being investigated for their biological properties (e.g., antibacterial, antiviral, antioxidant, anticancer, etc.).

This volume undertakes a comprehensive examination and elucidation of the cultivation, botanical and ethnobotanical characteristics, medicinal applications, composition, food and non-food uses, and pharmacological potential of hops. It was born out of a desire to explore hops in all their facets, going beyond their immediate association with craft beer, providing a tribute to a versatile and fascinating plant whose influence is not limited to the food and beverage industries.

The intention is to provide a comprehensive contribution to the scientific community involved in food sciences, horticulture, clinical nutrition, health, and pharmacology, systematically reviewing the aspects that have led to recent advances in hops biochemistry, production, and functionality.

The editor hopes that this volume will be a valuable resource for researchers, students, and developers of related disciplines, and expresses sincere gratitude to all the authors for their excellent contributions and collaboration, and to the staff of Springer Nature for their assistance and support.

Rome, Italy
26 February 2025

Katya Carbone

Contents

Editor and Contributors

About the Editor

Katya Carbone is a Senior Scientific Researcher at CREA, Research Centre for Olive, Fruit and Citrus Crops (CREA OFA). She holds a degree in Industrial Chemistry (*summa cum laude*), a PhD in Applied Biotechnology, and a PhD in Food Biotechnology. She is currently Head of the Food Chemistry and Biotechnology Laboratory at CREA OFA. Her research focuses largely on hops and medicinal plants as well as on food quality and circular economy applied to the recovery and enhancement of agro-food waste and by-products. She has been involved in national and European-funded projects on hops and has collaborated with universities and industries to advance hop science. She is also active in scientific dissemination, giving invited talks, publishing research articles, and contributing to policy discussions on the development of the Italian hop supply chain.

Contributors

Elisabetta Bazzaro Department of Physiology and Pharmacology "V. Erspamer", Sapienza University of Rome, Rome, Italy

Katya Carbone CREA – Research Centre for Olive, Fruit and Citrus Crops, Rome, Italy

Barbara Čeh Department of Plants, Soil and Environment, Slovenian Institute of Hop Research and Brewing, Žalec, Slovenia

Erika Celi CREA – Research Centre for Olive, Fruit and Citrus Crops, Rome, Italy

Andreja Čerenak Department of Plants, Soil and Environment, Slovenian Institute of Hop Research and Brewing, Žalec, Slovenia

Federica Cisilino CREA – Research Centre for Agricultural Policies and Bio-Economy, Rome, Italy

Alessia D'Andrea CREA – Research Centre for Olive, Fruit and Citrus Crops, Rome, Italy

Silvia Di Giacomo Italian National Institute of Health – Unit of Human Nutrition and Health, Department of Food Safety, Nutrition and Veterinary Public Health, Rome, Italy

Antonella Di Sotto Department of Physiology and Pharmacology "V. Erspamer", Sapienza University of Rome, Rome, Italy

Marjeta Eržen Department of Plants, Soil and Environment, Slovenian Institute of Hop Research and Brewing, Žalec, Slovenia

Martina Galaverni Department of Food and Drug, University of Parma, Parma, Italy

Tommaso Ganino Department of Food and Drug, University of Parma, Parma, Italy

Institute of BioEconomy (IBE), National Council of Research – CNR, Sesto Fiorentino, Florence, Italy

David H. Gent Agricultural Research Service, United States Department of Agriculture, Corvallis, OR, USA

Forage Seed and Cereal Research Unit, United States Department of Agriculture–Agricultural Research Service, Corvallis, OR, USA

Scott J. Harper Department of Plant Pathology, Washington State University, Prosser, WA, USA

Jernej Jakše Department of Agronomy, Biotechnical Faculty, University of Ljubljana, Ljubljana, Slovenia

Ana Karničnik Klančnik Department of Plants, Soil and Environment, Slovenian Institute of Hop Research and Brewing, Žalec, Slovenia

Octave Lacroix Plant Protection Department, Slovenian Institute of Hop Research and Brewing, Žalec, Slovenia

Francesco Licciardo CREA – Research Centre for Agricultural Policies and Bio-Economy, Rome, Italy

Lucija Luskar Department of Plants, Soil and Environment, Slovenian Institute of Hop Research and Brewing, Žalec, Slovenia

Dario Macaluso CREA – Research Centre for Agricultural Policies and Bio-Economy, Rome, Italy

Ilaria Marchioni Department of Food and Drug, University of Parma, Parma, Italy

Timothy D. Miles Department of Plant, Soil and Microbial Sciences, Michigan State University, East Lansing, MI, USA

Boštjan Naglič Department of Plants, Soil and Environment, Slovenian Institute of Hop Research and Brewing, Žalec, Slovenia

Ester Percaccio Department of Physiology and Pharmacology "V. Erspamer", Sapienza University of Rome, Rome, Italy

Julija Polanšek Agricultural Institute of Slovenia, Ljubljana, Slovenia

Franček Poličnik Plant Protection Department, Slovenian Institute of Hop Research and Brewing, Žalec, Slovenia

Magda Rak Cizej Plant Protection Department, Slovenian Institute of Hop Research and Brewing, Žalec, Slovenia

Sebastjan Radišek Plant Protection Department, Slovenian Institute of Hop Research and Brewing, Žalec, Slovenia

Margherita Rodolfi Department of Food and Drug, University of Parma, Parma, Italy

Veronica Santarelli Department of Bioscience and Technology for Food, Agriculture and Environment, University of Teramo, Teramo, Italy

Siniša Srečec Križevci University of Applied Sciences, Križevci, Croatia

Annabella Vitalone Department of Physiology and Pharmacology "V. Erspamer", Sapienza University of Rome, Rome, Italy

Inga Zasada Agricultural Research Service, United States Department of Agriculture, Corvallis, OR, USA

Chapter 1
A Spotlight on the Economic and Statistical Data of the International and Italian Hop Sector

Francesco Licciardo, Dario Macaluso, Federica Cisilino, and Katya Carbone

1.1 Introduction

Hop cultivation has historically been concentrated in the humid temperate zones of the northern hemisphere, with Germany and the United States being the largest producers, accounting for the vast majority of the cultivated area and global volume of hops. However, in recent years, the cultivation of this plant has also entered the agricultural landscape of countries that do not have a brewing tradition, representing niche markets with interesting growth potential.

In Italy, the craft beer movement emerged in the early 1990s and has continued to grow steadily ever since, becoming a strong magnet for new entrepreneurs (particularly young people and women). Over the years, this phenomenon has been supported by a series of regulatory measures aimed at fostering the development and consolidation of an Italian product with a strong local identity. (Carbone et al., 2021; Van Holle et al., 2021). In this regard, public policies have played a crucial role in shaping the sector, encouraging innovation, supporting small-scale production, and enhancing the value of local raw materials and transitions. Law no. 154 of 28 July 2016, commonly referred to as the "agricultural annex" (Tarangioli et al., 2020), introduced for the first time a legal definition of craft beer in Italy. The law also tasked the Ministry of Agriculture, Food Sovereignty and Forestry with supporting the development and promotion of raw materials, particularly hops, through targeted initiatives and funding for research and process improvement. The legislation introduced was meant to be a driving force for the development of hop cultivation in Italy, a sector that currently represents a small niche, but which has

F. Licciardo (✉) · D. Macaluso · F. Cisilino
CREA – Research Centre for Agricultural Policies and Bio-Economy, Rome, Italy
e-mail: francesco.licciardo@crea.gov.it

K. Carbone
CREA – Research Centre for Olive, Fruit and Citrus Crops, Rome, Italy

K. Carbone (ed.), *All About Hops: The Crop, its Cultivation, and its Uses in Brewing and Beyond*, https://doi.org/10.1007/978-3-031-96606-4_1

considerable potential for development, partly because of the high adaptability of the species, its broad distribution throughout the country and its dual aptitude.

To fully evolve from a niche phenomenon into a structured agricultural sector, it is essential to gather comprehensive information on both production characteristics and business structures. Such data is crucial to effectively design and implement development and support policies tailored to the sector's needs.

In light of these considerations, the current chapter aims to provide a comprehensive overview of the international and national hop sectors from a statistical and economic perspective, establishing a foundation for understanding trend dynamics across different scales. To achieve this goal, the study explores both global and local dynamics of the brewing sector, drawing on a wide range of data sources beyond traditional statistics. The analysis is based on the most recent data available at the time of writing, providing stakeholders—particularly institutions—with practical tools to design measures that support and strengthen hop cultivation in Italy.

The chapter is organised as follows: it begins with an assessment of the potential and limitations of various statistical sources (Sect. 1.1.1). Then the global context (Sect. 1.1.2.1) and the European context (Sect. 1.1.2.2) are analysed, focusing on production areas and their development over time. This analysis utilises data from FAO and EUROSTAT, with the former updated to 2023 and the latter to 2022. These official statistics are supplemented with insights from unofficial sources, including data from the Barth-Haas Group, the world's leading provider of services for the hop industry.

Data from the International Hop Growers' Convention (IHGC) has been essential in analysing current trends in major hop-growing countries as of 2024 and in providing preliminary forecasts for 2025 (Sect. 1.1.3). Finally, Sect. 1.1.4 presents a detailed analysis of the Italian brewing industry over the past decade. This section exclusively refers to unofficial statistics from respected national and international organizations, allowing for an in-depth evaluation of medium-term trends.

All sources referenced are explicitly cited within the text, ensuring both transparency and reliability.

1.1.1 Challenges and Limitations of Utilizing Diverse Data Sources in Statistics

Statistics represent an indispensable tool for understanding, evaluating, and making decisions in the agri-food sector. Through the analysis of statistical data, it is possible to gain a clear insight into market trends, assess the effectiveness of policies, and develop informed and targeted strategies. The ability to use statistical data to make evidence-based decisions is crucial for the success and sustainability of the agri-food sector, allowing for the tackling of complex challenges and seizing of new opportunities. One of the main advantages of using statistics is the ability to transform raw data into useful information for decision-making purposes.

In this chapter, the use of statistical secondary data is crucial for analysing the hop and craft beer industry, as it provides important information for understanding market dynamics, production trends, and economic impacts. Secondary data or data collected from existing sources such as government institutions, international organisations, and research institutes are valuable resources for agribusiness analysis. They usually offer broader temporal and spatial coverage, allowing for comprehensive and detailed analyses.

Official sources used in this chapter such as FAO (FAOSTAT), the Italian National Institute of Statistics (ISTAT) and Eurostat (European Commission), and other sources such as The Brewers of Europe, AssoBirra, and the Italian Agricultural Payments Agency (AGEA) allow to track growth in production and yields by area, to understand which are the main producing countries, considering exports, consumption and employment. Such data is essential for evaluating both the hop cultivation and the industry's expansion. By leveraging statistical analyses, policymakers and stakeholders can develop strategies to enhance sector stability and competitiveness.

As noted by Watson et al. (2003) the reliability and comprehensive nature of secondary data make it indispensable for research; the value of secondary data is crucial in supporting evidence-based policy formulation and industry growth strategies. However, reliance on multiple data sources can present significant challenges, particularly concerning data consistency, compatibility, and reliability. One of the primary challenges of using secondary data from multiple sources is the risk of inconsistency and contradictions. Different organizations may employ varying methodologies, definitions, and data collection techniques, leading to discrepancies in the reported figures. For instance, the FAO might use global standards that do not align perfectly with the regional specifics captured by EUROSTAT or ISTAT. This inconsistency can complicate comparative analyses and hinder the development of coherent policies. Discrepancies in agricultural data often stem from differences in definitions and measurement techniques. For example, the definition of "agricultural area" can vary significantly between FAO and EUROSTAT, impacting the comparability of data on land use and crop production. Such discrepancies necessitate careful consideration and adjustment when analysing data from multiple sources. Temporal and spatial discrepancies are another significant challenge.

Data collection periods and geographical coverage can differ across organizations. FAO data might cover broader timeframes and geographical regions compared to more focused national datasets from ISTAT. These differences can lead to mismatches in temporal and spatial granularity, affecting trend analysis and regional studies. In the context of the hop and beer industry, these discrepancies can obscure important trends. For instance, FAO might report global hop production data, while ISTAT provides detailed regional statistics within Italy. Therefore, the interpretation of the data presented in this chapter must take this issue into account. Regarding the reconciling of different datasets, it would require sophisticated techniques to ensure temporal and spatial alignment. One effective solution to address data discrepancies is the harmonization of data standards across organizations. Establishing common definitions, methodologies, and reporting frameworks can significantly reduce inconsistencies. Initiatives like the Global Strategy to Improve Agricultural and Rural Statistics (GSARS) are already underway to harmonise agricultural data standards by developing standardised methodologies for agricultural statistics. Advanced data integration techniques can also help mitigate the challenges associated with using multiple data sources. Methods such as data fusion, statistical matching, and imputation can be employed to reconcile discrepancies and create cohesive datasets.

The use of secondary data from multiple sources is essential for comprehensive agricultural research and policymaking. However, the challenges of data inconsistency, temporal and spatial discrepancies, and methodological differences necessitate careful consideration and advanced techniques to ensure accurate and reliable analyses.

1.1.2 Situation of Hop Cultivation and Beer Production

1.1.2.1 World Scenario

According to the latest FAO estimates, the hop global area cultivated in 2023 was around 65,000 hectares, which is consistent with the average value for the period but shows a 1.5% decrease compared to the previous year. However, a dynamic analysis reveals a positive trend in the global area under hops over the observed period (Fig. 1.1).

It should be pointed out that the aggregate FAO data, referring to the macro-areas under consideration (Europe, Americas, Asia, Oceania), may include, in addition to official data, unofficial data, estimates and calculations. Data on the African

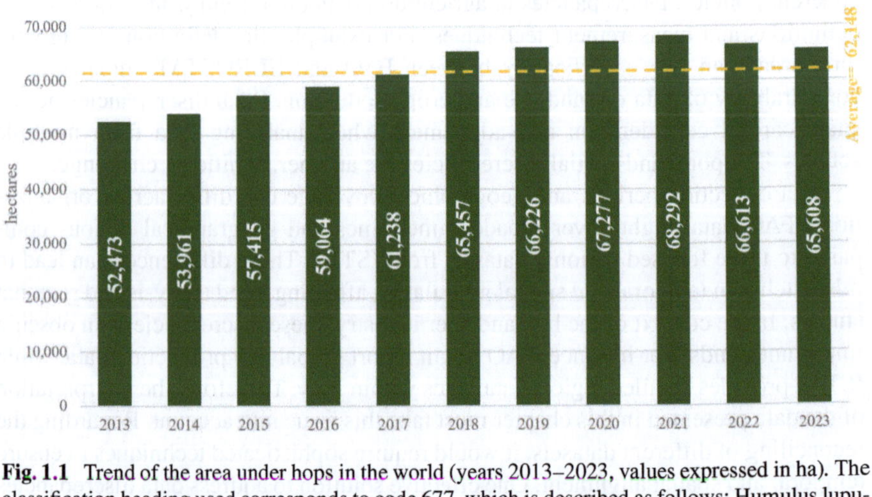

Fig. 1.1 Trend of the area under hops in the world (years 2013–2023, values expressed in ha). The classification heading used corresponds to code 677, which is described as follows: Humulus lupulus L., cones of hops, fresh or dried, whether or not ground, powdered, or in the form of pellets. (*Source*: our elaborations on FAO data (FAOSTAT consultation date 27 December 2024))

continent was omitted due to statistical inconsistencies (Licciardo et al., 2021). This may explain discrepancies with other sources.

The overall evolution was shaped by distinct trends across individual macro-regions. As shown in Fig. 1.2, there was a substantial stability in the production areas in Oceania and Asia between 2013 and 2023, while the growth of the cultivated area in the American continent followed an opposite trend (+53%). Europe is also on a growth trajectory, with a positive trend of +18%, although this is less intense than that of the Americas.

These trends can be examined more closely using indices based on a value of 100. An index value of 100 indicates that the situation has remained unchanged compared to the reference period (year 2000). Higher or lower values indicate an increase or decrease, respectively, compared to the reference year. As shown in Fig. 1.3, after a recessionary phase that—at the beginning of the last decade and particularly between 2011 and 2014—affected all production areas, albeit with different intensities, starting from 2015 the trend differed significantly. Specifically, there was a strong and continuous expansion of hop cultivation in the Americas, while the trend in other macro-regions was markedly different. In Asia and Oceania, a steady decline in cultivated areas was observed, whereas in Europe there was a brief recovery between 2017 and 2018 followed by a standstill in the most recent available year.

The latest FAO estimates for 2023 indicate a negative global trend (Table 1.1). Only the European continent and Oceania have shown growth in cultivated areas compared to the previous year, with increases of 3.6% and 2.1% respectively. However, in the case of Oceania, its relative contribution to the global hop area remains minimal at only 1%. Other regions of the world, notably the Americas, have experienced a decline in cultivated areas over the past year. Nevertheless, this did

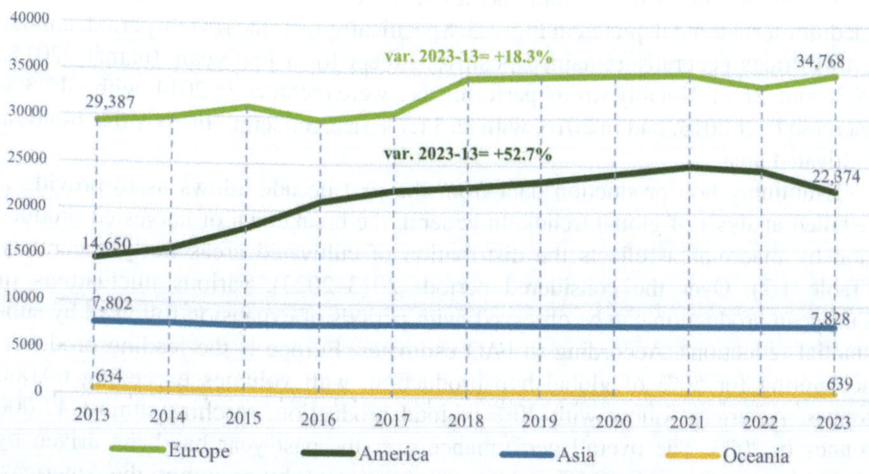

Fig. 1.2 Trend of the hop area per macro-area (years 2013–2023, values in ha). (*Source*: our elaborations on FAO data (FAOSTAT consultation date 27 December 2024))

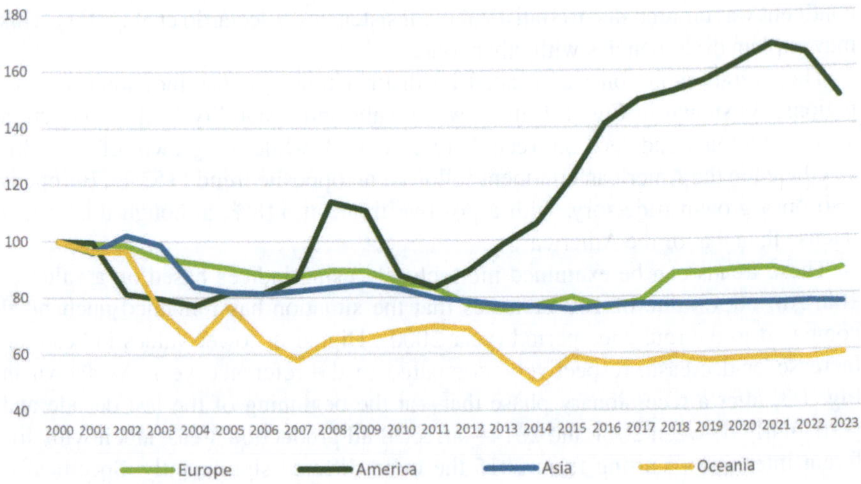

Fig. 1.3 Trend 2000–2023 of the area under hops per macro-area (index numbers base year 2000 = 100). (*Source*: our elaborations on FAO data (FAOSTAT consultation date 27 December 2024))

not affect the relative shares of the two main macro-regions: Europe (53%) and the Americas (34.1%).

In 2022, economic uncertainties, such as supply chain challenges and high production costs, hindered the growth potential of many global economic sectors. World hop production was similarly impacted by the economic downturn, resulting in an 18% decrease in output, equivalent to a reduction of 23,401 tonnes. Estimates for 2023 indicate a significant recovery in global production, which increased by approximately 10% within a year, exceeding 120,000 tonnes (Fig. 1.4).

The global production volume achieved in 2023 remained consistent with the medium-term trend depicted in Fig. 1.5. Specifically, over the review period, annual growth rates generally remained positive, except for a few years (namely 2015, 2020 and 2022). Notably strong performances were recorded in 2014, with a 15.3% increase over 2013, and in 2016, with an 11.7% rise, reflecting the positive trend in cultivated area.

Examining hop production data from the past decade allows us to provide a detailed analysis of global trends. In general, the breakdown of harvested production by macro-area reflects the distribution of cultivated areas fairly accurately (Table 1.2). Over the considered period (2013–2023), various fluctuations in European production can be observed, with periods of expansion followed by substantial reductions. According to FAO estimates, Europe is the leading producer, accounting for 50% of global hop production, with volumes exceeding 60,000 tonnes. Americas follow with 40% of total production, reaching almost 47,000 tonnes in 2023. The overall performance over the past year has been driven by Europe's growth rate (+18.5%) and, to a lesser extent, by growth in the Americas. Contributions from Asia and Oceania remain significantly lower. The American

Table 1.1 Area under hops per macro-area and share in total (2013–2023, values expressed in ha and in %)

	2013	2014	2015	2016	2017	2018	2019	2020	2021	2022	2023	Var. %		
												2021/20	2022/21	2023/22
Area under hops per macro-area (ha)														
Europe	29,387	29,917	31,072	29,703	30,546	34,338	34,521	34,714	34,817	33,563	34,768	0.3	−3.6	3.6
America	14,650	15,755	18,039	20,967	22,241	22,662	23,266	24,117	25,017	24,578	22,374	3.7	−1.8	−9.0
Asia	7,802	7,753	7,663	7,779	7,838	7,815	7,816	7,821	7,827	7,846	7,828	0.1	0.2	−0.2
Oceania	634	537	641	615	613	641	623	626	630	626	639	0.6	−0.6	2.1
Total	52,473	53,962	57,415	59,064	61,238	65,456	66,226	67,278	68,291	66,613	65,609	1.5	−2.5	−1.5
Incidence of area under hops per macro-area (%)														
Europe	56.0	55.4	54.1	50.3	49.9	52.5	52.1	51.6	51.0	50.4	53.0			
America	27.9	29.2	31.4	35.5	36.3	34.6	35.1	35.8	36.6	36.9	34.1			
Asia	14.9	14.4	13.3	13.2	12.8	11.9	11.8	11.6	11.5	11.8	11.9			
Oceania	1.2	1.0	1.1	1.0	1.0	1.0	0.9	0.9	0.9	0.9	1.0			
Total	100	100	100	100	100	100	100	100	100	100	100			

Source: our elaborations on FAO data (FAOSTAT consultation date 27 December 2024)

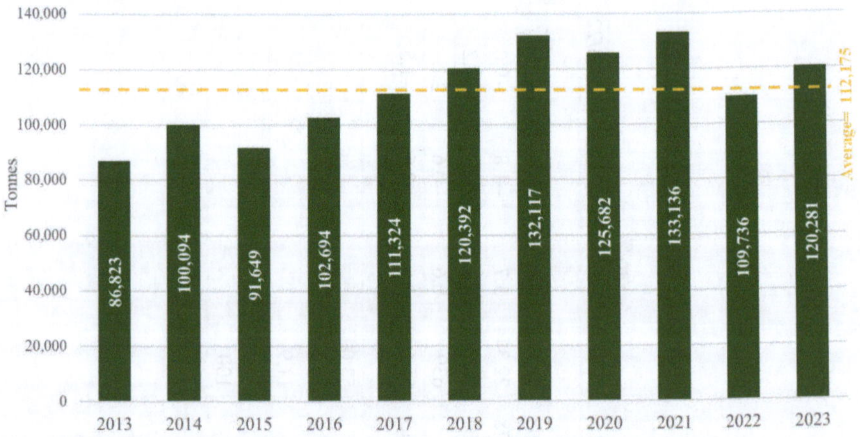

Fig. 1.4 Trend of hop production in the world (2013–2023, values expressed in tonnes). (*Source*: our elaborations on FAO data (FAOSTAT consultation date 27 December 2024))

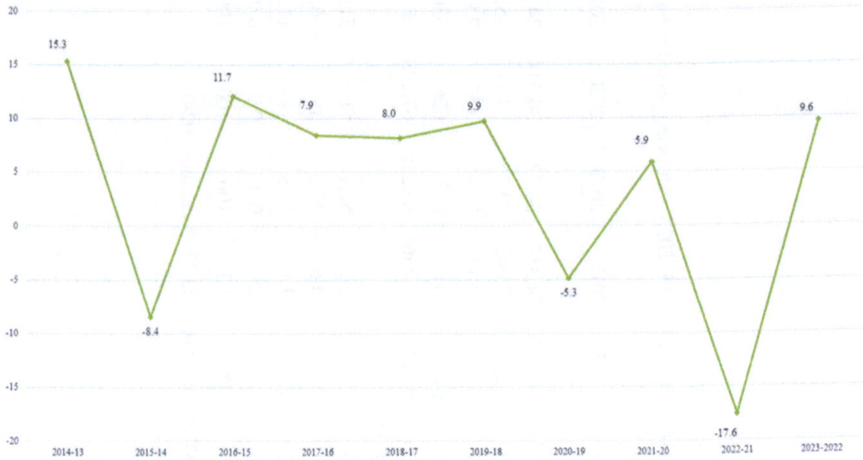

Fig. 1.5 Growth rates of hops production in the world (2013–2023, values expressed in %). (*Source*: our elaborations on FAO data (FAOSTAT consultation date 27 December 2024))

continent is clearly gradually increasing its global market share, often at the expense of Europe's historically dominant position.

The most recent trends focus on alpha-acid (AA) yield, which, in contrast to the decline observed in 2022 (−24%), showed a positive shift of 9.4% in 2023, in line with the increase in production (Fig. 1.6).

According to BarthHaas (2024), the proportion of bittering hops also rose in 2023, from 40.5% to 43.1%, while the share of aroma hops decreased from 59.5% in 2022 to 56.9% in 2023. This represents an increase of over 8,000 tonnes for bitter varieties (+19%), compared to approximately 4,200 tonnes for aroma varieties (+7%).

Table 1.2 Hop production by macro-area and share of the total (values in tonnes and in percentages for 2013–2023)

	2013	2014	2015	2016	2017	2018	2019	2020	2021	2022	2023	Var. %		
												2021/20	2022/21	2023/22
Hop production by macro-area (t)														
Europe	42,570	55,475	43,112	50,320	50,558	59,161	68,464	65,771	67,876	50,961	60,399	3.2	−24.9	18.5
America	31,861	32,656	36,229	39,997	48,388	48,970	51,293	47,564	52,925	46,416	47,673	11.3	−12.3	2.7
Asia	10,842	10,652	10,749	10,879	10,901	10,733	10,838	10,830	10,804	10,825	10,655	−0.2	0.2	−1.6
Oceania	1,550	1,311	1,559	1, 498	1,477	1,529	1,522	1,518	1,532	1,534	1,554	0.9	0.1	1.3
Total	86,823	100,094	91,649	102,694	111,324	120,392	132,117	125,682	133,136	109,736	120,281	5.9	−17.6	9.6
Incidence of area under hops per macro-area (%)														
Europe	49.0	55.4	47.0	49.0	45.4	49.1	51.8	52.3	51.0	46.4				
America	36.7	32.6	39.5	38.9	43.5	40.7	38.8	37.8	39.8	42.3				
Asia	12.5	10.6	11.7	10.6	9.8	8.9	8.2	8.6	8.1	9.9				
Oceania	1.8	1.3	1.7	1.5	1.3	1.3	1.2	1.2	1.2	1.4				
Total	100	100	100	100	100	100	100	100	100	100				

Source: our elaborations on FAO data (FAOSTAT consultation date 27 December 2024)

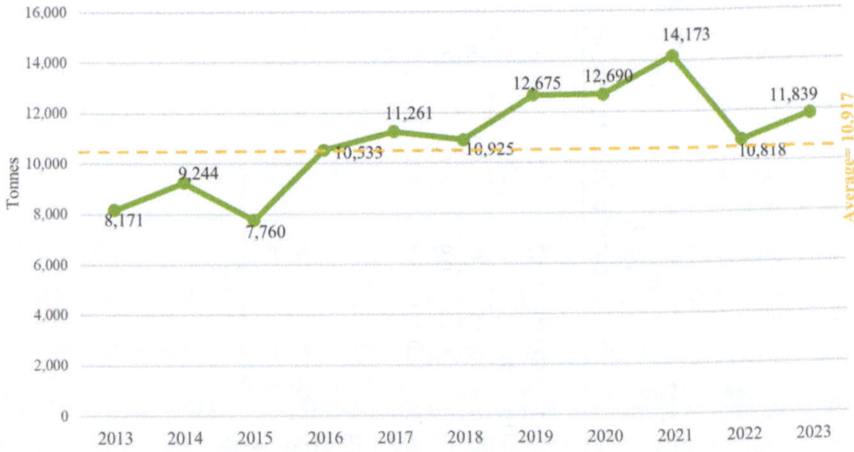

Fig. 1.6 Trend in hop production in alpha-acids (2013–2023, values in tonnes). (*Source*: our elaborations on BarthHaas Report (different years))

The growth in the share of bitter varieties can be attributed to contrasting trends in the two leading hop-producing countries, the United States and Germany. The global production of alpha hops in these countries has remained largely stable at 81%. However, within the aroma category, the United States experienced a decline in its share to 60% (2022: 70%), while Germany saw an increase to 33% (2022: 12%).

1.1.2.2 European Union Scenario

At the end of 2022 (European Commission—DG AGRI, 2023), there were 2,127 hop-growing farms in the EU-27, an increase of 1.9% compared to 2021 (Fig. 1.7), with a total growing area of around 30,030 hectares, representing 50% of the world's hop production area. The total quantity produced was about 44 thousand tonnes, with an average price between €5.48 and €8.78 per kilogram. It should be noted that these data are based on Member States' communications submitted by 15 April each year, as per Regulation (EC) No. 2017/1185. Not all Member States submit this communication, so the data may be underestimated compared to other statistical sources.

Hops are grown in more than 13 EU countries (Table 1.3). The Member States with the greatest vocation for hops are concentrated in central Europe, i.e. Germany, the Czech Republic and Slovenia. Another important producer is Poland (1,652 ha), while France, Spain, Romania and Austria have much smaller growing areas than the Central European countries, respectively 726, 580, 275 and 269 ha. Finally, the Netherlands have a much smaller production area (5 ha).

In general, the sudden climatic variations of recent years have had a significant impact on hop cultivation in EU countries. Periods of drought (2022–2023) have alternated with particularly rainy periods (2024), creating conditions that

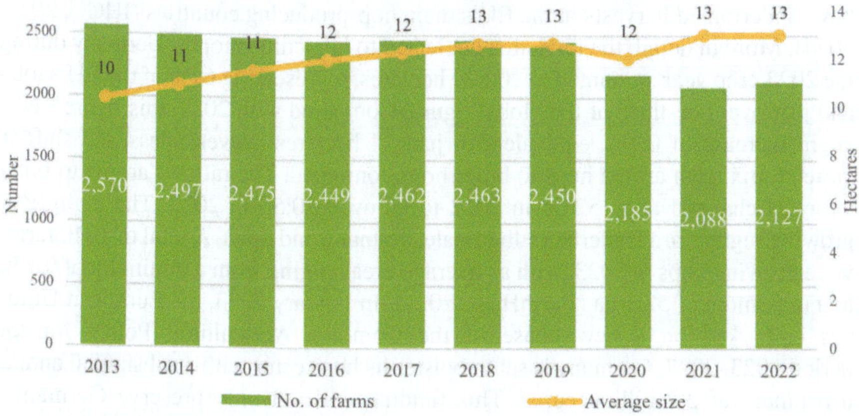

Fig. 1.7 Number of hop-growing farms and average size in leading EU-27 producing countries (2013–2022). Legend: The data from 2013 to 2019 also includes the UK. (*Source*: our elaboration on European Commission—DG AGRI, 2023)

Table 1.3 EU-27 acreage and crop (2023, values in ha, tonnes and in %)

	Acreage (ha)	Var. % 2022/21	Production (tonnes)	Var. % 2022/21	Ø Alpha (%)	Alpha (tonnes)	Var. % 2022/21
Germany	20,629	0.1	41,234	19.8	9.5	3,932	13.7
Czech Republic	4860	−1.7	6997	57.2	3.6	249	47.3
Slovenia	1676	3.1	2735	19.8	6.9	189	65.8
Poland	1652	−4.4	3505	2.3	9.2	321	2.9
France	726	7.9	940	22.2	4.9	46	58.6
Spain	580	−2.8	851	−19.6	10.8	92	−22.0
Romania	275	1.9	255	37.8	9.3	24	41.2
Austria	269	1.1	381	−15.3	7	27	−12.9
Belgium	185	0.0	309	27.7	9	28	75.0
Slovakia	38	5.6	20	−9.1	3.1	1	0
Bulgaria	33	−15.4	60	3.4	10.1	6	0
Italy[a]	20	–	20	–	5.8	1	–
Portugal	12	0	10	25.0	10	1	0
Netherlands	5	0	2	0	12.2	0	0
European Union	30,960	−0.1	57,319	21.0	8.6	4,917	15.1
Rest of Europe	1208	−8.6	1525	21.8	6.9	105	25.0
Europe	32,168	−0.4	58,843	21.0	8.5	5,022	14.8

Source: BarthHaas GmbH & Co. KG, 2024
[a]No data are available for Italy in 2022

have undermined harvests in the EU's main hop-producing countries (IHGC, 2023, 2024). More in detail, the total area dedicated to hop cultivation in Germany during the 2023 crop year amounted to 20,629 hectares, representing 67% of the EU's total and just over one-third of the global figure. Compared with 2022, this figure saw a slight increase of 0.1%, equivalent to just 24 hectares. Nevertheless, the shift in varietal mix from aroma hops to bitter hops continued. The ratio of aroma to bitter varieties changed from 52:48 in 2022 to an even 50:50 in 2023. The main hop-growing regions are Hallertau, Elbe-Saale, Tettnang and Spalt. A total of 1040 farms were growing hops in 2023, with an average area ranging from a minimum of 9.2 ha to a maximum of 52.1 ha (BarthHaas, 2024). In January 2023, the European Union has embarked on a new phase of the Common Agricultural Policy for the period 2023–2027. Germany is set to bolster its hop sector with a substantial annual investment of 2.2 billion euros. This funding will not only preserve Germany's global leadership in hop production but also promote more sustainable and potentially climate-neutral cultivation practices in future.

The Czech Republic is another significant hop producer. With a cultivated area of less than 5,000 hectares, it is the second-largest hop producer in the EU, accounting for 16% of the total, and the third-largest globally, contributing 8% to the world's total hop-growing area (6% in terms of production volume). Slovenia is the EU's third-largest hop producer, accounting for 5% of cultivated area and total production. Together, Germany, the Czech Republic, Slovenia, and Poland account for 46% of global hop production, highlighting the importance of hop cultivation in these countries.

As illustrated in Fig. 1.8, the overall trend in hop cultivation across the EU from 2013 to 2023 was marked by relative stability in terms of area under cultivation. In

Fig. 1.8 Evolution of the EU-27 hop area and quantities (2013–2023, values expressed in ha and tonnes). Legend: The data from 2013 to 2019 also includes the UK. (*Source*: our elaboration on European Commission - DG AGRI, 2023 and BarthHaas Report (different years))

contrast, production exhibited a steady and gradual upward trend from 2016 onwards, interrupted by a sharp decline in 2022. During that year, EU hop production plummeted to 47,360 tonnes, representing a 26.5% decrease compared with the previous year. However, the 2023 crop year signalled a robust recovery, with production reaching 57,319 tonnes and returning to medium-term average levels (BarthHaas, 2023, 2024).

During the review period, the average AA production was slightly above 5,000 tonnes. The strong growth in AA volumes seen from 2019 to 2021 came to an abrupt halt in 2022. This decline was primarily attributed to unfavourable cultivation conditions in Germany and neighbouring regions during the summer of 2022 (IHGC, 2023). The season was characterised by extreme heat, insufficient rainfall during the critical months of June to September, and severe hailstorms. These adverse weather conditions also impacted the Czech Republic and other nearby hop-growing regions, including France and Slovenia, resulting in yields and AA levels that were well below average.

The EU and global hop markets are also facing a significant surplus of AA due to several strong harvests in recent years, particularly in 2021 (IHGC, 2023). This surplus led to an oversupply in 2022, largely driven by increased cultivation of traditional aromatic varieties such as Perle and Tradition, as well as premium aromatic varieties like Saazer and Tettnanger.

Following the already poor harvest of 2022, growing conditions in Central Europe during the summer of 2023 were again characterised by numerous hot days and prolonged periods of insufficient rainfall. The regions of Hallertau, Spalt, and Tettnang, as well as the Czech Republic and France, were particularly affected by these dry spells, resulting in AA values that were notably below average. In contrast, the Elbe-Saale region, Slovenia and Poland benefited from sufficient rainfall in June and July, which was crucial for the harvest. Climatic conditions significantly influenced AA yields across most varieties. Nevertheless, overall volumes in 2023 were higher than those recorded in the previous year (Fig. 1.8).

1.1.3 Main Hop-Producing Countries in 2024: Projections for 2025

Germany and the USA's dominance in hop cultivation and production is clearly illustrated in Fig. 1.9, which presents data from the IHGC on the leading hop producing countries. Together these five countries account for almost 86% of the world's hop area and 89% of world production.

In particular, the USA and Germany are the key players on the international stage, together accounting for three-quarters of global production and 69% of the area projected for 2024.

The ongoing decline in demand for hops, a trend that has been observed over the past 2 years, has resulted in a further reduction of global hop-growing areas in 2024.

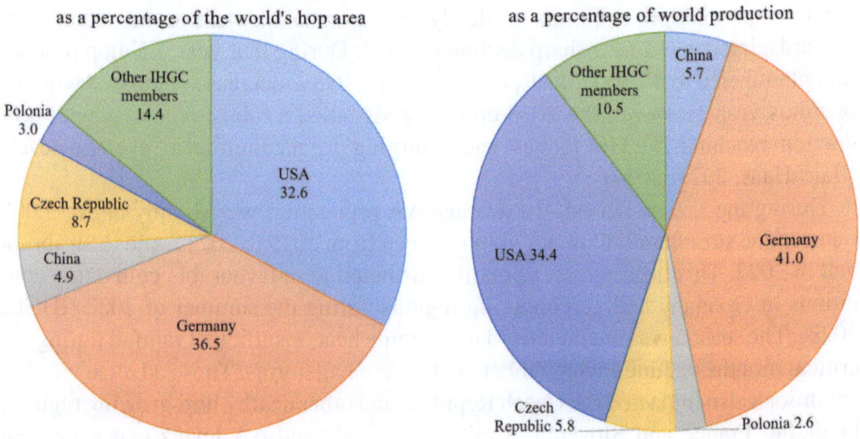

Fig. 1.9 Top five hop-producing countries (2024, values in %). (*Source*: IHGC, 2024)

This marks a reversal of the uninterrupted growth seen over the past decade, during which time the area dedicated to hop cultivation expanded by over 14,000 hectares. According to IHGC estimates (IHGC, 2024), the global hop area is expected to be 55,636 hectares (Table 1.3) for the 2024 crop year, representing a decline of over 4,000 hectares (−7%) compared to the previous year.

The latest projections estimate the 2024 global crop at 113,310 tonnes, with an alpha-acid production of 12,353 tonnes. This reflects a 4.3% decrease in harvested volumes compared to 2023, while alpha-acid production fell marginally by 0.5%. This situation is the result of contrasting trends in the two leading hop-producing nations, Germany and the USA.

In Germany, hop production increased by 13%, rising from 41,234 tonnes in 2023 to 46,497 tonnes in 2024, despite a 1.6% reduction in the cultivated area. The 2024 growing season in Germany was characterised by above-average rainfall and very hot temperatures, which negatively impacted traditional aromatic varieties, resulting in below-average alpha levels.

Conversely, the United States experienced a decline in both cultivated area (−17.5%, a reduction of 3842 hectares) and production, with volumes projected to drop by 18.2% compared to 2023 (Hop Growers of America, 2024; USDA, 2024). This decrease was primarily due to a contraction in the area allocated to aromatic varieties, although there was a compensatory increase in bittering varieties.

Globally, a significant surplus of hops remains in cold storage. This surplus, coupled with declining demand from craft breweries, has led American brewing experts to predict a further 5% reduction in acreage in 2025.

As is often the case with oversupply in agricultural markets, falling prices have generated international concern within the industry and beyond. Several factors have contributed to this surplus, including the decline in beer sales, the contractual dynamics between supply and demand, and over-planting by farmers in response to the pre-pandemic boom in craft beer, which has since slowed considerably.

Demographic and lifestyle changes among younger generations in Western countries also play a significant role, as this group is becoming more aware of the consequences of alcohol consumption. Stricter alcohol advertising regulations, lower drink-driving thresholds, and weakening purchasing power across many nations further dampen global beer demand, exacerbating the decline in sales.

IHGC forecasts indicate that after the 2025 and 2026 harvests, there will be a significant reduction in acreage and farm closures in Europe. Estimates suggest that between 5,000 and 10,000 hectares would need to be reclaimed globally in the short term to restore market balance (Table 1.4).

1.1.4 Evolution of the Hop Market in Italy During the Last Decade

Over the last few decades, Italy has witnessed the rise and the establishment of a new business model in the brewing industry: the craft beer industry, which has gained great success and credibility, becoming one of the most significant phenomena in the Italian food and drink sector. In addition to this, Italy is the only Member State to have passed a law on the cultivation and initial processing of medicinal plants (Legislative Decree 75/2018). This law introduces hops to the list of cultivated and

Table 1.4 Area, production and α-acids in selected countries (November 2024, estimated data)

Country	Hop acreage (ha)					Hop production (in Tonnes = 1000 kg)			
	Aroma	Alpha	Area[a]	New	Total	Aroma	Alpha	Total	AA
Australia[b]	649	109	758	0	758	1343	165	1,508	247
China[b]	300	2400	2700	0	2,700	700	5,800	6,500	470
Germany	9,498	10,026	19,523	765	20,289	18,179	28,318	46,497	5,391
Great Britain	428	108	536	0	536	526	168	694	56
Polonia	756	855	1,612	41	1,652	1307	1679	2,974	283
New Zealand[b]	1,000	400	1,400	0	1,400	1400	630	2,030	204
Czech Republic	4,572	77	4,649	196	4,845	6400	150	6,550	235
Slovenia	1,554	18	1,572	72	1,644	2200	45	2,245	125
Spain	10	562	572	18	590	13	970	983	114
USA	12,698	5,442	18,140	0	18,140	24,500	14,456	38,956	4,863
South Africa[b]	87	317	404	11	404	137	545	682	87
Total IHGC members	33,224	21,064	54,290	1,372	55,636	59,227	54,093	113,310	12,353
Δ Σ 2024–2023	−4368	289	−4,076	−58	−4,145	−8,703	3,670	−5,043	12
Δ Σ 2023–2022	−3952	1,468	−2,475	84	−2,426	4,772	7,229	11,998	1,006

Source: IHGC (2024)
[a]New hop areas
[b]Data referring to July 2024

cultivatable species, providing farmers with new opportunities for multifunctionality and diversification of farm income.

The direct, indirect and induced contributions of the craft brewing sector to the Italian economy are becoming considerable. Over the last decade, the sector has grown steadily, with only a slight downturn in the 2020 due to the Covid-19 pandemic (Fig. 1.10). The craft beer phenomenon has expanded rapidly, leading to significant growth in production (+31.5% compared to 2013) and in the number of breweries. This has also triggered a process of development of the entire brewing industry and prompted reflection on the regulation of the sector (Amoriello et al., 2017; Carbone et al., 2017; Ruggeri et al., 2024). This positive trend was evident in the performance achieved in 2022, which recorded the highest beer production volume in the analysed period. The year also saw a growth rate of +3.3% compared to 2021, highlighting the consistent growth of the national brewing sector within the EU context (Fig. 1.11). However, according to recent data from AssoBirra, 2024, beer production in Italy declined to 17.4 million hectolitres in 2023, representing a 5.5% decrease compared to the 18.3 million hectolitres produced in 2022. As shown in Fig. 1.11, the negative peak in 2023 also impacted the other EU countries (Eurostat, 2023). Despite this downturn, the figure for 2023 surpassed pre-pandemic levels (17.3 million hectolitres in 2019) and remained significantly above the long-term average (15.855 million hectolitres).

The growth in domestic beer production between 2013 and 2023 was partly due to increased consumption, particularly from 2015 onwards: that year saw a rise of over six percentage points, propelling Italy ahead of member states such as Romania, the Czech Republic, and the Netherlands with production reaching 22.3 million hectolitres in 2022.

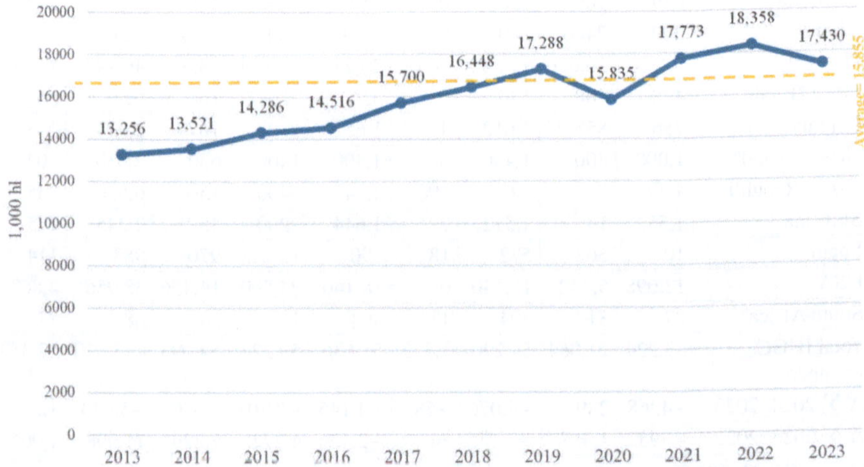

Fig. 1.10 Trend of beer production in Italy (2013–2023, 1000 hl). Legend: Microbreweries included. (*Source*: own elaboration on The Brewers of Europe, 2024 and AssoBirra, 2024)

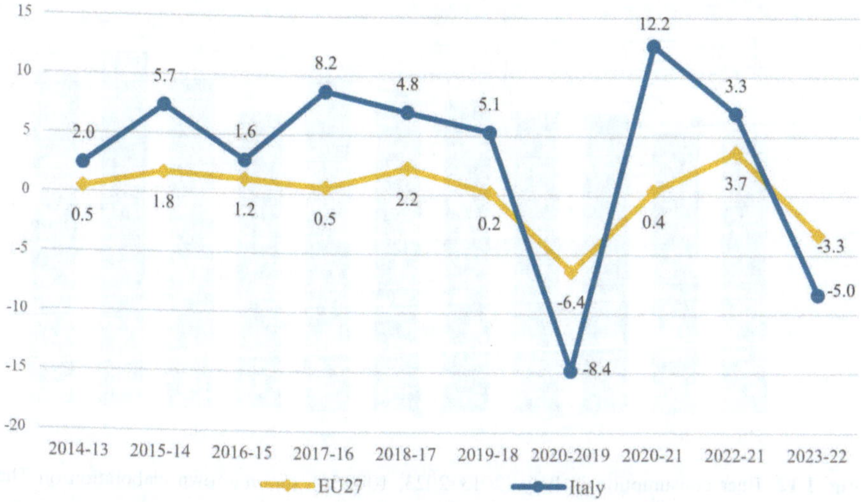

Fig. 1.11 Growth rates of beer production in Italy and EU27 (2012–2023, values in %). (*Source*: own elaboration on The Brewers of Europe (2024))

According to the latest data from AssoBirra, 2024, beer consumption in Italy reached 21.2 million hectolitres in 2023 (Fig. 1.12). While this represents a decline of nearly six percentage points from the 2022 record high, it equalled the previous 2019 peak of 21.2 million hectolitres and surpasses the 2021 figure. Over the past decade, consumption has grown by 21%, rising from 17.5 million hectolitres in 2013. This steady upward trend highlights the increasingly important role that beer plays in Italian food culture today.

The production of Italian beer is primarily associated with large industrial breweries. According to AssoBirra (AssoBirra, 2024), there were 12 such breweries in 2023, alongside numerous small and micro-breweries. Heineken Italia Spa maintained its leading position among the beers marketed and consumed in Italy in 2023, with a market share of 32%.

The increasing maturity of the domestic market and the growing appreciation for beer culture have had a significant influence on per capita consumption. In 2023, per capita beer consumption in Italy reached 36 litres (The Brewers of Europe, 2024), highlighting the sector's positive development. However, this figure remains well below the EU average of 69 litres and is closer to that observed in other Mediterranean countries such as France (33 litres) and Greece (38 litres). Notably, the rise in domestic consumption is largely sustained by domestic production, primarily for logistical reasons. This reliance underscores the considerable growth potential for Italian breweries (AssoBirra, 2021). Furthermore, as illustrated in Table 1.5, domestic consumption is increasingly characterised by a preference for low and zero-alcohol beers. These accounted for 1.86% of total consumption in 2023, which is a 4.5% increase compared to the previous year.

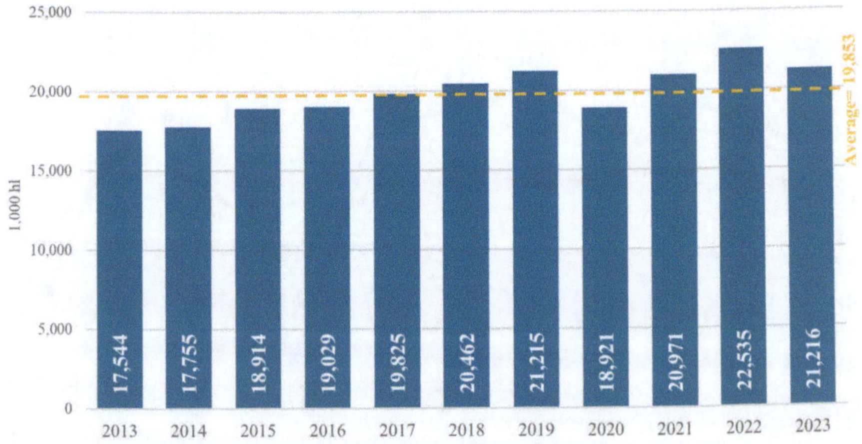

Fig. 1.12 Beer consumption in Italy (2013–2023, 1000 hl). (*Source*: own elaboration on The Brewers of Europe (different years) and AssoBirra (different years))

Table 1.5 Market segmentation in Italy (2013–2023, values in %)

Kinds of beer	2013	2014	2015	2016	2017	2018	2019	2020	2021	2022	2023
Low—non-alcoholic	1.62	2.58	2.63	2.03	1.86	1.75	1.60	1.30	1.43	1.78	1.86
Lager	92.94	91.16	89.95	89.59	88.15	86.56	85.21	84.18	81.29	82.85	82.73
Specials	5.44	6.26	7.42	8.38	9.99	11.69	13.19	14.52	17.28	15.37	15.41
Total	100	100	100	100	100	100	100	100	100	100	100

Source: AssoBirra (different years)

Over the past decade, the growth of the domestic beer sector has been largely driven by increasing demand for craft products (Tarangioli, 2020). This trend has been bolstered by advancements in research and innovation, the sector's ability to adapt to evolving consumer preferences, and supportive legislative measures. These factors have also been instrumental in driving and boosting exports, which increased by 87% between 2012 and 2023 (AssoBirra, 2024). In 2023, exports reached a record high of 3.6 million hectolitres, reflecting the global recognition of Italian quality, particularly in countries with a long brewing history such as the United Kingdom, Germany, and the Netherlands.

However, export data for 2023 show a year-on-year decrease of 5.4% compared with 2022. Notably, exports to the UK fell from 48.2% of the total in 2022 to 44.1% in 2023, equating to a decrease of 250,000 hectolitres. In contrast, exports to Albania and France recorded significant growth, with exports to France rising by an impressive 57% in 2023 (AssoBirra, 2024). Furthermore, Italy imports almost all its brewing raw materials, particularly hops (Gargani et al., 2018).

Like many other European countries, Italy has also witnessed a craft beer boom in recent years. However, this movement is not entirely new; Italy's craft beer journey began nearly three decades ago. The first Italian craft brewery was founded in

Lazio in the late 1980s, and was soon followed by others in Piedmont and Lombardy (Tarangioli & Carbone, 2020). The craft movement links the added value of its products to the territory and raw materials used, aiming to offer consumers a wide range of unique tastes and styles. Craft beers often draw inspiration from the natural resources, distinctive flavours, and unique characteristics of the regions in which they are produced. This approach fosters a stronger connection with the local community, gaining increasing popularity among beer enthusiasts seeking more authentic and sustainable experiences (Carbone & Licciardo, 2023; Laura Saggio et al., 2022). Licciardo and Carbone (2024) attribute the success of Italian craft breweries to their ability to demonstrate the value of raw materials and their strong links to local areas. These characteristics have bolstered the credibility of the movement, making it one of the most significant developments in the Italian food sector over the past decade. Several authors (Aquilani et al., 2015; Donadini et al., 2016; Fastigi & Cavanaugh, 2017; Garavaglia, 2009) have emphasised that the rapid growth of microbreweries in Italy is part of a wider global trend driven by competitive forces and the increasing demand for quality and quantity-focused products (Bernetti et al., 2020).

Currently, hops grown in Italy are used almost exclusively to fulfill the requirements of the craft and agricultural beer markets. The microbrewery sector has experienced substantial growth, with the number of licences increasing from 825 in 2017 to 975 in 2022 (Agenzia delle Dogane e dei Monopoli). Nearly half of these licences are concentrated in just five regions, with Lombardy leading the way with 131 registrations (Fig. 1.13).

Despite these successes, the craft beer segment appears to have entered a mature phase in the last 3 years, with its share of total sales volumes remaining below 4%. Recently, new entrepreneurial ventures have emerged within the food sector,

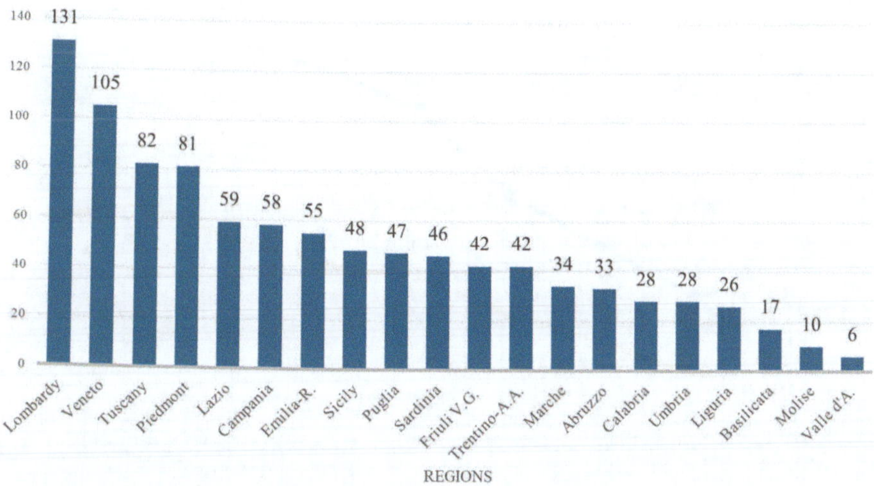

Fig. 1.13 Italian regions by number of microbreweries in 2022. (*Source*: own elaboration on The Customs and Monopolies Agency data)

particularly in the production of *bruscandoli* (hop sprouts), as well as in cosmeceuticals, including creams, hydrosols, and essential hop oil.

The cultivation of hops in Italy can be considered an alternative supply chain that continues to face challenges in gaining significant traction (Licciardo et al., 2021; Licciardo & Carbone, 2024; Ruggeri et al., 2024). Data provided by the Italian Agricultural Payments Agency (AGEA) for the marketing years from 2015 to 2023 reveal a phenomenon that, while expanding, remains small in scale. Similar to trends observed among major global producers, the sector experienced a contraction in 2023, with unsold quotas from previous years still present.

In general, the trend recorded over the considered period is positive for the two examined variables: the invested areas have risen constantly, from 4.1 ha in 2015 to 89 ha in 2023, and the number of holdings has increased, too (Fig. 1.14). As a consequence of these dynamics, the average farm size has also increased (around 5000 m² in 2022), which confirms the strong national interest in hop cultivation. The costs for constructing a specialised plant (0.5 ha), including expenses related to soil preparation, plant construction, planting, and irrigation, range from € 16,000 to € 20,000 (including labour). These costs vary depending on the specific adopted solutions.

This is still a limited crop for Italy, and the produced quantities cannot cover our companies' hop requirements. However, the current trend is encouraging and demonstrates the development of the agricultural phase, partly driven by the demand for craft beer. In today's scenario, the hops grown along the peninsula are used to meet the needs of the craft and agricultural beer market.

As far as territorial distribution is concerned, observation of the data reveals a clear concentration of cultivated areas in the northern regions, due to soil and

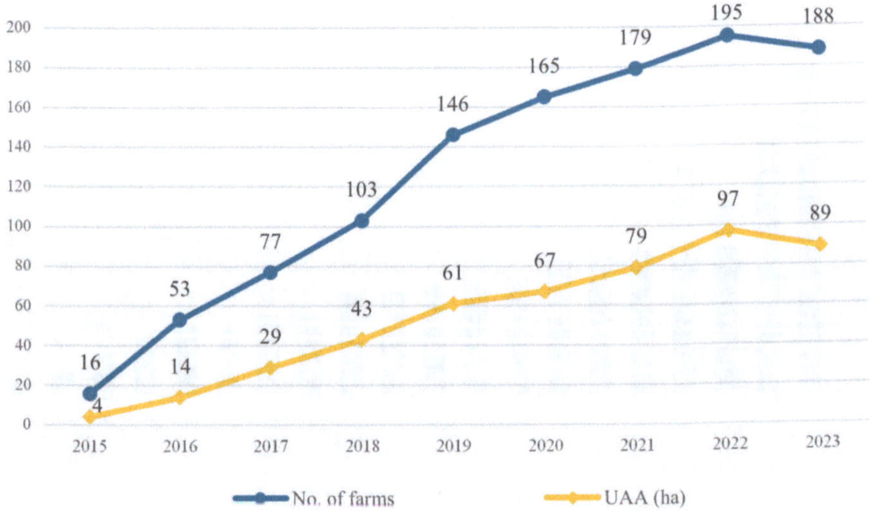

Fig. 1.14 Number of farms with hop and Utilised Agricultural Area in Italy (2015–2023). (*Source*: own elaboration on AGEA data)

climate conditions but also, in many cases, to brewing productivity. Emilia-Romagna, Veneto, and Piedmont accounted for half of the Italian hop-growing area in 2023 (Fig. 1.15), with 21, 17 and 10 hectares respectively. In terms of farms, a greater concentration in northern Italian regions is evident, with four regions (Veneto, Lombardy, Emilia-Romagna, and Piedmont) accounting for 52%. However, the propensity of central Italian holdings towards hop cultivation is also worth highlighting, particularly in Tuscany, Lazio, Umbria, and Marche.

Although the Italian hop market offers opportunities for farmers such as income diversification, farm multifunctionality, potential market growth driven by rising national demand and the prospect of higher selling prices, several critical issues remain. These include:

- Small production size;
- Instability of supply in terms of quality and quantity;
- Lack of specialised technical training for supply chain operators;
- Lack of specialised processing centres;
- Lack of supply-demand alignment.

These challenges are deeply interconnected and limit the development and competitiveness of the Italian hop sector.

The sector's small production scale is a key issue, limiting its ability to meet demand and making it overly dependent on imports, which increases costs and reduces resilience to external market fluctuations. Supply instability, both in terms of quality and quantity, due to climate variability or non-standardised cultivation practices further complicates the situation. A salient issue in this regard is the lack of technical expertise among farmers, which means they are unable to manage critical processes such as determining the optimal harvest time. This, in turn, adversely impacts the aromatic profile of hops, potentially yielding products that fail to meet

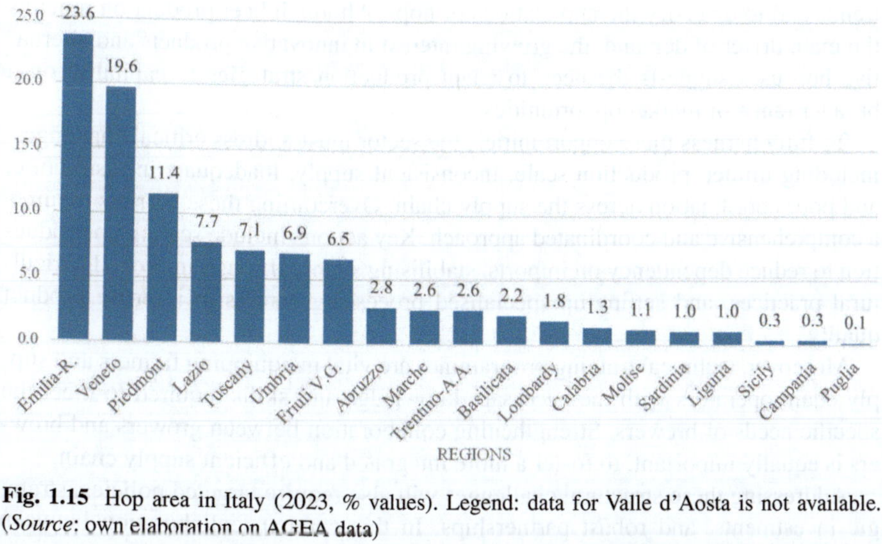

Fig. 1.15 Hop area in Italy (2023, % values). Legend: data for Valle d'Aosta is not available. (*Source*: own elaboration on AGEA data)

the exacting standards of master brewers. The paucity of access to specialised technical training for operators along the supply chain further highlights the need for improved knowledge transfer and technical capacity to implement optimal practices. The absence of specialised processing facilities further exacerbates the situation, making producers dependent on basic or external facilities, which often compromises the quality and market value of the final product. The sector's potential is also limited by weak coordination between supply and demand, highlighting the need for a more structured system that aligns brewers' requirements with cultivation outputs. Brewers must play a pivotal role in fostering the growth of hop cultivation and promoting collaboration to ensure that the supply chain aligns with market demands.

1.1.5 Closing Remarks

This chapter presents an analysis that highlights the dynamic nature of the global hop sector, which is characterised by significant variations in production and consumption across countries. The Americas, led by the United States, continue to play a pivotal role despite recent reductions in cultivated areas, largely due to their substantial production capacity and influence on market dynamics. The European Union is underpinned by a policy framework that is strongly oriented towards sustainability, remains a key player, with Germany, the Czech Republic, Slovenia, and Poland being its most active member states in the hop sector.

Within this broader context, Italy has significant potential, particularly in the niche craft beer market, which could enhance its role in the local economy and generate considerable direct, indirect, and induced benefits. However, realising the sector's full potential requires rethinking strategies to address emerging consumer trends and to diversify the applications of hops. Although beer production remains the main driver of demand, the growing interest in innovative products and alternative hop uses suggests the need to adapt production strategies to capitalise on a broader range of market opportunities.

To fully harness these opportunities, the sector must address critical challenges, including limited production scale, inconsistent supply, inadequate infrastructure, and poor coordination across the supply chain. Overcoming these barriers requires a comprehensive and coordinated approach. Key actions include scaling up production to reduce dependency on imports, stabilising supply through improved agricultural practices, and setting-up specialised processing centres to improve product quality.

Moreover, technical training programmes are vital in equipping farmers and supply chain operators with the necessary knowledge and skills required to meet the specific needs of brewers. Strengthening collaboration between growers and brewers is equally important, to foster a more integrated and efficient supply chain.

Addressing these structural challenges will also require targeted policies, strategic investments, and robust partnerships. In this context, collaboration between public institutions, farmers, and brewers is vital to design and implement measures that are aligned with market demands and sustainability objectives. Such efforts will enable the sector to capitalise fully on the growing demand for beer and other hop-based products and innovations, thereby strengthening its economic resilience and adaptability to changing market dynamics.

References

Amoriello, T., Carbone, K., Monteleone, A., Pagano, M., & Tarangioli, S. (2017). *Criticità e opportunità per lo sviluppo sostenibile della filiera brassicola.*

Aquilani, B., Laureti, T., Poponi, S., & Secondi, L. (2015). Beer choice and consumption determinants when craft beers are tasted: An exploratory study of consumer preferences. *Food Quality and Preference, 41*, 214–224. https://doi.org/10.1016/j.foodqual.2014.12.005

AssoBirra. (2021). *Annual Report 2020.*

AssoBirra. (2024). *Annual Report 2023.*

BarthHaas. (2023). *Report 2022/2023.*

BarthHaas. (2024). *Report 2023/2024.*

Bernetti, I., Alampi Sottini, V., Cipollaro, M., & Menghini, S. (2020). A survey on the performance of the Italian brewing companies. *Italian Review of Agricultural Economics, 75*(1), 37–50. https://doi.org/10.13128/REA-11741

Carbone, K., & Licciardo, F. (2023). Luppolo made in Italy, una filiera in (lenta) crescita, *31*, 48–50.

Carbone, K., Amoriello, T., Pagano, M., Sperandio, G., Assirelli, A., Tarangioli, S., & Monteleone, A. (2017). Prospettive interessanti per il luppolo italiano. *Informatore Agrario, 20*, 49–51.

Carbone, K., Bianchi, G., Petrozziello, M., Bonello, F., Macchioni, V., Parisse, B., De Natale, F., Alilla, R., & Cravero, M. C. (2021). Tasting the Italian terroir through craft beer: Quality and sensory assessment of cascade hops grown in Central Italy and derived monovarietal beers. *Food, 10*(9), 2085. https://doi.org/10.3390/foods10092085

Donadini, G., Fumi, M. D., Kordialik-Bogacka, E., Maggi, L., Lambri, M., & Sckokai, P. (2016). Consumer interest in specialty beers in three European markets. *Food Research International, 85*, 301–314. https://doi.org/10.1016/j.foodres.2016.04.029

European Commission - DG AGRI. (2023). *Hop report for the harvest year 2022*. https://agriculture.ec.europa.eu/farming/crop-productions-and-plant-based-products/hops/hops-harvest-reports_en

Eurostat. (2023). *Beer production back to pre-pandemic level*. https://ec.europa.eu/eurostat/web/products-eurostat-news/w/ddn-20230803-1#:%E2%88%BC:text=Germany

Fastigi, M., & Cavanaugh, J. R. (2017). Turning passion into profession: A history of craft beer in Italy. *Gastronomica, 17*(2), 39–50. https://doi.org/10.1525/gfc.2017.17.2.39

Garavaglia, C. (2009). Struttura di mercato e concorrenza nell'industria birraria in Italia. *L'industria, 3*, 523–550. https://doi.org/10.1430/30006

Gargani, E., Faggioli, F., & Haegi, A. (2018). A survey on pests and diseases of Italian Hop crops. *Italus Hortus, 24*, 1–17. https://doi.org/10.26353/j.itahort/2017.2.117

Hop Growers of America. (2024). *2023 Statistical Report*. https://www.usahops.org/news/2023-hga-statistical-report

IHGC. (2023, novembre). *Economic Commission—Summary Reports 2023*. www.usahops.org/img/blog_pdf/466.pdf

IHGC. (2024, novembre). *Economic Commission—Summar Reports 2024*. www.usahops.org/img/blog_pdf/492.pdf

Licciardo, F., & Carbone, K. (2024). Il luppolo non manca ma calano i brindisi. *Il luppolo non manca ma calano i brindisi, 30*, 45–49.

Licciardo, F., Carbone, K., Ievoli, C., Manzo, A., & Tarangioli, S. (2021). *Outlook economico-statistico del comparto luppolo*. https://doi.org/10.13140/RG.2.2.10805.81120.

Ruggeri, R., Rossini, F., Roberto, S. R., Sato, A. J., Loussert, P., Rutto, L. K., & Agehara, S. (2024). Development of hop cultivation in new growing areas: The state of the art and the way forward. *European Journal of Agronomy, 161*, 127335. https://doi.org/10.1016/j.eja.2024.127335

Saggio, L., Carbone, K., Licciardo, F., Perinotto, M., & Tarangioli, S. (2022). Tutto il potenziale del luppolo. *Tutto il potenziale del luppolo, 11*, 50–51.

Tarangioli, S. (2020). La filiera brassicola in Italia. In *Annuario dell'agricoltura italiana 2018* (Vol. LXXII). CREA.

Tarangioli, S., & Carbone, K. (2020). Il comparto delle birre artigianali e le attività di ricerca sul luppolo. In *Annuario dell'agricoltura italiana 2018*. CREA.

Tarangioli, S., Monda, M., & Manzo, A. (2020). La normativa di riferimento della filiera brassicola. In *Annuario dell'agricoltura italiana 2018* (Vol. LXXII). CREA.

The Brewers of Europe. (2024). *European beer trends. Statistics report, 2023 edition*. https://brewersofeurope.eu/european-beer-trends-2024-edition-and-previous-years/

USDA. (2024). *National Hop Report 2024*.

Van Holle, A., Muylle, H., Haesaert, G., Naudts, D., De Keukeleire, D., Roldán-Ruiz, I., & Van Landschoot, A. (2021). Relevance of hop terroir for beer flavour. *Journal of the Institute of Brewing, 127*(3), 238–247. https://doi.org/10.1002/jib.648

Watson R. T., Crawford M., & Farley S. (2003). *Strategic approaches to science and technology in development. Policy Research Working Paper Series 3026*. World Bank Group. http://documents.worldbank.org/curated/en/422921468739154192/Strategic-approaches-to-science-and-technology-in-development

Part I
Hops: Botany, Biodiversity and Cultivation

Chapter 2
Hop Botany and Physiology

Tommaso Ganino, Martina Galaverni, Ilaria Marchioni, and Margherita Rodolfi

2.1 Hop Botany

2.1.1 Botanical Classification

The species *Humulus lupulus* L., commonly called hops, belongs to the *Cannabaceae* family (Table 2.1).

Two other species belong to the genus *Humulus*: *Humulus japonicus* and *Humulus yunnanensis*. China appears to be the geographical region of origin of the genus *Humulus* (Murakami et al., 2006). Subsequently, it spread widely in Japan, America, and Europe. Today, hops are cultivated in all temperate regions of the world.

The species *Humulus lupulus* L. has been classified into five varieties according to their morphological characteristics and geographical region: variety *lupulus* for European hops, variety *cordifolius* Small for Japanese hops, variety *neomexicanus* Nelson & Cockerell, variety *pubescens* Small and variety *lupuloides* Small for American hops (Zanoli & Zavatti, 2018).

The etymology of the name is complex and there are many hypotheses. Pliny the Elder, the Roman writer and naturalist, describes this plant as "*lupus salicarius*", in the sense of weed. *Lupulus* would be the diminutive of *lupus*, wolf, due to a supposed aggressive behaviour of the plant. Just as the wolf attacks and strangles its

T. Ganino (✉)
Department of Food and Drug, University of Parma, Parma, Italy

Institute of BioEconomy (IBE), National Council of Research – CNR, Sesto Fiorentino, Florence, Italy
e-mail: tommaso.ganino@unipr.it

M. Galaverni · I. Marchioni · M. Rodolfi (✉)
Department of Food and Drug, University of Parma, Parma, Italy
e-mail: martina.galaverni@unipr.it; ilaria.marchioni@unipr.it; margherita.rodolfi@unipr.it

K. Carbone (ed.), *All About Hops: The Crop, its Cultivation, and its Uses in Brewing and Beyond*, https://doi.org/10.1007/978-3-031-96606-4_2

Table 2.1 Hop botanical classification

Domain	*Eukariota*
Kingdom	*Magnoliophyta*
Class	*Magnoliopsida*
Order	*Urticales*
Family	*Cannabaceae*
Genus	*Humulus*
Species	*Humulus lupulus* L.

prey, the hop is also capable of suffocating the plants on which it grows. The term *Humulus* is derived from Latin, but the meaning is dubious to say the least. *Humulus* is said to derive from the Latin *humus* meaning ‘damp earth’ or it could be a reference to the plant’s prostrate habit. However, there are also those who refer to a Germanic origin from the word *humall*, but in this case the perplexity is even greater (Zanoli & Zavatti, 2018).

2.1.2 *Growth Habit*

Wild hop (*Humulus lupulus* L.) is a dioecious, perennial, climbing plant; it is indigenous to north-temperate areas (between latitudes 35° and 55°) (Small, 1978; Bocquet et al., 2018), which frequently grows in moist thickets, slopes, river banks, alluvial woods, or along fences and hedges, often in sandy soils (Small, 2016; Jastrombek et al., 2022). This species prefers open sunny locations, but it can tolerate moderate shade (Small, 2016). The climbing habitus ensure to the plant the leaves exposure to sunlight, twining the bines spirally over the tall vegetation, especially shrubs and trees (Small, 2016). A proper support is, hence, pivotal for a quick and vigorous growth (30 cm per day in the peak of the vegetative period, reaching heights of 7–10 m, occasionally 15 m) (Gutierrez et al., 2023).

Hop plants cultivated for industrial purposes are grown under a trellis system (vertical supporting strings suspended from a trellis wirework and anchored in the ground), which supports the natural climbing habits (bining) of plants (Kenny, 2004; Adams, 2021). Depending upon the trellis system, either two or four bines are allowed to grow from a single plant, and no more should be trained to avoid to excessively weak the rhizome (Adams, 2021). Moreover, the latter determine the useful commercial life of hop plant, which is between 12 and 20 years (De Keukeleire et al., 1999).

2.2 Vegetative Morphology and Anatomy

When analysing the hop plant from a morphological and anatomical point of view, a distinction between the underground and aboveground portions of the plant must be made basically for their different ability of surviving winter temperatures. The hypogeal portion of the plant cope with cold temperature thanks to a secondary thickening, contrastingly with the epigeal one that die at the onset of the cold period. Besides, hop perennation is possible thanks to rhizome dormient buds that are annually renewed. These buds can survive up to 4 years and develop under favourable conditions into new aboveground stems (Rybáček, 1990). Anatomical and morphological differences also exist between organs, whether they are hypogeal or epigeal, thus resulting in different functions. The underground organs system mainly consists of the root system and the rhizome, while vegetative and reproductive organs form the aboveground portion (Rybáček, 1990).

2.2.1 *Rhizome*

The underground organs system is composed of the root system and the rhizome (vulgarly and erroneously called "rootstock"), with the latter including all the modified stem organs below the soil level. These modified stems can be distinguished in *true rhizomes* and *suckers* depending on whether they grow vertically or horizontally, respectively. The rhizome primary function is to preserve the meristematic tissue during the cold period either for the continuous growth of the underground system and the rapid growth of the above-ground one at the onset of the vegetative period. Furthermore, underground stems can also be successfully used for vegetative propagation. From an anatomical point of view, at the centre of an internode section all these modified stems are characterised by the presence of primary xylem bundles with many vessels of secondary xylem outwards. Then, placed externally of the secondary xylem, a ring of secondary phloem, used as a storage tissue and made up with multiple vessels and parenchyma, can be found (Rybáček, 1990). Similarly, in autumn organic substances are stored in the secondary xylem. Underground stems externally present a secondary bark (periderm) with visible lenticels, which make gas exchange and tissue respiration possible. Therefore, both rhizomes and suckers can be distinguished into *one-year old organs* or *older organs* based on the type of wood they are made of. *Young wood* comprises suckers and modified stems up to one-year old that develop from buds on old wood. Young wood primary role is to allow the development of new aboveground shoot in the next vegetative season. More specifically, underground modified stems (also called *new wood*) grow vertically towards the ground and are characterised by short and thick internodes, whereas new suckers have several tillering nodes with dormant buds (Rybáček, 1990). In addition, new suckers are characterised by the presence of a top buds with a strong apical dominance. However, branches up to 1 m length can develop from those buds in case of interrupted apical dominance. *Well-grown wood,* instead, comprehends both rhizomes and suckers with more than one-year of age. Their maturity can be determined by the number of annual rings, which are thicker (2–6 mm thick) in old wood than in old suckers.

2.2.2 Roots

The hop root system is a complex mixture of roots that all have as primary purpose the absorption of water and nutrients from the soil and their transport to the other plant organs for a further transformation. Furthermore, roots also exert a storage function especially the so-called *tuber roots*. Roots can be easily distinguished from other underground organs for the absence of buds. They are regenerated every spring from buds located on the branched stem tissue close below the ground and forms the upper side of the rhizome (Neve, 1991).

The root system can be described taking into consideration multiple approaches, such as the root age, growth direction and function. According to Rybáček (1990), the hop root system is made up of two main types of roots depending on their maturity, namely *skeletal roots* and *terminal rootlets,* both developed from the rhizome. The skeletal roots represent the basis of the root system and constitute its skeleton. The main function of this root category is to anchor the plant to the soil and distribute plant liquid (Rybáček, 1990). Skeletal roots are big, perennial, brown and lignified. They generate an annual ring and get longer every year without any limiting factor. Skeletal roots can be divided into *horizontal* and *vertical* based on their spatial position. Terminal rootlets, also named active, are fine, young, white roots that can often be found as horizontal root ending (Graf et al., 2014). They present a primary anatomical structure and, starting from the tip, three main portions can be identified: the root carp meristem, the elongation cell zone and the root hair zone (Rybáček, 1990). The adsorption of water takes place in the root hair zone, while salts are adsorbed in the zone of cell elongation (Meyer & Anderson, 1952). Besides, a distinction between *thin* end-rootlets (0.2–0.6 mm) and *thick* end-rootlets (about 1 mm) can be made according to their thickness, with the first usually being much more in number than the latter. In addition, in a recent overview of the root system given by Graf et al. (2014), a third root type in relation to age was described including white and young roots developed out of the shoots above the rootstock in the tilled soil within the row. The hop root system can also be examined in its three-dimensional state, considering separately *vertical* and *horizontal* roots. Vertical roots refer to all skeletal roots and their lateral branches that develop vertically; they are the first roots developed from a new hop plant, and their number depends on whether they develop from one germinating seed or by vegetative propagation, going from just one root to a multiple, respectively. Furthermore, vertical roots can also grow from horizontal roots and are designated as *anchor roots.* Their primary role is to anchor the plant to the soil, but they also help taking water from the deeper soil layers. With regards to horizontal roots, they can be produced by all the rhizome organs, and more specifically they are called "*rootstock roots*" when are produced from the rootstock (rhizome) and *floor roots* when are grown from vertical roots. According to Graf et al. (2014), the root system can be subdivided into three main zones based on the prevalence of horizontal and vertical roots. Starting from the first layers of the soil, a prism-shape root system consisting of both horizontal and vertical white root ramifications within the hilled rows can be found. Deeper into the soil, the root system becomes cylindrical as it mainly consists of horizontal roots and then become cuboid due to the prevalence of vertical roots. Considering the ways in which the root system is organised to maximise the absorption of water and salts, it is likely that horizontal roots develop close below the soil surface for the absorption of rainwater, while vertical ones are used as assurance in periods of drought to exploit deeper regions of the underground. However, the length of hop roots and the expansion of the root system are still discussed issues as many factors such as the hop cultivar, soil conditions, and agronomic practises can play a role. Beard (1943) found out differences in the root system and the proportion between vertical and horizontal roots in relation to the soil conditions. He reported of differences due to the soil type and, therefore of larger and deeper root system under well-drained conditions compared to poorly drained soil. Similarly, Burgess (1964) reported root depth of 150 cm and a lateral extension ranging from 180 to 240 cm in well-drained soil. The distribution of the root system is also affected by the row management with limitations in the lateral growth within regularly cultivated rows as stated by Neve (1991) and Brant et al. (2020). Among the agronomic practises, irrigation plays a key role in the length of vertical roots. According to Sobotik et al. (2018), roots in poorly irrigated soils can reach 3.7 m in depth, compared to 1.4 m in well-irrigated soils. Furthermore, Koie et al. (2018) reported for the first time of a varietal effect on the hop root expansion.

From an anatomical point of view, hop root is a typical dicotyledonous root and it can be described according to their stage of development. In this sense, it is possible to identify the primary root structure and the secondary root structure.

In primary structure, the first root tissue, from the outside towards the inside, is the epidermis. This tissue consists of a single layer of cells with the presence of absorbent root hairs. Immediately below the epidermal layer, there is the cortical layer or cortex. This consists of vacuolated cells with large intercellular spaces. In young hop roots, no cells belonging to the mechanical tissue are visible in this zone (Miller, 1958). Various authors have indicated that some of these cells may contain a tetraploid number of somatic chromosomes (Miller, 1958). In the inner part of the root, it possible to observe the central cylinder. The endoderm separates the cortex from the central cylinder. Endoderm is composed by a single layer of cells, and it is formed by the Casparian strips, primary parietal structures surrounding the endodermis cells. The Casparian stripes contain lignin, while suberin is found in the cell wall of the endodermis cells after the formation of the Casparian stripes. In the primary hop root, the Casparian stripe only appears when the primary structure is differentiated. Proceeding towards the inside of the structure, after the endodermis there is a single layer of cells that forms the pericycle. This part is important for the formation of lateral roots. In the innermost part of the central cylinder, we find the conducting tissues: xylem and phloem. Primary xylem and primary phloem are arranged in separate and alternate rows in a radial way. According to the number of rows, the primary hop root is usually diarch, and rarely triarch.

The primary xylem consists of protoxylem and metaxylem. The first has a considerably narrower diameter than the latter. The protoxylem consists of tracheids whose cell walls are thickened in an annular, helical or spiral manner. The metaxylem consists mainly of trachea with scalariform cell wall thickening.

The primary phloem also consists of protophloem and metaphloem, which differ in cell diameters and speed of formation. The first to form is the protophloem. In each case, both are formed by sieve-tube elements and companion cells.

During growth, the root forms its secondary structure, where a vascular cambium originates from the parenchyma cells located between the diarch primary xylem and primary phloem. The cambium of the root of hops is composed of both ray and fusiform initials. The ray initials form the multiseriate rays, while the fusiform initials form the conduction system. Anatomically, hops roots can be divided into two types: fibrous roots and fleshy storage roots (Miller, 1959). The fibrous roots are the smaller ones with a reduced viability (in terms of days of functionality). Although they are present as lateral roots of larger ones and are found at various depths, these small, thin, fibrous and 'absorbent' roots tend to stand out in the upper layers of the soil (Miller, 1959). Fleshy storage roots are roots that have the function of storing reserve substances, mainly in the cortical zone. In these roots, fibres can be found only outside the xylem. In the older xylem it is not uncommon to find cells with abundant tillae (Fig. 2.1), just as it is not uncommon to find calcium oxalate crystals (druse) in the parenchyma zones.

2.2.3 *Buds*

Aboveground organs grow from buds on the meristem tissue, in the upper area of the rhizome, just below the soil surface. In early winter, with chilling temperatures and short daylengths, buds go into dormancy whose duration is necessary for the meristem tissue to change from a vegetative to a floral state (Neve, 1991). In early spring, warmer temperature and longer daylength bring the buds out of dormancy and they start to produce new shoots. Depending on the hop variety, rhizome buds

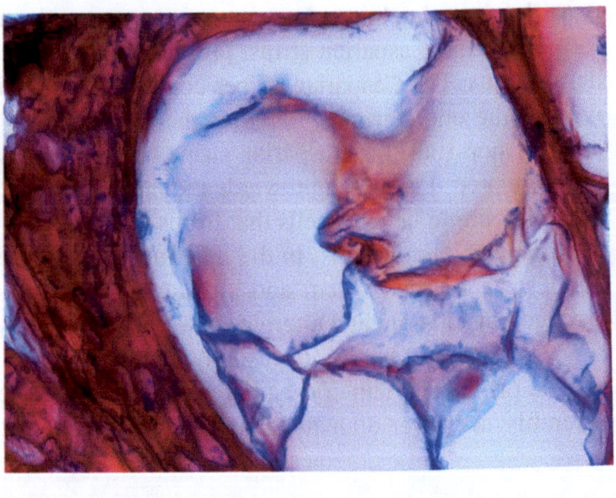

Fig. 2.1 Transverse section of the xylem cell in which the presence of tillae occluding the vessel can be observed. (400× magnification—photo: Tommaso Ganino)

can have a range of colour varied from reddish to light green or nearly white (Burgess, 1964). Different types of aboveground buds can be identified in the hop plant. The *terminal bud* on the tip of emerged shoots from the rhizome is characterised by numerous bracts enveloping its apex; those bracts are turning green as bud emerges from the ground. Normally, two or three shoots out of the multiple emerging from the rhizome are hand-trained on strings, while the others are removed (Haunold, 1980). According to Thomas and Schwabe (1970), the apical meristem tissue constantly changes its dimensions, but great variations in shape and size occur in conjunction with the beginning of the lateral bud development. Besides, the differentiation of individual tissues in the apex occur further in time. In the axils of bracts *leaf buds* can be found, while three *lateral dormant buds* are usually positioned in the axils of leaves, but only the central one gets activated by the interruption or suppression of the apical dominance, while the other two are suppressed. Lateral branches develop from axillary buds on the main stem of male and female plants. However, the development of such buds due to the damage of the terminal one is undesirable since plants then often fail to climb properly (Haunold, 1980). When stems and branches are fertile, generative organs grow from their terminal buds. By contrast, in case of stem non-fertility, buds atrophy and the same occur to the other both active and dormant buds downwards (Rybáček, 1990).

2.2.4 Sprout and Stem

Hop stem, commonly called "hop bine", constitutes the main part of the aboveground system and it consists of a main bine and its lateral branches. In early spring, new sprouts (or shoots) begin to emerge from the ground. Their colours range from reddish purple with different shades of mottling to green depending on hop variety. Each new shoot grows vertically for the first four to seven nodes, then its apex starts to rotate clockwise allowing the bine to climb around a support. According to Bell (1958), a simple helix is described by the bine around the support to which it clings, whose radius of curvature is constant probably due to differences in cell elongation between the parenchymatous cells in the convex and concave portions of the bine. Furthermore, the same Author reported that the lateral branches development occurred as the curvature of the main bine decreased. The twining of a hop bine around a support is enhanced by tiny stout hook hairs whose dimension and morphology vary depending on their position along the stem. According to Burgess (1964), hairs in the lower, non-twining internodes are characterised by a small pillar-like base consisting of a single cell embedded in the epidermal tissue that branches in its upper part into two needle-like portions, whose walls are strengthened making them rigid. On the contrary, hairs in the upper-climbing part of the bine are smaller with the base reduced to a small protuberance and shorter branches. One of the two-needle like portions is strictly adpressed to the surface of the bine, while the other one is raised to make an angle with the bine creating a small downward claw. Therefore, without limitation in space and assisted in by stout hairs, hop bines reach heights of 8–9 m (up to 30 cm per day), occasionally extending up to 15 m. However, several factors like mechanical perturbation, can affect its growth. According to Bauerle (2019), thigmomorphogenic cues (i.e. touching and/or bending) within the active elongation zone of hop bines resulted in a limited internode elongation, and an increase in bine compactness and node density per unit area, without any effect on the number of female flowers. Therefore, such mechanical stimuli can be useful methods to reduce the length of hop bine in case of confined cultivation area (e.g. the greenhouse). During the growing season the elongation occurs only in the last 3 or 4 internodes of the bine, but in the meanwhile the underlying ones continue to thicken until their cortex is torn. Therefore, as only nodes are entire, the anatomical structure of a well-grown hop bine should be studied only through section at the centre of a node.

Anatomically, young hop shoots have irregular and variable cross-sections (Fig. 2.2), a characteristic that is highly cultivar-dependent (Rodolfi et al., 2022). The epidermis of sprouts consists of a monolayer of cells. Different types of accessory organs, typical of the species, can be observed on the epidermis: non-glandular and glandular hairs, lupulin glands and climbing hairs (Metcalfe & Chalk, 1957). The cortical zone consists of parenchyma and mechanical tissues. The latter tissues are lamellar-type collenchyma (Rodolfi et al., 2022), which are characterised by elongated cells with an irregularly thickened cell wall composed predominantly of cellulose and pectin. The amount of this type of tissue seems to be cultivar-dependent (Rodolfi et al., 2022). The inner zone of the structure, adjacent to the cortex, is occupied by vascular strands of the open collateral type. This term denotes bundles of

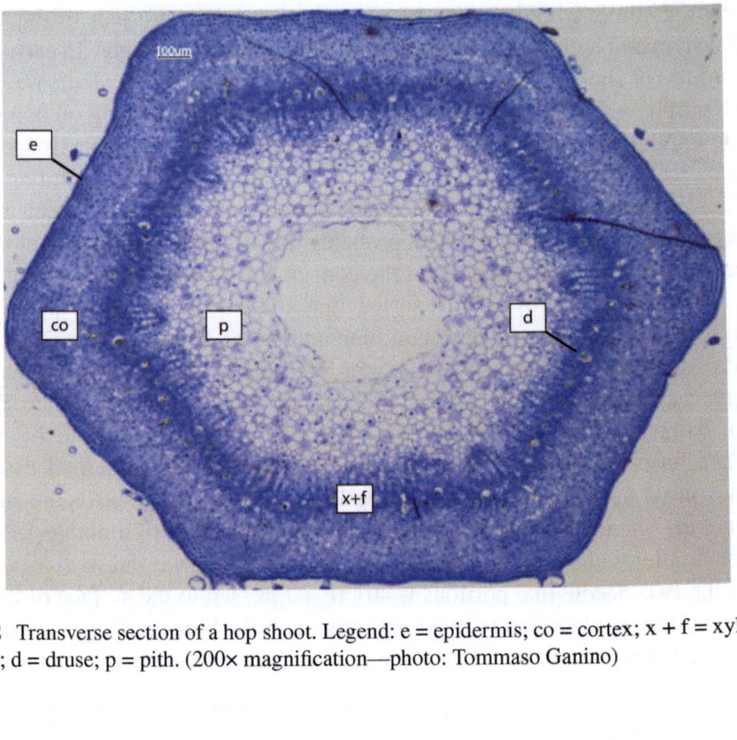

Fig. 2.2 Transverse section of a hop shoot. Legend: e = epidermis; co = cortex; x + f = xylem and phloem; d = druse; p = pith. (200× magnification—photo: Tommaso Ganino)

xylem and phloem separated by a meristematic ("pro-cambium") tissue. Near the phloem and in parenchymatic tissues, calcium oxalate crystals (druse) are frequently present, as confirmed by Schweingruber et al. (2011) (Fig. 2.3). The xylem is characterised by being made up of tracheae and tracheids and in this area bundles of fibres (called xylem fibres) are present. According to Gloser et al. (2011), vessels of secondary xylem can conduct solutes more powerfully compared to those of the primary xylem, as they are longer and wider; this characteristic, makes hop one of the most efficient plants in water conducting among several angiosperms and tropical lianas.

The central area of the shoot consists of a pith zone. This part is generally composed of reserve parenchyma with the presence of starch granules. The size of the pith zone is cultivar dependent. The presence of druses and sand crystals in the pith tissue of the hop stem was reported by Konyar et al. (2014). Furthermore, it was observed in a study by Rodolfi et al. (2022) that the shoot tissues are rich in tannin inclusions.

During the growing season, the shoot lignifies and its internal structure changes (Fig. 2.4a). The anatomical structure of the stem in cross-section appears greatly modified. The primary protection tissue (epidermis) is replaced by a secondary protection tissue, the cork. Below the cork layer is a secondary meristem (the cork cambium) and a phelloderm composed of living parenchyma cells. The phelloderm is in continuity with the cortex. This zone is characterised by parenchymatous tissue and many extra-xylar fibres (mechanical tissue). The fibres are grouped in bundles and are dispersed throughout the cortex (Fig. 2.4a). The central cylinder consists of

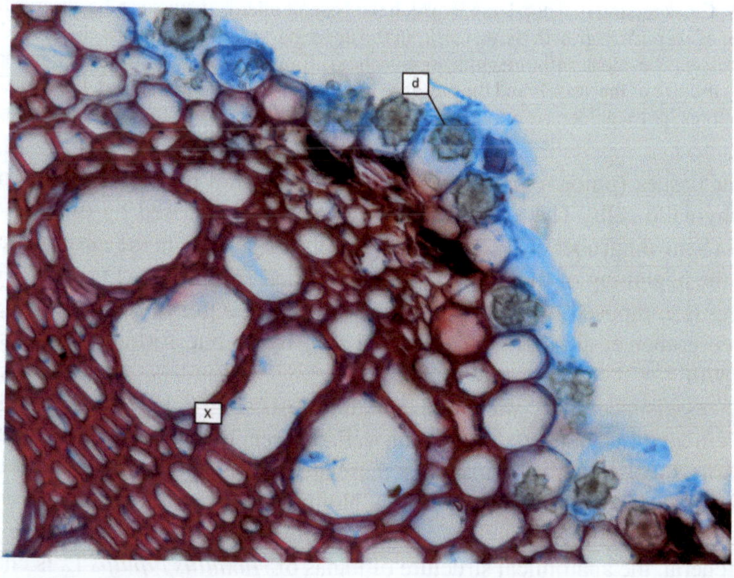

Fig. 2.3 Transverse section of a hop shoot in which druse can be observed near the vascular bundle (above, irregularly shaped grey inclusions within the parenchyma cells). Legend: d = druse; x = xylem. (200× magnification—photo: Tommaso Ganino)

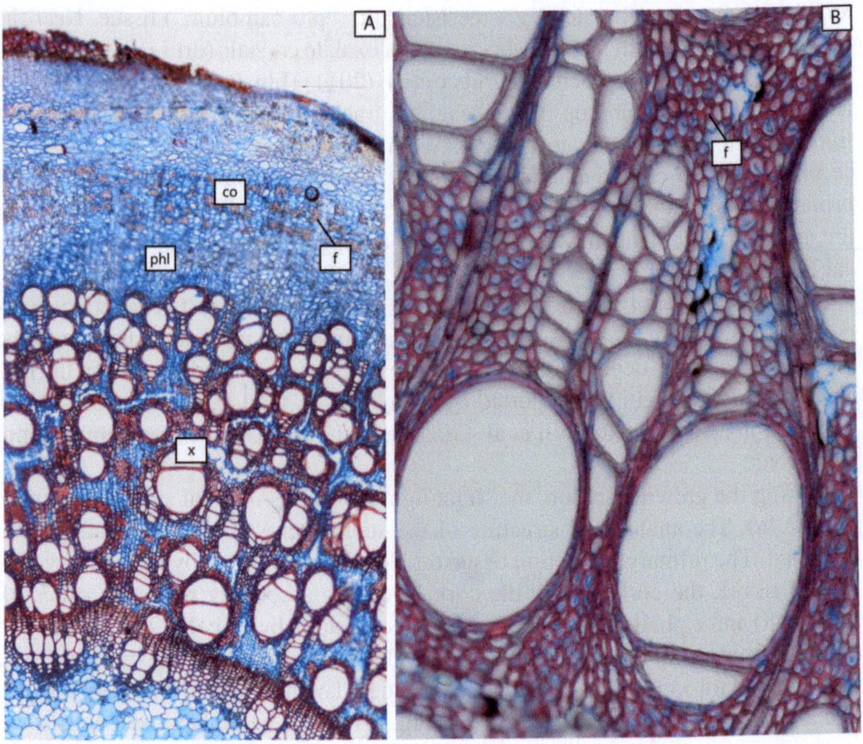

Fig. 2.4 Cross section of a hop liana in which the general microstructure can be observed: (**a**) the presence of secondary growth tissue (cork, in the upper part of the image) and the second ring of xylem vessels is evident. (40× magnification—photo: Tommaso Ganino); (**b**) detail of hop xylem in which the size of the vessels and the presence of xylem fibres can be observed. Legend: co = cortex; f = fibres; phl = phloem; x = xylem. (magnification 200×—photo: Tommaso Ganino)

vascular tissues (phloem and xylem) and, in the central part, of a pith formed by parenchymatic cells. The xylem of hops stem is characterised by two systems of vessels (Schweingruber et al., 2011): the vessels are smaller in the initial xylem ring and at the beginning of the formation of the second xylem ring (Fig. 2.2). The second ring is composed by many large vessels with 200 μm of diameter (Fig. 2.4a) (Schweingruber et al., 2011). This vessel ring is the one formed during secondary growth.

The vessels have thick-walls and contain tylosis (Fig. 2.1). The wide vessels of the xylem are often surrounded by thin-walled parenchyma cells (Fig. 2.4a). These cells form so-called parenchymal rays; these are usually between 3 and 8 cells wide (Fig. 2.4a) (Schweingruber et al., 2011). Numerous xylem fibres can be observed within the xylem (Fig. 2.4b).

In general, the anatomical structure of lianas of *Humulus lupulus* L. is characterised by: large vessels, ring-porosity, thin-walled and unlignified strands of parenchyma (Schweingruber et al., 2011).

Another important characteristic of the hop liana (and indeed of almost all organs of the plant) is the presence of secretion structures. Two secretory structures, namely *laticifer system* and *secretory idioblasts*, appear along the hop bine. According to Ramos et al. (2024), the *laticifer system* in hop is articulated, unbranched and it is present in almost all the plant organs, excepts for the roots and the anthers of staminate flowers. Latex is produced and stored transparent, but oxidation eventually changes the colour into amber or orangish due to the contact with the environment, and contains several substances, such as phenolic compounds, terpenes, tannins, and starch. The *laticifer system* originates from procambial cells, ranging in number from three to five, close to the phloematic elements, and grows among vascular cells or between laticifer elements basically due to the dissolution of the terminal walls of precursors cells by cellulase and pectinase. Like the *laticifer system*, hop *secretory idioblasts* produce a transparent substance rich in several metabolites, among which phenolic compounds, tannins, and starch. These structures are usually referred as *phenolic idioblasts,* as their structure looks very similar to that of phenolic-secreting cells. The cells of these secretion structures are cube-shaped, with pecto-cellulosic walls, and they can be easily recognised by the presence of numerous small vacuoles containing phenolic droplets. *Phenolic idioblasts* are characteristics of all the hop vegetative and floral organs, primarily within the sub-epidermis of leaves and the stem parenchyma, thus suggesting their potential role in the plant defence.

As hop bine is the main residual biomass of hop cultivation after the cones harvesting, a great interest has recently grown for its potential applications in new productive areas, such as in the paper industry. According to Haunreiter et al. (2021), hop bine has an amount of total lignin, hemicellulose and ash comparable to that of industrial hemp, but a significantly lower cellulose content (Table 2.2). Differences in the carbohydrate composition of the fibres are reported in Table 2.3.

Table 2.2 Comparison of composition of hop stem with industrial hemp (Haunreiter et al., 2021)

	Hop bine	Industrial hemp
Cellulose (% w/w)	32.9	42.5–44.0
Hemicellulose (% w/w)	26.0	22.5–25.1
Lignin (% w/w)	25.9	23.4–24.8
Ash (% w/w)	3.4	1.2–3.3
Extractives (% w/w)	12.0	1.1

Table 2.3 Comparison of the carbohydrate composition of hop stem with industrial hemp (Haunreiter et al., 2021)

	Hop bine	Industrial hemp
Arabinose (% w/w)	9.0	0.4
Galactose (% w/w)	1.2	1.0
Glucose (% w/w)	32.9	41.7
Xylose (% w/w)	15.2	14.6
Mannose (% w/w)	0.6	0.0

2.2.5 Leaf

Leaves in hop are decussate or occasionally tricussate and arise from nodes on the main bine and lateral branches in pairs opposite each other. However, towards the end of lateral branches they can be alternate. Hop leaves are simple, with a variable shape going from cordate to deeply lobed, except for those near the end of lateral branches and on the tip of the main bine that are entire. According to the hop variety, the number of lobes on well-grown leaves of the main bine usually ranges from three to five, but 7-lobed can sometimes be found (Fig. 2.5). Leaves on lateral branches are contrastingly compact or with three to five lobes. A U-shaped sinus occurs between two consecutive lobes and its depth is a varietal feature; the American cultivars and their hybrids usually present deeper sinuses compared to those in European and Asiatic varieties. The margin of the leaf is serrated, and variations in the number of teeth in the central lobe are related to the hop cultivar. Hop leaves are usually characterised by equal length and width whose size can vary, depending on the position and the vigour of growth. The leaf venation is palmate and is characterised by the main vein protruding on the lower surface. The colour of the leaves is a varietal factor and encompasses shades of green, yellow, and purple. Similarly, the colour of the petiole, also called leafstalk, is cultivar dependent and

Fig. 2.5 Examples of hop leaf morphology. Top: tri-lobate leaves of the cultivar Columbus (green petiole); Bottom: penta-lobate leaves of an Italian wild genotype (purple petiole). (Photo: Tommaso Ganino)

varies from red to pale green, with the upper face being usually the most strongly coloured. The leafstalk is slightly fleshy, its length is three fifths that of the leaf and is characterised by a shallow furrow on the upper side. Both the leaf and its petiole are hairy. Hairs in the upper surface of the leaf are small, stout, hooked and margin-directed, making the leaf rough. In addition, finer, softer and three times longer hairs compared to the hooked ones can be found in a variable extent either on the upper and lower surface of the leaf and on the lower surface of the veins. According to the hop variety, the amount of hair in the abaxial surface of the leaf can vary.

Different types of secretory structures can be identified on hop leaves as shown in Table 2.4. The leaf laticifer system is distributed in the petiole and near the vascular elements, as well as the phenolic idioblasts. According to Oliveira et al. (1988), the glandular trichomes on hop leaves and petioles can be distinguished into two types: peltate (lupulin) and bulbous. Peltate trichomes originate from a protodermal cell by two anticlinal divisions in perpendicular planes, followed by two periclinal ones that constitute the initials of the head cells, the basal and the stalk cells. Head cells divide successively in radial and irregular planes. Fully developed peltate trichomes are built of a glandular head consisting of 30–72 cells, four stalk cells and four basal cells. On the other hand, bulbous trichomes are also formed from a protodermal cell by an anticlinal division followed by two periclinal ones that produce the initials of the glandular head cells, and the basal and stalk cells. Fully developed bulbous trichomes consist of four (occasionally eight) head glandular cells, two stalk cells and two basal cells (Oliveira et al., 1988). Both the types produce secretions, whose colours are yellow and transparent for the peltate and bulbous type, respectively, and are stored in the subcuticular space. In peltate trichomes the accumulation of the secretion is followed by the separation of the cuticle from the cell without any rupture, unlike that of the bulbous type that breaks after distension. Besides, the density of lupulin glands on leaves decline during leaf extension, therefore it is likely that their final number is reached in leaf development. According to Naraine and Small (2017), the glandular trichomes in the abaxial side of the leaf blades can be found in a higher density near the juncture between the leaf and its petiole compared to other peripherical parts to protect the basal region from damage by insects. The gland density also seems to be related to the node age with higher values in young leaves, suggesting a protective function against the attack of

Table 2.4 Presence of secretory structures in the hop leaf according to Ramos et al. (2024)

Plant organ		Laticifer system	Glandular trichomes	Phenolic idioblasts
Leaf (adaxial side)	Petiole	+	+	+
	Leaf blade	–	+	+
	Central vein	+	+	–
	Parallel veins	+	+	+
Leaf (abaxial side)	Leaf blade	–	+	+
	Central vein	+	–	–
	Parallel veins	+	–	+

Symbols: (+): Presence; (–): Absence

herbivores that is yet to be confirmed. The density of leaf glands can be variable depending on the taxonomic varieties of *Humulus lupulus* L. as reported by Small (1978). In addition, Naraine and Small (2017) described the European hop varieties as much more susceptible to the attack of herbivores than the American ones. Morcol et al. (2021) confirmed the differences in the leaf gland density and, thus, in the bitter acids and prenyflavonoids content by comparing the southwestern American *H. neomexicanus* with some European *H. lupulus* L. cultivars and obtaining higher values with the first. They also reported similar differences in the content of non-glandular phenolic compounds produced by the green leaf tissue (Morcol et al., 2021).

Anatomically, the leaf has a cross-sectional anatomical structure consisting of a large central vein (Fig. 2.6) and a thin leaf mesophyll (110–140 μm) (Fig. 2.7). The central vein consists of a large cribro-vascular bundle (Fig. 2.6a). Surrounding the cribro-vascular bundle is a parenchymatic tissue. The latter is surrounded, adaxially and abaxially, by collenchymatic tissue (Fig. 2.6b).

The foliar mesophyll (Fig. 2.7) is composed by two principal tissues: palisade tissue (photosynthetic tissue) and spongy tissue. Palisade is composed by 2 or 3 layers of cells (the thickness ranged from 65 to 80 μm). Below the palisade layer is the spongy layer. The cells of the spongy layer are characterised by large intercellular spaces. This arrangement allows the leaf to store gas. The spongy tissue is on average 20–40 μm thick. The palisade tissue represents about 70%, while the spongy mesophyll is about 25% of the entire mesophyll/leaf blade (Rodolfi et al., 2021). According to Konyar et al. (2014), calcium oxalate crystals in the druse shape can be found in hop leaves (Figs. 2.6 and 2.7), specifically around the vascular tissue and in the palisade and spongy tissue of the mesophyll.

The upper epidermis is composed by large cells that appear elongated longitudinally (Rodolfi et al., 2021). On the upper epidermis, it is common to find segregation tissue (Fig. 2.7): glandular and non-glandular trichomes. The non-glandular trichomes consist in cells with intracellular mineralised inclusions (cystoliths,

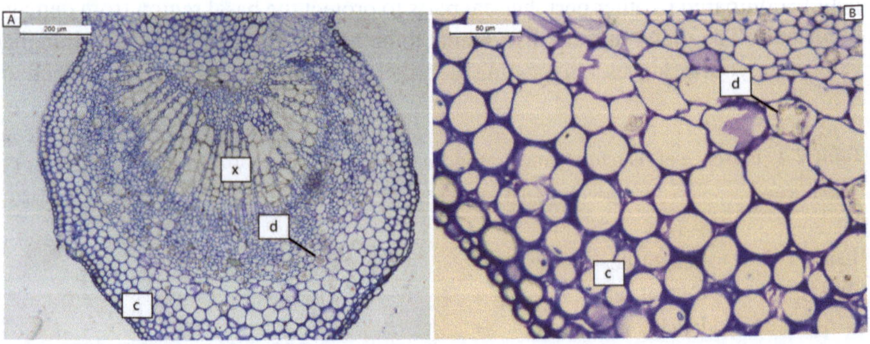

Fig. 2.6 Cross section of a vein of hop leaf (**a**) in which the general microstructure can be observed. A detail of the collenchyma can be seen in (**b**). In the same image, the inclusion of calcium oxalate crystals (druse) can be observed. Legend: c = collenchymatic tissue; d = druse; x = xylem. (Photo: Tommaso Ganino and Giancarlo Cangelli and Marcello Abbadati)

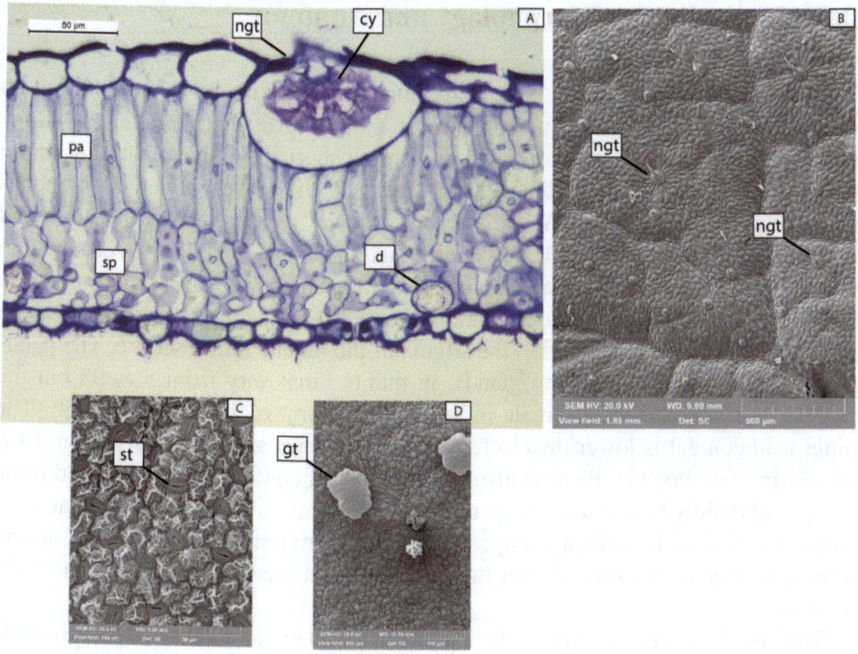

Fig. 2.7 In (**a**), a cross-section of the hop leaf blade (typical structure of dorsoventral leaves) can be observed. On the upper epidermis occasionally appear "non-glandular trichomes" containing cystoliths (intracellular mineralised inclusions). On the inferior epidermis are the stomata. The mesophyll consists of a layer of palisade tissue and a spongy tissue. In (**b**), the upper epidermis observed with scanning electron microscope (SEM). In figure it is possible to observe the spread of non-glandular trichomes over the entire leaf surface. On the lower page (**c** and **d**), stomata, and trichomes (**c**), and the lupulin glands (**d**) can be observed. Legend: ngt = non-glandular trichomes; cy = cystoliths (intracellular mineralised inclusions); pa: palisade tissue; sp. = spongy tissue; d = druse; st = stomata; gt = glandular trichomes. (Photo: Tommaso Ganino and, the photograph (**a**) by Tommaso Ganino and Giancarlo Cangelli and Marcello Abbadati)

Rodolfi et al., 2021). These inclusions consist of calcium carbonate, and a small amount of magnesium (Setoguchi et al., 1989). In general, two types of glandular trichomes were identified on hop leaves (Sugiyama et al., 2006): large peltate-type (lupulin glands—Fig. 2.7d) on the abaxial side, and stalked capitate glands (bulbous trichomes, composed by a stalk and four head cells) on both adaxial and abaxial side, near the leaf veins.

On the lower epidermis are stomata, which open into substomatous chambers that are connected to the intercellular air spaces between the cells of the spongy mesophyll and those of the palisade mesophyll. Both the upper and lower epidermis are protected by a layer of epidermal wax. The epicuticular wax layer of hop leaves is composed by homologous series of hydrocarbons, wax esters, benzyl acyl esters, aldehydes, primary alcohols, fatty acids, and the triterpenoids (Ülz et al., 1993). Primary alcohols are the main wax component, and no single class of lipids is predominant (more than 80%).

2.3 Reproductive Morphology and Anatomy

As hop is a dioecious specie, male and female reproductive organs are present on different plants. The male inflorescence is a densely branched, greenish-yellow cymose panicle with small individual flowers on short stalks that emerge from the upper portion of the main bine and the axil of the leaves of lateral branches (Fig. 2.8a) (Burgess, 1964). When in bloom, flower measure 2 mm in length and 6 mm in width (Mez, 1969). The five lobes that make up the yellow-green sepaloid perianth are separated by five delicate, short filaments at their bases that finish in lengthy anthers. The fine golden pollen (Fig. 2.8b), that anthers freely hang from flowers, is carried by the wind to the stigma of the female inflorescence. The petals are also endowed with lupulin glands, in number that vary from 5 to 98 but the number is smaller than in female inflorescence (Burgess, 1964; Mez, 1969). The bitter acid content is lower than in female inflorescence and could vary from 0.1% to 1%. In male flowers, there is also an essential oil content that was settled to be composed mainly by the sesquiterpenes humulene and caryophyllene, that can constitute the 75% of the total oil composition. Instead, monoterpenes like myrcene and farnesene, that are almost obliquus in female inflorescence, are not revealed in male flowers.

The buds found in the axils of the leaves on lateral branches gave rise to pistillate flowers; the top of the main bine formed short branches with rounded terminal buds. The spherical buds have short stalks that end in "burrs," which are the female inflorescence. Green, with a short central axis, female inflorescences measure 5 mm in length and 6 mm in diameter. They display alternating pairs of stipular bracts. With the exception of the hop variety 'Northern Brewer', where a leaf occasionally forms in between the stipular bracts; these bracts are vestigial of a leaf that disappeared during the ancestral growth of the plants (Burgess, 1964). However, under the strobile bracts, other varieties may also grow leaves in response to environmental or nutritional stressors.

Fig. 2.8 Male flowers grouped in cymose panicle (**a**), and pollen of *H. lupulus* L. (**b**). (Photo: Margherita Rodolfi)

Stipular bracts have tiny branches that start in the axils and end in four tiny protuberances. Each of these protuberances has a female flower enfolded at the base of a minute bracteole (Fig. 2.9). Every single flower has a tiny green perianth, in which, the single-ovule bicarpelate ovary is enclose (Rybacek, 1991). There are two filamentous stigmas on the ovary, and they lack styles. The stigmas protrude, giving the inflorescences a brush-like look that helps them catch pollen that is floating in the air (Burgess, 1964; Rybacek, 1991). During flowering, if a dry spell follows a period of rain, plants that have previously flowered rebloom, creating a second inflorescence (Rybacek, 1991).

Occasionally, plants with both male and female flowers can be observed. Flowering is strongly influenced by the environment and, under stress conditions (such as non-appropriate light and temperature), a female plant may produce a few male flowers. The intensity of this event often varies from season to season. Even cultivars can present this problem with varying intensity: for example, the cultivar 'Cascade' and the cultivar 'Columbus' are particularly notorious for this phenomenon. Genetic analyses have shown that the plants affected by this phenomenon are all genetically (at the chromosome and DNA level) female. Moreover, male flowers produced by female plants are sterile. Truly hermaphrodite plants, capable of producing viable seeds, do exist, but they are extremely rare.

Fig. 2.9 Female flowers grouped in a strobile inflorescence. (Photograph kindly donated by Paolo Focherini, MO, IT)

Because stigmas are atrophic and white, they turn brown and lose their ability to capture pollen relatively rapidly. During the transformation of the flower into a strobile (cone), several modifications occur: the axis lengthens and thickens, the stem changes into a rachis and bract, and the bracteoles enlarge, transform into covering bracteolets, and finally become real bracts (Fig. 2.10a, b). Fine hairs are rare in bracts and bracteoles. The lupulin, which is composed of multicellular lupulin glands, is the hop cone's most commercially significant component. At flower bud stage, lupulin glands are present in rudimentary form, which take on a cup-shaped or globular form upon maturation. These glands grow quickly and secrete more resins and essential oils, which are in charge of the golden yellow colour.

Sometimes, the pressure exerted by the buildup of lupulin, causes the cuticles of overripe hops to break. Nonetheless, lupulone and bracteoles are brittlely attached, and lupulone may be lost in small amounts due to wind or during the picking and drying processes. A single ovule in the ovary develops into two coiled embryos upon fertilization, and the pericarps harden and turn brown as the fruit ripens. The stigmas of unfertilised blooms elongate and eventually die off. The seedless cone develops more slowly and is smaller than the fertilised one, but it yields more lupulone. The bracteole first covers the seeds or remnants of the unfertilised blooms.

Fig. 2.10 Strobile of hops (also known as hop cones) (**a**), composed of bracts (**b**) each containing a fruit (achenes, **c**). (Photo: Margherita Rodolfi)

One seeded achene is the hops' fruit (Fig. 2.10c). Mature hop strobiles may contain up to 18% of their weight in seeds. The occurrence of a male plant that flowers at the same time as the female plant, the direction in which the wind moves while carrying pollen, the environment, and the timing of flowering are all factors that affect pollination and consequently the development of seeds. Hops variety affects seed output, but female flower fertilization has a particular control on it. In the commercial realm, seeds are frequently deemed undesirable. According to Regulation No. 1850/2006 EC, hops must declare on their label that they are "seeded hops" if the percentage of seeds in their weight surpasses 2%. In general, female hop seedlings were more common than male ones, which could be explained by parthenogenetic seed production. Furthermore, female seeds possess the tendency to germinate earlier than male (Neve, 1991).

2.4 Vegetative and Reproductive Physiology

2.4.1 Annual Cycle and Phenological Growth Stages

Hop plant is characterised by a perennial overwintering rhizome and annual climbing bines that die back each autumn (McAdam et al., 2014). The annual cycle of this species is marked by a succession of nine principal phenological growth stages, namely (0) sprouting; (1) leaf development; (2) formation of side shoots; (3) elongation of bines; (5) inflorescence emergence; (6) flowering; (7) development of cones; (8) maturity of cones; (9) senescence and entry into dormancy (Rossbauer et al., 1995; Rossini et al., 2021). According to this numbering and the BBCH-scale system, Rossbauer et al. (1995) accurately described each hop secondary growth stage, to which a proper two-digit code has been assigned (Table 2.5). The document was originally written in German, later translated in English by Meier in his monograph (Meier, 2001).

The first phenological stage (stage 0) includes the dormancy (or cryptovegetation) period of the hop rhizome (approx. between the second half of October and the beginning of the spring in the northern hemisphere), thanks to which this species is able to survive during the cold winter (Acosta-Rangel et al., 2024). With the break of dormancy, the activation of the buds on the upper part of the perennial wood occurs, followed by their underground growth and finally the shoot sprouting above the soil surface (beginning the vegetative phase) (Rybáček, 1990) (Fig. 2.11). In this phase, the brittle shoots grow linearly until the first three visible nodes are formed (Adams, 2021). Shortly afterwards, hop shows its nature of climbing plant as soon as the tip of the growing primary shoots move in space looking for a proper support to which twist (BCCH 12) (Rybáček, 1990).

The vegetative growth continues (BCCH 13-39) during the late spring and throughout the summer, with the thickening of the main bines diameter and the vigorous and fast growth in its upward climb (Fig. 2.12a); at the same time, an

Table 2.5 Phenological growth stages and their respective BBCH-scale code (Rossbauer et al., 1995; Meier, 2001)

Growth stage	Code	Description	Growth stage	Code	Description
0: Sprouting	0	Dormancy: rootstock without shoots (uncut)	6: Flowering	61	Beginning of flowering: about 10% of flowers open
	1	Dormancy: rootstock without shoots (cut)		62	About 20% of flowers open
	7	Rootstock with shoots (uncut)		63	About 30% of flowers open
	8	Beginning of shoot-growth (rootstock cut)		64	About 40% of flowers open
	9	Emergence: first shoots emerge at the soil surface		65	Full flowering: about 50% of flowers open
1: Leaf development	11	First pair of leaves unfolded		66	About 60% of flowers open
	12	2nd pair of leaves unfolded (beginning of twining)		67	About 70% of flowers open
	13	3rd pair of leaves unfolded		68	About 80% of flowers open
	1.	Stages continuous till ...		69	End of flowering
	19	9 and more pairs of leaves unfolded	7: Development of cones	71	Beginning of cone development: 10% of inflorescences are cones
2: Formation of side shoots	21	First pair of side shoots visible		75	Cone development halfway: all cones visible, cones soft, stigmas still present
	22	2nd pair of side shoots visible		79	Cone development complete: nearly all cones have reached full size
	23	3rd pair of side shoots visible	8: Maturity of cones	81	Beginning of maturity: 10% of cones are compact
	2.	Stages continuous till ...		82	20% of cones are compact
	29	Nine and more pairs of side shoots visible (secondary side shoots occur)		83	30% of cones are compact

(continued)

Table 2.5 (continued)

Growth stage	Code	Description	Growth stage	Code	Description
3: Elongation of bines	31	Bines have reached 10% of top wire height		84	40% of cones are compact
	32	Bines have reached 20% of top wire height		85	Advanced maturity: 50% of cones are compact
	33	Bines have reached 30% of top wire height		86	60% of cones are compact
	3.	Stages continuous till ...		87	70% of cones are compact
	38	Plants have reached the top wire		88	80% of cones are compact
	39	End of bine growth		89	Cones ripe for picking: cones closed; lupulin golden; aroma potential fully developed
5: Inflorescence emergence	51	Inflorescence buds visible	9: Senescence, entry into dormancy	92	Overripeness: cones yellow-brown discoloured, aroma deterioration
	55	Inflorescence buds enlarged		97	Dormancy: leaves and stems dead

The first digit of the code refers to the principal growth stage, while the second digit refers to the secondary growth stage which corresponds to an ordinal number or percentage value

Fig. 2.11 Shoots sprouting above the soil surface after winter. (Photograph kindly donated by Paolo Focherini, MO, IT)

Fig. 2.12 Vigorous and fast growth of the shoot (**a**) and subsequent formation of lateral branches (**b**). (Photograph kindly donated by Paolo Focherini, MO, IT)

intense development of lateral branches occurs (Rybacek, 1991; De Keukeleire et al., 1999) (Fig. 2.12b).

In contrast with Rossbauer et al. (1995) and Meier (2001), Dodds (2017) propose in his guide the necessity of reverse the order of the BBCH growth stages 2 (formation of side shoots) and the BBCH growth stages 3 (elongation of bines), since side shoots generally grow after the elongation of bines: when the shoots reach their maximum height, the apical dominance is reduced. Therefore, the numbering and the growth stages order, according to Dodds (2017), are: (1) sprouting; (2) leaf development; (3) elongation of bines; (4) formation of side shoots; (5) flower emergence; (6) flowering; (7) development of cones; (8) maturity of cones; (9) senescence (start of dormancy).

Hop vegetative growth stops when first reproductive structures make their appearance (BBCH 51): in particular, male inflorescences develop on the terminal portion of the main stem, while female inflorescences grow on the upper terminal one (both of them on secondary and tertiary branches) (Haunold, 1980; Gutierrez et al., 2023). Before anthesis, male inflorescences resemble a panicle; the flowers have filaments and anthers completely formed, and the latter are ready to release pollen for dehiscence (Haunold, 1980). Female inflorescences develop through a "pin" stage (stigmas just visible) and the following "burr" stage (young receptive inflorescence with stigmas fully extended) (Shephard et al., 2000). In this latter stage, the synthesis of resins begins (Skinner et al., 1974). Flowering occurs in summer (BBCH 61–69), followed by a very rapid growth of the strobili (cones) on female plants, which last at least a month (BBCH 71–79) (Gutierrez et al., 2023). At full development (BBCH 79), the cones reached their maximum size, their bracts are open and the lupulin is poorly coloured (Skomra & Koziara-Ciupa, 2020). Successful pollination produces an achene (fruit) enclosed loosely within the vestigial perianth whorl (Shephard et al., 2000).

Harvesting of unfertilised strobili for industrial purposes has to be performed at fully maturity, approximately 5–6 week after the "burr" stage (BBCH 89) (Skinner et al., 1974). At this phenological stage (mid-to-late August thought mid-to-late September), strobili green colour loses vividness, and bracts and bracteoles edges start to became slightly browned; the amount of cone humidity reduces, the powdery dark-golden lupulin reach the centre of the strobilus and the peculiar hop aroma is dominating over all other perceivable aromatic notes (Adams, 2021). Over-ripeness (BBCH 92) occur with the embrittlement of hop strobili, and the physical losses of lupulin during harvesting and kilning (with a consequent decrease in resin and essential oil yield) (Skinner et al., 1974).

Lastly (BBCH 97), all the above-ground organs die back; this process of decay start with the top of bines and the upper shoots, continuing downwards to reach the part of the bine near the soil; underground organs receive nutritional reserves by senescent bines, getting ready to face the period of dormancy (Rybáček, 1990; Gutierrez et al., 2023).

2.4.2 *Physiology of the Phenological Growth Stages*

2.4.2.1 Photosynthetic Activity

The vegetative growth of hops begins when the dormancy of the rhizome is broken. Shoot sprouting is initially sustained by the nutritive reserves (mainly polysaccharides converted in monosaccharides) stocked in the underground organs at the end of the annual cycle (Rybáček, 1990). The photosynthetic processes are fairly limited in young expanding leaves (BCCH 12), because of their small leaf area (and hence the reduce ability to capture solar radiation), of the low amount of photosynthetic pigments and the poor carbon dioxide (CO_2) fixation (the latter is also duc to low stomatal conductance by small and underdeveloped stomata) (Kenny, 2004; Pokorný et al., 2011; Hejnák et al., 2014). In these conditions, the respiration and the capacity for non-radiative thermal dissipation are high (Hejnák et al., 2014). Therefore, the photosynthetic rate is closely linked to leaf development, reaching its full potential with an expansion of at 75% or greater (Kenny, 2004). Hereafter, the photosynthetic activity rises progressively throughout the subsequent phenological stages (drastically between BCCH 32 and BCCH 37), and the highest values have been reported during the reproductive phase (Rybáček, 1990; Pokorný et al., 2011, 2012; Hejnák et al., 2014). Indeed, the photosynthetic capacity is particularly high during the flowering (BCCH 61–69) and hop cone formation phases (BBCH 71–79) (Rybáček, 1990), even if relevant values were also observed during the maturity of strobili (BCCH 81–89) (Pokorný et al., 2011, 2012; Hejnák et al., 2014). In this regard, it is important to note that the leaves that develop in those specific phenological phases grow on branches of different levels, contributing to the overall plant photosynthetic rate once they reach their full maturity. In particular, the leaves of the first and second node reached their photosynthetic maturity from the 61 BBCH phase, while the ones of third and fourth level reached the same developmental stage from the 81 BBCH phase (Hejnák et al., 2014). Noteworthy is also the contribution of hop strobili, which have photosynthetically active bracts that act as a source of carbon assimilation, also pivotal for lupulin biosynthesis (Bauerle, 2023). Nevertheless, a decrease in the photosynthetic activity occurs with the overripeness of the strobili (yellow-brown discolouration) and leaf senescence (Rossbauer et al., 1995; Meier, 2001).

As previously reported, the hop vegetative stage is characterised by a fast and vigorous growth that, undoubtedly, require high amount of photoassimilates to sustain plant climbing and the side shoot formation. An increase in the photosynthetic activity is thus necessary to cope with this requirement. The activity of the photosynthetic apparatus rises again during the generative stage, since inflorescence and strobili increase the plant energy demand acting as carbon sinks (Brazel & Ó'Maoiléidigh, 2019). Nevertheless, green reproductive organs (mainly strobili) give a remarkable contribution to the overall plant photosynthetic activity, particularly noticeable after a certain stage of development and when leaves begin to senescence (Brazel & Ó'Maoiléidigh, 2019; Bauerle, 2023).

Considering the intrinsic growth habit of hop plants, some additional consideration worth to be mentioned. First of all, the plant reaches considerable heights and its canopy is quite dense due to numerous lateral and sublateral branches; this cause uneven irradiance in the height profile that affect the photosynthetic efficiency of the leaves, which could be thus reduced in the shaded ones (Kenny, 2004; Hniličková et al., 2007). Nevertheless, all leaves throughout the canopy have a high photosynthetic potential, which is promptly express by shaded leaves when they are exposed again to the direct sunlight (Kenny, 2004).

Lastly, hop plants are characterised by heterophylly: young leaves are heart-shaped, while mature leaves have tree, five to seven lobes; the offshoot leaves are often compact or with three to five lobes (Hejnák et al., 2014). Differences in leaf area can mirror differences in photosynthetic activity at full development (Hejnák et al., 2014).

2.4.2.2 Growing Degree Days (GGD)

Growth and development of hop plants are strongly dependent on temperature, and both parameters can be estimated, during the growing season, by Growing Degree Days (GDD), also known as heat units or thermal time (Rossini et al., 2021). GDD are used to describe the heat energy received by a specific crop over a given time period, relating it to progress in development or growth processes (McMaster & Wilhelm, 1997); this is due to the fact that plants require a specific amount of heat to develop from one point in their lifecycle to the next one (Miller et al., 2001). Therefore, GDD can be used retrospectively to calculate the current growth stage of a crop, or to help forecast the date that a crop will reach a predetermined growth stage or phenology models (Miller et al., 2001).

Until now, very few investigations have been performed to calculate GDD for hop's main phenological phases (Rossini et al., 2016; Ruggeri et al., 2018; Marceddu et al., 2020). All studies have been conducted in Italy, highlighting a high variability of GGD according to hop genotypes, maturity timelines, and environmental and growth conditions (greenhouse or open field).

As regards shoot emergence (BBCH 09), earliest sprouting cultivars ('Cascade', 'Hallertauer Magnum', 'Yeoman') recorded a mean thermal time need of less than 160 GGD (119.3, 130.0, and 154.2 GGDs, respectively), while the latest ones ('Tettnager', 'Hallertauer Bitter', 'Perle') have surpassed the 230 GGD (233.5, 240.2, and 278.73 respectively) (two-year open field study, 2013–2014; Central Italy, 42°26 N, 12° 04 E, altitude 310 m a.s.l.) (Ruggeri et al., 2018). These values change significantly if the third consecutive year (2015) has taken into account: cv 'Cascade' shoots need 243.5 GGD to emergence, while those of cv 'Yeoman' required a mean thermal time of 286.4 GGD. Differences have been also highlighted for the latest sprouting cv 'Perle', which shoot emergence occurred with 395.4 GGD (Rossini et al., 2016). This high variability in shoot emergence time, observed between consecutive years, is probably due to the fact that the hop plant reaches its maturity after three growing seasons, so in the first few years after planting the plant is not yet fully adapted to the new conditions. The plants yield will also not be stable until the fourth year after planting.

The thermal time for the whole hop growing period (sprouting, BCCH 09; flowering, BBCH 65; development of cones, BCCH 71; maturity of cones, BBCH 89) has been investigated for 20 hop cultivars in a single scientific publication, which reported heat units ranging between 1739.0 GGD of early-maturing cv 'Pheonix' and 2168.8 GGD of late-maturing cv 'Hallertauer Spat' (Rossini et al., 2016).

The correlation between GGD and hop yield and quality is of particular interest for industrial purposes. Until now, a negative correlation has been observed between GGDs and both cones yield (grams of dry matter, g DM) and hop biomass production (dry plant weight, g DM). On the contrary, dry cones weight showed to be positively correlated to GDDs accumulation along the reproductive stages (from the onset of flowering to full ripening) (Marceddu et al., 2020). Moreover, hop biomass production seems to be maximised with smaller accumulation of thermal units, suggesting that the most productive genotypes might show a more efficient response to the growth environment's conditions, by having a rapid and vigorous vegetative development (Marceddu et al., 2020).

2.4.2.3 Photoperiod

Photoperiod, namely the amount of light and darkness in a daily cycle of 24 hours (h), regulates many aspects of plant development. Daylength modulate plant growth in response to seasonal changes and synchronise key developmental transitions with favourable conditions to sustain their ecological fitness (i.e., survival of adult organisms able to produce progeny before their death) (Osnato et al., 2022). In northern latitudes, for example, shortening daylength in autumn is used as a cue by many trees and perennial species for the induction of cold hardiness and bud dormancy in anticipation of the freezing winter temperatures, whereas ground-level woodland plants may use short days to induce flowering in early spring, enabling them to complete seed production before the leaf canopy fully forms and limits the available light (Jackson, 2009).

As concerns hop, the rhizome dormancy require a daylength less than 10.5 h, the budbreak induction is triggered by more than 11.2 h of daylight, and the vegetative growth occurs in a great extent with a photoperiod above 13 light hours (Gutierrez et al., 2023; Acosta-Rangel et al., 2024). Regarding flowering time, hop is considered a short-day plant (SDP), which transition from the vegetative to regenerative growth stage is induced when the photoperiod is shorter than the critical daylength (CDL) that, for this species, is 16.5 h of daylight (Jackson, 2009; Gutierrez et al., 2023). The optimal flowering takes place at near-critical photoperiods, which they are short enough to induce flowering and still long enough to prevent premature buds' dormancy (Faragó et al., 2009). Despite this general information, the optimum light pattern for hops flowering time may change between different cultivars, as the CDL is variable and usually genotype specific (Shephard et al., 2000).

2.4.2.4 Phenotypical Factors Affecting Flowering Time

Hop flowering time is also affected by the physical height and size of the main bine; indeed, this specie has to reach a genotypic minimum maturity requirement before a reproductive inductive photoperiod can be signalled (Thomas & Schwabe, 1969). This size requirement, known as juvenility, ensures the plant is an adequate developmental stage to properly sustain flowering (Adams, 2003).

2.5 Reproductive Biology

As previously mentioned, hop is dioecious (female and male flowers develop in separate plants), although occasionally monoecious individuals may naturally occur (Edwardson, 1952). These latter, definitely very rare, produce terminal female flowers on inflorescences of male plants (Ainsworth, 2000).

In most wild populations, female plants are about twice as numerous as males, and unisexual colonies are often established by extensive spread of the rhizomes (vegetative propagation, individuals with the same genetical information) (Small, 2016). However, crossbreeding, i.e. sexual reproduction, is a powerful means for plants (and higher organisms in general) to introduce genetic variability (Haunold, 1980).

Hop is normally diploid ($2n = 2x = 20$) with nine autosomal bivalents (A) and two distinct sex chromosomes (X and Y) (hop genome size: 2.8 pg—2.7 Gigabase pairs) (Grabowska-Joachimiak et al., 2006; Škof et al., 2012). Female and male plants have homomorphic (XX) and heteromorphic (XY) sex chromosomes, respectively (Neve, 1991; Shephard et al., 2000). X and Y chromosomes are known to share rather low homology, and the latter is smaller (Divashuk et al., 2011; Škof et al., 2012).

Sex determination is dependent on an X/A balance system, according to which a ratio of the number of X-chromosomes to the number of sets of autosomes of 0.5 gives rise to a male plant; a ratio of 1.0 gives a female plant, while intermediate ratios give rise to intersexual, monoecious individuals (Shephard et al., 2000; McAdam et al., 2013). Triploids are mainly male with a few terminal female inflorescences ($2n$ = 27XXY; X/A ratio = 0.67), while tetraploids generally carry approximately equal numbers of male and female flowers ($2n$ = 36XXY; X/A ratio = 0.75) (Shephard et al., 2000). Polyploid hop plants occasionally occur in the progeny of crossbreed plants by human intervention (Škof et al., 2012).

The X/A balance system suggest that the sex expression in hop plants may be regulated by genes on the autosomes or may be X-linked; indeed, the Y chromosome is not essential to the development of the male phenotype, even if it is required for pollen maturation, thus affecting its fertility (Škof et al., 2012; McAdam et al., 2014).

For many plants cultivated for industrial purposes, triploids are preferred over individuals with other sets of chromosomes. Indeed, in comparison with diploid hop plants, triploids show:

1. a higher shoot twist index (namely the ability to climb up on wires or trellis);
2. greater vigour and higher yields;
3. a more rapid closure of cones after their formation (reducing the penetration of the pathogenic fungi);
4. more robust and less brittle cones when harvested (Trojak-Goluch et al., 2021).

Moreover, triploids are sterile and fruit production is, therefore, avoided (Trojak-Goluch et al., 2021).

2.5.1 Pollination and Fruit Development

Hop male plants produce staminate flowers, containing large amounts of pollen (Wodehouse, 1971). Pollen grains (Fig. 2.8b) are approximately 20 μm in size, oblate (equatorial view) to circular (polar view) in shape, with a thin, scabrata-granulata exine and three slightly protrusive pori (Kallajxhiu et al., 2018; Liberatore et al., 2019). Pollen is released by dehiscence of the anthers, which mainly occur under favourable atmospheric conditions (warm and dry air) (Wormald, 1915; Burgess, 1964). The pollination is anemophilous (with the help of the wind), and pollen may travel over distances of several kilometres without loss of viability (Haunold, 1980). The length of the pollen shedding period is usually associated with the plant vigour and it may last from 4 to 52 days (Brooks, 1963).

Before pollination, the bracts of female strobili are small and only the basal ones are visible (Edwardson, 1952). Stigmas receptivity to pollen last up to 3 weeks, at the end of which they begin to drop off from the inflorescences (Haunold, 1980). If the pollination takes place, the bracteoles enlarge, the indehiscent fruit (achene) starts to develop and, in a short time, overall strobilus size increase, taking on the characteristic "fir-cone" appearance (Edwardson, 1952; Haunold, 1980). Fruits grow enfolded by the bracteoles and surrounded by the resin glands; inside them, the coiled embryo with two cotyledons and good amount of food reserves (mainly lipids and proteins) are present (Haunold, 1980). At maturity (after 45–50 days after pollination), the fruit is small (4–7 mg of weight), hard waterproof, exhibiting a long dormancy and slow germination (Haunold, 1980; Korpelainen & Pietiläinen, 2021). Achenes dissemination rely on wind, which waft them far away from the mother plant, while they are still attached the bracteoles (Small, 2016).

In hop fields, fruits production is not desirable, since they can produce off-flavours in beer, due to the oxidation of the fatty acids (Harrison, 1971). For this reason, male plants are promptly removed to prevent the pollination of female inflorescence (Adams, 2021). Unfertilised strobili are comparatively small, but generally richer in essential oils and resins, characteristic the enhance their brewing value (Almaguer et al., 2014).

References

Acosta-Rangel, A., Agehara, S., & Rechcigl, J. (2024). Double-season production of hops (*Humulus lupulus* L.) with photoperiod manipulation in a subtropical climate. *Scientia Horticulturae, 332*, 113177. https://doi.org/10.1016/j.scienta.2024.113177

Adams, S. R. (2003). Regulators of growth: Juvenility. In *Encyclopedia of applied plant sciences* (pp. 1051–1056). Elsevier - Academic Press

Adams, S. A. (2021). *Cultivating hops for cone production in Nebraska*. Nebraska Extension. EC3050. https://digitalcommons.unl.edu/agronomyfacpub/1438

Ainsworth, C. (2000). Boys and girls come out to play: The molecular biology of dioecious plants. *Annals of Botany, 86*(2), 211–221. https://doi.org/10.1006/anbo.2000.1201

Almaguer, C., Schönberger, C., Gastl, M., Arendt, E. K., & Becker, T. (2014). Humulus lupulus–a story that begs to be told. A review. *Journal of the Institute of Brewing, 120*(4), 289–314.

Bauerle, W. L. (2019). Disentangling photoperiod from hop vernalization and dormancy for global production and speed breeding. *Scientific Reports, 9*(1), 16003.

Bauerle, W. L. (2023). *Humulus lupulus* L. strobilus photosynthetic capacity and carbon assimilation. *Plants, 12*(9), 1816.

Beard, F. H. (1943). Root studies. X. The root-systems of hops on different soil types. *Journal of Pomology and Horticultural Science, 20*(4), 147–154. https://doi.org/10.1080/03683621.1943.11513608

Bell, P. R. (1958). Twining of the hop (*Humulus lupulus* L.). *Nature, 181*, 1009–1010. https://doi.org/10.1038/1811009a0

Bocquet, L., Sahpaz, S., Hilbert, J.L., Rambaud, C. & Rivière, C. (2018). Humulus lupulus L. a very popular beer ingredient and medicinal plant: Overview of its phytochemistry, its bioactivity, and its biotechnology. *Phytochemistry Reviews, 17*, 1047–1090. https://doi.org/10.1007/s11101-018-9584-y

Brant, V., Krofta, K., Kroulík, M., Zábranský, P., Procházka, P., & Pokorný, J. (2020). Distribution of root system of hop plants in hop gardens with regular rows cultivation. *Plant, Soil and Environment, 66*(7), 317–326. https://doi.org/10.17221/672/2019-PSE

Brazel, A. J., & Ó'Maoiléidigh, D. S. (2019). Photosynthetic activity of reproductive organs. *Journal of Experimental Botany, 70*(6), 1737–1754.

Brooks, S. N. (1963). Relation of training date to pollen shedding in male hops, *Humulus lupulus* L. *Crop Science, 3*, 275–277.

Burgess, A. H. (Ed.). (1964). *Hops: Botany, cultivation, and utilisation*. Interscience Publishers.

De Keukeleire, D., De Cooman, L., Rong, H., Heyerick, A., Kalita, J., & Milligan, S. R. (1999). Functional properties of hop polyphenols. In *Plant polyphenols 2: Chemistry, biology, pharmacology, ecology* (pp. 739–760). Springer, Boston, MA.

Divashuk, M. G., Alexandrov, O. S., Kroupin, P. Y., & Karlov, G. I. (2011). Molecular cytogenetic mapping of Humulus lupulus sex chromosomes. *Cytogenetic and Genome Research, 134*(3), 213–219.

Dodds, K. A. (2017). *Hops: A guide for new growers*. New South Wales Department of Primary Industries.

Edwardson, J. R. (1952). Hops—Their botany, history, production and utilization. *Economic Botany, 6*(2), 160–175. https://doi.org/10.1007/bf02984875

Faragó, J., Pšenáková, I., & Faragová, N. (2009). The use of biotechnology in hop (*Humulus lupulus* L.) improvement. *Nova Biotechnologica, 9*(3), 279–293.

Gloser, V., Baláž, M., & Svoboda, P. (2011). Analysis of anatomical and functional traits of xylem in Humulus lupulus L. stems.

Grabowska-Joachimiak, A., Sliwinska, E., Pigula, M., Skomra, U., & Joachimiak, A. J. (2006). Genome size in *Humulus lupulus* L. and *H. japonicus* Siebold & Zucc. (Cannabaceae). *Acta Societatis Botanicorum Poloniae, 75*, 207–214.

Graf, T., Beck, M., Mauermeier, M., Ismann, D., Portner, J., Doleschel, P., & Schmidhalter, U. (2014). *Humulus lupulus* - The hidden half. *BrewingScience, 67*, 161–166.

Gutierrez, R. M., de Oliveira, R. R., Ribeiro, T. H. C., de Oliveira, K. K. P., Silva, J. V. N., Alves, T. C., do Amaral, L.R., de Souza Gomes, M., de Souza Gomes, M., & Chalfun-Junior, A. (2023). Unveiling the phenology and associated floral regulatory pathways of *Humulus lupulus* L. in subtropical conditions. *bioRxiv*, 2023-09. https://doi.org/10.1101/2023.09.15.557988.

Harrison, J. (1971). Effect of hop seeds on beer quality. *Journal of the Institute of Brewing, 77*, 350–352.

Haunold, A. (1980). Hop. In: Hybridization of crop plants, Book Editor(s): Walter R. Fehr, Henry H. Hadley, pp. 393–406. doi:https://doi.org/10.2135/1980.hybridizationofcrops.c27.

Haunreiter, K. J., Dichiara, A., & Gustafson, R. (2021). Structural and chemical characterization of hop bine fibers and their applications in the paper industry. *Industrial Crops and Products, 174*(114217). https://doi.org/10.1016/j.indcrop.2021.114217

Hejnák, V., Hniličková, H., & Hnilička, F. (2014). Effect of ontogeny, heterophylly and leaf position on the gas exchange of the hop plant. *Plant, Soil and Environment, 60*(11), 525–530.

Hniličková, H., Hnilička, F., & Krofta, K. (2007). Determining the saturation irradiance and photosynthetic capacity for new perspective varieties of hop (*Humulus lupulus* L.). *Cereal Research Communications, 35*(2), 461–464.

Jackson, S. D. (2009). Plant responses to photoperiod. *New Phytologist, 181*(3), 517–531.

Jastrombek, J. M., Faguerazzi, M. M., de Cássio Pierezan, H., Rufato, L., Sato, A. J., da Silva Ricce, W., Vieira Marques, V., Rodrigues Leles, N., & Roberto, S. R. (2022). Hop: An emerging crop in subtropical areas in Brazil. *Horticulturae, 8*(5), 393.

Kallajxhiu, N., Kapidani, G., Naqellari, P., Pupuleku, B., Turku, S., Dauti, A., & Jançe, A. (2018). An allergopalynologic comparative study of the Humulus Lupulus' pollen grains (Cannabaceae) in different habitats. *Thalassia Salentina, 40*(2), 75–82.

Kenny, S. T. (2004). Photosynthetic measurements in hop (Humulus). In *I International Humulus symposium 668* (pp. 241–248).

Koie, K., Myoda, T., Yoshida, H., & Itoh, H. (2018). Quantitative evaluation of the hop (Humulus lupulus L.) root system based on wall profile method. *Plant Root, 12*, 11–15. https://doi.org/10.3117/PLANTROOT.12.11

Konyar, S. T., Öztürk, N., & Dane, F. (2014). Occurrence, types and distribution of calcium oxalate crystals in leaves and stems of some species of poisonous plants. *Botanical Studies, 55*(1). https://doi.org/10.1186/1999-3110-55-32

Korpelainen, H., & Pietiläinen, M. (2021). Hop (*Humulus lupulus* L.): Traditional and present use, and future potential. *Economic Botany, 75*(3), 302–322.

Liberatore, C. M., Calabuig-Serna, A., Rodolfi, M., Chiancone, B., & Seguí-Simarro, J. M. (2019). Phenological phases of flowering in hop (*Humulus lupulus* L.) and their correspondence with microsporogenesis and microgametogenesis. *Scientia Horticulturae, 256*, 108639.

Marceddu, R., Carrubba, A., & Sarno, M. (2020). Cultivation trials of hop (*Humulus lupulus* L.) in semi-arid environments. *Heliyon, 6*, e05114.

McAdam, E. L., Freeman, J. S., Whittock, S. P., Buck, E. J., Jakše, J., Čerenak, A., Javornik, B., Kilian, A., Wang, C. H., Andersen, D., Vaillancourt, R. E., Carling, J., Beatson, R., Graham, L., Graham, D., Darby, P., & Koutoulis, A. (2013). Quantitative trait loci in hop (Humulus lupulus L.) reveal complex genetic architecture underlying variation in sex, yield and cone chemistry. *BMC Genomics, 14*(1), 360. https://doi.org/10.1186/1471-2164-14-360

McAdam, E. L., Vaillancourt, R. E., Koutoulis, A., & Whittock, S. P. (2014). Quantitative genetic parameters for yield, plant growth and cone chemical traits in hop (*Humulus lupulus* L.). *BMC Genetics, 15*, 1–18.

McMaster, G. S., & Wilhelm, W. W. (1997). Growing degree-days: One equation, two interpretations. *Agricultural and Forest Meteorology, 87*(4), 291–300.

Meier, U. (2001). BBCH-monograph: Growth stages of mono-and dicotyledonous plants. In *Technical report* (2nd ed., p. 158). Federal Biological Research Centre for Agriculture and Forestry.

Metcalfe, C. R., & Chalk, L. (1957). *Anatomy of the dicotyledons* (Vol. 1 & 2, p. 806). Oxford at the Clarendon Press.

Meyer, B. S., & Anderson, D. B. (1952). *Plant physiology*. Van Nostrand, Ed. (2nd ed.).

Mez, E. (1969). Luppolo. In: Enciclopedia Agraria Italiana. Ed. Ramo Editoriale Degli Agricoltori, *Roma, 6*, 1064–1077.

Miller, R. H. (1958). Morphology of *Humulus lupulus*. I. Development of the primary root. *American Journal of Botany, 45*(5), 418–431. http://www.jstor.org/stable/2439643

Miller, R. H. (1959). Morphology of Humulus lupulus. II. Secondary growth in the root and seedling vascularization. *American Journal of Botany, 46*(4), 269–277.

Miller, P., Lanier, W., & Brandt, S. (2001). Using growing degree days to predict plant stages. *Ag/Extension Communications Coordinator, Communications Services, Montana State University-Bozeman, Bozeman, MO, 59717*(406), 994–2721.

Morcol, T. B., Matthews, P. D., & Kennelly, E. J. (2021). Differences in leaf chemistry and glandular trichome density between wild southwestern American hop (*Humulus neomexicanus*) and commercial hop cultivars. *Journal of Agricultural and Food Chemistry, 69*(27), 7798–7814. https://doi.org/10.1021/acs.jafc.1c02710

Murakami, A., Darby, P., Javornik, B., Pais, M. S. S., Seigner, E., Lutz, A., & Svoboda, P. (2006). Molecular phylogeny of wild hops, *Humulus lupulus* L. *Heredity, 97*(1), 66–74.

Naraine, S. G. U., & Small, E. (2017). Germplasm sources of protective glandular leaf trichomes of hop (Humulus lupulus). *Genetic Resources and Crop Evolution, 64*(7), 1491–1497. https://doi.org/10.1007/s10722-017-0540-2

Neve, R. A. (1991). *Hops* (1st ed.). Chapman and Hall.

Oliveira, M. M., Salomk, M., Pais, S., Oliveira, M. M., & Pais, M. S. S. (1988). Glandular trichomes of Humulus lupulus var. Brewer's Gold: Ontogeny and histochemical characterization of the secretion. *Nordic Journal of Botany, 8*(4), 349–359.

Osnato, M., Cota, I., Nebhnani, P., Cereijo, U., & Pelaz, S. (2022). Photoperiod control of plant growth: Flowering time genes beyond flowering. *Frontiers in Plant Science, 12*, 805635.

Pokorný, J., Pulkrábek, J., Štranc, P., & Bečka, D. (2011). Photosynthetic activity of selected genotypes of hops (*Humulus lupulus* L.) in critical periods for yield formation. *Plant, Soil and Environment, 57*(6), 264–270.

Pokorný, J., Krivánek, J., Jezek, J., & Pulkrábek, J. (2012). Characterization of changes in photosynthetic rate, transpiration and chlorophyll content during the hop growing season. In *III International Humulus Symposium 1010* (pp. 163–168).

Ramos, F. P., Iwamoto, L., Piva, V. H., & Teixeira, S. P. (2024). Updating the knowledge on the secretory machinery of hops (*Humulus lupulus* L., *Cannabaceae*). *Plants, 13*(6), 864. https://doi.org/10.3390/plants13060864

Rodolfi, M., Barbanti, L., Giordano, C., Rinaldi, M., Fabbri, A., Pretti, L., Casolari, R., Beghé, D., Petruccelli, R., & Ganino, T. (2021). The effect of different organic foliar fertilization on physiological and chemical characters in hop (*Humulus lupulus* L., cv Cascade) leaves and cones. *Applied Sciences, 11*(15), 6778.

Rodolfi, M., Rinaldi, M., Caligiani, A., Paciulli, M., Lolli, V., Chiancone, B., & Ganino, T. (2022). Hop green sprouts preservation and valorisation as semi-finished and finished products: Impact of different treatments on microstructural, physical and chemical traits. *European Food Research and Technology, 248*(5), 1203–1215.

Rossbauer, G., Buhr, L., Hack, H., Hauptmann, S., Klose, R., Meier, U., Stauss, R., & Weber, E. (1995). Phänologische Entwicklungsstadien von Kultur-Hopfen (*Humulus lupulus* L.). *Nachrichtenbl. Deut. Pflanzenschutzd., 47*, 249–253.

Rossini, F., Loreti, P., Provenzano, M. E., De Santis, D., & Ruggeri, R. (2016). Agronomic performance and beer quality assessment of twenty hop cultivars grown in Central Italy. *Italian Journal of Agronomy, 11*(3), 180–187.

Rossini, F., Virga, G., Loreti, P., Iacuzzi, N., Ruggeri, R., & Provenzano, M. E. (2021). Hops (*Humulus lupulus* L.) as a novel multipurpose crop for the Mediterranean region of Europe: Challenges and opportunities of their cultivation. *Agriculture, 11*(6), 484.

Ruggeri, R., Loreti, P., & Rossini, F. (2018). Exploring the potential of hop as a dual purpose crop in the Mediterranean environment: Shoot and cone yield from nine commercial cultivars. *European Journal of Agronomy, 93*, 11–17.

Rybáček, V. (Ed.). (1990). *Hop production*. Elsevier Science Publishers.

Rybacek, V. (1991). *Hop Production*, 1st ed. Elsevier Science: Amsterdam, The Netherlands, ISBN 9780444987709.

Schweingruber, F. H., Börner, A., & Schulze, E. D. (2011). *Atlas of stem anatomy in herbs, shrubs and trees*. Springer Science & Business Media.

Setoguchi, H., Okazaki, M., & Suga, S. (1989). Calcification in higher plants with special reference to cystoliths. In *Origin, evolution, and modern aspects of biomineralization in plants and animals* (pp. 409–418). Springer US.

Shephard, H. L., Parker, J. S., Darby, P., & Ainsworth, C. C. (2000). Sexual development and sex chromosomes in hop. *New Phytologist, 148*(3), 397–411.

Skinner, R. N., Kavanagh, T. E., Hildebrand, R. P., & Clarke, B. J. (1974). Hop maturation and resin formation: Field trials. *Journal of the Science of Food and Agriculture, 25*(9), 1121–1133.

Škof, S., Čerenak, A., Jakše, J., Bohanec, B., & Javornik, B. (2012). Ploidy and sex expression in monoecious hop (*Humulus lupulus*). *Botany, 90*(7), 617–626.

Skomra, U., & Koziara-Ciupa, M. (2020). Stability of the hop bitter acids during long-term storage of cones with different maturity degree. *Polish Journal of Agronomy, 40*, 16–24. https://doi.org/10.26114/pja.iung.406.2020.40.03

Small, E. (1978). A numerical and nomenclatural analysis of morpho-geographic taxa of Humulus. In *Systematic botany* (pp. 37–76). American Society of Plant Taxonomists.

Small, E. (2016). Hop (*Humulus lupulus*)—A bitter crop with sweet prospects. *Biodiversity, 17*(3), 115–127. https://doi.org/10.1080/14888386.2016.1199327

Sobotik, M., Graf, T., Himmelbauer, M., Bodner, G., Bohner, A., & Loiskandl, W. (2018). In-situ root system characterization of hop and maize via soil profile excavation. *Bodenkultur, 69*(2). https://doi.org/10.2478/boku-2018-0011

Sugiyama, R., Oda, H., & Kurosaki, F. (2006). Two distinct phases of glandular trichome development in hop (*Humulus Lupulus* L.). *Plant Biotechnology, 23*, 493–496.

Thomas, G., & Schwabe, W. W. (1969). Factors controlling flowering in the hop (*Humulus lupulus* L.). *Annals of Botany, 33*(4), 781–793.

Thomas, G. G., & Schwabe, W. W. (1970). Apical morphology in the hop (*Humulus lupulus*) during flower initiation. *Annals of Botany, 34*(4), 849–859. https://doi.org/10.1093/oxfordjournals.aob.a084416

Trojak-Goluch, A., Kawka-Lipińska, M., Wielgusz, K., & Praczyk, M. (2021). Polyploidy in industrial crops: Applications and perspectives in plant breeding. *Agronomy, 11*(12), 2574. https://doi.org/10.3390/agronomy11122574

Ülz, P. G. G., Üller, E. M., Errm, T. H., Lösel Bayer, P., & Monheim, P. (1993). Epicuticular leaf waxes of the hop (*Humulus lupulus*). Chemical composition and surface structures. *Zeitschrift für Naturforschung, 48*, 689–696.

Wodehouse, R. P. (1971). *Hayfever plants* (2nd revised ed.). Hafner Publishing.

Wormald, H. (1915). Variation in the male hop, *Humulus lupulus* L. *The Journal of Agricultural Science, 7*(2), 175–196.

Zanoli, P., & Zavatti, M. (2018). Pharmacognostic and pharmacological profile of Humulus lupulus L. *Journal of Ethnopharmacology, 116*(3), 383–396.

Chapter 3
The Evolution of Hop Breeding: Integrating Molecular Markers and Genomic Insights

Andreja Čerenak, Jernej Jakše, Lucija Luskar, Marjeta Eržen, Margherita Rodolfi, and Tommaso Ganino

3.1 Introduction

Plant breeding is the science of enhancing plants to meet human needs while taking agriculture and food production requirements into account. Regardless of plant species, the primary stages include selecting plant parents with desirable properties and breeding them to create new genotypes that combine the superior traits of parents. The primary properties that are enhanced are yield quantity and quality, disease resistance, and abiotic stress tolerance. It combines conventional methods like crosspollination with modern techniques called molecular breeding. Although hops belong to minor crop, in recent decades, molecular breeding has contributed to new discoveries in hop genetics and the application of effective methods for enhanced variety development.

Hops have been used in brewing since the Early Middle Ages, with the first documented use being during the seventh and ninth century in Bohemia, Slovenia and Bavaria (Neve, 1991). The first stages of hop breeding began in the eighth century when female plants were selected from feral populations for cultivation (Kralj & Sušnik, 1983). The first efforts to develop varieties were made in the fifteenth century when superior genotypes were isolated, from feral populations through

A. Čerenak (✉) · L. Luskar · M. Eržen
Department of Plants, Soil and Environment, Slovenian Institute of Hop Research and Brewing, Žalec, Slovenia
e-mail: andreja.cerenak@ihps.si; lucija.luskar@ihps.si; marjeta.erzen@ihps.si

J. Jakše
Department of Agronomy, Biotechnical Faculty, University of Ljubljana, Ljubljana, Slovenia
e-mail: jernej.jakse@bf.uni-lj.si

M. Rodolfi · T. Ganino
Department of Food and Drug, University of Parma, Parma, Italy
e-mail: margherita.rodolfi@unipr.it; tommaso.ganino@unipr.it

K. Carbone (ed.), *All About Hops: The Crop, its Cultivation, and its Uses in Brewing and Beyond*, https://doi.org/10.1007/978-3-031-96606-4_3

mass selection (Henning, 2006). Saazer, Hallertauer, Spalter and Hersbrucker evolved as a result of local genotype selection and the use of vegetative reproduction cultivars. Neve (1991) provided an excellent summary of the historical evolution of breeding programs, but the rise of private breeding programs throughout the last thirty+ years of development has brought about new developments.

3.2 Overview of Classical Breeding

The key phase in classical breeding is hybridisation or sexual reproduction, which is accomplished by crossing selected female and male plants. Natural crosses between female populations and wild male plants did not result in significant genetic progress. Salmon's crossing of American and European germplasm in the early twentieth century signaled a significant shift in hop breeding (Neve, 1964). One of the simplest processes is open pollination of selected female plants where seeds are developed after successful pollination with unknown male pollinators. As a result, the effectiveness of breeding is limited because half of the genome inherited by male plant is unknown. Early released varieties such as Brewers Gold and Bullion were selected from open pollinated (OP) female plants. Later, new varieties registered as OP were also introduced from both public and private breeding programs.

Another innovation was the use of male test-crosses to determine the value of male parents. Analysing progenies to determine the breeding value of a parent takes a significant amount of time and money. Consequently, the method is rarely employed in hop breeding programs due to a long perennial plant's juvenile phase. In general, hop breeders observe male plants and record their vigour, disease resistance, the occurrence of the phenophases, and characteristics associated with male flowers such as their quantity, distribution on the plant, and duration of pollen shedding.

Nowadays, hop breeders pick parent lines based on phenotype, chemical or molecular analysis. Understanding the genealogy of the chosen material is crucial. They are primarily focused on yield quantity and quality (alpha-acid content, aroma characteristics), and disease resistance (Nesvadba et al., 2022). One of the most crucial aspects of integrated plant protection is plant resistance to diseases, which also poses a major challenge in breeding new varieties. As a result, pathogen selection is extremely important in the development of hop varieties. In general, breeding resistant plants requires the application of appropriate selection methods and understanding of plant-pathogen interactions (Radišek et al., 2007). Diseases included in selected breeding programs are prioritized based on how important they are for a specific cultivation area. Downy mildew (*Pseudoperonopora humuli*), powdery mildew (*Podosphaera macularis*), verticillium wilt (*Verticillium nonalfalfae, V. dahliae*), and fusarium canker (*Fusarium sambucinum*) are the primary concerns. Furthermore, with the increased awareness of climate change and the importance of sustainable use of natural resources, research into how different abiotic conditions affect hop growth and development is becoming increasingly important. As a result, breeding programs in different countries include the monitoring of different variety responses to abiotic factors (Kolenc et al., 2016; Contin et al., 2023; Nesvadba, 2022).

Some techniques in the selection process are technically demanding and time-consuming, there are chances to improve the current selections by utilizing different marker systems (MAS; Marker Assisted Selection), of which molecular markers are particularly helpful.

Primarily, crossing combinations mostly comprise diploid parents (having two distinct sets of chromosomes, one from each parent), but triploid crosses (tetraploid x diploid parent; having three homologous sets of chromosomes) are also helpful. Triploid varieties can be more vigorous, they generally do not produce fertile seeds, and the impact of the diploid parental component on the offspring is smaller. Tetraploids (containing four complete sets of chromosomes) can be developed with colchicine treatment, and in recent decades, they have also been developed in tissue cultures (Škof et al., 2007). When a tetraploid female is crossed with a diploid male, the majority of the seeds contain triploid embryos (around 90%), with approximately 70% of the offspring being female (Kralj, 1989). The number of monoecious individuals in triploid progeny tends to rise (Kralj, 1973; Beatson et al., 2003).

Since seeded hop plants are not often valued in the market, selection fields planted with male plants are typically separated from female plants. Rarely, 'polycross' hop fields are maintained. This method requires a special spatially isolated field to be planted with a larger number of male and female genotypes selected for crossbreeding. This crossing method is suitable for population breeding (with recurrent selection), which occurs in cycles. For each new cycle, a new planting trial with the best plants from the resulting offspring has to be established. Consequently, superior materials selected during a cycle are planted in the next stage, and desirable alleles concentrate in the population.

The use of backcrossing strategies to integrate different commercially important features into a single genetic background is not well documented (Henning, 2006). There are more than 300 different hop varieties according to the International Hop Growers' Convention (IHGC, 2023). Most were developed in private breeding programs where breeders are primarily required to limit the free use of their produced varieties by others. Developing new varieties demands substantial resources, including investment, time, materials, and labour. However, because hop plants multiply vegetatively, once these varieties are released, they can be easily replicated by others, making it difficult for the creators to recoup their investment. In terms of intellectual property (IP), cultivars can be safeguarded by Plant Breeders' Rights in EU and plant patents in the USA. Unlike patents, genes can be combined freely, allowing for additional crosses and the development of new varieties during the protection period (Guimaraes et al., 2021). Between 2010 and 2020, the preponderance of proprietary varieties gave rise to unprecedented cartel-like powers, creating opportunities for controlling supply and pricing (Pavlovič et al., 2022; MacKinnon & Pavlovič, 2022). Private varieties are affecting not only the hop market, but also breeding programs. Rarely are patented varieties given to stakeholders who are not the owners. Patented varieties are rarely made readily available to breeders for use as parent components in subsequent crossings. As a result, genetic diversity is being reduced, particularly in public breeding programs that were the driving force behind the development in hop production decades ago.

3.3 Molecular Markers

3.3.1 Introduction

The advent of molecular marker technology in the 1980s revolutionized plant breeding by introducing a novel approach to germplasm evaluation, genetic selection and improvement. Molecular markers are specific nucleotide sequences that exhibit polymorphisms, which are variations in the DNA sequence among individuals and segregate according to Mendelian laws. These polymorphisms can result from a variety of genetic alterations, including insertions, deletions, translocations, substitutions and duplications (Mondini et al., 2009). Different types of molecular markers, categorized as hybridization-based, polymerase chain reaction (PCR)-based, or sequencing-based, have been developed alongside advancements in molecular methods. According to the principles of genetics, they may be co-dominant or dominant (the latter cannot distinguish from heterozygote loci), or, they could have paternal or maternal organelle inheritance, bi-parental nuclear inheritance or maternal nuclear inheritance (Semagn et al., 2006). One of the main benefits of molecular markers is that they are not affected by environmental factors and are based only on variations in DNA which can distinguish between very closely related genotypes (Hartl & Seefelder, 1998).

The field of molecular markers was pioneered by Restriction Fragment Length Polymorphism (RFLP), a hybridization-based technique developed in the 80s. The discovery of PCR in the 90s revolutionized marker development, resulting in the creation of PCR-based markers like RAPD (Randomly Amplified Polymorphic DNA), AFLP (Amplified Fragment Length Polymorphism), SSR (Simple Sequence Repeats / microsatellites) and SCAR (Sequence Characterized Amplified Regions). Advancements in sequencing technology further expanded the repertoire with sequence-based markers like SNP (Single Nucleotide Polymorphisms) and DArT (Diversity Arrays Technology) (Nadeem et al., 2018).

The development of molecular markers has significantly contributed to more efficient germplasm diversity studies and breeding efforts for several important crop cultivars. Hops are one of the most important industrial crops in countries like the USA, Germany, Czech Republic, China, Poland, Slovenia, Australia, New Zealand, Spain and Great Britain. Due to the ongoing demand for hop quality in the global market, there is always a need for new varieties (higher alpha acids, different aroma profiles, new sustainable cultivation technologies, resistances to diseases and pathogens) (Javornik, 2019; Henning et al., 2016, 2017; Luskar et al., 2022). However, hop molecular research is hindered by a number of issues, for example genome size, high heterozygosity, dioecy, a long dormant period and the continued scarcity of excellent reference genomic sequence, to name a few (Howard et al., 2011).

3.3.2 *Overview of the Markers Systems Used in Hop Research*

Hop varieties were first classified using min-max models based on biochemical markers such as essential oils and flavonoids (Kralj et al., 1991). With the development of new nucleic acid-based techniques new approaches to identification emerged, like RAPD, Sequence Tagged Sites (STS), AFLP markers and microsatellites (Čerenak et al., 2004), which will be discussed in the next paragraphs. Markers were mainly used to conduct diversification studies and to determine between different hop varieties (Javornik, 2005). Due to comparatively rapid genome analysis without prior nucleotide sequence knowledge, the initially utilized RFLP markers were replaced with RAPD or AFLP markers (Abbott & Fedele, 1994; Hartl & Seefelder, 1998). Molecular marker evolution continued with extensive research utilizing short repeats of a few nucleotides in the genome sequences – microsatellites in various hop studies, including hop cultivar varietal authenticity, sex determination and disease markers (Javornik, 2005). The development of DArT markers was prompted by the need for a high-throughput method. The first study to use DArT markers for hop genotyping was conducted by Howard et al. (2011). With the advent of high-throughput sequencing (HTS), also known as next-generation sequencing (NGS), genotyping-by-sequencing technologies have allowed for accurate studies of the hop genome, revealing thousands of polymorphisms (Henning et al., 2016, 2017).

3.3.2.1 Randomly Amplified Polymorphic DNA (RAPD) Markers

RAPD markers application was first documented already in 1990 (Williams et al., 1990). They are based on the amplification of genomic DNA segments using single usually 10-mer long primers that are shorter than classic primers. They are referred to as random, since the primers are not selected based on any genome sequence similarities. RAPD primers anneal to different regions in the genome and are typically visualized on agarose gel after amplification (Butler, 2012). Polymorphisms can only be detected if they are present at or between primer binding sites. The disadvantages of RAPD markers are that they are dominant (Nadeem et al., 2018) and have lower interlaboratory reproducibility, as RAPD reaction is far more sensitive to component variation than the classical PCR reaction due to shorter primers which amplify anonymous locations in the genome (Bardakci, 2001). RAPD reaction requires a small amount of total genomic DNA, making superior to hybridization assays. RAPD markers can be utilized in a variety of applications, including genetic mapping as genetic markers associated with a desired trait in population and evolutionary genetics (Bardakci, 2001).

RAPD markers were the first markers utilized in hop molecular studies shortly after their introduction (Abbott & Fedele, 1994; Pillay & Kenny, 1996), however they exhibited minimal genetic diversity. RAPD polymorphisms were screened for six closely related hop cultivars and showed specific bands for one of the cultivars, but Sequence-Tagged Sites (STS) converted polymorphic RAPD bands could not distinguish between analysed cultivars (Brady et al., 1996). Šuštar-Vozlič and Javornik (1999) were the first to report that RAPD study was able to distinguish between American and European hop cultivars.

Polley et al. (1997) examined ten male and ten female plants using RAPD markers for Y chromosome (male specific markers) sex determination. They screened 1000 different RAPD primers, and 900 of them produced amplified DNA patterns. Thirty-two RAPD primers revealed specific fragments associated with male traits. These thirty-two RAPD primers were then amplified in a larger sample of different hop varieties from different countries yielding three primer patterns capable of amplifying male specific fragments. One of them derived from primer OPJ9 was successfully transformed to an STS marker. This was the first trait-specific marker in hop. RAPD markers associated with hop sex have also been detected in the study of Buck et al. (2009).

3.3.2.2 Amplified Fragment Length Polymorphism (AFLP) Markers

AFLP markers were the upgrade of molecular markers integrating the concepts of RFLP and RAPD/PCR markers developed by company Keygene (Vos et al., 1995). AFLP markers were more cost effective, more technologically advanced and showed better reproducibility. They are also dominant, though attempts to score them as codominant based on bands intensities have been made. They have higher reproducibility and are highly polymorphic. AFLP markers have high genome abundance and are usually visualized on denaturating polyacrylamide gels (Nadeem et al., 2018). AFLPs are generated from a genomic DNA cut employing two sticky-end restriction enzymes, often frequent (4 base pairs (bp) recognition site) and rare cutter (6 bp recognition site). *Eco*RI and *Mse*I were originally used, however enzymes used are also species specific. Adapters specific for enzyme-generated ends are then ligated, which introduces the known sequence sites for annealing the amplification primers (Butler, 2012). One disadvantage of AFLP markers is the use of polyacrylamide gels (PA) and radioactivity for visualisation (Nadeem et al., 2018), which was eventually solved with real-time fluorescence detection by capillary electrophoresis (Hartl & Seefelder, 1998).

Genetic diversity of eight major hop aroma and bitter varieties from a breeding program in Hüll using AFLP markers was studied by Hartl and Seefelder (1998). It was found that the combination of EcoRI-MseI restriction enzymes with three additional selective bases in the primers could distinguish between studied hop varieties. Their work was upgraded in 2000 when they used seven AFLP primer pair combinations to analyze six German experimental lines and 84 cultivars from different parts of the world to determine their genetic relatedness. With 130 polymorphic fragments, accessions were divided in two main clusters: European aroma-type accessions and lines which were developed from wild American accessions incorporated into European varieties (Seefelder et al., 2000).

In the 2001 study conducted by Jakše et al. (2001) AFLP markers polymorphisms of 41 cultivars were compared to polymorphisms found by microsatellites. AFLP dendrogram showed good clustering of closely related accessions and, additionally, isolated two geographically distinct hop germplasms. Genetic dendrograms showed low correlation between the two marker systems.

Potential parental combinations for studying heterosis in hops were performed using AFLP markers (Henning & Townsend, 2005) analysing 80 male and 26 female hop genotypes. Six primers amplified 550 fragments, 89.1% of which were polymorphic, revealing two major genetic clusters: European hops and European–wild American hybrids. Further subdivision into 13 smaller clusters highlights potential parental combinations for studying heterosis in hops. In further study, the same authors (Townsend & Henning, 2009) analysed 60 hop accessions from cultivated and native gene pool by AFLP markers. Based on the results the authors concluded that incorporating the studied native American hop accessions could significantly enhance genetic diversity in hop breeding programs.

In a study by Henning et al. (2010), AFLP marker polymorphisms and AFLP-derived genetic distance estimates among 16 cultivars and 31 male accessions was shown that they could be used to predict superior offspring performance.

3.3.2.3 Microsatellites

The next DNA molecular markers were microsatellites (synonyms: short tandem repeats or STRs, simple sequence repeats or SSRs, variable number of tandem repeats or VNTRs, simple sequence tandem repeats or SSTR), composed of short nucleotide repeat units, usually 1–6 base pairs long and abundantly present in the genome, chloroplasts and mitochondria. They could also be present in protein-coding genes and in expressed sequence tags (ESTs) commonly referred to as expressed microsatellites or EST-SSRs. Microsatellites are co-dominant with high reproducibility and high level of polymorphism. They have medium genome abundance and can also be visualized on agarose gel, although alleles with small differences cannot be resolved (Nadeem et al., 2018). They are cost-effective, multi-allelic, and low quantity and low-quality DNA can be used for amplification (Abdurakhmonov, 2016). The use of microsatellites polymorphisms was also incorporated in the technique called inter simple sequence repeats (ISSR), where primers based on microsatellite repeats amplify loci between adjacent microsatellites (Marwal et al., 2014). They can be effectively employed in plant mapping studies, genetic diversity studies, genotyping, population analysis and marker assisted selection. They are highly informative and polymorphic within and between species (Štajner et al., 2005). The disadvantages of microsatellites include the occurrence of homoplasy and the presence of null alleles (Nadeem et al., 2018), as well as the need for genomic libraries, primer synthesis, and clone sequencing for their development (Hartl & Seefelder, 1998). However, the availability of sequencing data and the accessibility of next-generation sequencing (NGS) for most species have helped overcome these challenges. The first study which isolated microsatellites in hops was published in 1996 by Brady et al. (1996). In the study 57 microsatellite sequences were isolated from the genomic library using four of them for PCR amplification and hop genotyping. With these four microsatellites, a wide range of cultivated and wild hops were successfully identified and proved to be an asset.

To differentiate between as many accessions as possible, several primer pair combinations for detecting and amplifying microsatellites were tested. Most varieties (44 out of 63) included in the study of Čerenak et al. (2004) can be differentiated by using five primer pairs (5-2, 11a59, 3a88, 7a82, HlGT4). The microsatellite loci 5-2 and 11a59 detected the largest number of alleles (nine), so they were chosen in the first stage of the identification. An example of hop sample identification is shown in Fig. 3.1. Beyond differentiation, 124 wild and cultivated accessions were screened using primer pairs for four microsatellite loci to assess diversity. High variability was noticed between North American wild hops, whereas European wild hops differed less due to a smaller gene pool (Jakše et al., 2004). To expand the reservoir of discriminating loci, the isolation and characterization of microsatellites in hop was also done by Štajner et al. (2005). They developed six genomics libraries which were enriched with different microsatellite motifs (GA, GT, ACA, AGA, CAG and ACTC). They selected 25 microsatellites out of 60 primer pairs to assess polymorphism in 67 hop accessions.

In the Koelling et al. (2012) research, 1007 unique SSR markers (di-, tri-, tetra- and pentanucleotide repeats) were identified out of 10,581 previously published EST library from hop var. Hallertauer Taurus and Nugget. Markers were amplified

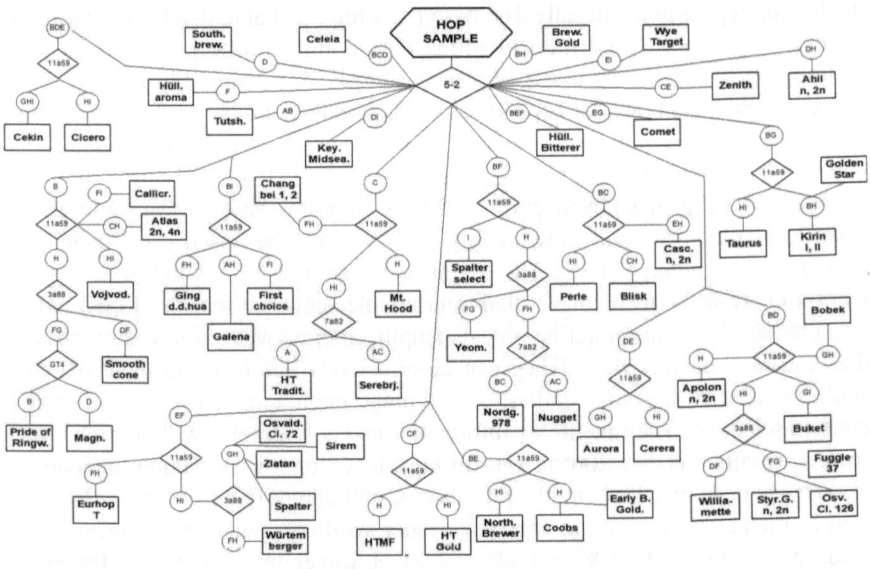

Fig. 3.1 Identification flowchart for 63 hop cultivars. Letters in circles indicate obtained phenotypes and marks in rhomboids loci used in analysis (Čerenak et al., 2004)

on eight different hop cultivars (Aurora, Super Alpha, Cluster, Hallertauer Magnum, Columbia, Hallertauer Taurus, Vanguard, and Brewers Gold). 95.8% of markers were successfully amplified for each cultivar. This was the first study on hops using such a large number of SSR markers. All the studies before used only a small number of SSRs: Jakše et al. (2002) used 11 new microsatellites, Štajner et al. (2005) used 25 microsatellites, Murakami et al. (2006) used 11 SSRs, Bassil et al. (2008) used 8 SSRs and Jakše et al., (2008a) used 24 SSRs.

Microsatellites were not used only to study genetic diversity, but also to link them to some specific traits like sex-linkage. Jakše et al. (2008b) identified microsatellite marker HlAGA7 specific for male hop plants.

3.3.2.4 Single Nucleotide Polymorphism (SNP) Markers

SNPs are single base-pair changes in the genome. There can be nucleotide substitutions, transversions, transitions, insertions or deletions. SNPs are highly distributed across the genome both in the coding and non-coding regions. SNPs can be used in genotyping techniques such as genotyping by sequencing (GBS) and next generation sequencing (NGS) (Nadeem et al., 2018), for linkage map construction, diversity analysis, DNA fingerprinting, trait mapping, quantitative trait locus (QTL) analysis and association studies (Young & Vivier, 2021). SNPs are co-dominant, with high reproducibility, they are highly polymorphic and have very high genome abundance. SNPs are cost-effective, and don't need prior sequence information (Nadeem et al., 2018). In comparison to microsatellites, SNPs are genetically more stable and have a higher frequency of occurrence. SNPs can be visualized using microarrays or other high-throughput techniques (Young & Vivier, 2021).

Matthews et al. (2013) were the first researchers who used NGS to determine SNPs in hops. With GBS they identified 17,128 SNPs in 178 hop genotypes. They found out that it is sufficient to use a subset of 3068 SNPs for differentiation between all studied genotypes.

SNP markers were also used in the study by Van Holle et al. (2019) for differentiation of hop batches in the brewing industry to verify biochemical divergences. They selected 56 different commercial hop varieties from Australia, Europe, Japan, New Zealand, North America and South Africa. They used the GBS approach to determine 1830 SNP markers which generated 48 unique genetic fingerprints.

Genetic diversity of wild Portuguese hops compared to commercial was analyzed by Machado et al. (2022). They used 33 varieties and 110 wild genotypes which were characterized using high-resolution melting analysis of short amplicons targeting SNP loci. They could differentiate between 75% of all selected genotypes which were classified into three main clusters: wild accessions, and other two mixed with both wild and commercial genotypes.

SNP markers were used to determine sex in seedlings. Seven hundred and sixty-five hop genotypes with known sex phenotypes were selected. GBS and genome-wide association studies (GWAS) revealed highly significant markers for sex determination. The most significant marker, SM1 was then validated in a separate population of 94 samples from 15 families (Clare et al., 2024).

3.3.2.5 Diversity Array Technology (DArT) Markers

DArT markers are microarray hybridization technology used for genotyping of polymorphic loci distributed over the chromosome. With these markers several thousand genomic loci could be genotyped. They are dominant, highly reproducible and have a high polymorphism level with a high marker index. Polymorphism could be visualized using microarrays. The only disadvantages of DArT markers are high developmental costs and marker dominance (Nadeem et al., 2018). DArT markers are fragments obtained with a reduction procedure of genome complexity where the use of methylation sensitive endonuclease (PstI) is used (Gawroñski et al., 2016). DArT markers can be used in many different studies such as linkage mapping, marker-assisted selection, genetic diversity, genetic identity testing, directed breeding of superior cultivars (Howard et al., 2011).

Whittock et al. (2009) were the first who used DArT markers in hop diversity study. Later on, Howard et al. (2011) developed 730 high quality polymorphic markers from 6144 random genomic hop clones of 92 hop accessions. According to hop accessions, they could be divided into two groups; North American and European whereas hybrid accessions were genetically somewhere in between both groups but closer to European accessions.

Ten specific DArT markers were also used in the study by Čerenak et al. (2019) where new male specific sequences were selected to determine the sex of young hop seedlings. DArT markers showed complete correlation with male sex trait in three different crossed families. Four out of ten DArT markers were converted into PCR markers. They also developed a multiplex PCR approach for their use (Luskar et al., 2023).

DArT markers and microsatellites were also used in diversity studies for constructing linkage maps since other markers such as AFLP and RAPD were not as suitable. 511 polymorphic markers were identified among 6439 DArT clones and used for linkage maps in QTL analysis (McAdam et al., 2013).

3.4 Diversity Studies, Mapping and Marker Assisted Selection—Overview

3.4.1 Diversity Studies in Hop Plant

Species biodiversity can depend on how well they can adapt to a certain environment. Different growing conditions and ecosystems in which plants grow can produce genotypes that are biologically better suited to that environment, owing to the process of natural selection (Gienapp et al., 2008). This concept highlights the importance of hop biodiversity in traditional (USA, Germany, Czech Republic, Poland, Slovenia…) and non-traditional hop growing regions. The regions where hops have not been cultivated for centuries are a rich source of natural diversity and of beneficial traits that can be exploited in breeding programs, such as finding genotypes with biotic or abiotic stress tolerance traits to establish a more sustainable hop cultivation. Furthermore, hop cultivation is still in its infancy in some of these regions, so it would be interesting to study and understand the biodiversity that is already present in order to select and identify new genotypes that might be interesting for brewing or medicinal purposes that are undoubtedly better suited to the territory. With this in mind, it would be possible to meet the market demand for more sustainable production that is more suited to the local regions (Mafakheri et al., 2020).

Biodiversity knowledge is the foundation for crop development in new production regions, and various research is conducted to delve deeper and find intriguing new resources. These studies have historical background, showing the hop production in some regions in ancient times before being discontinued for various reasons (e.g., in Italy, France, Scandinavia). In countries where hop is a well established crop, like in different European states or the USA, breeding programs contribute to the depletion of hop agricultural biodiversity, resulting in cultivars with more homogenous characteristics (Patzak et al., 2010a, Patzak et al., 2010b). Furthermore, protected varieties are very rarely included in research studies because of their intellectual properties (MacKinnon and Pavlovič, 2023).

Therefore, by crossing and combining beneficial genes from wild hops, it is possible to create new elite varieties (Carron et al., 2022). The common trend that will be shared by all countries in the future is the rediscovery of hop biodiversity richness, as well as the tendency to reintroduce and valorise local production. It is already possible to observe a "spring" of the hop production in all these areas.

3.4.1.1 Diversity Studies in Traditional Hop-Growing Countries

Hop breeders rely heavily on genetic variation or diversity to better understand, conserve and expand their genetic reservoir (Howard et al., 2011). European cultivars were used as a breeding resource because of the flavour characteristics that brewers prefer. Wild germplasm provides new genetic resources for breeding, helping to overcome the limited genetic variation found in modern hop breeding programs.

Three major hop genetic groups were identified by the majority of genetic studies with some variations due to the selection of included accessions: North American, European and hybrids (Šuštar-Vozlič & Javornik, 1999; Seefelder et al., 2000; Jakše et al., 2004; Murakami et al., 2006; Howard et al., 2011). Additionally, genotypes were divided into five subgroups within the European groups, which corresponded to regional geographic adaptation (Šuštar-Vozlič & Javornik, 1999). Using microsatellites, a study by Jakše et al. (2004) that examined genetic diversity in 59 wild hop accessions and 65 cultivated hop accessions from all over the world was also able to distinguish between North American and European hops. 23 STS and 46 SSR markers used in the study by Patzak et al. (2010a) successfully distinguished between Caucasian and European wild hops. Murakami et al. (2006) discovered close genetic relationships among European wild hops (Czech Republic, Germany, Austria, Hungary, Italy, Portugal), classifying Eurasian wild hops as broadly spread group. Wild hops from Georgia/Caucasus and the Altai region were somewhat different. This structure of European, Altai, and Caucasus germplasm was later confirmed by Bassil et al. (2008) and Štajner et al. (2008). In the study by Patzak et al. (2010a), European cultivated noble aroma hops were clustered together with European wild hops. Many older European varieties were developed through clonal selection and would most likely contain lesser amounts of detectable polymorphism in them. As a result, the male and female genotypes selected from the European collection may have originated from a narrower genetic base than their counterparts in the United States Department of Agriculture Agricultural Research Service (USDA-ARS) collection, therefore they represent a more homogeneous gene pool (Henning et al., 2004). High genetic differences between North American wild accessions, constituting a single distant cluster, and all the other accessions were identified by microsatellite analysis (Jakše et al., 2004). Similar findings were reported by Bassil et al. (2008). The USDA-ARS hop genetics and breeding program has historically utilized a wide variety of wild American hops, both male and female, in the development of new germplasm. Thus, the males in the USDA-ARS collection represent a wide range of the potential gene pool present in wild American hops (Henning et al., 2004). AFLP analysis carried out by Seefelder et al. (2000) proved that introducing American wild hop accessions into European cultivars creates a genetically distinct group that differs from the traditional European aroma types. Similarly, the genetic classification of 135 world cultivars using 276 polymorphic markers (Patzak & Henychová, 2018) demonstrates the development of cultivars through hybridization with European landrace germplasm with wild germplasm from North America (Fig. 3.2). The cultivars are categorized by two components of the principal coordinate analysis (PCoA) and divided into four quadrants based on origin: Late Cluster, Continental Landrace, Fuggle and Northern Brewer and Brewers Gold.

3.4.1.2 Diversity Studies in Non-traditional Hop-Growing Countries

Several countries have conducted research to analyse the genetic diversity of their indigenous wild hop populations. Rodolfi et al. (2018) studied wild Italian hops, comparing them to European and American cultivars using microsatellite markers. Similarly, Grdiša et al. (2021) investigated wild hops in Croatia and found high genetic diversity and low population differentiation. The biodiversity observed on the French island of Corse demonstrated genetic (SSRs analysis) and chemical variability between wild hops and commercial cultivars Cascade, Smaragd and Tettnanger (Dabbous-Wach et al., 2021). Furthermore, the interest in hop

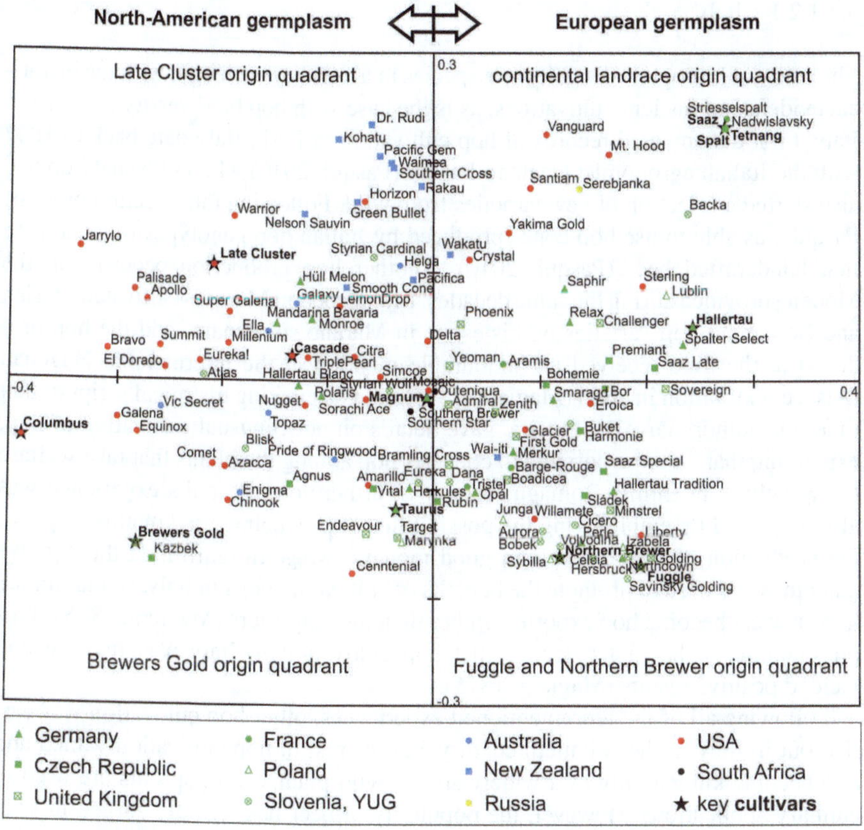

Fig. 3.2 Principal coordinate analysis of 135 hop cultivars with country of origin revealed by DARwin v. 5.0.155 (Dissimilarity Analysis and Representation for Windows, http://darwin.cirad.fr/darwin) based on 276 polymorphic molecular markers; *x*, *y*—the first and the second principal coordinate, respectively; colours and signs represent the country of cultivar origin (Patzak & Henychová, 2018)

production in the French territories highlighted novel raw materials and local wild hop biodiversity (Paguet et al., 2023, 2024).

In Portugal, Mafakheri et al. (2020) evaluated the genetic diversity and structure of wild hops in Northern Iran using both morphological and molecular characteristics. Additionally, Machado et al. (2022) assessed wild hop diversity using SNP genotyping. These studies collectively contribute to a better understanding of hop genetic diversity on a global scale. In terms of plant diversity, some of the non-traditional hop countries are discussed in greater detail below.

3.4.1.2.1 Italy

The study of biological diversity of a species in a territory can begin with the historical evidence of ancient cultivations, as is the case with hop biodiversity research in Italy. First documented records of hop cultivation in Forli, Italy date back to 1847 with the Italian agronomist Gaetano Pasqui (Pasqui, 2010) who cultivated hop and also started a selection of new varieties from wild. Following the selection process, Pasqui was able to use hop cones produced by Italian hop genotypes to launch the first handcrafted beer (Pasqui, 2010). Another hop production occurred in the Modena province during the same decades: Montecuccoli Marquis cultivated Styrian and Bohemian hop varieties on his estate in Marano sul Panaro, and the hop produced at the time received an honourable mention at the international Hagenau (Alsace) exhibition in 1860 (Magiera, 1875). In a fascinating historical script of that time, the author, namely Mageira, gave details on new unusual cultivation of hop, explaining that "it to be possible to consider hop among the plants that take welfare in agriculture in Emilia Romagna region". Furthermore, Pasqui's experience was also described by emphasizing the possibilities of producing good quality hops for the production of "good beer and good money". Mageira reaffirmed the positive outcomes and tried to promote the benefits of cultivating hops in Italy, so that Italian hop growers become hop exporters rather than just importers (Magiera, 1875). This presentation is the first evidence that hop cultivation in Italy was feasible and yielded positive results (Magiera, 1875).

Following all of the aforementioned experiences, other hop cultivation was carried out in Italy in the twentieth century, but interest in hop cultivation waned and hop become known only as a widely spread wild plant whose sprouts are used in culinary applications. However, the popularity of beer has recently surged in Italy and a large number of small breweries have sprouted up, sparkling a renewed interest in local ingredients. Furthermore, introducing hops as a new crop for Italian farmers might be a great way to add high-value production and differentiate output. For these reasons, Italian germplasm is being studied from both genetic and phytochemical perspectives.

In two studies, Mongelli and associates examined the inherent genetic and phytochemical variations in Italian wild germplasm (Mongelli et al., 2015, 2016). In the first study 22 wild hops and 5 cultivars were compared. High variability in the population studied was observed, as well as some chemical similarities with Italian germplasm. The analysis of bitter acids and essential oil fraction revealed noteworthy characteristics in the analysed samples: α-acid and β-acid content ranged between 2.98 g/100 g and 11.11 g/100 g, showing higher content than previously reported for European wild hops (Patzak et al., 2010a, Patzak et al., 2010b), and the volatile fraction varied with certain genotypes characterized by high α- and β-selinene and β-myrcene content. Notably, in some samples *trans*-β-farnesene (typical molecule of the so called 'European noble hops') was detected in some samples, emphasizing the possibility to exploit some ecotypes from the Italian wild biodiversity panorama in breeding programs. In 2018, the application of SSR genetic markers enabled the study of genetic heterogeneity across 80 wild hop accessions and 43 hop cultivars from the United States and Europe (Rodolfi et al., 2018). The findings were fascinating since they showed a genetic differentiation between cultivar and Italian wild germplasm, with some varieties having wild European germplasm in their progeny that is genetically closer to the Italian wild population. Furthermore, the study of the allelic frequency has revealed the greater number of alleles and unique alleles in the Italian population, a high level of biodiversity. The differences in allelic composition between wild and cultivated Italian hops indicate the existent genetic separation. These differentiations were also outlined in 2006 by Murakami and associates (2006) in a study on the genetic diversity (using 11 SSR markers) of 133 hop samples collected in Asia, North America and Europe, which included four Italian wild hop accessions. The results displayed in a dendrogram showed the four Italian accessions positioned in the European germplasm cluster, but with a higher genetic distance. More recently, in 2022, Italian biodiversity was studied in two northern regions from the genetic and phytochemical perspective (Rodolfi et al., 2022). The observation of 60 wild genotypes, using eight SSR markers and analysis of the aromatic profile and the bitter acids fractions indicated once again high variability present in the Italian hop population. In the north Italian genotypes, there are significant genetic differences among northern Italian genotypes, with α-acid values ranging from 0 to 4.12%, which is consistent with European wild hops studies. These Italian accessions are also characterised by interesting essential oil yield, which reach values above 0.7% and the volatiles composition, showing some genotypes characterized by important spicy fraction, and others by fruity and floral notes. Other studies focused on the central southern part of Italy, and Ghiselli et al., (2022) analysed the biodiversity of 14 wild hop accessions, comparing them to Cascade and Hallertau Taurus cultivars grown in the same conditions. They discovered variability in bitter acids composition in Italian accessions, in particular in alpha and beta acids, which are quite low in comparison with the results obtained in other studies on wild European hops. Interestingly, the polyphenol and flavonoid levels, as well as anti-radical power were high and generally comparable to commercial varieties. Twelve wild hop populations were compared to 6 cultivars to determine the genetic polymorphism using 13 SSR loci (Riccioni et al., 2021). A high level of genetic polymorphism in the population was discovered in the biodiversity study of Southern Italy (Calvi et al., 2023).

The variability in the chemical composition was found to be high throughout the Italian peninsula, especially in the essential oil composition, yet several common features appear to be individuated all over Italy: an important presence of α-humulene and selinene isomers in most of the studied populations, together with multiple samples containing a significant quantity of farnesene isomers, a chemical that together with α-humulene, appears to be distinctive of noble hops (Rodolfi et al., 2022, 2018).

3.4.1.2.2 Portugal

In Portugal hop cultivation is present but only commercial cultivars are grown. Despite this, wild hop is widely grown throughout the country, and is traditionally used as a medicine. In the 1960s it was used in brewing. In 1997, Portugal launched a project named “Cultivate Cultures: Ecologies of Hops” to preserve and study Portuguese hop biodiversity, which went on until 2000. In order to achieve this, more than 160 genotypes were gathered and stored in a collection field (installed at the Portuguese Genebank—BPGV) for study and preservation. During this project, hops were found in the country’s northern and central parts but not, in the regions bellow the Tagus River. Morphological studies revealed high variability in the recovered accessions compared to some commercial cultivars, thus a genetic characterization of the indigenous germplasm was performed using SNP (Single Nucleotides Polymorphism) molecular markers. Although wild germplasm has an interesting intraspecific variability, the genetic structure of wild Portuguese hops is characterized by a common genetic variance different from commercial varieties. Portuguese wild accessions demonstrated high polymorphism and genetic diversity when compared to commercial cultivars, indicating the importance of the area's genetic resources for breeding (Alfaro-Saiz et al., 2024). Nonetheless, the lack of data on the phytochemical characters of these genotypes prevented a thorough characterization of the biodiversity present in the Portuguese region.

3.4.1.2.3 Northern Europe

Norway, Sweden, Denmark and Finland have a similar past in terms of hop culture. Evidence of hop growing in these areas dates back to the Middle Ages, when hop gardens grew close to monasteries and large households, though modest cultivation experiences were likely there much earlier. From the fifteenth century onward, farmers on the Scandinavian peninsula were obliged to cultivate a specific quantity of hop plants. Although this requirement was later lifted, it demonstrated the plants’ historical significance on the peninsula. In particular, in Sweden legislative requirement to cultivate hops began in 1414 and remained in effect until 1860 (Hornsey, 2003; Korpelainen & Pietiläinen, 2021).

Despite the period of flourishing hop production on the Scandinavian peninsula, where hops were also grown on the Lofoten Islands, hop production decreased, from 1103 ha of cultivated land to around 200 ha from 1103 to 1881. In the twentieth century, Scandinavian hop cultivation virtually died in the twentieth century due to the economic motives such as breweries being centralized and hops being imported from Germany at a lower price. Nowadays hops are becoming increasingly important owing to a growing interest by New Nordic Food and local foods, as well as a growing interest in local raw materials, which means that brewing could benefit from these new trends. For these reasons, and to study and preserve hops, the vestiges of ancient cultivations in Denmark, Norway and Sweden have recently been rediscovered. The genetic identity of the Scandinavian germplasm is most likely linked to ancient, now unknown cultivars, as well as local naturalized clones and populations. A clonal archive has been established to preserve and study the Scandinavian population (Solberg et al., 2014). From the morphological characterisation of biodiversity a some characteristics that Danish and Norwegian hops share. Interestingly, Danish germplasm have green shoots, but Norwegian hop accessions are characterized by red shoots (Solberg et al., 2014). Furthermore, the chemical characterization of Norwegian hops, has revealed a good cohumulone content, a bitter acid content above the average compared to European wild hops, and alpha-acid values above 6% (Solberg et al., 2014).

Thanks to historical documentation, it was possible to determine where ancient hopyards were located in Sweden, and a genetic study on the population that is still present allowed us to conclude that within the populations the genotypes present demonstrated high clonality and a low rate of sexual reproduction in feral germplasm (Karlsson Strese et al., 2014). That means that population recombination and natural expansion are uncommon or sluggish in Sweden, and the hop plants now present, may have originated from ancient cultivation (Karlsson Strese et al., 2014). Furthermore, there is no evidence of similarity between those feral populations, with the genotypes present in the European genebank clones, revealing the genetic difference between the current population and the nineteenth century cultivated accessions; instead, it appears that the previous cultivations based on Swedish hops rather than European cultivars were replaced and persisted in the region because they better adapted to the Swedish soil (Karlsson Strese et al., 2014).

3.4.1.2.4 Brazil

Brazil is the world's third largest beer producer (Duarte et al., 2024), and breweries rely heavily on hops as a raw material. Although the brewing industry is one of the most important sectors of the Brazilian economy, nearly all of the hops used in the beer production are imported from the United States and Germany. There is a considerable public and private commitment to hop production in Brazil as a result of the significant growth in craft breweries that demand raw materials with unique notes and character for the beer. Even though Brazil is located in a hemisphere unconventional for hop cultivation, modern technologies combined with genetic enhancement will make it possible to achieve intriguing yields. Crop adaptability to climate conditions is one of the biggest challenges in Brazil's latitude. Daylight and rainfall needed for hop growing are the key elements that limited the hop experience in Brazil. Brazilian hop cultivation dates back to the nineteenth century, when hops were sourced from European seedlings. Hop production and commerce were abandoned when local breweries discovered it was more convenient to purchase hop from other countries (Jastrombek et al., 2022). In the 2000s, more unsuccessful attempts to grow hops using plants originating from Canadian seeds were made. Nevertheless, plants remained in the field and spread, and as a result, the naturally selected plants were registered as a new variety under the name of 'Mantiqueira' (Jastrombek et al., 2022). The use of new technologies such as LED lamp to regulate photoperiods, as well as plant flowering and irrigation systems, allows for better hop cultivation also at various latitudes in Brazil. The varieties cultivated in Brazil are mainly North American and German cultivars, which appear to have distinctive characteristics appreciated by Brazilian brewers cultivating hops in Brazil (Jastrombek et al., 2022).

The registered Brazilian hop varieties are the previously mentioned 'Mantiqueira', followed by 'Canastra' and 'Brasylijsk' (Jastrombek et al., 2022). Canastra cultivar had the highest total phenolic content (35.10 mg GAE/g dry weight), and a study comparing Brazilian cultivars to European cultivars in Brazil suggests that Brazilian cultivars contain higher quantities of flavonoids than European cultivars. 'Mantiqueira' contains 4.9% α-acids, 4.1% β-acids, and 49.8% cohumulone: 'Canastra' is high alpha cultivar and 'Brasylijsk' contains 5.9% of alpha acids, 15.18% β-acids and 57.8% cohumulone (Jastrombek et al., 2022).

3.4.2 Genetic Mapping

Genetic mapping, also known as linkage mapping or meiotic mapping, is one of the important genetic methods in plant breeding for identification and isolation of key genes with qualitative and quantitative properties. It is based on the assumption that genes on the same chromosome are inherited through linkage. The recombination frequency between two loci depends on their distance and represents a distance measure between two loci, and is measured in centimorgans (cM). One of the aims of genetic mapping is to find a strong link between marker and phenotypic trait and their mutual inheritance (Čerenak & Javornik, 2008). Genetic map construction allows us to pinpoint genes or quantitative trait loci, as well as do comparative mapping between different species based on gene similarities and function in the phenotype expression (Temesgen, 2021). Linkage mapping in bi-parental F1 populations has been the primary method for quantitative trait locus (QTL) analyses in hop. The maps made using older marker techniques were less dense (Seefelder et al., 2000; Koie et al., 2005; Čerenak et al., 2006; Patzak et al., 2012; Henning et al., 2011; Jakše et al., 2013; McAdam et al., 2013) than the new ones, created by using thousands of SNP markers (Henning et al., 2015, 2017; Havill et al., 2023). A lower average distance suggests denser coverage, meaning markers are more closely spaced and provide a more detailed picture of the genome. For instance, markers used by Jakše et al. (2013) resulted in 5.88 cM average marker distance while Henning et al. (2017) reported 0.34 cM distance using SNP markers. This significant reduction in average marker distance demonstrates the benefits of adopting newer marker technologies like SNP genotyping, which can produce a significantly higher density of markers across the genome than the previous methods. An example of high density map created with 60 K SNP markers is presented in Fig. 3.3 (Padgitt-Cobb et al., 2020).

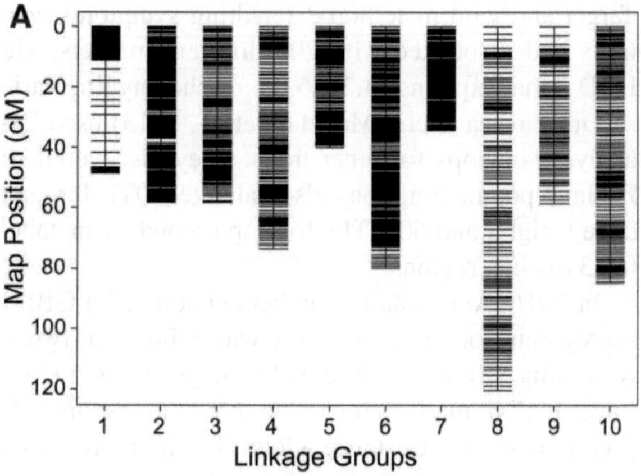

Fig. 3.3 Linkage groups and size across ten linkage groups (9 autosomes and 1 pseudo-autosomal region) (Padgitt-Cobb et al., 2020)

3.4.2.1 Quantitative Trait Loci (QTL) Mapping

Quantitative trait loci (QTL) mapping was first described by Sax (Sax, 1923). QTL mapping is a combination of genetic mapping and traditional quantitative genetics. QTLs are genes or loci/specific regions in the genome linked to quantitative traits (Čerenak & Javornik, 2008). It gives us information about the contribution to the trait variation in a population. For QTL mapping, information about the number of polymorphic genetic markers and quantitative trait values for individual samples is required (Zeng, 2001). Compared to other crops, only a small number of QTL studies have been undertaken in hops. The desirable hop traits are sex identification, resistance to diseases, increased yield capacity and improved bittering, flavour and aroma chemistry (Henning et al., 2011; McAdam et al., 2013). American researchers have primarily focused on resistance to powdery mildew (Henning et al., 2011; Padgitt-Cobb et al., 2020; Havill et al., 2023) and downy mildew (Henning et al., 2015, 2016, 2017), as these diseases cause the most economic damage in their region, whereas verticillium wilt is more common in Europe and Slovenian researchers have dedicated efforts to search for the wilt resistance linked QTL (Jakše et al., 2013).

Genetic QTL linkage map of hop was created by Koie et al. (2005). They developed 212 AFLP markers for variety Chinook and male parent SaM. QTLs were detected for 7 chemical components across 4 linkage groups. QTLs were also determined for alpha-acids and yield where AFLP and microsatellite markers were employed. In the population of 111 offspring (Magnum x wild male hop 2/1) a total of 13 putative QTLs for alpha acid content, 13 for dry cone weight and 18 for harvest index were identified on the two maps over the years (Čerenak et al., 2009).

The first study to identify molecular markers linked to the QTLs associated with disease susceptibility was conducted by Henning et al. (2011). They identified QTLs associated with powdery mildew in crosses of Perle x USDA 19058M where three QTLs in three linkage groups were identified. In their study they created the most comprehensive map of the hop genome. In the F1 mapping population of Wye

Target and wild male hop 2/1 wilting symptoms were evaluated along with other traits and associated with 203 different markers. The study identified QTL with LOD 7 that explains 24.2–26.0% of phenotyping variance (Jakše et al., 2013).

Other researchers (McAdam et al., 2013) used DArT markers to conduct QTL analyses of hops for other traits. They discovered sex-linked marker in the New Zealand population. They also validated QTL for essential oil content and one dry cone weight, and 60 QTLs for hop secondary metabolites were identified, mapped to 13 distinct regions.

In 2016, American researchers conducted a GBS analysis on a hop population segregating for resistance to downy mildew in two different environments - field (Corvallis, OR and Yakima, WA) and greenhouse (Henning et al., 2016). They identified 39 SNP markers related to infection response, 17 of which were further analysed to discover resistance alleles. Furthermore, researchers found differential sets of SNP markers related to the environment, suggesting two different of disease response. Their research identified four highly selective SNP markers and resistance alleles, however no validation report has been published to date.

In respect of medicinal use of hop metabolic compounds, QTL molecular markers for xanthohumol content, desmethylxanthohumol content and bitter acid content were detected by Patzak et al. (2012) and evaluated across 68 world hop cultivars. STS, SSR, EST-SSR and AFLP molecular markers were used to analyse 116 individual female F1 hop genotypes. The genetic linkage maps for the female parent Taurus and male parent H06/14 were created using 106 markers in total.

The development of the genetic map requires a complete whole genome chromosome assembly, which was not available for hop until 2023 (Padgitt-Cobb et al., 2023). The hop genome is large (2.8 Gb), heterozygous and complex, and early assembly efforts resulted in fragmented assemblies. Scaffold based assemblies were available earlier for Shinshuwase (Natsume et al., 2015) and Cascade (Hill et al., 2016) and later the Cascade draft haplotype-phased assembly (Padgitt-Cobb et al., 2021) and chromosome-level assembly (Padgitt-Cobb et al., 2023). The estimated repeat content is approx. 64% with of long terminal repeats (LTRs). Additionally, genome complexity and heterozygosity pose challenges for assembly and annotation.

3.4.3 Marker-Assisted Selection (MAS)

Marker-assisted selection (MAS) is a technique in which genetic markers are used to select plants or animals with desirable traits. Once QTLs are identified, markers closely linked to them can be developed. QTL mapping often identifies relatively large chromosomal regions associated with a trait, making it challenging to pinpoint the exact causal genes. Further fine-mapping or functional studies are usually required to identify the specific genes underlying the QTL and should be validated across different populations or accessions with a known phenotype (Holland, 2004). Additionally, QTL mapping may have limited power to detect all contributing loci when examining highly complex traits controlled by multiple genes with minor effects. Numerous QTLs have been identified in hop through linkage mapping. However, the reproducibility of these QTLs across diverse hop genotypes has been limited due to the likely presence of genotype-specific genetic interactions and variations (Murakami et al., 2006). The majority of detected QTLs appear to be family specific, hindering their successful transfer and utilization in larger breeding initiatives (Čerenak et al., 2006; Patzak et al., 2012).

Sex selection is required in the early stages of the breeding program because only ripe female inflorescences are used in the brewing process and also because seeded female cones have lower market value. Hop plants reach reproductive phase in two years, therefore use of sex molecular markers is desirable for early detection of sex. Sex related markers are among the most studied, using various marker techniques. Buck et al. (2009) developed four sex linked markers: three SCAR markers, and one high-resolution melting (HRM) curve analysis marker which has been effectively applied in the New Zealand hop breeding program (Beatson et al., 2016). Čerenak et al. (2019) used DArT male marker sequences (Howard et al., 2011) and developed multiplex PCR for MAS. Four PCR markers were selected and tested on the global hop population. The method was upgraded with qPCR TaqMan probes where a single male marker was able to detect all male plants (Luskar et al., 2023). The method is commonly employed in the Slovenian hop breeding program.

QTLs associated with pathogen resistance have not yet been translated into molecular markers for MAS. Despite identifying potential markers for downy mildew resistance, the study by Henning et al. (2016) lacks validation, making its broader applicability questionable. Similarly, KASP markers for powdery mildew resistance have shown promise, although their current applicability is limited to accessions associated with the source pedigree (Havill et al., 2023). It is worth noting that markers for powdery and downy mildew resistance are reportedly being utilized in MAS by private breeding companies, suggesting possible improvements in this field that have not yet been made public.

To develop more widely applicable and effective MAS markers in hop breeding, future research should prioritize large-scale, field-based GWAS studies on a variety of genotypes with well-characterized phenotypes. As pointed out by Henning et al. (2016), SNP markers obtained from field phenotyping are likely to be most relevant in practical breeding applications. The integration of such markers into MAS programs has the potential to considerably accelerate hop cultivar development, overcoming the challenges posed by the crop's long generation time, dioecious nature, and demanding phenotyping requirements. This might eventually streamline the breeding process, allowing for more efficient and cost-effective creation of new hop cultivars with desirable traits.

References

Abbott, M. S., & Fedele, M. J. (1994). A DNA-based varietal identification procedure for hop leaf tissue. *Journal of the Institute of Brewing, 100*(4), 283–285. https://doi.org/10.1002/j.2050-0416.1994.tb00825.x

Abdurakhmonov, I. Y. (2016). Introduction to microsatellites: Basics, trends and highlights. In *Microsatellite markers* (Vol. 1). https://doi.org/10.5772/66446

Alfaro-Saiz, E., Cámara-Leret, S., González-González, M., Fernández-Álvarez, Ó., Rodríguez-Fernández, S., López-López, D., et al. (2024). The memory of hops: Rural bioculture as a collective means of reimagining the future. *Sustainability, 16*(6), 2470. https://doi.org/10.3390/su16062470

Bardakci, F. (2001). Random amplified polymorphic DNA (RAPD) markers. *Turkish Journal of Biology, 25*, 185–196. https://journals.tubitak.gov.tr/biology/vol25/iss2/10

Bassil, N. V., Gilmore, B., Oliphant, J. M., Hummer, K. E., & Henning, J. A. (2008). Genic SSRs for European and North American hop (Humulus lupulus L.). *Genetic Resources and Crop Evolution, 55*(7). https://doi.org/10.1007/s10722-007-9303-9

Beatson, R. A., Ferguson, A. R., Weir, I. E., Graham, L. T., Ansell, K. A., & Ding, H. (2003). Flow cytometric identification of sexually derived polyploids in hop (Humulus lupulus L.) and their use in hop breeding. *Euphytica, 134*(2), 189–194. https://doi.org/10.1023/B:EUPH.0000003882.23615.c5

Beatson, R. A., Alspach, P. A., Stephens, M. J., Buck, E. J., Datson, P. M., & Ferguson, A. R. (2016). Breeding polyploid hop cultivars for New Zealand conditions. *Acta Horticulturae, 1127*, 9–14. https://doi.org/10.17660/ActaHortic.2016.1127.2

Brady, J. L., Scott, N. S., & Thomas, M. R. (1996). DNA typing of hops (Humulus lupulus) through application of RAPD and microsatellite marker sequences converted to sequence tagged sites (STS). *Euphytica, 91*(3). https://doi.org/10.1007/BF00033088

Buck, E. J., Wiedow, C., Carlisle, C., Chagné, D., & Beatson, R. (2009). Validation of new sex specific DNA markers for hops (Humulus lupulus L.). *Acta Horticulturae, 848*. https://doi.org/10.17660/ActaHortic.2009.848.33

Butler, J. M. (2012). Non-human DNA. In *Advanced topics in forensic DNA typing* (pp. 473–495). Elsevier. https://doi.org/10.1016/b978-0-12-374513-2.00016-6

Calvi, A., Aci, M. M., Lupini, A., & Preiti, G. (2023). Morphological and genetic analysis of wild hop (Humulus lupulus L.) germplasm from Calabria region in South Italy. *Agronomy, 13*, 252. https://doi.org/10.3390/agronomy13010252

Carron, C. A., Fuchsmann, P., Tena Stern, M., Moix, A., Corsi, S., Droz, E., et al. (2022). Genetic, phytochemical and agronomic characterization of wild hop ecotypes from Switzerland. In *XXXI international horticultural congress (IHC2022): International symposium on medicinal and aromatic plants: Domestication, 1358* (pp. 11–18). https://doi.org/10.17660/ActaHortic.2023.1358.2

Čerenak, A., & Javornik, B. (2008). The basis of genetic mapping of agricultural plants. *Hop Bulletin, 15*, 23–31. https://www.cabidigitallibrary.org/doi/pdf/10.5555/20093037986

Čerenak, A., Jakše, J., & Javornik, B. (2004). Identification and differentiation of hop varieties using simple sequence repeat markers. *Journal of the American Society of Brewing Chemists, 62*(1), 1–7. https://doi.org/10.1094/asbcj-62-0001

Čerenak, A., Šatovic, Z., & Javornik, B. (2006). Genetic mapping of hop (Humulus lupulus L.) applied to the detection of QTLs for alpha-acid content. *Genome, 49*(5), 485–494. https://doi.org/10.1139/g06-007

Čerenak, A., Satovic, Z., Jakse, J., Luthar, Z., Carovic-Stanko, K., & Javornik, B. (2009). Identification of QTLs for alpha acid content and yield in hop (Humulus lupulus L.). *Euphytica, 170*, 141–154. https://doi.org/10.1007/s10681-009-9920-9

Čerenak, A., Kolenc, Z., Sehur, P., Whittock, S. P., Koutoulis, A., Beatson, R., Buck, E., Javornik, B., Škof, S., & Jakše, J. (2019). New male specific markers for hop and application in breeding program. *Scientific Reports, 9*(1). https://doi.org/10.1038/s41598-019-50400-z

Clare, S. J., King, R. M., Tawril, A. L., Havill, J. S., Muehlbauer, G. J., Carey, S. B., et al. (2024). An affordable and convenient diagnostic marker to identify male and female hop plants. *G3: Genes, Genomes, Genetics, 14*(1), jkad216. https://doi.org/10.1093/g3journal/jkad216

Contin, D. R., Habermann, E., de Souza, B. C., de Oliveira, E. A. D., Martinez, C. A., Vieira, P. C., & Da Costa, F. B. (2023). Exploring the tropical acclimation of European and American hop cultivars (Humulus lupulus L.): Focus on physiology, productivity, and chemical composition. *European Journal of Agronomy, 151*, 126990. https://doi.org/10.1016/j.eja.2023.126990

Dabbous-Wach, A., Rodolfi, M., Paolini, J., Costa, J., & Ganino, T. (2021). Characterization of wild Corsican hops and assessment of the performances of German hops in Corsican environmental conditions through a multidisciplinary approach. *Applied Sciences, 11*(9), 3756. https://doi.org/10.3390/app11093756

Duarte, P. F., do Nascimento, L. H., Bandiera, V. J., Fischer, B., Fernandes, I. A., Paroul, N., & Junges, A. (2024). Exploring the versatility of hop essential oil (Humulus lupulus L.): Bridging brewing traditions with modern industry applications. *Industrial Crops and Products, 218*, 118974. https://doi.org/10.1016/j.indcrop.2024.118974

Gawroński, P., Pawełkowicz, M., Tofil, K., Uszyński, G., Sharifova, S., Ahluwalia, S., Tyrka, M., Wędzony, M., Kilian, A., & Bolibok-Brągoszewska, H. (2016). DArT markers effectively target gene space in the rye genome. *Frontiers in Plant Science, 7*, 01600. https://doi.org/10.3389/fpls.2016.01600

Ghiselli, L., Marini, L., Taiti, C., Calamai, L., Paffetti, D., & Masi, E. (2022). Hops germplasm: phytochemical characterization of wild humulus lupulus of central and northern Italy. *Plants, 11*(12), 1564.

Gienapp, P., Teplitsky, C., Alho, J. S., Mills, J. A., & Merilä, J. (2008). Climate change and evolution: Disentangling environmental and genetic responses. *Molecular Ecology, 17*(1), 167–178. https://doi.org/10.1111/j.1365-294X.2007.03413.x

Grdiša, M., Šatović, Z., Liber, Z., Jakše, J., Varga, F., Erhatić, R., & Srečeć, S. (2021). High genetic diversity and low population differentiation in wild hop (Humulus lupulus L.) from Croatia. *Applied Sciences, 11*(14). https://doi.org/10.3390/app11146484

Guimaraes, B. P., Nascimento, P. G. B. D., & Ghesti, G. F. (2021). Intellectual property and plant variety protection: Prospective study on hop (Humulus lupulus L.) cultivars. *World Patent Information, 65*, 102041. https://doi.org/10.1016/j.wpi.2021.102041

Hartl, L., & Seefelder, S. (1998). Diversity of selected hop cultivars detected by fluorescent AFLPs. *Theoretical and Applied Genetics, 96*(1). https://doi.org/10.1007/s001220050716

Havill, J. S., Richardson, B. J., Rohwer, C. L., David, G. H., Henning, J. A., & Muehlbauer, G. J. (2023). Identification of quantitative trait loci associated with R1-mediated resistance to powdery mildew and sex determination in hop (Humulus lupulus L.). *Theoretical and Applied Genetics, 136*, 154. https://doi.org/10.1007/s00122-023-04399-7

Henning, J. (2006). The breeding of hop. In *Brewing* (pp. 102–122). Woodhead Publishing. https://doi.org/10.1533/9781845691738.102

Henning, J. A., & Townsend, M. S. (2005). Field-based estimates of heritability and genetic correlations in hop. *Crop Science, 45*(4), 1469–1475. https://doi.org/10.2135/cropsci2004.0360

Henning, J. A., Steiner, J. J., & Hummer, K. E. (2004). Genetic diversity among world hop accessions grown in the USA. *Crop Science, 44*(2), 411–417. https://doi.org/10.2135/cropsci2004.4110

Henning, J. H., Townsend, M. S., & Matthews, P. (2010). Predicting offspring performance in hop (Humulus lupulus L.) using AFLP markers. *Journal of the American Society of Brewing Chemists, 68*(3), 125–131. https://doi.org/10.1094/ASBCJ-2010-0706-01

Henning, J. A., Townsend, M. S., Gent, D. H., Bassil, N., Matthews, P., Buck, E., & Beatson, R. (2011). QTL mapping of powdery mildew susceptibility in hop (Humulus lupulus L.). *Euphytica, 180*, 411–420. https://doi.org/10.1007/s10681-011-0403-4

Henning, J. A., Gent, D. H., Twomey, M. C., Townsend, M. S., Pitra, N. J., & Matthews, P. D. (2015). Precision QTL mapping of downy mildew resistance in hop (Humulus lupulus L.). *Euphytica, 202*, 487–498. https://doi.org/10.1007/s10681-015-1356-9

Henning, J. A., Gent, D. H., Twomey, M. C., Townsend, M. S., Pitra, N. J., & Matthews, P. D. (2016). Genotyping-by-sequencing of a bi-parental mapping population segregating for downy mildew resistance in hop (Humulus lupulus L.). *Euphytica, 208*(3), 545–559. https://doi.org/10.1007/s10681-015-1600-3

Henning, J. A., Gent, D. H., Townsend, M. S., Woods, J. L., Hill, S. T., & Hendrix, D. (2017). QTL analysis of resistance to powdery mildew in hop (Humulus lupulus L.). *Euphytica, 213*(4). https://doi.org/10.1007/s10681-017-1849-9

Hill, S. T., Coggins, J., Liston, A., Hendrix, D., & Henning, J. A. (2016). Genomics of the hop pseudo-autosomal regions. *Euphytica, 209*(1), 171–179. https://doi.org/10.1007/s10681-016-1655-9

Holland, J. B. (2004). Implementation of molecular markers for quantitative traits in breeding programs—Challenges and opportunities. In *Proceedings of the 4th international crop science congress* (Vol. 26, pp. 1–13). Regional Institute. https://www.ars.usda.gov/ARSUserFiles/60701500/Publications/Holland/Holland4thInt'lCropSciCong04.pdf

Hornsey, I. S. (2003). *A history of beer and brewing* (Vol. 34). Royal Society of Chemistry.

Howard, E. L., Whittock, S. P., Jakše, J., Carling, J., Matthews, P. D., Probasco, G., Henning, J. A., Darby, P., Cerenak, A., Javornik, B., Kilian, A., & Koutoulis, A. (2011). High-throughput genotyping of hop (Humulus lupulus L.) utilising diversity arrays technology (DArT). *Theoretical and Applied Genetics, 122*(7), 1265–1280. https://doi.org/10.1007/s00122-011-1529-4

International Hop Growers' Convention. (2023). *2023 IHGC hop variety list* [PDF]. https://www.ihgc.org/wp-content/uploads/2023_IHGC_HopVarietyList.pdf

Jakse, J., Cerenak, A., Radisek, S., Satovic, Z., Luthar, Z., & Javornik, B. (2013). Identification of quantitative trait loci for resistance to Verticillium wilt and yield parameters in hop (Humulus lupulus L.). *Theoretical and Applied Genetics, 126*(6), 1431–1443.

Jakše, J., Kindlhofer, K., & Javornik, B. (2001). Assessment of genetic variation and differentiation of hop genotypes by microsatellite and AFLP markers. *Genome, 44*(5), 773. https://doi.org/10.1139/gen-44-5-773

Jakše, J., Bandelj, D., & Javornik, B. (2002). Eleven new microsatellites for hop (Humulus lupulus L.). *Molecular Ecology Notes, 2*(4). https://doi.org/10.1046/j.1471-8286.2002.00309.x

Jakše, J., Šatović, Z., & Javornik, B. (2004). Microsatellite variability among wild and cultivated hops (Humulus lupulus L.). *Genome, 47*(5). https://doi.org/10.1139/G04-054

Jakše, J., Luthar, Z., & Javornik, B. (2008a). New polymorphic dinucleotide and trinucleotide microsatellite loci for hop Humulus lupulus L. *Molecular Ecology Resources, 8*(4). https://doi.org/10.1111/j.1755-0998.2007.02053.x

Jakše, J., Štajner, N., Kozjak, P., Èerenak, A., & Javornik, B. (2008b). Trinucleotide microsatellite repeat is tightly linked to male sex in hop (Humulus lupulus L.). *Molecular Breeding, 21*(2). https://doi.org/10.1007/s11032-007-9114-x

Jastrombek, J. M., Faguerazzi, M. M., de Cássio Pierezan, H., Rufato, L., Sato, A. J., da Silva Ricce, W., et al. (2022). Hop: An emerging crop in subtropical areas in Brazil. *Horticulturae, 8*(5), 393. https://doi.org/10.3390/horticulturae8050393

Javornik, B. (2005). Genetic research on hops. *Hop Bulletin, 12*, 71–78. https://www.cabidigitallibrary.org/doi/full/10.5555/20063057035

Javornik, B. (2019). Genetic studies of hop and its pathogens in Slovenia—A retrospective. *Hop Bulletin, 26*, 5–22. http://www.dlib.si/stream/URN:NBN:SI:DOC-A4F24XQF/76638dfb-9d09-40ee-9ae1-55cc9de6457a/PDF

Karlsson Strese, E. M., Lundström, M., Hagenblad, J., & Leino, M. W. (2014). Genetic diversity in remnant Swedish hop (Humulus lupulus L.) yards from the 15th to 18th century. *Economic Botany, 68*, 231–245. https://doi.org/10.1007/s12231-014-9273-8

Koelling, J., Coles, M. C., Matthews, P. D., & Schwekendiek, A. (2012). Development of new microsatellite markers (SSRs) for Humulus lupulus. *Molecular Breeding, 30*(1), 479–484. https://doi.org/10.1007/s11032-011-9637-z

Koie, K., Inaba, A., Okada, Y., Kaneko, T., & Ito, K. (2005). Construction of the genetic linkage map and QTL analysis on hop (Humulus lupulus L.). *Acta Horticulturae, 668*, 59–66. https://doi.org/10.17660/ACTAHORTIC.2005.668.7

Kolenc, Z., Vodnik, D., Mandelc, S., Javornik, B., Kastelec, D., & Čerenak, A. (2016). Hop (Humulus lupulus L.) response mechanisms in drought stress: Proteomic analysis with physiology. *Plant Physiology and Biochemistry, 105*, 67–78. https://doi.org/10.1016/j.plaphy.2016.03.026

Korpelainen, H., & Pietiläinen, M. (2021). Hop (Humulus lupulus L.): Traditional and present use, and future potential. *Economic Botany, 75*(3), 302–322. https://doi.org/10.1007/s12231-021-09528-1

Kralj, D. (1973). Di x tetraploid crossing in hop (Humulus lupulus L.). *Research Reports of the Biotechnical Faculty, University of Ljubljana. Agricultural Issue, 21*, 155–174.

Kralj, D. (1989). Breeding of fertile aromatic hop (Humulus lupulus L.) with polyploidisation method. In *Research reports of VI hop symposium (Žalec, 1989)* (pp. 25–69). Slovenian Institute of Hop and Brewing—Zalec and Institute of Field and Vegetable Crops—Novi Sad.

Kralj, D., & Sušnik, F. (1983). Germplasm transfer in hops. In *Eucarpia workshop on pre-breeding in relation to gene banks* (pp. 169–179). Serbian Genetics Society.

Kralj, D., Zupanec, J., Vasilj, D., Kralj, S., & Pšeničnik, J. (1991). Variability of essential oils of hops, Humulus lupulus L. *Journal of the Institute of Brewing, 97*(3), 197–206. https://doi.org/10.1002/j.2050-0416.1991.tb01066.x

Luskar, L., Čerenak, A., & Jakše, J. (2022). Interaction of hop (Humulus lupulus L.) and Verticillium nonalfalfae. *Hop Bulletin, 29*, 5–22.

Luskar, L., Eržen, M., Čerenak, A., & Jakše, J. (2023). Sex determination in hop (Humulus lupulus L.) by detection of male markers using multiplex qPCR. In *The Plant & Animal Genome Conference, PAG 30, January 13–18, 2023, San Diego*. https://plan.core-apps.com/pag_2023/abstract/963d7d6c9252da3647857a6b5d0dd46d

Machado, J. C., Faria, M. A., Barata, A. M., da Silva, I. G., Čerenak, A., & Ferreira, I. M. P. L. V. O. (2022). Portuguese wild hop diversity assessment by fast SNP genotyping using high-resolution melting. *Journal of Applied Genetics, 63*(1), 103–114. https://doi.org/10.1007/s13353-021-00668-7

Mackinnon, D., & Pavlovič, M. (2022). Price dynamics in the market of hops. *Hop Bulletin*, (29), 65–73.

Mackinnon, D., & Pavlovič, M. (2023). The hops surplus influence on global market events. *Hmeljarski Bilten*, (30), 54–63.

Mafakheri, M., Kordrostami, M., Rahimi, M., & Matthews, P. D. (2020). Evaluating genetic diversity and structure of a wild hop (Humulus lupulus L.) germplasm using morphological and molecular characteristics. *Euphytica, 216*, 1–19. https://doi.org/10.1007/s10681-020-02592-z

Magiera, A. (1875). *Della coltivazione del luppolo*. Tipografia Sociale.

Marwal, A., Sahu, A. K., & Gaur, R. K. (2014). Molecular markers: Tool for genetic analysis. In *Animal biotechnology: Models in discovery and translation* (pp. 289–305). https://doi.org/10.1016/B978-0-12-416002-6.00016-X

Matthews, P. D., Coles, M. C., & Pitra, N. J. (2013). Next generation sequencing for a plant of great tradition: Application of NGS to SNP detection and validation in hops (Humulus lupulus L.). *BrewingScience, 66*(11–12), 185–191.

McAdam, E. L., Freeman, J. S., Whittock, S. P., Buck, E. J., Jakše, J., Čerenak, A., Javornik, B., Kilian, A., Wang, C. H., Andersen, D., Vaillancourt, R. E., Carling, J., Beatson, R., Graham, L., Graham, D., Darby, P., & Koutoulis, A. (2013). Quantitative trait loci in hop (Humulus lupulus L.) reveal complex genetic architecture underlying variation in sex, yield and cone chemistry. *BMC Genomics, 14*(1). https://doi.org/10.1186/1471-2164-14-360

Mondini, L., Noorani, A., & Pagnotta, M. A. (2009). Assessing plant genetic diversity by molecular tools. *Diversity, 1*(1). https://doi.org/10.3390/d1010019

Mongelli, A., Rodolfi, M., Ganino, T., Marieschi, M., Dall'Asta, C., & Bruni, R. (2015). Italian hop germplasm: Characterization of wild Humulus lupulus L. genotypes from northern Italy by means of phytochemical, morphological traits and multivariate data analysis. *Industrial Crops and Products, 70*, 16–27. https://doi.org/10.1016/j.indcrop.2015.02.036

Mongelli, A., Rodolfi, M., Ganino, T., Marieschi, M., Caligiani, A., Dall'Asta, C., & Bruni, R. (2016). Are Humulus lupulus L. ecotypes and cultivars suitable for the cultivation of aromatic hop in Italy? A phytochemical approach. *Industrial Crops and Products, 83*, 693–700. https://doi.org/10.1016/j.indcrop.2015.12.046

Murakami, A., Darby, P., Javornik, B., Pais, M. S. S., Seigner, E., Lutz, A., & Svoboda, P. (2006). Microsatellite DNA analysis of wild hops, Humulus lupulus L. *Genetic Resources and Crop Evolution, 53*(8). https://doi.org/10.1007/s10722-005-7765-1

Nadeem, M. A., Nawaz, M. A., Shahid, M. Q., Dogan, Y., Comertpay, G., Yildiz, M., Hatipoglu, R., Ahmad, F., Alsaleh, A., Labhane, N., Özkan, H., Chung, G., & Baloch, F. S. (2018). DNA molecular markers in plant breeding: Current status and recent advancements in genomic selection and genome editing. In *Biotechnology and biotechnological equipment* (Vol. 32, Issue 2, pp. 261–285). Taylor and Francis Ltd.. https://doi.org/10.1080/13102818.2017.1400401

Natsume, S., Takagi, H., Shiraishi, A., Murata, J., Toyonaga, H., Patzak, J., et al. (2015). The draft genome of hop (Humulus lupulus), an essence for brewing. *Plant and Cell Physiology, 56*(3), 428–441. https://doi.org/10.1093/pcp/pcu169

Nesvadba, V., Donner, P., & Charvátová, J. (2022). Hop breeding in The Czech Republic. In *Proceedings of the scientific-technical commission*, IHGC, Lugo, Galicia, Spain (pp. 3–7).

Neve, R. A. (1964). New hop varieties: Their production and evaluation. In *Report of the Department of Hop Research, Wye College for 1963* (pp. 7–9). Wye College.

Neve, R. A. (1991). The cultivated hop. In *Hops*. Springer. https://doi.org/10.1007/978-94-011-3106-3_2

Padgitt-Cobb, L. K., Kingan, S. B., & Henning, J. A. (2020). Genomic analysis of powdery mildew resistance in a hop (*Humulus lupulus* L.) bi-parental population segregating for "R6-locus.". *Euphytica, 216*(1), 1–15. https://doi.org/10.1007/s10681-019-2543-x

Padgitt-Cobb, L. K., Kingan, S. B., Wells, J., Elser, J., Kronmiller, B., Moore, D., Concepcion, G., Peluso, P., Rank, D., Jaiswal, P., Henning, J., & Hendrix, D. A. (2021). A draft phased assembly of the diploid Cascade hop (*Humulus lupulus*) genome. *Plant Genome, 14*(1). https://doi.org/10.1002/tpg2.20072

Padgitt-Cobb, L. K., Pitra, N. J., Matthews, P. D., Henning, J. A., & Hendrix, D. A. (2023). An improved assembly of the "Cascade" hop (*Humulus lupulus*) genome uncovers signatures of molecular evolution and refines time of divergence estimates for the Cannabaceae family. *Horticulture Research, 10*(2). https://doi.org/10.1093/hr/uhac281

Paguet, A. S., Siah, A., Lefèvre, G., Moureu, S., Cadalen, T., Samaillie, J., et al. (2023). Multivariate analysis of chemical and genetic diversity of wild *Humulus lupulus* L. (hop) collected in situ in northern France. *Phytochemistry, 205*, 113508.

Paguet, A. S., Siah, A., Lefèvre, G., Vandenberghe, M., Lutun, D., Degardin, N., et al. (2024). Phytochemical characterisation and aromatic potential for brewing of wild hops (*Humulus lupulus* L.) from Northern France: Towards a lead for local hop varieties. *Food Chemistry, 433*, 137302.

Pasqui, U. (2010). *L'uomo della birra*. Ed. Cartacanta, Forlì.

Patzak, J., & Henychová, A. (2018). Evaluation of genetic variability within actual hop (*Humulus lupulus* L.) cultivars by an enlarged set of molecular markers. *Czech Journal of Genetics & Plant Breeding, 54*(2) https://cjgpb.agriculturejournals.cz/pdfs/cjg/2018/02/07.pdf

Patzak, J., Nesvadba, V., Krofta, K., Henychova, A., Marzoev, A. I., & Richards, K. (2010a). Evaluation of genetic variability of wild hops (*Humulus lupulus* L.) in Canada and the Caucasus region by chemical and molecular methods. *Genome / National Research Council Canada = Génome / Conseil National de Recherches Canada, 53*(7). https://doi.org/10.1139/G10-024

Patzak, J., Nesvadba, V., Henychová, A., & Krofta, K. (2010b). Assessment of the genetic diversity of wild hops (*Humulus lupulus* L.) in Europe using chemical and molecular analyses. *Biochemical Systematics and Ecology, 38*(2), 136–145. https://doi.org/10.1016/J.BSE.2009.12.023

Patzak, J., Henychova, A., Krofta, K., & Nesvadba, V. (2012). Study of molecular markers for xanthohumol and DMX contents in hop (*Humulus lupulus* L.) by QTL mapping analysis. *BrewingScience, 65*, 96–102.

Pavlovič, M., Mackinnon, D., & Naglič, B. (2022). Study of market supply statistics in trading of hops. *Hop Bulletin, 29*, 73–85.

Pillay, M., & Kenny, S. T. (1996). Random amplified polymorphic DNA (RAPD) markers in hop, Humulus lupulus: Level of genetic variability and segregation in F 1 progeny. *Theoretical and Applied Genetics, 92*, 334–339. https://doi.org/10.1007/BF00223676

Polley, A., Ganal, M. W., & Seigner, E. (1997). Identification of sex in hop (Humulus lupulus) using molecular markers. *Genome, 40*(3), 357–361. https://doi.org/10.1139/g97-048

Radišek, S., Čerenak, A., & Javornik, B. (2007). *Hop disease resistance breeding: Procedures and techniques in selection of seedlings.* https://www.cabidigitallibrary.org/doi/full/10.5555/20083035391

Riccioni, C., Belfiori, B., Sileoni, V., Marconi, O., Perretti, G., Bellucci, M., & Rubini, A. (2021). High genetic and chemical diversity of wild hop populations from Central Italy with signals of a genetic structure influenced by both sexual and asexual reproduction. *Plant Science, 304*, 110794. https://doi.org/10.1016/j.plantsci.2020.110794

Rodolfi, M., Silvanini, A., Chiancone, B., Marieschi, M., Fabbri, A., Bruni, R., & Ganino, T. (2018). Identification and genetic structure of wild Italian *Humulus lupulus* L. and comparison with European and American hop cultivars using nuclear microsatellite markers. *Genetic Resources and Crop Evolution, 65*, 1405–1422. https://doi.org/10.1007/s10722-018-0622-9

Rodolfi, M., Marieschi, M., Chiancone, B., & Ganino, T. (2022). Assessment of the genetic and phytochemical variability of Italian wild hop: A route to biodiversity preservation. *Applied Sciences, 12*(11), 5751. https://doi.org/10.3390/app12115751

Sax, K. (1923). The association of size differences with seed-coat pattern and pigmentation in *Phaseolus vulgaris*. *Genetics, 8*(6), 552.

Seefelder, S., Ehrmaier, H., Schweizer, G., & Seigner, E. (2000). Genetic diversity and phylogenetic relationships among accessions of hop, Humulus lupulus, as determined by amplified fragment length polymorphism fingerprinting compared with pedigree data. *Plant Breeding, 119*(3), 257–263. https://doi.org/10.1046/j.1439-0523.2000.00500.x

Semagn, K., Bjørnstad, Å., & Ndjiondjop, M. N. (2006). An overview of molecular marker methods for plants. *African Journal of Biotechnology, 5*(25). https://www.ajol.info/index.php/ajb/article/view/56080

Škof, S., Bohanec, B., Kastelec, D., & Luthar, Z. (2007). Spontaneous induction of tetraploidy in hop using adventitious shoot regeneration method. *Plant Breeding, 126*(4), 416–421.

Solberg, S. Ø., Brantestam, A. K., Kylin, M., & Bjørn, G. K. (2014). Genetic variation in Danish and Norwegian germplasm collections of hops. *Biochemical Systematics and Ecology, 52*, 53–59. https://doi.org/10.1016/j.bse.2013.12.014

Štajner, N., Jakše, J., Kozjak, P., & Javornik, B. (2005). The isolation and characterisation of microsatellites in hop (*Humulus lupulus* L.). *Plant Science, 168*(1). https://doi.org/10.1016/j.plantsci.2004.07.031

Štajner, N., Satovic, Z., Cerenak, A., & Javornik, B., (2008). Genetic structure and differentiation in hop (Humulus lupulus L.) as inferred from microsatellites. *Euphytica, 161*(1–2), 301–311. https://doi.org/10.1007/s10681-007-9429-z

Šuštar-Vozlič, J., & Javornik, B. (1999). Genetic relationships in cultivars of hop (*Humulus lupulus* L.) determined by RAPD analysis. *Plant Breeding, 118*(2), 175–181.

Temesgen, B. (2021). Genetic mapping in crop plants. *Open Journal of Plant Science*. https://doi.org/10.17352/ojps.000028

Townsend, M. S., & Henning, J. A. (2009). AFLP discrimination of native North American and cultivated hop. *Crop Science, 49*(2), 600–607.

Van Holle, A., Muylle, H., Ruttink, T., Van Landschoot, A., Haesaert, G., Naudts, D., De Keukeleire, D., & Roldán-Ruiz, I. (2019). Single nucleotide polymorphisms and biochemical markers as complementary tools to characterize hops (*Humulus lupulus* L.) in brewing practice. *Journal of Agricultural and Food Chemistry, 67*(13), 3761–3771. https://doi.org/10.1021/acs.jafc.9b00816

Vos, P., Hogers, R., Bleeker, M., Reijans, M., Lee, T. V. D., Hornes, M., et al. (1995). AFLP: A new technique for DNA fingerprinting. *Nucleic Acids Research, 23*(21), 4407–4414. https://doi.org/10.1093/nar/23.21.4407

Whittock, S., Leggett, G., Carling, J., Kilian, A., Probasco, G., Darby, P., et al. (2009). Use of diversity array technology (DArT) for genotyping of *Humulus lupulus* L. In D. De Keukeleire et al. (Eds.), *Proceedings of the IInd international Humulus symposium* (Acta Horticulturae, 848) (pp. 59–63). International Society for Horticultural Science.

Williams, J. G. K., Kubelik, A. R., Livak, K. J., Rafalski, J. A., & Tingey, S. V. (1990). DNA polymorphisms amplified by arbitrary primers are useful as genetic markers. *Nucleic Acids Research, 18*(22). https://doi.org/10.1093/nar/18.22.6531

Young, P. R., & Vivier, M. A. (2021). Genetics and genomic approaches to improve grape quality for winemaking. In *Managing wine quality: Volume one: Viticulture and wine quality*. https://doi.org/10.1016/B978-0-08-102067-8.00030-0

Zeng, Z.-B. (2001). QTL mapping. In S. Maloy & K. Hughes (Eds.), *Brenner's encyclopedia of genetics* (2nd ed., pp. 8–12). Academic. https://doi.org/10.1016/B978-0-12-374984-0.01248-1

Chapter 4
Hops Cultivation and Production

Siniša Srečec, Margherita Rodolfi, Boštjan Naglič, and Tommaso Ganino

4.1 A Short Historical Overview of Hop Growing

The oldest historical documents about hop growing date back to 768 AD. Namely, in the will of Pepin the Short (714–768) the hop gardens in today's Hallertau region in Bavaria were left to the Cloister of Saint-Denis (Hornsey, 2003). Some sources named Freising as a first location of hop garden mentioned in Pepin the Short's will (Barth et al., 1994). However, despite the valuable historical documents some archaeological findings of hop cones fragments date from different ages and from different locations in Europe, such as Develier in Switzerland from the sixth to the eighth century, Serris-Les Ruelles in France from the seventh to the ninth century, Mikulčice in Czech Republic from the eighth to the ninth century, Lübeck in Germany thirteenth century finds, etc. (Verberg, 2020). These findings completely correspond with the historical outline of a thousand years of world hop growing (Barth et al., 1994). Also, the St. Hildegard or Abbess Hildegard of Bingen (1098–1179) was the first to describe the beneficial properties of hops in her book "Physica" in the twelfth century (written between 1150 and 1160) and she was very explicit in her advice to readers of that time "*if one intends to make beer from oats, it is prepared with hops*" (Kišgeci et al., 1984; Hornsey, 2003).

S. Srečec (✉)
Križevci University of Applied Sciences, Križevci, Croatia
e-mail: ssrecec@vguk.hr

M. Rodolfi · T. Ganino
Department of Food and Drug, University of Parma, Parma, Italy
e-mail: margherita.rodolfi@unipr.it; tommaso.ganino@unipr.it

B. Naglič
Department of Plants, Soil and Environment, Slovenian Institute of Hop Research and Brewing, Žalec, Slovenia
e-mail: bostjan.naglic@ihps.si

K. Carbone (ed.), *All About Hops: The Crop, its Cultivation, and its Uses in Brewing and Beyond*, https://doi.org/10.1007/978-3-031-96606-4_4

Nevertheless, the fateful connection between hops and beer was sealed on 23rd April 1516, when Bavarian Duke Wilhelm IV declared the "Purity Law" (*germ.* Reinheitsgebot) during the annual meeting of merchants in Ingolstadt (Schattenhofer, 1993). According to Barth et al. (1994) the spread of hop cultivation in other regions of Europe, in particular in the Danube region and in overseas countries, was mainly conditioned by the arrival of new residents in those areas. The new residents arrived:

1. from the hop-growing areas of Central Europe, mainly Germans and Slovaks, who migrated to the Danube region of the south-eastern European countries, and
2. from England who migrated to the area of New England in the United States (Neve, 1983; Kišgeci et al., 1984; Barth et al., 1994; Srečec, 2004).

One coloured woodcut from the 1880s shows the hop garden constructed with wooden poles whose tops are connected by thin ropes (Barth et al., 1994). So, it is possible to conclude that until the nineteenth century the hop gardens were constructed with poles. The poles were prepared from pine, because of their long life in the soil. The poles were then removed from the hop garden every year after the harvest. Such type of hop gardens construction was widespread in central Europe until the end of the second world war, when was successively replaced by modern trellis, as we know today (Barth et al., 1994; Virant, 2002).

4.1.1 Responses of Hop Growers on Some Historical Challenges

Until the 1830s usual method of preservation, not only of hop cones but also of wine and vegetables, was sulfuring. But this treatment was banned in 1830 because the sulfuring of hop cones might make appear old hop cones as fresh, making frauds easy to do. On 1st September 1846 the first official hop market was set up in Nuremberg, but because of strict regulations on brewers and traders it was quietly closed in 1860 and replaced by a free hop market, which lasted until the beginning of the twentieth century (Barth et al., 1994).

Nevertheless, the biggest historical challenge for hop production sustainability was the hop disease known as downy mildew, caused by the fungus *Pseudoperonospora humuli* (Miyabe & Takah.) G.W. Wilson (1914), which was noted for the first time in Japan in 1905, in England in 1920 and in Germany in 1924. This disease sporadically causes damage on hops, even today (Barth et al., 1994; Moir, 2000; Purayannur et al., 2021); damage may go from a reduction of cone yield and quality to plant death. On the other hand, downy mildew caused the establishment of a strong link between hop research and production (Barth et al., 1994; Virant, 2002). The spread of downy mildew led to the setting up of the Hop Research Institute at Hüll, Germany, in 1926. Also, till the end of the second world war, Slovenian hop production was almost completely shattered, partially because of war devastation, but mainly due to the *Pseudoperonospora humuli* fungus. The leading Slovenian hop cultivar at the time was German landrace hop cultivar Würtenberger, which was very sensitive to *Pseudoperonospora humuli* infections. Thus, in 1947 the Slovenian Institute for Hop Research and Brewing was founded, according to a decision of the General Assembly of the Hop cooperative—Hmezad. At the beginning, the Laboratory for hop growing was founded and through the years it was transformed and increased into the present Institute (Virant, 2002). It is very important to emphasise that this is not an isolated case in the world. In Žetac, Czech Republic, for instance, an agricultural research station was transformed into hop research station in 2024. Namely, in every hop growing country the initiative for organisation of hops research always started from hop producers and brewers. The scope of organised hop research was initially the development of new cultivars resistant or tolerant to *Pseudoperonospora humuli* and later, since the 1930s, to Verticillium wilt caused by *Verticillium albo-atrum* (Cooke) Wint. (Neve, 1986). Year after year, the new improved hop cultivars replaced traditional hops landraces. Eventually, even the traditional hops landraces have been improved by re-selection within phenotypically heterogenous landrace populations to improve the chemical properties of old hop cultivars (Neve, 1986, 1991). Improved knowledge of the chemistry of brewing allowed for progresses and changes in hop breeding programmes (Moir, 2000). According to chemical characterization of hops, all hop cultivars are divided into the groups of classic (traditional) aromatic cultivars, new aromatic cultivars, improved aromatic cultivars and bitter cultivars (Forster & Schmidt, 1994). This approach in hop breeding was strongly influenced by the innovations in hop processing. Namely, soon after invention of whirlpool separator in 1969 the use of hop powders and pelletised powders were introduced very soon and, already in 1975, approximately 20% of hop yield was processed into hop pellets. Moreover, in 1977 the first results of CO_2 supercritical extraction of hops were published, 9 years after discovery of the high solvent potential of liquid CO_2 for extraction of natural plant compounds (Pekhov, 1968; Laws et al., 1978; cited by Moir, 2000 as secondary source). Today CO_2 supercritical extraction is very important method in hop processing and the quality standards for hops are nowadays almost incomparable with those some 20 years ago. This is quite a big challenge for hop producers (Pavlovič, 2014).

4.2 The World Hops Growing Area

Geographical zone optimal for hops growing lays between 33 and 55° latitudes on north and south hemispheres (Kišgeci et al., 1984; Barth et al., 1994; Srečec, 2004; Kubeš, 2019; Mozny et al., 2023). According to official statistics of the European Commission hops are grown in 14 EU countries. Some 20,000 hectares are used for hop cultivation in Germany, accounting for 60% of the EU's hop-growing acreage and about one third of the surface area devoted to hop cultivation worldwide. The other main EU hop producers are the Czech Republic, Poland and Slovenia. The full list of hop production areas in EU is exposed in Table 4.1.

Table 4.1 List of European hop production areas (2017)

Country	Regions
Germany	Hallertau, Spalt, Tettnang, Rottenburg—Herrenberg—Weil der Stadt, Rheinpfalz, Bitburg, Elbe-Saale
Czech Republic	Žatec (Saaz)—Úštěk (Auscha)—Tršice (Trschitz)
Poland	Lubelski region, Wielkopolski region, Dolnośląski region, Opolski region, Śląski region, Kujawsko-pomorski region, Mazowiecki region, Podkarpacki region, Świętokrzyski region
Slovenia	Slovenia
UK	England
Spain	Castilla y León, Galicia, La Rioja, Navarra
France	Grand Est, Hauts-de-France
Austria	Oberösterreich, Niederösterreich, Steiermark
Belgium	West-Vlaanderen, Oost-Vlaanderen, Vlaams-Brabant, Hainaut
Portugal	Braga, Bragança
Slovakia	Piešťanská výrobná oblasť
Bulgaria	Pazardzhik region, Velingrad municipality, location town Velingrad
Hungary	Közép Magyarország

However, in recent years, new countries have started growing hops; for example, in Italy in 2023 there were 100 ha (Licciardo & Carbone, 2023). In this country, cultivation is widespread in the Centre and North of Italy, while the Regions where it is most concentrated are Emilia-Romagna and Veneto.

According to National Hop Report of the National Agricultural Statistics Service, Agricultural Statistics Board, United States Department of Agriculture (2023) the biggest region of hops production in US is Washington state, particularly Yakima Valley, followed by Idaho and Oregon.

In China the hop production is concentrated in two provinces, Gansu and Xinjiang (Pavlovič et al., 2006) and the centre of hops growing in Japan is Tono, in Iwate prefecture, placed near the east coast of the Honshu Island.

Speaking about the hops growing regions of the Southern hemisphere, the New Zealand's hops production is located on the South Island, primarily in Moutere Valley, while Australian hop production areas are located on the island of Tasmania and in the state of Victoria (Barth et al., 1994; Kopp, 2014). The largest hops growing area in Argentina is also the most southern hops production area in the world and it is in the Andean Patagonia region with currently 146 ha of hops (Testa et al., 2023).

4.3 Climatic and Soil Conditions for Hops Growing in a Climate Changing Era

Hops is a multipurpose crop whose utility primarily depends on its chemical characteristics (Wang et al., 2008; Nagel et al., 2008; Srečec et al., 2012; Rossini et al., 2021). Nevertheless, the yield of hop cones, as well as the chemical composition of hop secondary metabolites, primarily depend on the characteristics of the growing region, which nowadays vary year after year with longer or shorter periods of suitable and/or unsuitable growing conditions; primarily, due to global climatic changes or global warming. Climatic changes may lead to a reduction or even cessation of hop production in many countries, within the climate zone suitable for hop production, which until recently had a strong tradition of hop cultivation (Srečec et al., 2004; Mozny et al., 2009; Pavlovič et al., 2012; MacKinnon et al., 2020; Mozny et al., 2023). Moreover, global warming is also known to be the cause of more frequent extreme events, such as storms and intense rainfalls, that can cause damage of hop gardens.

Besides, a lot of hops varieties are very sensitive to climate conditions. Therefore, the hop production is one of the most vulnerable agricultural productions due to drought and heat in many European hop regions is correct (Potopová et al., 2021; Mozny et al., 2023). On the other hand, there is evidence of efficient hop production even in semiarid areas which are, at first sight, unsuitable for hop production (Rodolfi et al., 2018). However, careful choice of hop cultivars adapted to semiarid conditions, use of suited growing techniques and careful water management for irrigation, may allow regular yields and sufficient hop quality even in not typical hop growing regions such as semiarid climates (Fandiño et al., 2015; Rossini et al., 2016; Marceddu et al., 2020; Carrubba et al., 2022).

However, considering all these facts and the real threat of global climate changes, there is a logical question on what the global hop production will look like in a maybe not so far future: is it possible to grow hops in areas other than the usual hops growing zones? The answer is yes, provided that hop growers can ensure the sufficient water supply for irrigation, or are able to organise the production in humid areas. In this sense it should be reminded that dormancy and vernalization, are not necessary for the growth of hop flowers. Thus, it is possible to distribute hop production in a wider area than the 35–55° north and south latitudes, by selecting and developing new hop cultivars, through breeding and selection programmes, able to resist in the incoming climate condition and in world widespread areas, to alleviate risk in the supply chain and the risk for hop producers (Bauerle, 2019). As the role of research in the last centuries was essential to face hop diseases, and was the driving force to give new hope to the hop sector, now there is another time a strong need to develop new research to deal with climate change.

4.3.1 Requirements for Temperatures and Rainfalls

According to Kralj (1962), the climatologically ideal hop year should have the following weather conditions during the vegetative period in the northern hemisphere:

- From April to the first decade of May—dry and warm weather. This period corresponds to the BBCH[1] phenological growth stages of hops: rhizome with shoots (uncut), beginning of shoot-growth (rhizome cut), and emergence of first shoots at the soil surface (Kralj, 1962; Rossbauer et al., 1995 cited in Meier et al., 2001; Meier et al., 2009).
- From the first decade of May until the second half of June—rainy and not too hot weather. This period corresponds to the leaf development, formation of side shoots, and elongation of bines. In some years, depending on temperature inflorescence buds might be visible (Kralj, 1962; Rossbauer et al., 1995, cited in Meier et al., 2001; Meier et al., 2009).
- From the second half of June and during July and August—warm weather with sufficient precipitation and no severe drought. This period corresponds to the inflorescence emergence, flowering, development of cones, and cones maturity (Kralj, 1962; Rossbauer et al., 1995 cited in Meier et al. 2001; Meier et al., 2009).

Today, more than 60 years later, this description of the ideal hop year is still valid, but the periods have been shifted 1 month earlier, approximately. Namely, the minimal temperatures for the beginning of hop growth or, precisely, for the first shoots emerging at the soil surface, is 5 °C (Johnson, 1991; Zmrzlak & Kajfež-Bogataj, 1996; Srečec et al., 2004, 2008, 2013; Rossini et al., 2016).

However, the number of days is usually calculated using the Eq. 4.1 (Srečec et al., 2004, 2008, 2013; Rossini et al., 2016).

$$\mathrm{GDD}\sum_{days}\left(\frac{T_{\max}+T_{\min}}{2}\right)-T_b \tag{4.1}$$

Where:

GDD is the sum of effective temperatures (°C) in growing days
$T_{max.}$ is daily maximal temperature (°C)
T_{min} is daily minimal temperature (°C)
T_b is basic temperature for onset of vegetative growth which is 5 °C for hops.

It is very important to point out that hops are more sensitive to high temperatures than to high precipitations. High temperatures might slow down the growing of bines and speed up flowering at the same time thus, consequently, decreasing cone yield and quality, particularly on quantity of α-acids (Srečec et al., 2004, 2008; Mozny et al., 2009, 2023). The total loss of cone yield and alpha acids, corresponds to increase of average daily crop evapotranspiration (ET_0) during the flowering growth stages, and development and maturity of cones (Allan et al., 1998; Srečec et al., 2008; Fandiño et al., 2016). Initially, vegetative development takes place, then, after flowering, secondary metabolites are translocated from leaf mesophyll into the glandular trichomes which are organised on the epidermis of hop cone

[1] **B**iologische **B**undesanstalt für Land- und Forstwirtschaft, Bundessortenamt und **CH**emische Industrie

bracts (Wang et al., 2008); every heat stress decreases the number and volume of glandular trichomes (Srečec et al., 2011a).

From shoot emergence at the soil surface (after cut) till the end of cones ripening, as a rule the number of GDD required is between 1600 and 2000 °C. The rainfall needs vary from 200 to 400 mm per growing season, depending on hop cultivar (Srečec et al., 2008; Mozny et al., 2009; Rossini et al., 2016), but without considering extreme drought periods, which might increase the values of daily crop evapotranspiration (Srečec et al., 2008, 2013; Fandiño et al., 2016).

Considering the importance of alpha-acids accumulation, as the most important hop chemical constituent to determine hop price (Pavlovič, 2012; Pavlovič et al., 2012), it is understandable that development of mathematical models of alpha-acids yield prediction became very interesting for all stakeholders in hops supply chain (Kučera & Krofta, 2009; Srečec et al., 2013; Pavlovič et al., 2013; MacKinnon et al., 2020). Nevertheless, it is very difficult to establish the mathematical model for all hop cultivars, because of their different genetical and related physiological characteristics. The first model was developed by Kučera and Krofta (2009) for 'Saaz' aromatic hop cultivar grown in Czech Republic, the second one was developed for the 'Aurora' hop cultivar (Srečec et al., 2013), grown in Croatia by using artificial intelligence (Eq. 4.2).

$$y = \frac{5.38w - 453 - 1.33w^2}{x} : (-10) \leftrightarrow ET_0 \leq 4.5 \tag{4.2}$$

Where:

y—alpha-acids content in dry matter (%)
x—sum of effective temperatures (°C) from the emergence of shoots at the soil surface after the rootstock cut until hop cones maturity
w—sum of total rainfalls (mm) for the same period
5.38, 453 and 1.33 are the constants determined for cultivar Aurora by Eureqa AI software.

The authors set up a condition for accuracy of Eq. (4.2), according to which the daily reference crop evapotranspiration (ET_0) in July ≤4.5.

The accuracy of the empiric mathematical model (4.2) was estimated by using the linear regression, general linear regression, and Cobb-Douglas production model-function (Srečec & Scitovski, 2023). Therefore, it will be good to develop similar empiric mathematical models for the other hop cultivars grown in different agro-ecological conditions.

4.3.2 *Soil Requirements*

Literature about hop production explains that hops require extremely deep and structural soils with a neutral to slightly acidic reaction in the type of chernozems, eutric brown soils and alluvial soils (Kišgeci et al., 1984). However, it is well known that these soil types are the most fertile soil types. Additionally, according to FAO World reference base for soil resources, the share of these soil types in the world is quite small (IUSS Working Group WRB, 2007; Bašić, 2013). However, soil is certainly one of the agroecological factors that every grower can influence to some extent with certain interventions. Regardless of all the possibilities of soil repair, the cost of such procedures should not be forgotten. For hops production, the soils of heavy mechanical composition, as well as soils with too many skeletal particles are not the best suited for hop growing. The soils of heavy mechanical composition, i.e. hydromorphic soils must be avoided as they favour root asphyxia and might cause difficult or even questionable soil tillage in the wet phase of the year, usually between the end of winter and beginning of spring (Srečec, 2004; Bašić, 2013). Besides, the stagnant water or high-water level in these soils might cause death of seedlings or shorten the rootstocks lifespan (Srečec, 2004; Čeh, 2012). However, soils with rainwater stagnation may be suitable for hop production if their meso-relief ensures rapid rainwater runoff (Bašić, 2013), as is the case with hop cultivation on pseudogley soils in north-western Croatia (Srečec, 1998, 2004).

On the other hand, the colluvial skeletal, calcareous soil types (Bašić, 2013; Rossini et al., 2021) might be also suitable for hops growing, depending on their depth and share of skeletal material in multi-layered deposits and the same is with alluvial soil types in the river valleys (Knapič, 1996).

4.4 Preparations Before Planting the Hops

The cultivation of hops is unique if compared to the cultivation of other industrial crops as follows:

- Hops is a perennial crop, which means it does not 'forgive' any mistakes made by the grower in preparing the soil, establishing the hop garden and planting of seedlings.
- The commercial cultivation of hops is based on vegetative propagation, which means that the propagation material needs to be prepared in a timely manner.
- The vegetative growth of hops requires support due to the specific habit of the plant, which belongs to the climbing type. Before planting, it is necessary to erect the trellis system for the hop garden. This type of construction is a demanding and complex job that requires expertise and practical skills.

The most important features of these three aspects are given below.

4.4.1 Soil Preparation

The soil for hops growing must be well drained, without stagnant surface water in the ″wet″ phase of the year. Since hops are a perennial crop, it is recommended to take soil samples for pedochemical analyses 2–3 years before planting to opportunely improve the soil chemical characteristics. Moreover, it is also recommended to open one or several pedological profiles to determine the soil type, the depth of the impervious horizon and determine the soil texture of each horizon of the soil profile.

According to results pedophysical and pedochemical analyses, it is necessary to carry out all preparatory agrotechnical and agromeliorative works to prepare the production area for planting hops. That includes correction of soil acidity by carrying out calcification or implementation of soil acidification measures, depending on soil reaction. The optimal soil acidity for hop growing is pH between 6.0 and 6.7 (Mihelič, 2010; Čeh, 2012). In case of heavy soil texture, i.e. higher share of clay in the comparison with loam and sand particles, subsoiling is required, and it must be done below the depth of the clay rich horizon (Majer, 2002; Srečec, 2004; Čeh, 2012). It is recommended to do subsoiling in transverse direction to the future rows of hops as well as tillage direction (Majer, 2002). The speed of subsoiling must not exceed 2 km/h (Čeh, 2012).

4.4.2 *Hop Requirements for Nutrients*

During its growth and development, hops need macro- and micro- nutrients (Rodolfi et al., 2021). The nutrient requirements of hops depend on various factors: growing environment, soil type, cultivation practices, etc. In this section, the main nutrients and their importance will be analysed. Therefore, these requirements of hops on plant macronutrients and micronutrients, or macro and micro elements, must be carefully considered in calculations of soil macronutrients supplies as fertilisers in different formulations. Nevertheless, the uptake of macronutrients varies among the different hop cultivars (Čeh et al., 2019).

4.4.2.1 Nitrogen

As concerns the chemical characteristics of the soil, it is important to point out that hops annual uptake of nitrogen varies between 117 to 270 kg/ha.

Nitrogen is a fundamental element for plant growth and significantly influences the final yield. The plant's demand for nitrogen peaks at the beginning of June, when about 10% of the annual requirement is absorbed, and continues until July when assimilation is completed. Initial storage takes place in the leaves with subsequent, almost complete translocation into the cones (Gingrich et al., 2000). Deficiency of this element manifests itself in a general stunting of growth, yellowing of the plant, most noticeable in old leaves, and cones undersize. Conversely, excessive nitrogen fertilisation can increase the possibility of an attack by various diseases such as powdery mildew or Verticillium or phytophages such as mites or aphids (Gent et al., 2009).

The nitrogen has a very strong effect not just on yield of hop cones, but also on some quality parameters. Cone yield has a positive correlation with increasing rates of added nitrogen. On the other hand, the α-acids and total oil decreased as the nitrogen rate increased (Iskra et al., 2019) according to the following linear regression equations (Eqs. 4.3 and 4.4).

$$\alpha-\text{acids}(\%) = 22.0548 - 0.00545 \cdot (\text{Nitrogen dose in kg / ha}) \quad (4.3)$$

$$\text{Total oils}(\text{ml / 100g}) = 4.4026 - 0.00173 \cdot (\text{Nitrogen dose in kg / ha}) \quad (4.4)$$

Moreover, the accumulation of nitrates in hop cones increase with increasing doses of added nitrogen in fertilizing (Eq. 4.5, Iskra et al., 2019).

$$\text{Nitrate}(\text{ppm}) = -160.90 + 0.00701 \cdot (\text{Nitrogen dose in kg / ha}) \quad (4.5)$$

Since 1978, when for the first time nitrosamines, which are cancerogenic substances, were detected in beer, fertilizing with nitrogen is under special attention (Forster et al., 1989). The synthesis and the consequent accumulation of nitrosamines in hop cones increase with increasing doses of nitrogen fertilizing (Gent et al., 2022). Therefore, these facts must be considered in calculations of nitrogen supplies for hops fertilization.

4.4.2.2 Phosphorus

Hops have a low phosphorus requirement, absorbing between 38 to 74 kg/ha (expressed as P_2O_5) per year; 20–30% of this element is found in the cones, while the remainder is redistributed in leaves and lianas (Keller & Magee, 1954). In an experiment carried out by Boawn and Rasmussen (1969), no observable effects on cone colour, vigour or general growth were found after the application of phosphorus, but a reduction in the concentration of zinc in cone and leaf tissues was observed, without these showing deficiency symptoms. Furthermore, it has been shown through studies conducted in England (Gent et al., 2009) that phosphorus deficiency initially manifests itself in the lower leaves, which curl, their colour turns a dark green and their appearance appears withered. Lianas may also show some signs of phosphorus deficiency, appearing thin and weak, while cones may appear light brown.

4.4.2.3 Potassium

Hops potassium annual uptake is 110–256 kg/ha (K_2O). K is an important element for plant life; in hops 25% of the plant's potassium is stored in the cones, the rest is normally divided between leaves and vines. The plant removes an average amount of this element of 90 kg/ha per year. The optimum concentration of K in the hop garden soil is 200 mg/kg, below which a fertiliser supply is recommended, bearing in mind that it can be almost completely replenished at the end of the season by burying crop residues (Gingrich et al., 2000). Below the concentration threshold of 150 mg/kg, in the plant bronze chlorosis may occur between the veins of older leaves, which may develop into ash-grey colouration with subsequent phylloptosis (Gent et al., 2009).

Potassium, if given in excess, can compete with magnesium and nitrogen, making it difficult for the plant to absorb these elements (Brooks & Keller, 1960).

4.4.2.4 Calcium, Magnesium, Manganese, and Sulfur

The hops annual uptake of these elements from the soil is 150–180 kg/ha for calcium (CaO), about 40 kg/ha for magnesium (MgO) and 15–20 kg/ha of sulfur (Mihelič, 2010).

Calcium (Ca) is a key element from both a catalytic (activation of many enzyme systems) and structural (fundamental component of cell walls) point of view. It is important for the growth of leaf and root tissues, facilitates the transport and accumulation of carbohydrates, and the hydrolysis of starch. Calcium deficiency is manifested by internerval and marginal chlorosis of young leaves and subsequent necrosis of the leaf margins. Adult leaves, in case of calcium deficiency, show punctiform necrosis and yellow colour. An excess of this element causes ferric chlorosis and reduced growth of roots, shoots and leaves.

Magnesium (Mg) and Manganese (Mn) are essential for proper nutrition and optimal plant health even though they are absorbed in small amounts when compared to N or K. Their content in most soils is sufficient for the needs of the entire production cycle and any deficiencies or excesses can be attributed to soil pH (Gent et al., 2009). In fact, it has been ascertained that pH influences the availability of assimilable Mg and Mn. Moreover, excesses of these two elements can give rise to toxicity symptoms. This is why the optimal pH for the crop and for the availability of this element is in the 6–7 range (Gingrich et al., 2000). Higher pH values would lead to an immobilization of Mn without creating assimilation problems for Mg, while lower pH values would cause poor assimilation of Mg and a toxic and excessive availability of Mn (Brooks & Keller, 1960).

Sulfur is absorbed by the roots as SO_4^- and by the leaves as SO_2. This element is important for ensuring the structure and proper functioning of cells; in fact, it is an important element in some amino acids and coenzymes. It is also involved in energy transfer processes in the plant. In the case of a deficiency of this element, the leaves take on a lighter colouring, especially at the apical level, and are reduced in size. Normally there is no evidence of damage from excess of this element, except for the accumulation of the element in the soil and consequent soil acidification.

4.4.2.5 Iron, Boron, and Zinc

Iron (Fe) is undoubtedly the meso-element that the plant needs most, especially in the early periods of development, from shoot emergence to flowering. Deficiency is often due to the basic pH of the soil and not to a lack of the element itself (Brooks & Keller, 1960). Deficiency of this element is manifested by yellowish internerval chlorosis in the leaves. It is not uncommon for temporary iron deficiency symptoms to occur in young hop plants in the spring months, when the soil is moist or immediately after fertilisation with nitrogen or phosphorus (Gent et al., 2009).

Hops take up about 0.21 kg/ha of boron (B), so the required and suitable value of boron in soil is between 0.3 and 0.7 mg of boron per 1 kg of soil. Studies carried out in Washington State, Western Europe and New Zealand have observed a positive response from the crop when it was grown in soils containing at least 1.5 ppm of boron (B) (Cripps, 1956). Below this threshold, delays in shoot emergence with yellowing and wrinkling of younger leaves or, in shoots that have already emerged, yellowing of the shoot apex with subsequent necrosis, may occur (Gingrich et al., 2000).

Hops uptake of zinc (Zn) from the soil is only about 0.24 kg/ha (Mihelič, 2010). Zn is a very important element for cone growth and sizing. The assimilation of Zn depends on the pH of the soil; if the pH is above 7.5, Zn assimilation is complicated. Zn fertilization trials in the Willamette Valley and Yakima Valley (USA, 1994) found no significant changes in growth or yield of the crop (Gingrich et al., 2000). When the soil has a very low concentration of this element, virosis-like symptoms may occur with plant weakening and leaf chlorosis. These problems have been observed in soils containing an assimilable zinc concentration of less than 1 ppm (Gent et al., 2009). Extensive and prolonged deficiency can lead to qualitative and quantitative losses or, in extreme cases, to plant death (Brooks & Keller, 1960).

4.4.2.6 Manure

The distribution of manure is an important operation both for correct management of the crop and for the correct management of the soil. It is also recommended to use the manure 2 years before hop planting. Namely, only 20 t/ha of completely fermented, i.e. completely 'mature' manure is a valuable source of phosphorus, potassium and micronutrients. Also, with these 20 t/ha of manure about half of the annual nitrogen requirement of hops is added and may be released gradually. Also, in the case of calcareous soil types it is recommended to use the slurry (liquid manure) to modify soil acidity. Moreover, manuring is also important from the aspects of overall soil health (Elmqvist, 2022), due to organic matter added to the soil, not only before hop planting, but also during the years of hops growing.

4.4.3 Preparing the Construction of Hopyard

Since the wood poles were replaced with trellises the establishment of hopyards have been improved through the decades. Generally, there are two main types of hop garden trellis.

The first one, and at the same time the most common, is the wooden trellis, made with straight wood poles. The distance between individual columns are 9.6 m × 9.1 m, the value 9.6 m determines the distance between the columns in the transverse direction of the cable car and the number 9.1 m determines the distance in the longitudinal direction of the cable car. With an inter-row planting distance of 2.4 m, there is a four inter-row spaces, or three rows without poles, in the middle fields with the mentioned placement of the posts. There may be other variants, such as 9.6 m × 10.4 m and 12.0 m × 10.4 m (Čeh, 2012). Nevertheless, preparing of the hopyard trellis is a considerable investment and requires not just technical competence and skills but also good organizational attitude of the farmer. The hopyard trellis erection must be made one year before planting. That means that most materials must be ready at the right time. In Table 4.2 the requirements of materials are shown for the construction of a 5 ha hopyard (Srečec, 2004).

The second type of hopyard trellis construction implies concrete trellis. In that type of trellis, the wooden columns are replaced with concrete ones. These are placed on flat terrain and are always rectangular in shape. Due to the economy, the concrete columns are placed on larger surfaces. A concrete column has a longer service life compared to a wooden one, which is closely related to the quality of maintenance due to its structural design (Čeh, 2012). It is also important to point out that the anchors of concrete trellis (2.2 m) must be placed much deeper in comparison with the anchors of wooden trellis (1.6 m) (Majer, 2002). However, because of their high price they are not always preferred to wooden trellis.

Finally, before setting up the hopyard trellis, it is very important to consider some issues about future organisation of hop harvest. Namely, the hopyards must not be far away from facilities with stationary hop cones picking machines and hop cones driers. That's because in the hop harvest the whole hop plants are cut off and transported in special trailers to the facility where the hop cones will be separated from stalks and leaves and dried as soon as possible (Fig. 4.1).

Table 4.2 Materials required for the construction of a 5 ha hopyard trellis (Srečec, 2004)

Item	Quantity
Acacia poles 8–10 m long and ø = 18–30 cm	100 m^3
Cables 7 × 3.7 mm (7 wires of ø = 3.7 mm each)	12 km
Cables ø = 2 × 2.5 mm or 2 × 2.7 mm	24 km
Various metal accessories: tension bolts, couplings, profiles on the bends of "heart" ropes, bolts for anchors, etc.	3.2 tonnes
Virus-free seedlings	13,000 pieces (approx.)

Fig. 4.1 Ideal placement of hopyards nearby facility. (Photo: S. Srečec)

4.5 Hops Growing

Hops is a labour-intensive culture. The justification of this statement is best explained by an old German proverb, which says: "Hops must see their master every day". Very often, in conversation or at gatherings, hop growers state something along the lines of: "*Well, if we had done this or that...*". Though, the price of hops is extremely high compared to other industrial plants. On the other hand, investments in production are also extremely high. That is why any loss of harvest or loss of quality entails a reduction in total income, and the total financial result of production depends on this. The timely performance of all necessary work and the implementation of all agrotechnical measures of fertilization, care and protection of plantations depend on the personal commitment of the hop grower, his family members, and the provision of a sufficient number of seasonal workers. Success in production will not be absent if all works are carried out on time.

4.5.1 Planting

When all the mentioned preparatory work is done, planting is started. There are two types of hop seedlings i.e. of A- and B-certificate. The seedlings of A-certificate are completely virus free plant material, produced by tissue culture methods, known as meristem method. That supposes a selection of mother plants of typical phenotypes within the genetically pure hop cultivar. Afterward, these selected plants are tested for plant virus infection by serological method or using ELISA-test. If the test is negative, such plants are multiplied again and maintained in a protected area. From a larger number of mother plants, explants are cut from the meristem zone, because plant viruses translocate slowly to the zones of primary growth; in addition, the regeneration of explants into proto-callus and callus proceeds much faster if the explants are taken from the meristem region. Young plants are developed on a nutritive substrate in a climate chamber, and afterwards they are transplanted into special container pots and placed in the greenhouse under strictly controlled conditions of humidity and temperature for further development. When the young plants are completely developed and rooted, they are taken out of the greenhouse in special vegetation pots for outdoor acclimatization. At the end of the acclimatization time, they are transplanted into degradable container packaging that is previously filled with peat and planted in the hop garden. The price of the seedlings obtained in this way is very high. On the other hand, a high percentage of young plants die. That means the need to replant or underplant to fill empty places in the garden.

However, the hop garden planted with seedlings of A-certificate receives the certificate of the parent hop garden to produce hop seedlings of the B-certificate. Every year, tests are carried out to control possible infection with plant viruses. It is also important to emphasise that it is necessary to prune such hop garden using a machine that hasn't been used on an infected plantation but will only be used for a virus-free hop garden. Before and after pruning, it is good to wash the disc knives of the machine with hot water under pressure and then disinfect them with a formaldehyde solution. This greatly reduces the spread of plant viruses and viroids. On the other hand, B-certificate seedlings are nothing, but cuttings obtained by pruning the garden of hop mother plants based on A-certificate planting material. Each cutting should have two rows of visible buds, whitish at the base and reddish at the top and that is the basic condition for the cutting to be used for rooting. For the hop grower there is another important difference between hop gardens based on A-certificate planting material and plantations based on B-certificate planting material: hop gardens based on B-certificate plants achieve the productive stage already in the first year of planting and enter in commercial exploitation in the second year, which is not the case with plantations based on A-certificate planting material (Fig. 4.2).

When choosing the plant distance between the rows and within the rows, the characteristics of the habitus of the hop cultivar, soil fertility and ecological conditions must be considered, as they determine the spacings for planting (Table 4.3). Generally, every hop breeder is recommending the number of plants per hectare for the hop cultivar created by him/her.

A-certificate seedlings come in ready-made degradable packaging known as "containers". Such packaging is inserted into the planting holes and the soil is simply poured up to the collar region of the young plant.

The B-certificate plantlets are usually planted at depths of 8 cm in heavy soils, or at depths of 10–15 cm in lighter soils. It is recommended to put a layer of completely fermented (i.e. "mature") manure at the bottom of each planting place, to speed up the growing of young plant roots. However, the zone of manure must be separated from the seedling by a tiny layer of soil, approximately 5 cm (Srečec, 2004).

Fig. 4.2 Hop garden on the first growing year. The plants are grown on one bine per each plant. (Photo: S. Srečec)

Table 4.3 The most common spacings between rows and within rows for hop plantlets (Majer, 2002; Čeh, 2012)

Distance between the rows (cm)	Distance within the row (cm)	Vegetation space (m^2) per each plant	Number of plants per hectare
240	100	2.40	4166
240	110	2.64	3787
240	120	2.88	3472
240	130	3.12	3205
240	140	3.36	2976
280	110	3.08	3246
280	120	3.36	2976
280	130	3.64	2747
300	110	3.30	3030

4.5.2 Agrotechnical Practices During Hop Cultivation

From the second year and then each year, until the end of the hop garden exploitation, the cutting back of hops must be done every spring. It is important to point out that the primarily reason for such agrotechnical measure is decreasing the number of shoots primarily infected with downy mildew, caused by the fungus *Pseudoperonospora humuli* (Miyabe & Takah.) G.W. Wilson (Wilson, 1914). The primary infection with downy mildew always starts at early spring. The infected hop shoots are easily recognizable. They have very short internodes, their leaves are smaller than those of healthy shoots, and very curly. Moreover, all infected shoots are curly if compared to healthy ones which are straight with fully expanded young leaves. The spring cutting of hops starts at phenological growth stages of sprouting at the phase of beginning of shoot-growth (Rossbauer et al., 1995; Meier et al., 2001). The cutting may be done manually with hop knives, but only in the small hop gardens, or most often, with hop cutting machines. The cut must be about 2–3 cm above the head of the rhizome at the speed of 2.5 to maximum 4 km/h (Čeh, 2012). The cuttings that the cutting machine ejects left and right can be prepared for planting and used as B-certificate plantlets, but only if the hops are planted with A-certificate plants. Immediately after planting the plantlets or after cutting the rhizomes, the hops bines must be set up. The young plants must be directed or trained in clockwise in the northern hemisphere or counterclockwise on southern hemisphere. The hop twines must be set up every year. In the first year of hop growing only one bine is left per each plant (Fig. 4.2), but in the second year and in the remainder years of the culture, two bines are grown per each plant after the V-shaping and, on each of bines usually three, up to four plants are trained.

The other cultivation practices include green mulching between the rows and wrapping the rows by discing machines. Also, in case of plant disease infection and attack of hop pests, the protection of hops must be implemented using special sprayers for hopyards.

It is also recommended to keep a ground cover between the rows to avoid evapotranspiration and loss of water from the soil (Fandiño et al., 2015). It is important to point out that in case of water deficit, irrigation must be implemented soon after the young hop plants are directed on the bines (training). Every delay of irrigation after training causes significant losses on yield of hop cones and quality, depending on the cultivar, and on water deficit (Gent et al., 2021).

4.6 Hop Irrigation

To produce the highest yields, the hop plants require a good supply of water. Irrigation as an agri-technical measure was quickly established in hop production especially in the drier regions of the world (Neve, 1991). With irrigation practices plants water shortage is prevented and therefore high and good quality yields are provided each year. In many areas of the world and in Europe there is a shortage of irrigation water, excessive use of irrigation water and sub-optimal dosing of irrigation water. Climate change is one of the environmental factors that will significantly affect the development of agriculture in the future (Blanco et al., 2017). The demand for irrigation water will increase with climate change, which, with (already) unchanged water consumption, would mean an increase in pressure on water systems.

Hop irrigation technologies have evolved. In the first half of the twentieth century, furrow irrigation (flood irrigation) was developed, recommended and used in the USA and the EU. In the second half of the twentieth century, it was replaced by sprinkler irrigation (sprinklers placed over hop plants and hose pull irrigators) (Rybacek, 1991; Zepp et al., 1995). In parallel, the practice of underground drip irrigation of hops (pipes laid 40 cm below the plants) developed (Rybacek, 1991), and later the drip irrigation from the top of the hop wirework structure (Slavik & Kopecky, 1994) and the addition of nutrients using fertigation, which increased nutrient utilisation (Wample & Farar, 1983).

Improper irrigation of hop (causing insufficient amount of water in the soil) is reflected in lower yields, uneven ripening, uneven size of cones and a small amount of lupulin, which represents the main market value of hops (Evans, 2003; Srečec et al., 2008). The consequences of drought events can be reflected in lower yields and lower quality of the hop yield in the next two growing seasons.

Today two basic technologies of hop irrigation are widely used. First technology is a sprinkler type of irrigation or, more specifically, hose pull irrigators and second one is different varieties of drip irrigation. The hose-pull irrigator consists of a rain gun mounted on a sledge (Fig. 4.3). The sledge is connected to its own hose which is pulled to a large drum (hose reel) which is located on the other end of the field. The drum is rotating and therefore pulling the hose (FAO, 2001). Sprinkler technology of irrigation, which has been in use for more than three decades, has poor irrigation efficiency (estimated water losses are 40%) and quite many disadvantages and limitations. This technology of irrigation is very labour demanding, it has high energy requirements, is increasing chances of nutrient leaching into the groundwater and is not suitable for irrigation of new hop plantations (baby hops) because of the high pressures of sprinkling water.

Fig. 4.3 Hose-pull irrigator irrigating hop field. (Photo: B. Naglič)

Higher irrigation efficiency can be achieved by using the latest drip irrigation technologies which have recently become very popular in hop production industry. This technology is gradually replacing the sprinkler irrigation. The great potential of drip irrigation lies in increased water use efficiency, improved water management (because a smaller soil surface area is wetted the evaporative losses are lower than for surface, sprinkler or micro-sprinkler irrigation), improved crop yield and quality, decreased energy requirements (because of lower operating pressures) and low labour intensity. There are three different versions of drip irrigation used in hop production. First one is subsurface drip irrigation where a dripline is placed in the soil beneath the plants (Fig. 4.4), second one is surface drip system placed on the top of the soil surface (top of the ridge) along the hop row (Fig. 4.5) and the third version represents the dripline placed at the top of the hop wirework structure (Fig. 4.6). When the driplines are placed at the top of the wirework structure (represents the growing support for hop), at the height of 6–7 m, the irrigation hose is attached to the optional trailing wire which runs in the direction of hop row. This option is favoured by many European hop producers because the irrigation pipes, in comparison with the drip system where pipes are placed on the soil surface, do not need to be removed and stored every year before the hop growing season and after harvest. Irrigation pipes on the soil surface outside the growing season would prevent necessary agricultural operations, such as pruning and ploughing which are carried out annually.

Fig. 4.4 Hop drip irrigation where a dripline is placed in the soil beneath the plants. (Photo: B. Naglič)

Fig. 4.5 Hop surface drip pipes placed on the top of the soil surface (top of the ridge). (Photo: B. Naglič)

Fig. 4.6 Driplines placed at the top of the hop wirework structure. (Photo: B. Naglič)

The type of drip irrigation technology in hop plantations when driplines are placed at the top of the hop wirework structure substantially alters the basic characteristics of conventional surface drip irrigation methods. Surface drip irrigation is particularly suitable for grapevines, apple trees and olive trees, because water can be applied without interfering with cultural operations and without wetting the plants and their fruits. Therefore, with conventional surface or subsurface type of drip irrigation very good water efficiency (estimated water losses are 10%) can be achieved. Also, other advantages, as insensitivity to wind, very low risk of the disease development caused by wetting of plants and no impact on soil erosion speak in favour of drip irrigation (Dasberg & Or, 1999; Lamm et al., 2007). When drip irrigation system is placed at the top of the wirework structure (on the height of 6–7 m) or above plants, irrigation efficiency is unknown—larger area of soil surface is wetted because droplets rebound from the soil which is causing significant losses of water because of evaporation. Water dripping from such height also causes soil erosion and in windy conditions, when water drops are wetting the leaf surface, it can also affect the development of the fungi caused diseases.

Published scientific contributions and research projects in the field of hop irrigation relate primarily to the hop plants water needs, the impact of irrigation on quantity and quality of hop yield, different irrigation technologies, the impact of irrigation on the leaching of nutrients (nitrate) into groundwater and decision support about hop irrigation with sprinkler irrigation (Naglič et al., 2016).

It is important that the growing season of hops begins with a good supply of soil water, but experience shows that excessive early irrigation can damage the roots, causing hop downy mildew and root rot (Evans, 2003; Zepp et al., 1995). The first critical period is during flowering (first half of July) and the second during cone growth (from the second half of July to the first half of August) (Rybacek, 1991; Nakawuka et al., 2017). If the groundwater level is high (i.e. 120–150 cm below the surface), it can meet most of the water needs of hops during the growing season and significantly reduce the need for irrigation (Evans, 2003; Kišgeci, 1974). Hops can also absorb some water through the leaves or above-ground parts, i.e. 5000 plants receive 3465–8870 l of water with dew overnight, covering 20% of the daily water requirement. For 1 kg of dry matter hops need 300–500 l of water (Kišgeci, 1974).

Several studies have examined hop evapotranspiration (ET). Wagner and Veronek (1977, 1978) studied ET in the Aurora hop variety and found two peaks in May and July. Majer (2000) reported that crop coefficients (kc) increased with plant growth, ranging from 1.2 (ET = 0.97 mm/day) at early stages to 2.1 (ET = 4.7 mm/day) during cone growth. Hop crop factors vary by region: 0.3–1.05 (Allen et al., 1998), 0.33–1.00 on light soils, and 0.43–1.07 on heavier soils in Slovenia (Matičič, 1977). For the Nugget variety, average kc values were 0.69 (early), 1.02 (mid), and 0.84 (late) between 2012 and 2014 (Fandiño et al., 2019).

In Europe, Czech hop producers were the first who started with introduction of drip irrigation in hop production industry (Pintar & Knapič, 2001). Preliminary results showed that implementation of a subsurface drip irrigation system in already established hop fields has an negative impact on the yield of hops in the first year. The results also showed that all variants of drip irrigation had a positive influence on the content of alpha acids in hops cones when compared to sprinkler irrigation (Knapič & Pintar, 1998). Results from a 12-year hop irrigation study in the Czech Republic where the drip irrigation was placed on the top of the wirework structure, showed that this method of irrigation positively affects microclimate in the hop plantation under their given conditions. Irrigation decreased the temperature for 4 °C, which has, in hot summer days, positive effect on the biochemical and physiological processes in hop plants. Their results also showed that irrigation on average, over the 12 years, increased hop yield for 22%. In irrigated hops higher content of alpha acids, which is the most important quality parameter the hops, were determined (Kopecký & Ježek, 2007). A positive impact of drip irrigation in hop in the Czech Republic was also reported by Svoboda et al. (2008). The results of their research, which was based on a three-year drip irrigation experiment (2005–2007) showed that the yield of irrigated hop was on average higher by 21% throughout the entire duration of the experiment compared to non-irrigated hops. On average 7% increase in alpha acids was observed for irrigated hop. Positive effect of hop irrigation has also been observed by Slavik and Kopecky (1997) (cit. by Svoboda et al., 2008) and Delahunty et al. (2011), where the hop yield increase from 20% to 26% was observed. Some studies indicate that irrigation of hops does not affect the quality of yield (Slavik & Kopecky, 1994; cit. after Svoboda et al., 2008).

In Germany research was carried out on sandy and clay soils, where the quantity and quality of non-irrigated hops were compared with the quantity and quality of drip-irrigated hops (along the ridge, underground next to the ridge and underground between two rows). Two approaches were evaluated for the correct regulation of irrigation: measurement of soil water tension (gypsum blocks, Watermark sensors at a depth of 30 cm) and recording of the water balance. Net photosynthesis, transpiration and water use efficiency were also monitored. The results showed that irrigation had an effect on hop yield and quality only in the driest year, 2015, where yield increased by 35% with above-ground irrigation on the soil surface along the ridge and underground next to the row. A 30% increase in hop yield was observed with subsurface irrigation between two hop varieties. The alpha acid content increased significantly in all three cases (Graf, 2016).

Deficit irrigation of hop was also investigated. Researchers analysed the effect of deficit irrigation on hop yield and quality on heavy clay soils. In the experiment, water was added by subsurface drip irrigation. Three different amounts of water were added (60, 80 and 100% hop ET needs). The hop plants responded to the lack of water with a reduced yield. Compared to 100% irrigation, hop dry matter yield was 30, 33, 25 and 19% lower for Mt. Hood, Willamette Columbus and Chinook varieties over 2 years with 60% water addition. Eighty percent irrigation reduced yield by 14, 10 and 3% for Mt. Hood, Willamette and Chinook, respectively. On the other hand, the results showed that the effect of drought stress did not have a statistically significant effect on hop yield quality. Although the reduction in yield due to lack of water for year-round production was not statistically significant, the loss of revenue due to reduced yield was greater than the savings in production costs due to reduced water use. Economically, therefore, deficit irrigation of hops in the Yakima Valley (Washington) is not a recommended irrigation management tool in hop production (Nakawuka et al., 2017).

Studies based on soil water monitoring for hop irrigation management include the use of the gravimetric method (Kišgeci & Vučić, 1979; Kišgeci, 1974), neutron meters (Nakawuka et al., 2017), electrical conductivity measurement (Graf, 2016; Slavik & Kopecky, 1998) and TDR meters (Naglič et al., 2016). The effect of the three irrigation methods (drip, furrow and sprinkler) on hop yield was not statistically significant. It is assumed that drip irrigation did not give better results due to the limited wetting of the soil around the hop rhizomes, as a result of which the zone of the most active roots (60–140 cm from the rhizomes) remained insufficiently wetted (Kišgeci & Vučić, 1979). A comparison was made of the effect of different irrigation management methods (no irrigation, irrigation based on the amount of water in the upper 60 cm of the soil, irrigation based on critical periods of plant development, irrigation based on physiological indicators) on the quantity and quality of the crop. A 1 °C increase in average daily temperature increased evapotranspiration by 1.8 m^3/ha. Irrigation based on critical periods of plant development had a positive effect on yield. Irrigation based on soil moisture had a positive effect on alpha acid content. Irrigation based on physiological indicators had a negative effect on crop quality (Kišgeci, 1974).

Simulation models have also proved useful in the management of hop irrigation. In the case of hops, they can help to determine more accurately the plant coefficients (kc) of the plant and the relationship between transpiration and crop quantity and quality (Fandiño et al., 2015), as well as climate change simulations on potential evapotranspiration and alpha acid content in hops (Bárek et al., 2009; Mozny et al., 2009). Based on a three-year study (2012–2014) aimed at calibrating and verifying the SIMDualKc soil water balance model for nugget hops, the researchers proposed, among other things, plant coefficients for three growing seasons (early, mid and late) of 0.69, 1.02 and 0.85, respectively. They found that hop transpiration accounted for 92% of actual evapotranspiration in the middle of the hop growing season. The influence of water use and hop transpiration on the quantity and quality of hop yield was evaluated. A positive influence of hop transpiration on cone yield (coefficient of determination $r^2 = 0.92$) and concentration of alpha and beta acids was found (Fandiño et al., 2015). Using the DAISY programme, it was estimated that the potential evapotranspiration of hops in southern Slovakia will increase by 34% by 2075 under the GISSprep and CCCM climate scenarios (Bárek et al., 2009). Hops appear to be particularly vulnerable to climate change, as the CORAC model estimated that in Slovakia, the yield of the Saaz variety will decrease by 7–10% and the alpha acid content by 13–32% by 2100 if irrigation practices remain unchanged (Mozny et al., 2009).

4.7 Hops Harvest, Drying of Hop Cones and Storage

Harvesting hops is an extremely difficult and rather time-consuming activity that used to be done by hand only. Today, there are different models of hop picking machines (Gobor et al., 2013; Campbell & Pearson, 2019; Assirelli et al., 2020) they can be either mobile or stationary.

The term hop harvest in the technical sense includes four operations, which follow one another in a relatively short time, i.e.:

1. Cutting the hop bines in the hop garden and transporting them to the stationary machines for picking the cones. However, if the hop producers use mobile hop harvest machines the picking of the hop cones is done in the hop garden.
2. Cleaning the lot of hop cones from leaves and bines particles.
3. Drying and conditioning the dried hop cones.
4. Pressing of dried hop cones into hop bags and bales.

In planning the hop harvest every hop grower must consider the following issues:

- Determining the timely start of harvest.
- Organization of work and planning of work operations to "capture" the period of greatest accumulation of α-acids in hop cones.
- Prevention of undesirable oxidation processes due to excessively high drying temperature of hop cones and/or due to poor compaction of cones in hop bags or bales.

If harvesting is started too early, there is a real possibility that most of the crop will be harvested before the maximum accumulation of α-acids in the cones, so in case of premature harvesting, the total loss of α-acids in dried hop cones is high (Howard & Tatchell, 1956; Virant & Majer, 2003; Srečec et al., 2004, 2008, 2011a, b; Srečec, 2006). However, the biggest losses of α-acids are detected during the drying of hop cones. Namely, hop cones after separation from the bines have high water content (about 80%), so their drying must begin as soon as possible, otherwise they will suffer of self-heat. Drying is most often carried out in batch three- to four-floors gravity dryers with fluidization of dry and hot air that passes from the heat exchanger through the floors with hop cones exit on the roof of the dryer into the atmosphere. Optimum air temperatures for drying hop cones range from 62 to 65 °C, with a flow of 0.25–0.30 m/s, with the thickness of the cone layer ranging from 30 to 35 cm (Rossbauer & Münsterer, 2003; Srečec et al., 2009; Rybka et al., 2017). However, an increase in drying temperatures by 10 °C doubles the rate of oxidation processes that are inevitable in any drying process. Unfortunately, to shorten the drying time, hop producers raise the drying temperature to even 70 °C (Forster, 2003). Also, increasing the drying temperatures decreases the content of total essential oils in hop cones and reduces the intensity of their aroma, depending on the drying temperature (Rybka et al., 2018). The optimal water content in hop cones after drying is 9–10% but, due to the producers' fear of the possibility of self-heating, and in drastic cases, of self-ignition (which sometimes happens if the cones are not tightly packed in the hop bags), the cones are often 'dried out', so they contain less than 8% water, which leads to increased losses of α-acids (Srečec, 2006). Drying of cones in high temperatures causes shrivelling of peltate glandular trichomes and cracking of bulbous glandular trichomes (Srečec et al., 2011b) and increases the oxidation process of bitter and aromatic compounds.

After drying, hops are conditioned. This term refers to the balance of the water content within a batch of dry hop cones. Namely, all hop cones in the same batch are not and cannot be equally dry. In addition, due to the morphology of cones, their different parts release water differently. The cone bracts release water faster compared to rachis. Therefore, after drying, the cones are placed in conditioning chambers where they are cooled by active ventilation, and due to their own hygroscopicity, they are moistened with moisture from the circulating atmospheric air, until equilibrium is reached between the humidity of the cones and the relative humidity of the air. In doing so, the relationship between the humidity of the cones and the relative humidity of the circulating air should be considered (Table 4.4). Conditioning time is approximately 2–3 hours.

Pressing cones into hop bags or bales may also cause losses. Losses occur in two ways. First, they arise due to the manipulation of the cones at the time of filling the hop bags. Then, especially if the cones are too dried, the yellow flower powder (lupulin) falls off the bracts. Another way of causing losses is related to compressing of hop cones into hop bales. It is known that hop cones are a bulky product, so due to the reduction of bulk for the rational use of storage and transport space, in large hop plants, cones are most often compacted into cubic bales, in which more than 150 kg of cones/m^3 can be pressed (Srečec, 2006). With the increase in the compression force on the cones, the frequency of bursting of bracts glandular trichomes increases proportionally (Srečec et al., 2011b). The walls of lupulin grains provide limited protection against oxygen penetration, therefore hop cones with intact lupulin grains are less susceptible to oxidation during storage and consequently to losses of valuable chemical compounds (Forster, 2003; Srečec, 2006; Srečec et al., 2011b). Therefore, if the hop cones are dried at temperatures of 60–65 °C, baled at less than 90 kg, and wrapped in burlap the losses of bitter and aromatic substances decrease (Weber et al., 1979).

Not so long time ago, there were just no cold storage facilities in Europe where hop cones compressed into bales could be stored immediately after harvest. Most hop warehouses are of the floor type, without cooling systems, in which hop bags and bales are received and stored on several floors, until the moment of processing or shipping. Warehouses for hop cones are divided into three groups:

Table 4.4 Equilibrium of air humidity and moisture of hop cones (Rossbauer & Münsterer, 2003; Srečec, 2006)

Desired cones moisture (%)	Relative air humidity (%)
12	65
10	60
9	55
8	50

1. Warehouses of mediocre functionality where temperatures fluctuate between 5 and 30 °C.
2. Good hop warehouses, with temperature variations from 5 to 20 °C.
3. Refrigerated warehouses where the temperature does not exceed 5 °C.

The processes of oxidation can significantly increase as early as during the harvest itself, and if the hops are stored in inadequate hop warehouses they intensify, and cause decrease in the proportion of α-acids in total bitter substances, as well as a decrease in the freshness of hops. Thus, it often happens that older hops and/or hop pellets, produced in the previous growing year and stored in cold storage at temperatures from 0 to 4 °C, have greater freshness and a higher content of α-acids compared to fresh, non-properly stored hops, which is expressed by the hop storage index (HSI). The HSI has been used in evaluating varietal differences in storage, measuring permeability and antioxidant levels in lupulin glands, determining hopping rates in breweries, and evaluating grower and dealer handling practices (Nickerson & Likens, 1979; Forster, 2001, 2002; Srečec, 2006; Rutnik et al., 2023). The values of HSI might vary in interval of approximately 0.23 to close to 1 and the degradation process of α-acids increase as the values of HSI are closer to 1. However, the smaller hop producers with direct supply chain to brewery, might decrease the time and the costs of storage by processing the hop cones into hop pellets soon after their drying (Marić & Srečec, 2006; Srečec et al., 2009).

References

Allen, R., Pereira, L. S., Raes, D., & Smith, M. (1998). *Crop evapotranspiration: Irrigation and drainage paper No. 56*. Food and Agriculture Organization of the United Nations (FAO).

Assirelli, A., Carbone, K., & Ciccoritti, R. (2020). Mechanical hop-picking solutions in Italian cultivated areas. *Sustainability, 12*(5006). https://doi.org/10.3390/su12125006

Bárek, V., Halaj, P., & Igaz, D. (2009). The influence of climate change on water demands for irrigation of special plants and vegetables in Slovakia. In K. Strelcová (Ed.), *Bioclimatology and natural hazards* (pp. 271–282). Springer. https://doi.org/10.1007/978-1-4020-8876-6_23

Barth, H. J., Klinke, C., & Schmidt, C. (1994). *The hop atlas—The history and geography of cultivated plant*. Joh. Barth & Sohn.

Barth-Haas. (2023). *Barth-Haas report 2022/2023*. https://www.barthhaas.com/resources/barthhaas-report

Bašić, F. (2013). *The soils of Croatia*. Springer Science+Business Media. https://doi.org/10.1007/978-94-007-5815-5

Bauerle, W. L. (2019). Disentangling photoperiod from hop vernalization and dormancy for global production and speed breeding. *Scientific Reports, 9*(1), 16003. https://doi.org/10.1038/s41598-019-52548-0

Blanco, M., Ramos, F., Van Doorslaer, B., Martínez, P., Fumagalli, D., Ceglar, A., & Fernández, F. J. (2017). Climate change impacts on EU agriculture: A regionalized perspective taking into account market-driven adjustments. *Agricultural Systems, 156*, 52–66. https://doi.org/10.1016/j.agsy.2017.05.013

Boawn, L. C., & Rasmussen, P. E. (1969). Phosphorus fertilization of hops. *Agronomy Journal, 61*(2), 211–214.

Brooks, S. N., & Keller, K. R. (1960). Effect of time of applying nitrogen fertilizer on yield of hops. *Agronomy Journal, 52*, 516–518.

Campbell, S. M., & Pearson, B. J. (2019). Harvest techniques for hops (*Humulus lupulus*): ENH1314/EP578, 12/2019. *EDIS, 2019*(6), 4–4. https://doi.org/10.32473/edis-ep578-219

Carrubba, A., Marceddu, R., & Sarno, M. (2022). Hop (*Humulus lupulus* L.): Suitability of traditional cultivars to a low-trellis farming system in a semiarid environment. *HortScience, 57*, 1409–1415. https://doi.org/10.21273/HORTSCI16754-22

Čeh, B. (Ed.). (2012). *Hops from seedlings to cones*. Slovenian Institute for Hop Research and Brewing. https://www.ihps.si/wp-content/uploads/2016/08/Hmelj_URN-NBN-SI-DOC-KCSG6E7S.pdf

Čeh, B., Čremožnik, B., & Oset Luskar, M. (2019). Nutrient uptake with hop (*Humulus lupulus* L.) as the basis for fertilization rate determination and hop biomass after harvest related to variety. In *New challenges in agronomy 2019, proceedings of symposium* (pp. 63–70) https://www.cabidigitallibrary.org/doi/pdf/10.5555/20193168252

Cripps, E. G. (1956). Boron nutrition of the hop. *Journal of Horticultural Science, 31*(1), 25–34.

Dasberg, S., & Or, D. (1999). *Drip irrigation*. Springer. https://doi.org/10.1007/978-3-662-03963-2

Delahunty, K., Johnston, J., & Westfield, M. E. (2011). *An experiment on the effectiveness of irrigation and cover cropping to produce sustainable hops in Maine*. USDA Northeast SARE Final Report: FNE11–711.

Elmqvist, T. (Ed.). (2022). *Regenerative agriculture in Europe: A critical analysis of contributions to European Union Farm to Fork and Biodiversity Strategies* (EASAC policy report 44). German National Academy of Sciences Leopoldina. https://easac.eu/fileadmin/PDF_s/reports_statements/Regenerative_Agriculture/EASAC_RegAgri_Web_290422.pdf

European Commission. (2017). *List of hop production areas*. https://agriculture.ec.europa.eu/system/files/2019-08/hop-production-areas_en_0.pdf

Evans, R. (2003). *Hop management in water-short periods* (EM4816, Drought Advisory). Washington State University Extension. https://hdl.handle.net/2376/7026

Fandiño, M., Olmedo, J. L., Martínez, E. M., Valladares, J., Paredes, P., Rey, B. J., Mota, M., Cancela, J. J., & Pereira, L. S. (2015). Assessing and modelling water use and the partition of evapotranspiration of irrigated hop (*Humulus lupulus*), and relations of transpiration with hops yield and alpha-acids. *Industrial Crops and Products, 77*, 204–217. https://doi.org/10.1016/j.indcrop.2015.08.042

Fandiño, M., Martínez, E. M., Rey, B. J., Valladares, J., Olmedo-Nadal, J. L., Mirás-Avalos, J. M., & Cancela, J. J. (2019). Irrigation scheduling for *Humulus lupulus* (L.) cv. "Nugget": Climate and soil-plant relations. *Brewing Science, 72*, 188–195. https://doi.org/10.1016/j.indcrop.2015.08.042

Food and Agriculture Organization of the United Nations. (2001). *Irrigation manual: Planning, development monitoring and evaluation of irrigated agriculture with farmer participation (Module 8: Sprinkler irrigation systems planning, design, operation and maintenance)*. FAO Subregional Office for Southern and East Africa.

Forster, A. (2001). The quality chain from hops to hop products. In *Proceedings of the 48th IHGC Congress in Canterbury*.

Forster, A. (2002). What happens to hop pellets during unexpected warm phases? *Brauwelt International, 1*, 43–46.

Forster, A. (2003). Degradation of hop pellets in inert atmosphere with possible consequences caused by unusual warm phases. *Joh. Barth & Sohn Technical Information*.

Forster, A., & Schmidt, R. (1994). The characterisation and classification of hop varieties. In *EBC Symposium on Hops in Zoeterwoude* (*EBC Monograph XXII*) (pp. 251–270).

Forster, A., Beck, B., & Schmidt, R. (1989). Problematic substances of hops—Thoughts and investigations. In *Proceedings of the 76th October Meeting of the VLB in Berlin, 1989*.

Gent, D. H., Barbour, J. D., Dreves, A. J., James, D. G., Parker, R., Walsh, D. B., & O'Neal, S. (2009). *Field guide for integrated pest management in hops*. Washington Hop Commission.

Gent, D. H., Claassen, B. J., Massie, S. T., Phillips, C. L., Shellhammer, T. H., Trippe, K. M., & Twomey, M. C. (2021). Delayed early season irrigation: Impacts on hop yield and quality.

Journal of the American Society of Brewing Chemists, 80(1), 62–65. https://doi.org/10.1080/03610470.2021.1915053

Gent, D. H., Block, M., Massie, S., Phillips, C., Richardson, B., Shellhammer, T. H., Trippe, K., & Wiseman, M. (2022). Nitrogen and sulfur fertility practices: Influences on hop chemistry, aroma, and nitrate accumulation. *Journal of the American Society of Brewing Chemists, 82*(1), 50–60. https://doi.org/10.1080/03610470.2023.2204412

Gingrich, G. A., Hart, J. M., & Christensen, N. W. (2000). *Hop. Fertilizer guide 79*. Department of Crop and Soil Science, Oregon State University.

Gobor, Z., Fuß, J., Fröhlich, G., & Portner, J. (2013). Stationary fully automated hop picking machine—Concept, prototyping, and preliminary testing. In *International Commission of Agricultural and Biological Engineers, Section V. CIOSTA XXXV Conference, Billund, Denmark*. https://www.lfl.bayern.de/mam/cms07/ilt/dateien/ilt5_hopfenpfluecke_puplikation1.pdf

Graf, T. (2016). *Tröpfchenbewässerung im Hopfenbau - Feldversuche, Physiologie und Rhizosphäre [Drip irrigation in hop production - field trials, physiology and rhizosphere]*. [Doctoral dissertation, Fakultät Wissenschaftszentrum Weihenstephan].

Hornsey, I. S. (2003). *Brewing*. The Royal Society of Chemistry.

Howard, G. A., & Tatchell, A. R. (1956). Development of resins during the ripening of hops. *Brewing Industry Research Foundation, 62*, 251–256.

Iskra, A. E., Lafontaine, S., Trippe, K., Massie, S., Phillips, C., Twomey, M. C., Shellhammer, T. H., & Gent, D. H. (2019). Influence of nitrogen fertility practices on hop cone quality. *Journal of the American Society of Brewing Chemists, 77*(3), 199–209. https://doi.org/10.1080/03610470.2019.1616276

IUSS Working Group WRB. (2007). *World reference base for soil resources 2006: First update 2007*. World Soil Resources Reports No. 103. FAO. https://www.fao.org/fileadmin/templates/nr/images/resources/pdf_documents/wrb2007_red.pdf

Johnson, D. A. (1991). Two degree-day models for predicting initial emergence of hop shoots systemically infected with *Pseudoperonospora humuli*. *Plant Disease, 75*(3), 285–287.

Keller, K. R., & Magee, R. A. (1954). The effect of application rates of nitrogen, phosphorus, and potash on some chemical constituents in two varieties of hops. *Agronomy Journal, 46*(9), 388–391.

Kišgeci, J. (1974). *Vodni režim biljaka hmelja u različitim uslovima navodnjavanja i mineralne ishrane* [Water regime of hop plants under different irrigation and mineral nutrition conditions]. [Doctoral dissertation, University of Novi Sad].

Kišgeci, J., & Vučić, N. (1979). Reduced tillage in combination with herbicide treatments under different irrigation methods. In *The 8th Conference of the International Soil Tillage Research Organization (ISTRO), Bundesrepublik Deutschland* (pp. 217–221).

Kišgeci, J., Mijavec, A., Aćimović, M., Spevak, P., & Vučić, N. (1984). *Hops growing*. Faculty of Agriculture—Institute of Crop Science.

Knapič, M. (1996). Soil selection for hop gardens. *Hmeljar, 65*, 66–68. [In Slovenian].

Knapič, M., & Pintar, M. (1998). Implementation of subsurface drip irrigation in hop growing as a contribution to an environmentally acceptable production. *Fresenius Environmental Bulletin, 7*, 867–872.

Kopecký, J., & Ježek, J. (2007). Influence of weather conditions and irrigation on yield and quality of hops. In *Proceedings of the Scientific Commission of IHGC (International Hop Growers Convention)*.

Kopp, P. (2014). The global hop: An agricultural overview of the Brewer's Gold. In M. Patterson & N. Hoalst-Pullen (Eds.), *The geography of beer* (pp. 77–88). Springer. https://doi.org/10.1007/978-94-007-7787-3_8

Kovařík, M. (2020). Economic commission summary reports. *International Hop Growers' Convention*. Retrieved from https://www.ihgc.org/about/

Kovařík, M. (2023). Economic commission summary reports. *International Hop Growers' Convention*. Retrieved from https://www.ihgc.org/about/

Kralj, D. (1962). Influence of heat and precipitation on the development and growth of Savinjski Golding. In *Proceedings of the 1st Yugoslavian Symposium on Hop Production* (pp. 7–20) [In Slovenian].

Kubeš, J. (2019). Geography of world hop production 1990–2019. *Journal of the American Society of Brewing Chemists, 80*, 84–91. https://doi.org/10.1080/03610470.2021.1880754

Kučera, J., & Krofta, K. (2009). Mathematical model for prediction of alpha acid contents from meteorological data for 'Saaz' aroma variety. *Acta Horticulturae, 848*, 131–139.

Lamm, F. R., Ayars, J. E., & Nakayama, F. S. (2007). *Micro irrigation for crop production: Design, operation, and management*. Elsevier.

Laws, D. R. J., Peppard, T. L., Sharpe, F. R., & Pickett, J. A. (1978). Recent developments in imparting hop character to beer. *Journal of the American Society of Brewing Chemists, 36*, 69–72.

Licciardo, F., & Carbone, K. (2023). *Come sta la brassicoltura italiana? Stato attuale e prospettive di sviluppo*. Birra Nostra, (January–February 2023), 24–29. Retrieved from https://lobit.crea.gov.it/wp-content/uploads/2023/06/Carbone-Licciardo_BIRRA.pdf

MacKinnon, D., & Pavlovič, M. (2019). Global hop market analysis within the International Hop Growers' Convention. *Hop Bulletin, 26*, 99–108. Retrieved from http://www.dlib.si/stream/URN:NBN:SI:DOC-A4F24XQF/76638dfb-9d09-40ee-9ae1-55cc9de6457a/PDF

MacKinnon, D., Pavlovič, V., Čeh, B., Naglič, B., & Pavlovič, M. (2020). The impact of weather conditions on alpha-acid content in hop (*Humulus lupulus* L.) cv. Aurora. *Plant, Soil & Environment, 66*, 519–525. https://doi.org/10.17221/344/2020-PSE

Majer, D. (2000). *Vodni stres pri hmeljnih rastlinah ali O pomenu vode za življenje rastlin, o stresu pri rastlinah, o namakanju in o vplivu vodnega stresa na hmeljne rastline*. Hmeljarsko združenje Slovenije; Ministrstvo za znanost in tehnologijo; Probi MM. [In Slovenian].

Majer, D. (Ed.). (2002). *Handbook for hop growers*. Institute for Hop Research and Brewing. [Žalec, Slovenia]. [In Slovenian]. https://www.ihps.si/wp-content/uploads/2016/08/hmeljarski_prirocnik_2002.pdf

Marceddu, R., Carrubba, A., & Sarno, M. (2020). Cultivation trials of hop (*Humulus lupulus* L.) in semi-arid environments. *Heliyon, 6*(10), e05114. https://doi.org/10.1016/j.heliyon.2020.e05114

Marić, V., & Srečec, S. (2006). Integrated procedure of drying and processing of hop cones into hop pellets. *European Patent Office*, WO2006061665 (A2). https://worldwide.espacenet.com/publicationDetails/biblio?CC=WO&NR=2006061665&KC=&FT=E&locale=en_EP#

Matičič, B. (1977). *Evapotranspiration studies on different crops and irrigation water requirements (Final technical report, P. L. 480)*. Agricultural Engineering Department, Biotechnical Faculty.

Meier, U., Bleiholder, H., Buhr, L., Feller, C., Hacks, H., Hess, M., Lancashire, P. D., Schnock, U., Stauss, R., Boom, T., Weber, E., & Zwerger, P. (2009). The BBCH system for coding the phenological growth stages of plants—History and publications. *Journal für Kulturpflanzen, 61*, 41–52.

Mihelič, R. (Ed.). (2010). *Guidelines for expertly based fertilization*. Ministry of Agriculture, Forestry and Nutrition of the Republic of Slovenia. [In Slovenian]. https://www.gov.si/assets/ministrstva/MKGP/DOKUMENTI/KMETIJSTVO/RASTLINSKA-PRIDELAVA/GNOJILA/smernice-za-gnojenje_2011.pdf

Moir, M. (2000). Hops—A millennium review. *Journal of the American Society of Brewing Chemists, 58*(4), 131–146. https://doi.org/10.1094/ASBCJ-58-0131

Mozny, M., Tolasz, R., Nekovar, J., Sparks, T., Trnka, M., & Zalud, Z. (2009). The impact of climate change on the yield and quality of Saaz hops in the Czech Republic. *Agricultural and Forest Meteorology, 149*, 913–919. https://doi.org/10.1016/j.agrformet.2009.02.006

Mozny, M., Trnka, M., Vlach, V., Zalud, Z., Čejka, T., Hajkova, L., Potopová, V., Semenov, M., Semeradova, D., & Büntgen, U. (2023). Climate-induced decline in the quality and quantity of European hops calls for immediate adaptation measures. *Nature Communications, 14*, 6028. https://doi.org/10.1038/s41467-023-41474-5

Nagel, J., Culley, L. K., Lu, Y., Liu, E., Matthews, P. D., Stevens, J. F., & Page, J. E. (2008). EST analysis of hop glandular trichomes identifies an O-methyltransferase that catalyzes the biosynthesis of xanthohumol. *The Plant Cell, 20*, 186–200. https://doi.org/10.1105/tpc.107.055178

Naglič, B., Cvejić, R., & Pintar, M. (2016). Pregled objav s področja namakanja hmelja (*Humulus lupulus* L.) na porečju Savinje = Irrigation of hop (*Humulus lupulus* L.) in Savinja catchment: A review. *Hmeljarski Bilten, 23*, 41–55. [In Slovenian].

Nakawuka, P., Peters, T., Kenny, S., & Walsh, D. (2017). Effect of deficit irrigation on yield quantity and quality, water productivity, and economic returns of four cultivars of hops in the Yakima Valley, Washington State. *Industrial Crops and Products, 98*, 82–92. https://doi.org/10.1016/j.indcrop.2017.01.037

National Agricultural Statistics Service, Agricultural Statistics Board, United States Department of Agriculture. (2023). *National hop report*. https://www.usahops.org/enthusiasts/stats.html

Neve, R. A. (1983). Hop production 1958-1983. *Journal of the Institute of Brewing, 89*(3), 164–169. https://doi.org/10.1002/j.2050-0416.1983.tb04162.x

Neve, R. A. (1986). Hop breeding worldwide—its aims and achievements. *Journal of the Institute of Brewing, 92*(1), 21–24. https://doi.org/10.1002/j.2050-0416.1986.tb04370.x

Neve, R. A. (1991). *Hops*. Chapman & Hall. https://doi.org/10.1007/978-94-011-3106-3

Nickerson, G. B., & Likens, S. T. (1979). Hop storage index. *Journal of the American Society of Brewing Chemists, 37*(4), 184–187. https://doi.org/10.1094/ASBCJ-37-0184

Pavlovič, M. (2012). Production character of the EU hop industry. *Bulgarian Journal of Agricultural Science, 18*(2), 233–239. http://www.agrojournal.org/18/02-13-12.pdf

Pavlovič, M. (2014). *Hop industry—Quality management, decision support modeling*. Verlag Dr. Kovač.

Pavlovič, M., Luo, X., Košir, I. J., Virant, M., & Gu, F. (2006). Expansion of the Chinese hop and brewing industry. *Hop Bulletin, 13*, 61–69. https://www.cabidigitallibrary.org/doi/pdf/10.5555/20073014767

Pavlovič, V., Pavlovič, M., Čerenak, A., Košir, I. J., Čeh, B., Rozman, Č., Turk, J., Pazek, K., Krofta, K., & Gregoric, G. (2012). Environment and weather influence on quality and market value of hops. *Plant, Soil and Environment, 58*(4), 155–160. https://pse.agriculturejournals.cz/pdfs/pse/2012/04/01.pdf

Pavlovič, M., Pavlovič, V., Rozman, Č., Udovč, A., Stajnko, D., Deilang, W., Gavric, M., & Srečec, S. (2013). Market value assessment of hops by modeling of weather attributes. *Plant, Soil and Environment, 59*(6), 267–272. https://pse.agriculturejournals.cz/pdfs/pse/2013/06/06.pdf

Pavlovič, M., MacKinnon, D., & Naglič, B. (2022). Study of market supply statistics in trading of hops. *Hop Bulletin, 29*, 85–96. https://www.dlib.si/stream/URN:NBN:SI:DOC-H9BNX6X8/5e1a5e63-68df-44e6-9a8b-159d3c6104d9/PDF

Pekhov, A. V. (1968). Extraction of vegetable oils with condensed gases. *Maslozhir Prom, 34*, 25–28.

Pintar, M., & Knapič, M. (2001). Nekateri namakalni parametri in obremenitve okolja pri različnih tehnologijah namakanja. In T. Poje (Ed.), *Zbornik simpozija Trendi v razvoju kmetijske tehnike, Radenci, 14. in 15. junij 2001* (pp. 69–76). Društvo kmetijske tehnike Slovenije. [In Slovenian].

Potopová, V., Lhotka, O., Mozny, M., & Musiolková, M. (2021). Vulnerability of hop yields due to compound drought and heat events over European key-hop regions. *International Journal of Climatology, 41*(S1), E2136–E2158. https://doi.org/10.1002/joc.6836

Purayannur, S., Gent, D. H., Miles, T. D., Radišek, S., & Quesada-Ocampo, L. M. (2021). The hop downy mildew pathogen *Pseudoperonospora humuli*. *Molecular Plant Pathology, 22*(7), 755–768. https://doi.org/10.1111/mpp.13063

Rodolfi, M., Silvanini, A., Chiancone, B., Marieschi, M., Fabbri, A., Bruni, R., & Ganino, T. (2018). Identification and genetic structure of wild Italian *Humulus lupulus* L. and comparison with European and American hop cultivars using nuclear microsatellite markers. *Genetic Resources and Crop Evolution, 65*, 1405–1422. https://doi.org/10.1007/s10722-018-0622-9

Rodolfi, M., Barbanti, L., Giordano, C., Rinaldi, M., Fabbri, A., Pretti, L., Casolari, R., Beghé, D., Petruccelli, R., & Ganino, T. (2021). The effect of different organic foliar fertilization on physiological and chemical characters in hop (*Humulus lupulus* L., cv. Cascade) leaves and cones. *Applied Sciences, 11*(15), 6778. https://doi.org/10.3390/app11156778

Rossbauer, G., & Münsterer, J. (2003). Optimisation of hop drying and conditioning for the safeguarding of hop quality. *Proceedings of IHGC Technical Commission Meeting, Sofia.*

Rossbauer, G., Buhr, L., Hack, H., Hauptmann, S., Klose, R., Meier, U., Stauss, R., & Weber, E. (1995). Phänologische Entwicklungsstadien von Kultur Hopfen (*Humulus lupulus* L.). *Nachrichtenblatt des Deutschen Pflanzenschutzdienstes, 47*, 249–253. In U. Meier (Ed.), *Growth stages of mono- and dicotyledonous plants, BBCH monograph (2001).* Federal Biological Research Centre for Agriculture and Forestry.

Rossini, F., Loreti, P., Provenzano, M. E., De Santis, D., & Ruggeri, R. (2016). Agronomic performance and beer quality assessment of twenty hop cultivars grown in Central Italy. *Italian Journal of Agronomy, 11*, 746. https://doi.org/10.4081/ija.2016.746

Rossini, F., Virga, G., Loreti, P., Iacuzzi, N., Ruggeri, R., & Provenzano, M. E. (2021). Hops (*Humulus lupulus* L.) as a novel multipurpose crop for the Mediterranean region of Europe: Challenges and opportunities of their cultivation. *Agriculture, 11*, 484. https://doi.org/10.3390/agriculture11060484

Rutnik, K., Ocvirk, M., & Košir, I. J. (2023). The stability of hop (*Humulus lupulus* L.) resins during long-period storage. *Plants, 12*, 936. https://doi.org/10.3390/plants12040936

Rybacek, V. (1991). Hop production. In *Developments in Crop Science* (Vol. 16, pp. 211–219). Elsevier.

Rybka, A., Heřmánek, P., & Honzik, I. (2017). Theoretical analysis of the technological process of hop drying. *Agronomic Research, 15*, 859–865. https://www.cabidigitallibrary.org/doi/pdf/10.5555/20173199400

Rybka, A., Krofta, K., Heřmánek, P., Honzík, I., & Pokorný, J. (2018). Effect of drying temperature on the content and composition of hop oils. *Plant, Soil and Environment, 64*, 512–516. https://doi.org/10.17221/482/2018-PSE

Schattenhofer, M. (1989). *Hops from Germany*. CMA.

Slavik, L., & Kopecky, J. (1994). The effect of trickle irrigation on the hop yield. *Rostlinná výroba, 40*(1), 13–18.

Slavik, L., & Kopecky, J. (1997). Efficacy of hop irrigation under years with different precipitation levels. *Chmelařství, 70*, 44–46.

Slavik, L., & Kopecky, J. (1998). Influence of economical irrigation on the yield and quality of hops. *Rostlinná výroba, 44*(7), 293–297.

Srečec, S. (1998). Production characteristics of central European provenance hop cultivars (*Humulus lupulus* L.) in soil and climatic conditions of sub Kalnik region and possibilities for hop production in sub Kalnik region. *Sjemenarstvo, 15*, 169–177. [In Croatian]. https://hrcak.srce.hr/file/248338

Srečec, S. (2004). *Hop production*. Križevci College of Agriculture. [In Croatian].

Srečec, S. (2006). *Bitter compounds of hops during the vegetation, processing into hop pellets, storage, and beer production* (Doctoral dissertation). University of Zagreb, Faculty of Food Technology and Biotechnology.

Srečec, S., & Scitovski, R. (2023). Prediction of alpha-acids accumulation in hop cones according to empiric mathematical model, general linear regression, and Cobb-Douglas production model-function. In *Proceedings of the Scientific-Technical Commission of International Hop Growers Convention* (pp. 58–61) https://www.hopfenforschung.de/wp-content/uploads/2023/07/Proceedings_STC_2023.pdf

Srečec, S., Kvaternjak, I., Kaučić, D., & Marić, V. (2004). Dynamics of hop growth and accumulation of α-acids in normal and extreme climatic conditions. *Agriculturae Conspectus Scientificus, 69*, 59–62. https://hrcak.srce.hr/file/18891

Srečec, S., Kvaternjak, I., Kaučić, D., Špoljar, A., & Erhatić, R. (2008). Influence of climatic conditions on accumulation of alpha-acids in hop cones. *Agriculturae Conspectus Scientificus, 73*, 161–166. https://hrcak.srce.hr/file/42489

Srečec, S., Rezić, T., Šantek, B., & Marić, V. (2009). Hop pellets type 90: Influence of manufacture and storage on losses of α-acids. *Acta Alimentaria, 38*, 141–147. https://doi.org/10.1556/aalim.2008.0014

Srečec, S., Zechner-Krpan, V., Marag, S., Špoljarič, I., Kvaternjak, I., & Mršić, G. (2011a). Morphogenesis, volume, and number of hop (*Humulus lupulus* L.) glandular trichomes, and their influence on alpha-acid accumulation in fresh bracts of hop cones. *Acta Botanica Croatica, 70*, 1–8. https://doi.org/10.2478/v10184-010-0017-2

Srečec, S., Zechner-Krpan, V., Marag, S., & Mršić, G. (2011b). Hop pellets type 90: ESEM studies of glandular trichomes morphological and structural changes during the different phases of hop processing. *Acta Alimentaria, 40*, 282–290. https://doi.org/10.1556/aalim.40.2011.2.12

Srečec, S., Zechner-Krpan, V., Petravić-Tominac, V., Košir, I. J., & Čerenak, A. (2012). Importance of medical effects of xanthohumol, hop (*Humulus lupulus* L.) bioflavonoid in restructuring of world hop industry. *Agriculturae Conspectus Scientificus, 77*, 61–67. https://hrcak.srce.hr/file/123764

Srečec, S., Čeh, B., Savić Ciler, T., & Ferlež Rus, A. (2013). Empiric mathematical model for predicting the content of alpha-acids in hop (*Humulus lupulus* L.) cv. Aurora. *SpringerPlus, 2*, 59. https://doi.org/10.1186/2193-1801-2-59

Svoboda, P., Hnilicková, H., & Hnilicka, F. (2008). Changes in yield and quality of hop depending on the irrigation. *Cereal Research Communications, 36*, 891–894.

Testa, H., Cardozo, A., Chillo, V., Manescotto, M., & Saez, D. (2023). Improving hop plant nutrition and yield with plant growth-promoting rhizobacteria. In *Proceedings of the Scientific-Technical Commission of the International Hop Growers Convention* (pp. 47–53). Ljubljana. https://www.hopfenforschung.de/wp-content/uploads/2023/07/Proceedings_STC_2023.pdf

Verberg, S. (2020). From herbal to hopped beer: The displacement of regional herbal beer traditions by commercial export brewing in medieval Europe. *Journal of the Brewery History Society, 183*, 9–23.

Virant, M. (Ed.). (2002). *Half a century of the "House of Green Erudition"*. Institute of Hop Research and Brewing. [In Slovenian].

Virant, M., & Majer, D. (2003). Hop storage index—Indicator of brewing quality. *Proceedings of the IHGC Technical Commission Meeting, Sofia.*

Wagner, T., & Veronek, M. (1977). *Poročilo o rezultatih proučevanja evapotranspiracije in evaporacije na postaji Orla vas—Žalec v letu 1976* [Report on the results of studying evapotranspiration and evaporation at the Orla vas—Žalec station in 1976]. Žalec. [In Slovenian].

Wagner, T., & Veronek, M. (1978). *Preučevanje evpotranspiracije hmelja v letu 1977. Poročilo o evapotranspiraciji na ET postajo Orla vas v letu 1977* [Study of hop evapotranspiration in 1977. Report on evapotranspiration at the ET station Orla vas in 1977]. Žalec. [In Slovenian].

Wample, R. L., & Farar, S. C. (1983). Yield and quality of furrow and trickle irrigated hop (*Humulus lupulus* L.) in Washington State. *Agricultural Water Management, 7*, 457–470.

Wang, G., Tian, L., Aziz, N., Broun, P., Dai, X., He, J., King, A., Zhao, P. X., & Dixon, R. A. (2008). Terpene biosynthesis in glandular trichomes of hop. *Plant Physiology, 148*, 1254–1266. https://doi.org/10.1104/pp.108.125187

Weber, K. A., Jangaard, N. O., & Foster, R. T. (1979). Effects of postharvest handling on quality and storage stability of Cascade hops. *Journal of the American Society of Brewing Chemists, 37*, 58–60. https://doi.org/10.1094/ASBCJ-37-0058

Wilson, G. W. (1914). Studies in North American Peronosporales. VI. Notes on miscellaneous species. *Mycologia, 6*, 192–210.

Zmrzlak, M., & Kajfež-Bogataj, L. (1996). Phenological modeling of growth stages of the hop (*Humulus lupulus* L.). In *Proceedings of the 14th International Conference of Biometeorology* (p. 88).

Part II
Hops: Phytopathology and Entomology

Chapter 5
Hop Diseases: Characteristics and Management

Sebastjan Radišek, David H. Gent, Timothy D. Miles, Scott J. Harper, Jernej Jakše, and Inga Zasada

5.1 Introduction

The cultivation of hops has always been challenged by the occurrence of diseases, which, if not properly controlled, can completely destroy a crop or even lead to plant death. Among the most important diseases of hop in the majority of hop-growing regions worldwide are hop downy mildew (*Pseudoperonospora*

S. Radišek (✉)
Plant Protection Department, Slovenian Institute of Hop Research and Brewing, Žalec, Slovenia
e-mail: sebastjan.radisek@ihps.si

D. H. Gent
Agricultural Research Service, United States Department of Agriculture, Corvallis, OR, USA

Forage Seed and Cereal Research Unit, United States Department of Agriculture–Agricultural Research Service, Corvallis, OR, USA
e-mail: dave.gent@usda.gov

T. D. Miles
Department of Plant, Soil and Microbial Sciences, Michigan State University, East Lansing, MI, USA
e-mail: milesti2@msu.edu

S. J. Harper
Department of Plant Pathology, Washington State University, Prosser, WA, USA
e-mail: scott.harper@wsu.edu

J. Jakše
Department of Agronomy, Biotechnical Faculty, University of Ljubljana, Ljubljana, Slovenia
e-mail: jernej.jakse@bf.uni-lj.si

I. Zasada
Agricultural Research Service, United States Department of Agriculture, Corvallis, OR, USA
e-mail: inga.zasada@usda.gov

K. Carbone (ed.), *All About Hops: The Crop, its Cultivation, and its Uses in Brewing and Beyond*, https://doi.org/10.1007/978-3-031-96606-4_5

humuli Miyabe & Takah. G.W. Wilson), hop powdery mildew (*Podosphaera macularis* Wallr. U. Braun & Takam.), Verticillium wilt (*Verticillium nonalfalfae* Inderb., H. W. Platt, R. M. Bostock, R. M. Davis & K. V. Subbarao), and diseases caused by a complex of viruses and viroids. In addition to these, there are many more known pathogens on hop, representing a range of potential causes of economic damage. So far, over 30 different fungi, oomycetes, and bacteria, 12 viruses, 4 viroids, 12 plant-parasitic nematodes, and a phytoplasma from the group 16SrI-B have been described on hop. Among the known diseases are also some physiological disorders resulting from abiotic factors.

Hop cultivation has faced significant epidemic outbreaks of diseases in the past, greatly impacting global hop trade. The transmission of the hop downy mildew pathogen from Japan to Europe and the USA in the years 1926–1936 caused devastating outbreaks, compelling growers to cultivate less susceptible varieties and implement annual protection using fungicides. Similarly, in 1997, an outbreak of powdery mildew in Washington State destroyed more than 800 hectares of hop, prompting growers in the Pacific Northwest production areas to cease production of a highly susceptible cultivar and implement strict control measures to limit damage. In Europe, the lethal form of Verticillium wilt, a soil-borne disease, has caused extensive plant dieback, severely affecting regions such as England, Slovenia, and Germany. More recently, epidemic caused by viroids have caused substantial damage to hop production. A newly discovered viroid on hop, citrus bark cracking viroid (CBCVd), has severely affected more than 500 hectares in Slovenia since 2010, and its recent introduction to Germany has already affected several farms. Since 2004, the USA has reported the progression and spreading of hop stunt viroid (HSVd), significantly reducing growth and yield in several important varieties.

Continuous adaptation of pathogens to new hosts, the monoculture of hop cultivation, the introduction of contaminated planting material from other regions, the expansion of production into new areas, and climate change are, and will continue to be, risk factors for the emergence or re-emergence of disease. Reports from the last three decades show an increasing discovery of new pathogens of hop and the detection of new pathotypes and races, presenting future challenges in hop production. New detection methods for pathogens, epidemiological studies, resistance breeding, and new approaches in integrated pest management are crucial for future sustainable disease control in hop production. This chapter is therefore dedicated to reviewing diseases on hop, providing descriptions of the pathogens, disease symptoms, and management practices.

5.2 Fungal, Oomycete and Bacterial Diseases

5.2.1 Hop Downy Mildew

Downy mildew is one of the most important diseases of hop wherever it occurs, but especially in wet, humid production regions. The disease has a long history of research that is reviewed elsewhere (Kopp, 2016; Neve, 1991; Purayannur et al.,

2021; Royle & Kremheller, 1981; Salmon & Ware, 1925). Crop damage from downy mildew may be multifold, as the disease can substantially reduce yield and crop quality, approaching 100% in extreme cases. Soft resin content of cones can be reduced by 50% or more when cones suffer severe infection. The downy mildew pathogen causes systemic infection that may also cause a crown rot and plant death in highly susceptible cultivars. As such, downy mildew may have effects over multiple years and limit the cultivars that are economically viable in regions with climates favourable to the disease.

Downy mildew occurs broadly on hop throughout regions of the Northern Hemisphere. Most hop production regions in the Southern Hemisphere have been spared downy mildew because of geographic isolation pared with strict quarantine measures, although the disease is known to occur in South America. Downy mildew was reported in Argentina since the late 1950s, and reemerged as a serious problem in the early 2000s (Pérez et al., 2009). The disease was also recently confirmed in Brazil (Feltrin et al., 2024).

5.2.1.1 Casual Organism

Downy mildew is caused by the fungus-like organism *Pseudoperonospora humuli*. Among the five species presently recognised in the genus *Pseudoperonospora*, *P. humuli* is by far the most important pathogen of hop. Inoculation experiments under controlled conditions indicate that certain nettle (*Urtica*), *Cannabis*, and *Celtis* spp. may be infected by *P. humuli* (Hoerner, 1940; Jones, 1932b; Salmon & Ware, 1925). The closely related sister-species *P. cubensis* causes downy mildew on various cucurbits. Under artificial conditions, trace to slight infection of *P. humuli* on cucurbits and *P. cubensis* on hop have been recorded (Mitchell et al., 2011; Runge & Thines, 2012). Based on overlapping measurements and features of the sporangiphores and sporangia, and phylogeny based on the ITS, Choi et al. (2005) reduced *P. humuli* to a taxonomic synonym of *P. cubensis*. This proposal was challenged and not widely accepted because of the limited resolution of the ITS for delineating species boundaries in oomycetes. Later phylogenetic studies utilizing more loci, genome and transcriptome-based variants, and whole genome sequencing demonstrated conclusively that the species are distinct (Runge et al., 2011; Summers et al., 2015; Rahman et al., 2019; Thomas et al., 2017; Wallace & Quesada-Ocampo, 2017). *Pseudoperonospora humuli* populations in the U.S. and Europe are clonal and appear to be asexually reproducing or highly selfed (Bello et al., 2021; Gent, Adair, et al., 2019; Kitner et al., 2021; Wallace & Quesada-Ocampo, 2017). Multiple diagnostic PCR assays exist for *P. humuli* (Crandall et al., 2021; Gent et al., 2009; Higgins et al., 2022; Rahman et al., 2019).

5.2.1.2 Symptoms

Symptoms of downy mildew vary depending on the host tissue infected. A distinct symptom of the disease is systemically infected shoots that are colloquially termed “spikes”, so called because of their resemblance to wheat spikes. In general, spikes are stunted, have shortened internodes, and have brittle, downward-curled leaves (Fig. 5.1a). When such shoots occur on the basal portion of a plant, they are referred to basal spikes, in contrast to aerial spikes that occur on trained shoots or lateral branches. A terminal spike refers to infection of a trained, climbing bine. Spikes are further differentiated as primary versus secondary depending on the presumed time when the infection occurred. Primary spikes develop from infection of crown buds that emerge from winter dormancy, whereas secondary spikes develop following infection of the apical meristem from the secondary infection cycle.

Primary basal spikes begin to emerge from the perennial rootstock in late winter to early spring and characteristically have down-curled leaves with a distinctive chlorotic appearance (Fig. 5.1b). Provided favourable environmental conditions, the abaxial surface of leaves harbour the dark purple to black sporangiophores and

Fig. 5.1 Symptoms of downy mildew associated with systemic infection of shoots (**a–d**) and leaf lesions (**e, f**). (Photo: David H. Gent)

sporangia diagnostic for the disease (Fig. 5.1c). Aerial spikes similarly have shortened internodes, down-curled leaves, and partial chlorosis of leaves, but lack the complete chlorosis often observed on newly emerged primary spikes and generally there is a clear transition from apparently healthy tissue to the stunted and infected tissue. Lateral branches can be difficult to diagnose as downy mildew when infection occurs just after they emerge because the tissue may quickly desiccate. Infection of branches may occur at any time during their development, though, and later infection leads to the characteristic spike appearance (Fig. 5.1d). Infection of trained bines results in a cessation of their growth and causes bines to fall from the string.

Lesions form on leaves during periods of secondary spread, and appear angular and delimited by veins (Fig. 5.1e). Leaf lesions tend to desiccate quickly in warm, dry weather, and become necrotic.

Infection of cones by *P. humuli* can be quite damaging. Infected burrs turn dark brown, shrivel, and desiccate. Infected cones become dark brown, harden, and cease development. Bracteoles of affected cones often become discoloured more readily than bracts, and this gives cones a striped appearance; under high disease pressure though entire cones may become a dark brown (Fig. 5.2). Infected cones may support sporulation on the underside of bracts and bracteoles that is diagnostic for downy mildew.

Systemic infections can extend into roots, rhizomes, and crown tissues. Infected rhizomes may have reddish-brown to black flecks and streaks in the cortex (Fig. 5.3). The crown may be completely rotted, and plants may be killed in cultivars that do not tolerate crown infection. In rare instances in the U.S., a rot of stems due to

Fig. 5.2 Cones with dark brown discolouration due to downy mildew. (Photo: Sebastjan Radišek)

Fig. 5.3 Rhizome with downy mildew. Notice the dark reddish-brown spots and streaking. (Photo: Sebastjan Radišek)

downy mildew has been observed. Affected stems are swollen, either near the soil line or higher on the plant, brittle, and display the reddish-brown to black flecks and streaks usually observed in infected rhizomes, or a more generalised medium to dark brown discolouration throughout the cortex (Fig. 5.4). Wilting occurs above the portion of the affected stem. This situation appears to occur when a highly susceptible cultivar develops an extremely high level of systemic infection.

5.2.1.3 Disease Cycle and Epidemiology

A key aspect of the disease cycle and epidemiology of downy mildew is that *P. humuli* establishes systemic infections, enabling the pathogen to persist in the host through long periods of unfavourable weather and overwinter in infected dormant buds and rootstock. Detailed studies by Coley-Smith (1962, 1964, 1966), building upon earlier observations of internal mycelium of *P. humuli* occurring in various tissues (Salmon & Ware, 1925; Ware, 1926, 1929), demonstrated that diseased shoots originated from mycelial invasion of the rootstock and subsequent bud infection. Infections during late summer to early autumn may lead to mycelial invasion of the rootstock, from which *P. humuli* spreads into developing buds during the winter months. A portion of the infected buds give rise to basal spikes when shoots emerge in the spring. The timing of infection that leads to maximal levels of overwintering and basal spikes in the following year may vary by cultivar and environment. In England, plants inoculated in September routinely died from rootstock infection, whereas plants inoculated in October were more likely to survive and produce spikes (Coley-Smith, 1962). Conversely, foliar inoculations during June to October of Nugget hop plants in Oregon did not kill plants and inoculation in midsummer rather than early autumn tended to lead to the most spikes in the following year (Richardson & Gent, 2023).

Systemic infection of the rootstock may result from shoot infections that progress downward or by means of infection through natural openings at the base of stems. Some cultivars may support chronic infection depending on the sensitivity of

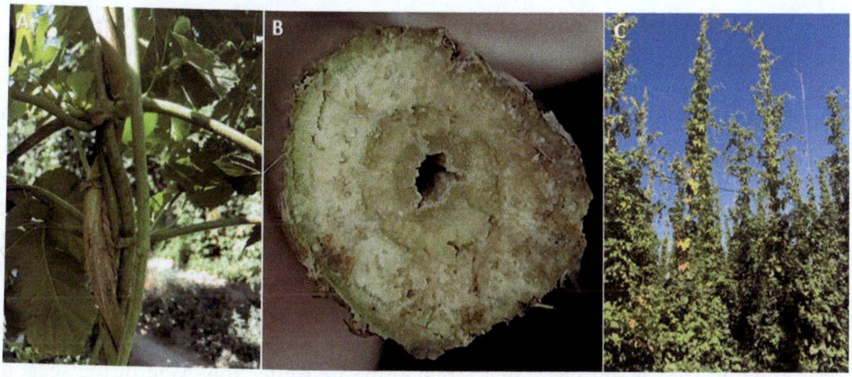

Fig. 5.4 Swollen and necrotic stems (**a**, **b**) and late-season wilting of plants (**c**) associated with severe downy mildew in a highly susceptible cultivar. (Photo: David H. Gent)

the cultivar to rootstock infection. Thus, a severe downy mildew outbreak in one year tends to lead to earlier and more severe outbreaks in the following season because of systemic infection of rhizomes and crown tissues (Gent et al., 2010; Johnson & Anliker, 1985). Indeed, there are hints that the founding population of *P. humuli* in a hop yard might persist over the life of the yard (Gent et al., 2019a). Carbohydrate reserves are reduced in systemically infected plants, resulting in a progressive weakening over time that reduces yield and could eventually kill plants.

Beginning in late winter to early spring, basal spikes begin to emerge from infected plants and provide primary inoculum to initiate the annual disease cycle. Emergence of diseased shoots is closely linked to plant growth and may proceed over a period of several months. Their initial emergence can be predicted using simple degree-day models (Gent et al., 2010; Johnson, 1991).

An unexplained oddity of the disease cycle is the biological function of oospores. *P. humuli* is homothallic (Gent et al., 2017a) and oospores form abundantly in diseased tissues, especially cones (Chee & Klein, 1998; Coley-Smith, 1962; Jones, 1932a; Mori, 1966). Germination and infectivity of oosporic inoculum has been reported in several instances (Arens, 1929; Bressman & Nichols, 1933; Jones, 1932a; Mori, 1966), but in multiple other studies similar attempts to demonstrate germination or infectivity have failed even though various assays suggest oospores are viable (Coley-Smith, 1962; Gent, Cohen, & Runge, 2017). These differences may point to oospores having a precise, but presently unknown, requirement for conditioning. Notwithstanding this uncertainty, oosporic inoculum does not appear to be essential for disease development (Skotland & Johnson, 1983).

The pathogen sporulates profusely on infected tissues when nighttime temperatures are greater than approximately 6 °C and humidity is greater than 90%. Sporangia are released daily in a cyclical pattern and are readily dispersed by wind, potentially over numerous kilometres.

Sporangia germinate indirectly to produce motile zoospores when the temperature is favourable and free water is present. The most severe disease outbreaks often are linked to daytime rain because zoospores enter hop tissues through open stomata

(Royle, 1970; Royle & Thomas, 1973). Infection is favoured by mild to warm temperatures (15 to 21 °C) when free moisture is present for at least 1.5 hours, although leaf infection can occur at temperatures as low as 5 °C provided leaf wetness persists for 24 hours or longer. Shoot infection requires at least 3 hours of wetness, with the rate of the infection process proportionate to temperature.

Various disease forecasting systems for hop downy mildew have been development in hop production regions in Europe and the U.S. (Skotland & Johnson, 1983; Royle & Kremheller, 1981). Most of these forecasting systems assess the risk of secondary infection events based on the combined effects of rain, temperature, and airborne spore density, although forecasting systems based solely on weather also exist. In Washington State in the U.S., a forecasting system was produced and used operationally for several years that targeted the first infection cycle of the season, with the goal of minimizing rootstock infection on highly susceptible cultivars sensitive to crown infection (Johnson et al., 1994; Skotland & Johnson, 1983).

5.2.1.4 Disease Management

No single management tactic provides satisfactory control of downy mildew. Rather, a strict set of sanitation measures, cultural practices, and timely fungicide applications are needed to manage the disease successfully. Cultivars vary widely in their susceptibility to downy mildew, although relatively few have extremely high levels of resistance (Feiner et al., 2021; Henning et al., 2015, 2016; Olatoye et al., 2023). Resistance to downy mildew is a polygenic trait with numerous loci having small, additive effects (Feiner et al., 2021; Henning et al., 2015, 2016; Olatoye et al., 2023). As a consequence, developing new cultivars with very high levels of resistance and the other traits essential for brewing is indeed a long, slow process. Where found, resistance traces back to materials bred in Europe and developed over decades of breeding efforts (Woods & Gent, 2016). Cultivars derived from North American germplasm tend to be among the most susceptible to the disease.

Because of the importance of systemic infection and the ease of introducing *P. humuli*, use of disease-free planting material is critically important when establishing new hop yards. When disease levels are relatively low within a yard, rouging and replacing heavily infected plants can further reduce primary inoculum. This practice becomes impractical in yards with extensive infection, though.

In well-established yards, thorough removal of all foliage during spring pruning substantially reduces later disease development (Gent et al., 2012). Pruning yards as late as possible and thoroughly removing all green tissue during the pruning operation, whether by mechanical pruning or using chemical desiccants, tends to reduce the severity of downy mildew. The optimal timing of pruning varies by cultivar and environment, and a potential trade-off with later pruning is reduced yield. Sanitation measures such as stripping the lower leaves from bines and hand removal of infected shoots removes both inoculum and reduces the favourability of the microclimate for disease. Similarly, management of weeds and other vegetation within or between hop rows can also decrease humidity and disease.

Timely fungicide applications are needed to manage downy mildew when weather is favourable to the pathogen. Fungicides are usually differentiated based on their suitability for targeting primary infection originating from systemically basal spikes versus protectant fungicides suitable for targeting secondary infection. The former includes highly systemic fungicides such as the phenylamides and phosphonates, or fungicides with local systemic activity such as fluopicolid. Numerous fungicides tend to be used for secondary infection, with the specific active ingredients varying by region based on local registration. The efficiency of fungicide applications can be improved substantially by timing their use to periods of high disease risk using one of the various disease warning systems. An explicit or implicit assumption of these systems is that systemic infection of the rootstock is minimal. If this assumption is not valid, growers tend to make scheduled applications. Resistance to multiple classes of fungicides is reported in *P. humuli* (Gent et al., 2020a; Hellwig et al., 1991; Klein, 1994), and future selection for fungicide resistance is expected given the fecundity of the pathogen and high potential for resistance in most modern fungicides. In situations where downy mildew has developed late in the season, yards can be harvested early to minimise cone infection. However, yields can be reduced in both the current and subsequent seasons when plants are harvested too early due to interruption of carbohydrate partitioning into the perennial root system.

5.2.2 *Powdery Mildew*

Powdery mildew is one of most damaging diseases of hop where it occurs. The disease can cause severe damage to both yield and quality, with the potential to render an entire crop unsaleable due to reductions in brewing qualities. The fungus that causes powdery mildew on hop is broadly distributed in production regions in Europe, the Pacific Northwest, Midwest, and eastern North America, and areas of Asia. The pathogen is also relatively common on feral and wild hop plants in the same regions. Geographic isolation and quarantines have largely prevented the establishment of the pathogen in the Southern Hemisphere, although a recent confirmation of the disease in subtropical areas of Brazil (Mendes Fagherazzi et al., 2021) highlights the potential for dissemination of the disease worldwide. Damage caused by powdery mildew depends heavily on susceptibility of a given cultivar and also varies by production region. Thorough reviews of the disease are provided elsewhere (Gent et al., 2008; Mahaffee et al., 2009; Neve, 1991; Royle, 1978).

5.2.2.1 Causal Organism

Powdery mildew is caused by the obligately biotrophic fungus *Podosphaera macularis*, with the basonym *Sphaerotheca macularis* appearing often in older literature (Braun & Takamatsu, 2000). The host range of *P. macularis* is known to include multiple species within the Cannabaceae including hop, the annual species Japanese hop (*H. scandens* syn. *H. japonicus)*, and *Cannabis sativa*. Although *C. sativa* is a host of *P. macularis*, most occurrences of powdery mildew on hemp are caused by another powdery mildew fungus, *Golovinomyces ambrosiae* (Rivedal et al., 2023). Natural infection of *C. sativa* by *P. macularis* under field conditions typically occurs late in the season near harvest, or when *C. sativa* is grown under protected cultivation. Older literature considered the related powdery mildew fungus, *Podosphaera aphanis*, on strawberry and other Rosaceae hosts as conspecific with *P. macularis*. These species have since been resolved and have distinct host ranges that do not overlap (Braun & Takamatsu, 2000). Various other powdery mildew fungi have been reported on hosts within the Cannabacecae, but not hop (Bradshaw et al., 2024).

Podosphaera macularis is heterothallic and requires each of two complementary mating types for sexual recombination and ascocarp formation (Wolfenbarger et al., 2015). Both mating types and the ascigerous stage are broadly distributed in Europe and eastern North America, whereas only the *MAT1–1* mating type has been detected in the western U.S. (Gent et al., 2020b; Weldon et al., 2021b; Wolfenbarger et al., 2015).

5.2.2.2 Disease Symptoms

Disease signs appear as powdery white colonies on leaves, buds, stems, and cones (Fig. 5.5). During periods of rapid plant growth, raised blisters often are visible before sporulation is observed.

Infection of burrs and young cones causes abortion or severe distortion of the cone as it develops (Fig. 5.6). When these tissues are infected, affected cones develop the characteristic white, powdery fungal growth. Signs of the pathogen are

Fig. 5.5 Characteristic colonies of *Podosphaera macularis* on a hop leaf. (Photo: W. F. Mahaffee)

Fig. 5.6 Severe powdery mildew due to infection of burrs and young cones. (Photo: David H. Gent)

less conspicuous when infection occurs later in cone development. In such instances, diffuse mycelium of *P. macularis* may be visible with the aid of magnification under bracts and bracteoles, but with limited sporulation. Affected cones become reddish-brown as tissues are killed (Fig. 5.7) or may turn a medium brown after kiln drying.

Where the ascigerous stage occurs, the dark pigmented chasmothecia often develop within colonies on leaves and cones (Fig. 5.8). Development of chasmothecia is most notable in later parts of the growing season, especially after early summer and into autumn. In severe instances, cones can become entirely blackened by chasmothecia.

5.2.2.3 Disease Cycle and Epidemiology

Podosphaera macularis persists over the winter months as hyphae within or on dormant infected buds, as chasmothecia in association with infested tissue, or on living host plants in protected areas such as greenhouses. The importance of each varies depending on whether both of the requisite mating types are present (Weldon et al., 2021a).

Crown buds infected in one season give rise to infected shoots the following spring. These are the so-called flag shoots, named because of their resemblance to a white flag against the contrast of otherwise green foliage in spring. Bud perennation leading to flag shoots is a rare event under field conditions. An extensive survey of hop yards in Oregon in the U.S. found flag shoots occurred in about 6% of yards,

Fig. 5.7 Reddish-brown discolouration of hop cones associated with late season occurrence of powdery mildew. (Photo: David H. Gent)

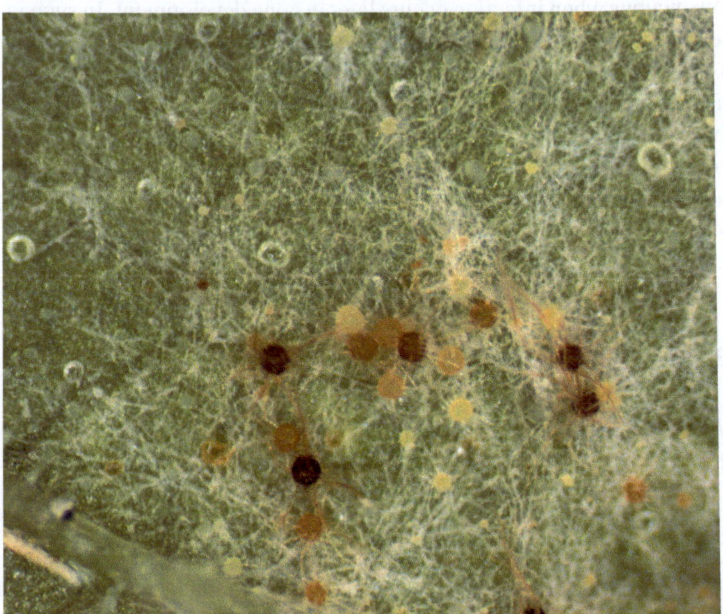

Fig. 5.8 Chasmothecia of *Podosphaera macularis* in varying stages of maturation on a hop leaf. (Photo: David H. Gent)

occurring on less than 1% of plants on average. Although relatively rare, bud perennation in a small proportion of yards is sufficient to initiate annual epidemics of powdery mildew. Flag shoots emerge in sync with plant growth, and the first infected shoots can be found as soon as healthy shoots begin to emerge in spring. Thus, the disease cycle may begin each year as soon as plant growth commences. The number of flag shoots produced in a given yard is related to cultivar susceptibility, prior occurrence of flag shoots in that yard, disease levels in the previous year, spring pruning practices, and winter temperature (Gent et al., 2018b; Gent et al., 2019c; Laurie et al., 2022).

Where both mating types occur, chasmothecia represent a second source of primary inoculum that may be present under entirely different conditions. Chasmothecia are durable structures that enable the fungus to persist in the absence of living host tissue. Thus, crop debris such as leaves and cones infected in the previous season become another inoculum source. Chasmothecia of *P. macularis* produce long, myceliod appendages that become entwined in the pannose mycelium of colonies, effectively anchoring chasmothecia to the infected tissue on which they were produced. This enables a large proportion of chasmothecia to remain adhered to the senescent host tissue after dormancy (Weldon et al., 2021c). Chasmothecia mature gradually during winter in proportion to temperature, and dehisce to release ascospores when wetted in spring. The optimal conditions for infection by ascospores are generally the same as those for the asexual conidia.

Greenhouse propagation of planting material during the winter months provides a third means for pathogen survival. This means of pathogen survival is particularly relevant and effective for disseminating *P. macularis*. Genetic studies with *P. macularis* suggest that one of the strains of the pathogen known to occur in the Pacific Northwest U.S. was broadly disseminated across commercial hop yards in the Midwest and eastern portions of the country in association with infected planting material (Gent et al., 2020b; Weldon et al., 2021b). Notably, the pathogen population in cultivated yards was, with only a few exceptions, distinct from the population found on wild or feral hop plants. Populations of *P. macularis* population in the U.S. appear to have a European origin, again pointing to probable dissemination of the pathogen in association with infected planting material.

The fungus is readily disseminated by wind and spreads from the initial disease focus to other leaves and green tissues, firstly within a given yard but soon afterwards to other nearby yards. Disease development is favoured by rapid plant growth, mild temperatures (8 to 28 °C), high humidity especially at night, and cloudy weather. Under ideal conditions of 18 to 21 °C, the generation time of the fungus can be as short as 5 days on highly susceptible cultivars. The short generation time of the fungus, copious sporulation, and its ability to disperse over long distance by wind results in fast-developing disease outbreaks within and between hop yards. Disease modeling suggests that the fungus may disperse over multiple kilometres given an adequate source strength of the inoculum and windy condition (Gent et al., 2019b).

Crop damage from powdery mildew can be quite difficult to estimate because the damage experienced depends sensitively on when infection occurs and the severity of the infection. Examples of severe disease outbreaks demonstrate that entire crops can be lost in highly susceptible cultivars with inadequate management, as was experienced in Washington in 1997 soon after the disease was first confirmed in the region (Mahaffee et al., 2003; Turechek et al., 2001). Burrs and young cones are quite susceptible and infection at these stages of crop development leads to the most substantial reductions in yield, alpha-acids content, colour defects, and premature ripening (Gent et al., 2014). Cones become somewhat less susceptible to powdery mildew with maturity, but unlike leaves (Turechek et al., 2001), never become fully resistant to the disease (Twomey et al., 2015). Infection during the later stages of cone development can lead to cone browning and hastened maturity in proportion to the amount of disease present on cones. Alpha-acids typically are not influenced by late-season attack of powdery mildew, but yield can be reduced by 20% or more because of shattering of overly dry cones during harvest due to early maturity. Several weak pathogens and secondary organisms can be found on cones attacked by powdery mildew, which can sometimes confuse diagnosis of the primary cause of cone discolouration.

5.2.2.4 Disease Management

Management of powdery mildew can be challenging in highly susceptible cultivars when environmental conditions are favourable to disease development. Disease management in susceptible cultivars involves an integration of crop sanitation practices, cultural practices to moderate canopy development, well-timed fungicide applications, and careful attention to the timing of harvest.

A key sanitation measure is spring pruning. Thoroughly removing all green tissue during spring pruning operations reduces or eliminates inoculum associated with bud perennation, reduces disease severity by delaying epidemics, and saves fungicide applications (Gent et al., 2012, 2019c; Laurie et al., 2022). Mechanical pruning tends to be more effective than chemical pruning in eliminating flag shoot, but logistics, consideration of other diseases, and costs often dictate use of the latter on many farms. Basal foliage removal is another important sanitation measure. Under high disease pressure, removal of basal foliage mid-season removes diseased tissue and delays disease development on leaves and cones (Gent et al., 2016). Established yards can tolerate some removal of basal foliage without reducing yield, but care is needed with first year plantings to avoid negatively affecting yield in the subsequent year.

In regions where powdery mildew occurs regularly, fungicide applications are essential for economic production of most susceptible cultivars. The timing of the first application of the season and later applications coincident with rapid plant growth are particularly critical in regions where conditions are highly favourable to the disease. The specific fungicides used vary based on local registrations and regulations, which makes it difficult to provide specific recommendations. In general, many fungicide programs can give adequate disease control on leaves when applied preventatively under low disease pressure (Nelson et al., 2015). On cones, however, differences in fungicide efficacy are substantial. A limited number of highly active fungicides provide the most effective disease control when targeted to the earliest stages of cone development (Nelson et al., 2015), provided that disease pressure is not unduly high and harvest is timely. Crop damage from powdery mildew can increase with later harvests, independent of prior fungicide applications. Research in the U.S. indicates that when powdery mildew is present, highly susceptible cultivars should be harvested by about 26% dry matter to maintain cone colour and minimise losses due to shattering during picking (Gent et al., 2014).

Where available, host resistance is the most efficient means of disease management. Multiple named R-genes have been described in hop, and other sources of uncharacterised resistance are known in cultivated varieties and germplasm (Gent et al., 2018a; Havill et al., 2023; Wolfenbarger et al., 2014). Huge efforts are expended to develop disease resistant cultivars. Perhaps unsurprisingly, most forms of resistance to powdery mildew have not been durable, as virulent strains of *P. macularis* have emerged repeatedly when a given form of host resistance is deployed broadly within a production region. Among the seven named R-genes, virulent isolates of *P. macularis* have emerged that can overcome all of these R-genes (Seigner et al., 2001; Wolfenbarger et al., 2016). The pathogen has also adapted to partial resistance, as found in the cultivar Cascade (Gent et al., 2017b). A long-term challenge for breeding for resistance remains identifying durable resistance, or combinations of resistance, a priori. In other hosts, combinations of R-genes or R-genes pared with quantitative resistance appear to provide longer lasting resistance than R-genes deployed individually (Mundt, 2014). The relatively long-lasting resistance in the cultivar Wye Target has been attributed to the combination of the major R-gene R2 and quantitative resistance (Darby et al., 1989). Whereas the concept is simple, selecting for multiple forms of resistance concurrently necessarily requires means to confirm multiple loci or genes have been combined. Molecular markers could play a key role in confirming multiple, distinct forms of resistance were indeed combined into an individual.

Research on the Mildew Locus O (*MLO*) in other plants affected by powdery mildew suggest that breeding for lack of susceptibility through *mlo* or other susceptibility gene null mutations could be an effective strategy for durable resistance (Kusch & Panstruga, 2017). Segregation of disease resistance and QTL studies indicate epistatic interactions among loci that confer susceptibility to powdery mildew in hop (Darby, 2013; Henning et al., 2011). The *MLO*-gene family occurs across diverse angiosperms, including hop. Mutations in *MLO* are associated with powdery mildew resistance in *C. sativa* (Pépin et al., 2021; Stack et al., 2023), suggesting that *mlo*-based resistance might be present and functional in hop. Research is ongoing to identify which members of the *MLO*-gene family condition powdery mildew susceptibility in hop and identify or develop *mlo* mutants with more durable resistance.

5.2.3 Verticillium Wilt

Verticillium wilt is an important vascular wilt disease that affects numerous dicot plant species in temperate and subtropical regions worldwide. The causal agents are soilborne fungi, which are members of the genus *Verticillium sensu stricto* (Inderbitzin & Subbarao, 2014). As a host plant, hop can be infected by the species *Verticillium nonalfalfae* and *V. dahliae*. Disease outbreaks on hop are predominantly caused by the fungus *V. nonalfalfae*, while infections by *V. dahliae* are comparatively rare and less severe (Neve, 1991). The disease can be particularly devastating in the case of infections of highly virulent strains of *V. nonalfalfae*, which induce rapid withering and dieback of the entire plant (Radišek et al., 2006). Verticillium wilt disease on hop was first recorded in 1924 in England, when a mild disease outbreak was caused by *V. nonalfalfae* (Harris, 1927). After 1933, more severe and progressive outbreaks appeared, resulting in significant economic damage to the English hop industry in the following decades. Pathogenicity tests of *V. nonalfalfae* isolates revealed that mild (fluctuating) and lethal (progressive) disease forms are associated with mild and highly virulent strains of *V. nonalfalfae* (Isaac & Keyworth, 1948). Until 1997, the lethal strains were geographically limited to England, but in that year, the first lethal disease outbreaks were found in Slovenia. Due to its rapid spread and the predominance of susceptible cultivars, a highly virulent pathotype (PV1) of *V. nonalfalfae* (genotype PG2) severely affected more than 200 hectares of Slovenian hop yards from 1997 to 2014 (Radišek et al., 2006; EPPO, 2014). Lethal strains of *V. nonalfalfae* were also reported from Germany (Seefelder et al., 2009), Belgium (Heungens et al., 2021), and most recently, from the Czech Republic (Svobodová, 2023).

5.2.3.1 Casual Organism

Verticillium wilt of hop is caused by *Verticillium nonalfalfae* Inderb., H. W. Platt, R. M. Bostock, R. M. Davis & K. V. Subbarao, and *Verticillium dahliae* Klebahn. Both fungi belong to a group of ten species in the genus *Verticillium* sensu stricto, which form a monophyletic clade in the Plectosphaerellaceae family of fungi associated with plants. Before the last taxonomic revision of the *Verticillium* genus, *V. albo-atrum* Reinke & Berthold was the species attributed to infection of hop, lucerne, and some other plant species. Phylogenetic studies split *V. albo-atrum* into three different species: *V. albo-atrum* sensu stricto, *V. nonalfalfae*, and *V. alfalfae*. The isolates pathogenic to hop are now classified as members of the species *V. nonalfalfae* (Inderbitzin et al., 2011; Giraldo & Crous, 2019). All three species share similar morphology, such as resting mycelium, which is the key morphological feature distinguishing them from *V. dahliae*, which forms microsclerotia. The host range and severity of disease caused by *V. nonalfalfae* and *V. dahliae* vary. In general, *V. dahliae* has more hosts and can infect over 200 host species. *V. nonalfalfae* has a narrower host range and, besides hop, can infect tomato, potato, eggplant, cucumbers, petunia, spinach, tree of heaven, cotton, *Fragaria* sp., and several weed species (Inderbitzin & Subbarao, 2014; Giraldo & Crous, 2019). Many of these plants can host both species, as seen in the case of hop, and the severity of the disease depends on fungal host adaptation. Hop is more susceptible to *V. nonalfalfae*, which has evolved highly virulent strains (termed PV1, PV2, PV3) that can be classified based on virulence testing on a differential set of hop cultivars (Sewell & Wilson, 1984). For molecular identification, species-specific PCR primers were developed by Inderbitzin et al. (2013). Additionally, Jakše et al. (2018) developed PCR primers capable of differentiating between lethal and mild strains, regardless of their geographical origin (Munda et al., 2020).

5.2.3.2 Disease Symptoms

Two Verticillium wilt syndromes occur on hop, consisting of a mild and a lethal form of the disease. The general disease symptoms found in both forms include yellowing and wilting of the leaves, beginning at the base of the plant and progressing upwards. Necrotic tissue appears on the affected leaves, especially on the margins and between the leaf veins (Fig. 5.9). Leaf margins exhibit upward curling and leaves drop very easily when lightly touched. Symptomatic bines develop corky epidermis and brown discolouration of vascular tissue (Fig. 5.9). which can be observed at the lower part of the bine. Unlike wilting caused by Fusarium canker, affected bines do not develop a restriction at the base (Neve, 1991; Radišek, 2009).

The characteristic symptoms of the mild form occur when susceptible cultivars are infected with less virulent strains of *V. nonalfalfae* or *V. dahliae*. Additionally, they manifest in the case of infection of tolerant cultivars by a highly virulent strain of *V. nonalfalfae* (Talboys, 1958). The first symptoms on leaves appear relatively late in the growing season, usually just before or after flowering when water demand is the greatest, and only some of the bines are affected. Symptomatic bines develop abnormal thickening from the base up to a height of 1.5 meters, along with limited browning of the vascular tissue, which is confined to the centre (Fig. 5.10). The thickening of the bines is linked to sparse mycelium presence in the xylem and the formation of intense tyloses (Cregeen et al., 2015). This process is accompanied by renewed cambial activity, resulting in xylem hyperplasia (Talboys, 1958). Mild wilt varies in intensity from season to season and rarely causes plant death. Thus, plants affected in one season may appear healthy the next. Affected plants are usually scattered throughout a hop yard, and the re-emergence of the disease is associated with excessive soil moisture and nitrogen fertilisation (Sewell & Wilson, 1967, 1974).

The lethal form of the disease occurs in susceptible cultivars infected with highly virulent strains of *V. nonalfalfae*. This form of the disease affects the majority of hop bines, which exhibit extensive and rapid dieback of leaves and lateral shoots, eventually leading to the death of the entire rootstock (Fig. 5.11). Bines rarely becomes swollen, and the entire vascular tissue develops brown discolouration (Fig. 5.9), which is a consequence of intense mycelium colonization. In highly susceptible cultivars, the first signs of the disease can be observed before flowering, progressing rapidly and culminating in wilting of the entire plant and leaf drop within 2–4 weeks (Fig. 5.11). Disease incidence is less influenced by ecological factors such as wetness and fertilisation, and its spread in hop yards is related to the cultivation pattern.

Fig. 5.9 Symptoms of Verticillium wilt. Note the interveinal chlorosis and necrosis on leaves (**a**), the swollen bines with a corky epidermis (**b**), and brown discolouration of vascular tissue (**c**, **d**). (Photo: Sebastjan Radišek)

5.2.3.3 Disease Cycle and Epidemiology

Both fungi overwinter in soil in the form of resting structures. In the case of *V. nonalfalfae*, these structures present as melanised swollen mycelium, whereas *V. dahliae* forms melanised microsclerotia. *V. nonalfalfae* may persist in soil free of host plants for up to 4 years, while *V. dahliae* can persist for more than 14 years (Sewell & Wilson, 1966; Schnathorst, 1981). The life cycle of both species is monocyclic and begins with root penetration near the root tips by hyphae that are formed on germinating resting structures, stimulated by root exudates. Infection progresses as conidia colonise the root cortex and invade xylem tissue. These conidia then travel

Fig. 5.10 Plants affected with the mild form of Verticillium wilt develop swollen bines (**a**) in which brown vascular tissue is limited to the centre (**b**). (Photo credits: Sebastjan Radišek)

Fig. 5.11 Plants affected with the lethal form of Verticillium wilt develop the first symptoms before flowering (**a**) and usually die before harvesting (**b**). (Photo: Sebastjan Radišek)

upward through the plant via transpiration flow. At the sites of vessel perforations, trapped conidia germinate and form new mycelium. Aided by hyphae and the secretion of hydrolytic enzymes, hyphae penetrate barriers and grow into new veins (Schnathorst, 1981). The occurrence of symptoms is associated with the clogging of xylem vessels by fungal biomass and the formation of tyloses within the vessel lumen (Cregeen et al., 2015). Both fungi also produce phytotoxins and cell wall-degrading enzymes, which contribute to the host cell death. When the plant tissue dies, the fungus breaks out of the xylem tissue and begins to form resting structures. The resting structures are returned to the soil in infected plant debris and remain dormant until induced to germinate and start the next disease cycle (Klimes et al., 2015).

As soil-borne fungi, spreading in contaminated hop yards primarily occurs through soil movement during cultivation and via remains of infected hop bines and leaves. Several weed species like *Chenopodium album*, *Senecio vulgaris*, *Solanum nigrum*, and *Urtica urens* can be symptomless carriers, helping to maintain or even increase soil infection potential (Sewell & Wilson, 1958). Between hop yards, the disease can be spread through uncleaned machinery and by returning fresh hop waste after harvesting to hop yards. Long-distance spreading is facilitated by infected planting material, which can introduce Verticillium wilt to new, previously uninfected areas. In the case of susceptible cultivars, the infection potential usually increases every year, especially if farmers use too much nitrogen fertilisation (Sewell & Wilson, 1974).

5.2.3.4 Disease Management

Verticillium wilt is difficult to control because there are no effective chemical treatments to cure infected plants or prevent the infection. Both fungi can survive in soil for several years, posing an additional challenge in preventing their spread, especially in the case of hop, which are grown as a perennial crop in monoculture. An integrated approach based on adapted cultural practices, planting resistant or tolerant cultivars, soil sanitation, and implementing hygiene measures are the only successful means of suppressing this disease. It is important to remove and destroy affected plants from the yard, as they present a potential source for inoculum buildup and disease spread. This measure is especially crucial in the case of lethal outbreaks to prolong the cropping life of a contaminated hop yard. In addition, conducting periodic disease monitoring for Verticillium wilt is essential to detect infected plants in the early stages to reduce the risk of further spread.

Adapted cultural practices, such as weed control, minimum cultivation, intercropping with non-host plants, and reduction of nitrogen fertilization, have proven to be important means of reducing disease incidence (Radišek et al., 2023; Sewell & Wilson, 1974). Hop waste from contaminated hop yards should be composted with a thermophilic stage and thoroughly mixed, as this presents an important sanitation measure. However, if the composting process is not carried out properly and high temperatures are not reached throughout the plant material, the fungi can survive (Sewell et al., 1962).

Crop rotation with non-host plants has been proven to be an effective soil sanitation method for reducing infection potential. A minimum four-year crop rotation with non-host plants or "grassing down" can efficiently eliminate or reduce the infection potential of *V. nonalfalfae* in the soil (Sewell & Wilson, 1966). Additionally, integrating non-host green manure plants such as sorghum and Sudan grass into the crop rotation process can shorten this period to two years (Radišek et al., 2023). These cover plants can also be used for interplanting during the hop vegetation period, while overwinter plants like barley and rye can be utilised during the winter period (Fig. 5.12). This practice positively affects microbiological soil parameters, suppresses weeds, improves soil structure, and contributes to soil health, thereby suppressing Verticillium wilt pathogens.

Resistant cultivars offer the most effective means of controlling this disease. However, even highly resistant cultivars can become infected, often exhibiting mild symptoms. Breeding for Verticillium wilt resistance commenced in England during the early 1930s, drawing upon two distinct sources of American wild hops (*H. lupulus var. neomexicanus*) identified as Y90 and AA 7 (Darby, 2001). These sources remain pivotal in resistance breeding. Progeny testing has revealed that Y90 contributes to polygenic resistance, whereas AA7 serves as a source of dominant resistance through two complementary genes (Jakse et al., 2013).

Hop relies on vegetative propagation, which favours the unnoticed transmission of pathogens through infected planting material. Therefore, the selected plant stock used for propagation should be disease-free, as demonstrated by testing. Controlled propagation is achievable in nurseries that are isolated from hop yards and utilise greenhouse production methods. In the case of production of planting material in hop yards, such plantations should be free from any symptoms of Verticillium wilt and with no history of Verticillium wilt. It is recommended that control of planting material be conducted within the framework of certification programs. In Europe, *V. nonalfalfae* and *V. dahliae* are subject to official control at the level of planting material, as outlined in Commission Implementing Regulation (EU) 2019/2072. Additionally, the European and Mediterranean Plant Protection Organization (EPPO) has developed standard PM 4/16(2) for the certification scheme concerning hop planting material.

Fig. 5.12 Sudan grass (**a**) and winter barley (**b**) as interplant cover plants suppress spreading of Verticillium wilt in an affected hop yard

5.2.4 Fusarium Canker

Fusarium canker is a disease frequently observed on hop worldwide, with reports dating back to at least the early 1900s. Usually, canker affects a sufficiently small percentage of bines as to not cause appreciable damage. However, there are reports of extensive crop damage from the disease scattered throughout the literature associated with specific environmental conditions or cultivars. In the extreme cases, 70% or more of plants may be affected. Wet soil conditions, co-infection by hop stunt viroid, and wounding of stems by insect pests have been implicated in the most serious occurrences of Fusarium canker (Davies, 1938; French et al., 1973; Salmon & Wormald, 1922).

5.2.4.1 Causal Organism

Fusarium canker has been attributed most often to *Fusarium sambucinum* (teleomorph *Gibberella pulicaris*). Reproducing the characteristic stems swelling and constriction at the base of stem requires specific conditions, and inoculation studies have confirmed that *F. sambucinum* is indeed capable of inducing canker symptoms. *F. sambucinum* also can cause a generalised root rot, wilting, and decay of plants derived from softwood cuttings or other propagative material. Whereas *F. sambucinum* appears to be the primary cause of canker, other *Fusarium* species may also be associated with cankered stems, namely *F. oxysporum*, *F. solani*, *F. equisti*, *F. meridionale*, and *F. redolens* (Augusto Moretti Ferreira Pinto et al., 2022; Solarska, 1981). *F. oxysporum* has been associated with wilted hop plants in Europe (Solarska, 1981). Studies are underway to determine if the other species can incite canker, or other disease symptoms, apart from *F. sambucinum*. A diagnostic quantitative PCR assay is available for detection and quantification of *F. sambucinum* (Thomas et al., 2022).

Fusarium sambucinum sensu stricto and other closely related species in the *Fusarium sambucinum* Species Complex are known to produce a number of mycotoxins of concern for human health (Munkvold et al., 2021). Hop-derived isolates of *F. sambucinum* contain the biosynthetic gene cluster required for production of certain trichothecene mycotoxins and produce diverse trichothecenes *in vitro*. Mycotoxin contamination in beer is rare and is routinely attributed to barley rather than hops. Given that *F. sambucinum* is also a pathogen of cones, there is a need to understand the diversity of mycotoxins produced by hop-derived isolates of *F. sambucinum* and whether the fungus could be a source of mycotoxins in hops in certain situations.

5.2.4.2 Disease Symptoms

Cankered bines can be found with careful inspection of plants in early to mid-spring on a small percentage of plants, with the incidence of the disease progressing throughout the summer. The most conspicuous aboveground symptoms of canker are usually first noticed just before or after bloom when plants are under water stress. Older leaves on the lower portion of affected bines may become generally chlorotic or display an interveinal chlorosis (Fig. 5.13). These bines soon wilt, and the foliage desiccates. Unlike with Verticillium wilt, the desiccated leaves remain attached to the dead bine. Bine wilting is often evident after cultivation or other field operations, or high winds (Fig. 5.14). The base of affected bines is swollen, tapering over a short distance (about 5 mm) near the point of attachment to the crown that may be attached by only a few vascular strands such that affected bines may break away from the crown with only a gentle tug (Fig. 5.15). Bines may also become entirely detached from the crown, particularly after high winds, cultivation, or other field operations. In severe instances, plants may be killed. In rarer instances, canker may appear soon after planting causing a rot of the crown that kills plants within the

Fig. 5.13 Marginal necrosis and interveinal chlorosis of older leaves preceding wilting caused by Fusarium canker. (Photo: David H. Gent)

Fig. 5.14 Wilted bines due to Fusarium canker causing detachment of the shoot from the crown. (Photo: David H. Gent)

Fig. 5.15 Hop stem with characteristic swelling just above an area of constriction near the soil line due to Fusarium canker. (Photo: David H. Gent)

first year. A root rot has also been observed on hop plants inoculated with *F. sambucinum* under controlled conditions.

5.2.4.3 Disease Cycle and Epidemiology

The disease cycle of Fusarium canker has not been investigated thoroughly and most of the information on the disease cycle is observational. *Fusarium sambucinum* is widespread in soil and is considered a soil inhabitant after introduction to a site. The fungus can be found in association with plant debris, diseased crowns, and apparently healthy planting materials. Although *F. sambucinum* is widespread and canker appears at a low incidence in most hop yards, severe occurrences of the disease are less common. Studies in the western U.S. indicate that *F. sambucinum* can be detected on the basal portions of stems soon after emergence, but with substantial variation between years or yards. These observations suggest that highly specific factors are critical for the disease, although these factors have not been defined precisely.

Wounds are thought to be important for pathogen ingress. These wounds may be created by mechanical or chemical damage (e.g., wind, cultivation, or potentially herbicide injury) around the soil line, cuts made to straps, or insect feeding.

Infection appears sensitive to wetness. Most, but not all, field observations of the disease in the literature suggest poor drainage as a contributing factor. Under experimental conditions, persistent high humidity or wetness near the base of stems favours infection. Experience also suggests that the disease tends to be more severe following wet winters. Relatively high temperatures (near 27 °C) appear to favour development of the characteristic stem canker.

Fusarium sambucinum may persist in association with diseased tissue on the rootstock or on the lower portions of stems infected in the previous year. Perithecia of the sexual stage may be found on the base straps (stems buried with soil during the growing season) with canker. In the western U.S., sporodochia of *F. sambucinum* (but rarely perithecia) are often observed on decaying hop stems affected by canker. The infested tissue of both types is a possible source of inoculum for perpetuating the disease. Diseased planting material has been implicated in canker, particularly when setts (small crowns) dry out during storage.

Fusarium sambucinum has a broad host range that may include potato, alfalfa, various grasses and cereals, and numerous woody plants. It is possible that inoculum produced on other host plants could incite canker on hop, although definitive evidence of this is missing. Recent evidence indicates host specialization in isolates of *F. sambucinum*, both on hop and other host plants. Isolates derived from hop caused the most severe disease on hop, whereas the isolates derived from potato were only weakly pathogenic on hop. The opposite was also true when the same isolates were inoculated into potato tubers.

5.2.4.4 Disease Management

Management of canker relies on sanitation measures to reduce inoculum potential, management of irrigation and drainage, and minimizing wounding. Cutting away necrotic tissue on the rootstock associated with canker is reported to reduce damage from canker in future years. Although reported to be effective, this practice may be impractical to implement at scale on large farms. In the U.S., the form of spring pruning influences the development of canker. In Washington, mechanical pruning using a rotary-type pruning implement consistently leads to a small but significant increase in incidence of bines wilting due to Fusarium as compared to using herbicides to chemically desiccate shoots. Presumably this is due to the more extensive wounding caused by this form of mechanical pruning. Tying strings on a hill together to prevent movement in the wind and from farm equipment operations may also reduce wounding at the base of stems and prevent cankered bines from completely detaching from the rootstock. Control of certain soil-dwelling insect pests that feed on hop is reported to reduce disease. Some reduction in wilting of bines is achieved by early-season hilling of soil near the base of plants. Following hilling, stems in the early stages of canker will develop roots that can prevent the entire wilting of the bine later in the season.

Fungicides are unavailable for canker. Research in England showed some promising results with application of thiabendazole directed to the bases of plants early in the season, but this is not a routine practice. Drench or dip treatments of setts with broad-spectrum fungicides such as maneb or benomyl have been suggested at times for mitigating the disease in young plants. Oomycete-specific fungicides such as mefenoxam have no effect on canker. The form and dose of nitrogen fertiliser applied can influence the severity of certain diseases caused by *Fusarium* pathogens. However, field studies in hop suggest that the development of stem cankers is little affected by nitrogen fertility practices.

5.2.5 Halo Blight

5.2.5.1 Causal Organism

Halo blight caused by the ascomycete *Diaporthe humulicola* is a recently described disease in the eastern United States and Canada (Allan-Perkins et al., 2020; Higgins et al., 2021; Hatlen et al., 2022; Sharma et al., 2023). Up to 50% yield damage has been observed due to cone shatter in high disease pressure situations (Higgins et al., 2021). With its ability to cause substantial damage to yield, an integrated management practice coupled with continued scientific investigation, will be key to mitigating the impact of this disease in locations with favourable climates for halo blight during the growing season.

5.2.5.2 Disease Symptoms

Halo blight presents various symptoms that significantly impact hop. Initial symptoms include lesions on the surface of hop leaves, surrounded by a chlorotic margin (Fig. 5.16). Lesions can be somewhat restricted but have the potential to cover more than 50% of the leaf area. Smaller lesions can often be mistaken for other foliar infections, such as those caused by *Colletotrichum fiorinae* or *Pseudoperonospora humuli* (Hatlen et al., 2023; Higgins et al., 2021). The disease is most detrimental when it affects hop cones, leading to their desiccation (Fig. 5.17). The desiccation can occur partially or fully and is often similar in appearance to Fusarium tip blight, powdery mildew, or grey mould. A notable symptom is the infection of the cone's strig, with potential white mycelia development post-desiccation. However, the appearance of these symptoms can vary, making visual diagnosis challenging.

5.2.5.3 Disease Cycle and Epidemiology

Diaporthe humulicola infects above-ground tissues of hop, including leaves, cones, and bines (Higgins et al., 2022). The fungus overwinters on dormant hop bines as asexual pycnidia (Hatlen et al., 2024, Fig. 5.18). This fungus likely initiates its cycle

Fig. 5.16 Halo blight caused by *Diaporthe humulicola* on a hop leaf in southwest Michigan, USA. (Photo: Timothy D. Miles)

Fig. 5.17 Halo blight caused by *Diaporthe humulicola* on developing hop cones in southwest Michigan, USA. Notice the dried bract tissue girdling the cone. (Photo: Timothy D. Miles)

Fig. 5.18 Pycnidia of *Diaporthe humulicola* on 'Centennial' hop tissue from the previous growing season. Note the small black dots on the tissue. (Photo: Timothy D. Miles)

in the spring when pycnidia release conidia that infect new growth. While the complete transmission pathway is still under investigation, it is suspected that the pathogen spreads through spores dispersed by wind, rain, or contaminated tools and plant material (Hatlen et al., 2024). The pathogen's ability to infect and persist in hop rhizomes and crowns remains unclear. Under artificial conditions, optimal growth of *D. humulicola* is 20 to 25 °C. Pycnidia of isolates form within 18 to 30 days on potato dextrose agar. They appear dark brown, smooth or covered in hyphae, and globose to ampulliform, with sizes approximately 1500 μm by 1200 μm. Alpha conidia under these conditions are unicellular, hyaline, and cylindrical to ellipsoid, measuring approximately 8 μm by 3 μm; no beta conidia were observed (Higgins et al., 2021).

5.2.5.4 Disease Management

Since its observation in 2018, effective management of halo blight has involved an integrated approach drawing on research from other important *Diaporthe*-caused diseases like Phomopsis stem and leaf spot of grape (Wilcox, 2017). Cultural practices such as sanitation, removing and destroying infected plant debris, and avoiding overhead irrigation can reduce the spread of the pathogen (Hatlen and Miles unpublished). The use of disease-free planting material is critical, given the potential for disease spread through rhizomes and cuttings. Fungicides may offer some control; efficacy can be inconsistent but applications targeting powdery mildew probably provide some control. Trifloxystrobin and pyraclostrobin have both been shown to restrict *D. humulicola* growth on artificial media (Allan-Perkins et al., 2020).

5.2.6 *Alternaria* Blight

Alternaria blight or Alternaria cone disorder on hop is primarily caused by the ascomycete *Alternaria alternata*, posing a significant concern for hop growers (Darby, 1988; Pethybridge et al., 2001a, b). Additionally, hop can be infected by *Alternaria humuli* (Simmons, 2002), and *A. humuli-scandens* has been reported on Japanese hop (*Humulus scandens*) in China (Zhang et al., 2009).

Whereas little is known about *A. humuli* and *A. humuli-scandens*, *A. alternata* is ubiquitous in the environment and affects a variety of crops, including apple, potato, sunflower, and wheat. A recent survey in Michigan found *Alternaria* spp. to be recovered through isolation approximately 50% of the time from cone tissue (Higgins et al., 2021). It has been reported in several locations where hops are grown (Twomey et al., 2015; Pethybridge et al., 2001a, b). Recently mycotoxin contamination in hops has been noted, which has raised the importance of this disease (Dodlek Šarkanj et al., 2024).

Symptoms initially appear as light reddish-brown discolourations on the tips of the bracteoles of young or mature hop cones (Fig. 5.19). As the disease progresses, these discoloured tips may cause the cones to have a distinctive green and brown striped appearance. The infection can spread to both bracteoles and bracts, especially when cones are damaged by wind or other mechanical factors (Twomey et al., 2015; Darby, 1988). In severe cases, the entire cone may turn dark brown, resembling symptoms of downy or powdery mildew.

Fig. 5.19 Alternaria blight observed in Michigan, USA. Notice the discolouration on the edge of the bract tissues. (Photo: Timothy D. Miles)

Under high magnification, fungal growth and spores, which appear dark brown, can be observed on the underside of infected tissues. Notably, in Australia and the UK, this pathogen has been observed to cause rapid necrosis of hop cones (Pethybridge et al., 2001a, b). *Alternaria alternata* is common in hop yards, often found on both diseased and healthy cones, making accurate diagnosis challenging as its presence alone does not confirm it as the primary cause of cone discolouration.

The disease cycle of *Alternaria alternata* begins with the pathogen overwintering on plant debris or as a weak pathogen on other plants (Darby, 1988). Infection typically occurs through mechanical damage, such as wind injury or wounds caused by insect feeding or other diseases. The pathogen thrives in conditions where there is mechanical injury to cones combined with extended periods of leaf wetness or high humidity. Temperatures above 18 °C during rain or moisture events favour the development of the disease. The spores are disseminated by wind, rain, or contaminated equipment, leading to new infections. Studies from the UK and Australia have shown that high humidity and mechanical damage accelerate infection and disease progression (Pethybridge et al., 2001a, b). *Alternaria alternata* also has a known association with powdery mildew, with cones affected by powdery mildew being more likely to also have *Alternaria* species at harvest (Twomey et al., 2015).

Managing Alternaria blight involves several cultural practices aimed at reducing mechanical damage and moisture on hop cones. To minimise mechanical injury, careful handling during field activities and chemical applications is essential. Reducing wind damage can be achieved by using windbreaks, altering hop training systems or planting hops in less exposed areas. Good cultural practices, such as stripping the lower bines to promote air circulation and timing irrigation to avoid prolonged wetness, may reduce disease incidence. Accurate diagnosis of the disease is crucial, as simply recovering the fungus from discoloured cones does not confirm it as the cause of the browning. Fungicides are not typically registered specifically for Alternaria blight control but fungicides belonging to the FRAC classes 3, 7, and 11 might provide disease suppression because they are broad spectrum and are often applied for powdery mildew control.

5.2.7 *Ascochyta Leaf Spot and Phoma Wilt*

Ascochyta leaf spot and Phoma wilt have been described that have similar symptoms in hop. Determining the etiology of these diseases is difficult due to several reclassifications of the *Didymellaceae* (Chen et al., 2015). Ascochyta leaf spot and Phoma wilt have been reported in several locations within Asia and continental Europe (Singh et al., 1985; Wormald, 1946; Radisek et al., 2007; Boerema et al., 2004). Additionally, the recent description of halo blight in North America could further complicate diagnoses and management (Allan-Perkins et al., 2020; Higgins et al., 2021).

Ascochyta leaf spot on hop has been attributed to the fungal pathogen *Ascochyta humuli* (Singh et al., 1985; Pidopličko, 1978). Phoma wilt is a disease caused by several species of fungal pathogens in the genus *Phoma*, including *Phoma herbarum* and *Boeremia exigua* (Radisek et al., 2007; Boerema et al., 2004). Initial symptoms for both diseases begin with leaf lesions that can grow larger and coalesce. Both diseases tend to develop small black pycnidia on the leaf surface in the centre of the lesion. Phoma wilt has been more characterised by chlorotic spots whereas Ascochyta leaf spot describes the development of grey margins at the lesion edge the lesions. Phoma wilt has also been described as typically having concentric circles rings (Radisek et al., 2007). Phoma wilt has been known also to affect cones leading to brown or red discolouration covering some or all of the bracts, indicating necrosis. In some cases, Phoma wilt has been described as causing stem and canker lesions leading to hop plants wilting. Since identification of this disease is further complicated by the recently described halo blight disease (Allan-Perkins et al., 2020; Higgins et al., 2021) diagnosis requires pathogen isolations and sequencing of marker genes. The current markers genes for identification within the *Didymellaceae* are LSU, ITS, RPB2 and TUB2 (Chen et al., 2015).

Conidia from pycnidia are likely the main method of spread. Mature pycnidia are usually small and black and will ooze conidia that are disseminated via rain splash, wind, or misting. For both diseases infection typically becomes evident in late summer, with infection occurring through stomata, wounds, or directly through the epidermis. Once inside the host, conidia bud, divide, and hyphae spread, initially without symptoms. Necrotic lesions appear as the pathogen becomes necrotrophic. Pycnidia formation begins with hyphae near the epidermis in lesions forming a cavity, followed by dark hyphal cell production. In the absence of a host, these fungi can survive saprophytically in plant debris or decaying material.

Currently, there are no specific management techniques for Ascochyta leaf spot or Phoma wilt in hop. However, several methods are proposed to reduce disease incidence. The removal of plant debris from hop yards can help minimise the spread by eliminating resting spores, pycnidia, or saprophytic pathogens. Diseased plants or plant material should be removed promptly. Ensuring adequate airflow around hop plants, avoiding misting or sprinkler systems, and practicing sterile techniques when pruning, transporting, or propagating stock are crucial. Crop rotation with nonhosts may also be beneficial. Although fungicides have shown action against *Phoma* species, none are specifically registered for Phoma wilt in hop. Further studies are needed to confirm the effectiveness of these treatments against these two diseases in hops.

5.2.8 Black Root Rot

Black root rot of hop, attributed to *Phytophthora citricola* sensu lato, has been reported broadly across hop production areas worldwide and can be a serious concern in susceptible cultivars when environmental conditions are favourable to the disease (Hoerner, 1949; Royle, 1968; Pichlmaier & Zinkernagel, 1992; Sonoda & Ogawa, 1968). The disease causes weak, unthrifty growth and wilting of affected bines, typically just before harvest or when plants experience water stress. The most serious damage to yards often results from plant death during winter dormancy, where 20 to 40% of plants may be killed (Roborgh, 1961; Sonoda & Ogawa, 1968). Affected roots and crowns are water-soaked and blackened, with a distinct boundary between diseased and healthy tissue (Fig. 5.20).

Since the original description of the disease, the identity of the causal organism has been revised and therefore the identity of the *Phytophthora* species that cause black root rot is uncertain. *P. citricola* was originally described as the cause of brown rot of citrus fruit (Sawada, 1927), but later misattribution of other species such as *P. cactorum* (Waterhouse, 1957) to *P. citricola* sensu lato resulted in a large and diverse species complex. At present, the former *P. citricola* species complex has been resolved into eight species and other taxa also may be present (Bezuidenhout et al., 2010; Brazee et al., 2017; Ginetti et al., 2014; Henricot et al., 2014; Scott et al., 2009; Tsykun et al., 2022). A consequence of this taxonomic revision is that earlier reports of the host range and distribution of *P. citricola* sensu lato are questionable. *P. citricola* sensu stricto is now thought to have a host range limited to

Fig. 5.20 Hop with dark discolouration and water-soaked rot from black root rot. (Photo: David H. Gent)

Citrus spp. and to occur in only a small number of countries (Jung & Burgess, 2009). It seems certain that black root rot on hop is caused by a member, or members, of the species complex other than *P. citricola* sensu stricto.

Recent evidence from the U.S. indicates that a causal agent of black root rot is *P. plurivora*; it is unknown if other members of the former *P. citricola* species complex also infect hop. *P. plurivora* has a host range spanning over 80 woody and trees species. *Rhododendron*, *Pieris*, *Acer* spp., *Alnus* spp., *Fagus sylvatica, Quercus* spp., and *Picea* spp. are some of the important hosts (Beal et al., 2021; Jung & Burgess, 2009; Parke et al., 2014; Schlenzig et al., 2015; Weiland et al., 2010). Hop-derived isolates of *P. plurivora* from Washington State and California represent a clonal lineage of the pathogen distinct from isolates derived from other common hosts. Isolates of *P. plurivora* from hop, and other isolates derived from certain other plants, also exhibit host preference. That is, isolates of the pathogen tend to cause the most severe disease on the host from which they were originally derived.

Most aspects of the biology and epidemiology of black root rot are unstudied on hop. Similar to plant diseases caused by other *Phytophthora* pathogens, black root rot is favoured by poor drainage or a high-water table. Wounding appears important for infection. The significance of black root has diminished with the development of resistant or tolerant cultivars. In New Zealand, for instance, a successful breeding program was initiated in 1952 with the specific goal of developing black-root-rot-tolerant cultivars to replace the highly susceptible California Cluster-type cultivar (Roborgh, 1961). Other cultivars, such as Fuggle, are considered tolerant despite not being bred specifically for tolerance to the disease (Christie, 1956). Certain phosphorous acid and phenylamide fungicides may suppress black root rot, but their efficacy on hop has not been documented well.

5.2.9 Cercospora Leaf Spot

Cercospora leaf spot is caused by the fungus *Phaeomycocentrospora cantuariensis* (E.S. Salmon and Wormald) Crous, H.D. Shin and U. Braun (Crous et al., 2013). The disease was initially observed in 1922 near Canterbury, England, and at that time, it was named *Cercospora cantuariensis* (Salmon & Wormald, 1923). While rarely reported in hop, significant damage was recorded in the Kärnten (Leutschach) region of Austria in 2005 and 2007, as well as in the nearby hop-growing area in Koroška (Radlje ob Dravi), Slovenia (Radišek, 2009). An assessment conducted on cones from Radlje ob Dravi in 2007 revealed disease incidence ranging from 7 to 26%, with severity ranging from 4.5 to 16.6% (Radišek et al., 2008). Additionally, reports from South Korea indicate that the fungus can infect Japanese hop (*Humulus scandens*), loofah (*Luffa cylindrica*), and Asian copperleaf (*Acalypha australis*) (Crous et al., 2013; Mycobank, 2024).

On hop, the disease initially manifests as small oval-shaped violet-brown spots on the leaves, which later expand to reach a diameter of up to one centimetre. With the progression of the disease, these spots coalesce, eventually covering the entire surface of the most affected leaves. On cones, irregularly shaped reddish-brown necrotic areas appear, which may spread and, in severe cases, cover the entire cone surface (Wormald, 1928; Radisek et al., 2009) (Fig. 5.21).

The epidemiological characteristics, along with the life cycle, of this disease remain largely unexplored. Most likely, the fungus, similar to related species, survives on dead remains of infected hop parts, forming stromata. Initial infections can be observed as early as the flowering stage on the lower leaves of plants. Warm and humid weather, often accompanied by frequent rainfall, accelerates the occurrence of larger outbreaks, mainly occurring late in the growing season, coinciding with cone infections. Since no forecasting systems have yet been established for this disease, prevention relies on regular scouting and timely application of fungicides. Effective fungicides for this purpose include azoxystrobin, trifloxystrobin, and myclobutanil, primarily registered for preventing hop downy mildew or powdery mildew (Radišek et al., 2008).

Fig. 5.21 Cercospora leaf spot on hop cones (**a**) and leaves (**b**, **c**). (Photo: Sebastjan Radišek)

5.2.10 Grey Mould

Botrytis cinerea Pers. Fr. (teleomorph *Botryotinia fuckeliana* (de Bary) Whetzel), is a necrotrophic fungus causing grey mould diseases in over 200 plants species. It is most aggressive on mature or senescent tissues of dicotyledonous hosts (Williamson et al., 2007). In general, grey mould presents a minor threat to hop; however, severe infections have been reported from Japan (Sakai, 1976) and Europe (Engelhard, 2006). Infections on hop can be observed during cone formation and are favoured by humid conditions. Certain cultivars, such as Hallertauer Magnum, Hallertauer Merkur, Hallertauer Taurus, and Nugget, tend to be susceptible to grey mould, particularly in Germany and Slovenia (Engelhard, 2006; Radišek unpublished).

Disease symptoms typically begin at the tip of the cone, where moisture tends to persist the longest. The tips of the bracts and bracteoles then turn light brown, with the browning later extending toward the strig. On the affected tissue, the fungus develops sporulating mycelia with conidiophores, which thrive in wet and humid conditions (Fig. 5.22). While the fungus can also infect leaves, such infections are rare and are often associated with injury.

Botrytis cinerea survives as a saprophyte on infected crop debris, where dormant resting structures (sclerotia) develop, enabling the fungus to persist in the soil. When conditions become favourable again, such as high humidity or moisture,

Fig. 5.22 Grey mould symptoms on hop cones. (Photo: Sebastjan Radišek)

sclerotia can germinate and give rise to new fungal growth. While *B. cinerea* primarily spreads via its asexual cycle, it also has the capacity to undergo a heterothallic sexual cycle, forming apothecia that release ascospores. However, the occurrence of the *B. cinerea* sexual cycle in nature is rare. The fungus predominantly spreads through airborne asexual conidia released from conidiophores (Williamson et al., 2007).

Disease management strategies primarily rely on fungicide application, with strobilurin fungicides being an effective option. Increasing row and plant spacing facilitates airflow within the canopy, reducing the duration of plant wetness and thereby mitigating disease development.

5.2.11 Cone Tip Blight

Cone tip blight is a disease described on hop in the Pacific Northwestern U.S. in 2004 (Bienapfl, 2003) but observed several years prior to the formal description. The disease is caused by a complex of *Fusarium* species. *Fusarium avenaceaum* and *F. sambucinum* were reported in the U.S. (Bienapfl et al., 2005) and Europe (Solarska, 2003), whereas *F. crookewellense* was reported in Australia (Bienapfl et al., 2005; Pethybridge et al., 2001a, b). These organisms have a nearly worldwide distribution, although cone tip blight has been reported only occasionally on hop. Other *Fusarium* species are occasionally isolated from hop cones and under controlled laboratory conditions these other *Fusarium* species may also cause necrosis on hop cones. Nonetheless, these various other species appear to be of secondary importance under natural conditions based on the infrequency of their isolation from affected cones.

Affected cones develop a characteristic necrosis of bracts and bracteoles, beginning at the distal end of the cone and progressing upward. Necrosis is apparent on the strig. Disease symptoms are conspicuous, but in most years and with most cultivars cone tip blight is a relatively minor issue that affects less than 1% of cones. However, more severe occurrences have been documented where nearly one third of cones may be affected (Bienapfl, 2003). The causes of these more severe occurrences are uncertain. *Fusarium sambucinum* and *F. avenaceum* can be isolated at the greatest frequency from burrs and stigmas, pointing to colonization of floral tissues well before the characteristic symptoms appear just before harvest. High humidity or rain near bloom is thought to be associated with more disease development later in the season.

Management approaches for cone tip blight have not been developed, and in most situations are not necessary. A natural question is whether suppression of canker could reduce inoculum to incite cone tip blight, as *F. sambucinum* is a causal agent of both diseases. Definitive data linking these diseases does not yet exist. Severe occurrence of the disease is often observed in certain cultivars, with Sorachi Ace being notable. However, there has not been a detailed examination of cultivar susceptibility. Certain fungicides or antifungal plant extracts applied at bloom or to young developing cones may reduce the frequency of colonization of the cone tip blight pathogens, but so far have had limited success in reducing the incidence of the disease.

5.2.12 Red Crown Rot

Red crown rot is a disease generally of minor importance in hop. The causal agent was described as a *Phacidiopycnis* sp. in Australia and later was attributed to *Phomopsis tuberivora*, the basonym of *Phacidiopycnis tuberivora* (Sutton, 1980), when a similar disease was detected in Oregon in the U.S. The pathogen is distributed worldwide, where it is causes stem end or dry rot on various woody and herbaceous plants. In Australia red crown rot was associated with a progressive decline in plant vigour and yield losses of up to 20%. In a severe case in Oregon, however, the progressive decline in vigor lead to eventual plant death of a small percentage of plants in a long-established yard of Fuggle hop.

The pith tissue of affected roots and crowns is orange to red, with a distinct boundary between diseased and healthy tissue (Fig. 5.23). A key point here is that roots and crowns of apparently healthy plants typically have a similar or identical appearance.

Damage from the disease becomes a concern as the severity of dry rot advances and yield becomes markedly affected. Given the weak pathogenicity of the causal agent, substantial yield damage associated with red crown rot seems to occur when

Fig. 5.23 Rot of the pith tissue of a root with red crown rot showing a distinct boundary between diseased and healthy tissue. (Photo: David H. Gent)

plants are attenuated by some other cause. Under such conditions, bine growth is weak, uneven, and lower leaves become generally chlorotic (Fig. 5.24). In extreme instances, severely affected plants can be killed.

The disease cycle of red crown rot is largely unknown. In Australia, the disease was attributed to planting poor quality rootstock, injury to crowns during spring pruning operations, and cultural practices that reduced carbohydrate reserves in the root system. Spatial patterns of the disease in an affected yard in Oregon did not indicate a directionality in aggregation of affected plants, casting doubts on the importance of mechanical wounding from spring pruning operation. Control measures usually are not warranted. Planting of high-quality planting material and agronomic practices that maximise and maintain carbohydrate reserves such as careful attention to harvest timing and ensuring adequate basal foliage after harvest likely render red crown rot only an occasional problem. Fumigation before planting is reported to improve establishment and yield hop yards affected by red crown rot, although fumigation should not be necessary on a routine basis if one was targeting only this disease.

5.2.13 Sclerotinia Wilt

Sclerotinia wilt tends to occur only occasionally on hop in wet, cool climates. The causal fungus, *Sclerotinia sclerotiorum*, has a plurivorous host range that spans hundreds of dicotyledonous plants. While damage on many hosts such as canola and

Fig. 5.24 Weak, uneven bine growth and chlorosis of older leaves associated with a severe occurrence of red crown rot in the cultivar Fuggle. (Photo: David H. Gent)

soybean can be quite damaging, the disease on hop usually occurs on and damages hop only when the soil and plants remain continuously wet and temperatures are mild. Under these conditions, the pathogen may infect the base of stems at or just below the soil surface. Infection may also occur higher on plants where stems are abraded from the tie used to anchor multiple strings together on a hill (Kropf et al., 2012). During wet weather, the pathogen causes water-soaked cankers on stems overlaid with a conspicuous fluffy, white mycelium over the infected tissue (Fig. 5.25). Dark, melanised sclerotia often are present on stems or in the pith tissue, especially as the disease progresses. Crop damage occurs when stem lesions expand to girdle and wilt a bine.

The lifecycle of *Sclerotinia sclerotiorum* is well studied in other pathosystems. Sclerotia are durable and long-lived in infested crop debris and soil. Myceliogenic or carpogenic germination of sclerotia may be important in certain crops, the later occurring only after conditioning by exposure to moist conditions and cool temperatures. In many host plants, ascosporic infection requires ascospores to colonise a senescent tissue as a nutrient source before invading a host. It is unclear whether myceliogenic or carpogenic germination of sclerotia is important in the infection cycle of hop, and both may be involved with the disease on hop. Carpogenic germination is suspected to be important given that infection of wounded stems may occur well above the soil surface. Myceliogenic infection is supposed because

Fig. 5.25 A stem canker caused by *Sclerotinia sclerotiorum*. Note the conspicuous white mycelium and round, black sclerotia associated with the diseased tissue. (Photo: David H. Gent)

severe epidemics of Sclerotinia wilt on hop reportedly are associated with hilling soil infested with sclerotia onto crowns and with frost injury of developing basal shoots (Salmon & Ware, 1936a, b). Routine control measures for Sclerotinia wilt usually are unnecessary, although special attention may be warranted if hop is planted after a susceptible host such as canola. Given the close relationship between the disease and wetness, practices that limit excessive foliage and promote drying of the soil surface limit disease development. Chemical and biological control products are used routinely for related diseases in many other crops, but are not developed or recommended in hop.

5.2.14 *Diseases of Minor Importance*

Several fungal and bacterial pathogens are described on hop, which are generally considered to be of minor importance to hop production. The diseases they cause are very rare, and usually do not cause severe crop damage. However, minor pathogens can become important in specific regions and may even become major pathogens in the future due to climatic changes and modified agricultural practices.

- *Armillaria Root Rot* (*Armillaria* spp.) starts with plant wilting and the presence of plaster-white sheets of the pathogen under the bark near the soil surface, progressing to a rot in the crown (Salmon & Ware, 1937). The pathogen persists in soil and decaying plant material, spreading through root contact or by rhizomorphs, and can remain viable for many years, infecting new plants when conditions are favourable. Prevention involves avoiding planting hop yards in areas previously occupied by susceptible tree crops unless all roots and stumps are removed, good soil drainage, and avoiding water stress.
- *Black Mould* (*Cladosporium* spp.) causes brown discolouration of bracts while bracteoles remain green, with dark spores visible under low magnification, primarily on shaded cones (Salmon & Ware, 1936a, b). The disease cycle details are not well-documented, but spores are spread by wind and water, infecting cones in shaded, humid conditions. Good air circulation and reducing humidity in hop yards can help prevent its development. Although the disease causes minimal damage, accurate diagnosis is crucial to avoid unnecessary treatments if black mould is mistaken for downy mildew.
- *Black Wilt* (*Diplodia seriata*) was first observed in August 2012 in Seneca Castle, New York, affecting 10 to 20% of the plants. Symptoms included dark stem discolouration, wilted leaves that remained attached, and dark brown to black pycnidia aggregated in the cortical tissue of affected stems. Although the disease has been confirmed through pathogenicity assays and gene sequencing, it currently appears to be a minor issue in New York State, possibly due to the long incubation period of the fungus which may limit symptom development on annually produced hop shoots (Twomey et al., 2016).
- *Colletotrichum fioriniae* has recently been reported to infect hop leaves and cones in the eastern United States (i.e., Michigan). In that report, the incidence and severity of the disease on leaves were approximately 20% and 5 to 10%, respectively (Hatlen et al., 2023). After incubation, lesions produced acervuli with orange spore masses. Cultures of the pathogen were identified as *Colletotrichum fioriniae* based on morphological characteristics and DNA sequencing. Pathogenicity tests confirmed the isolate's ability to cause similar lesions on 'Chinook' hop plants. Further investigation is needed to determine if management is warranted.
- *Crown Gall*, caused by the bacterium *Agrobacterium tumefaciens*, and has been reported on nearly every country where hops are commercially grown (Hoerner, 1945; Chen & Xiang, 1987). The disease manifests as fleshy to hard tumours (galls) on bines at or near the soil surface close to the crown and roots, leading to weak bine growth, wilting of affected bines, and, in severe cases, plant death. Susceptible hop cultivars include Fuggle, Late Cluster, and Southern Brewer. Crown gall is particularly damaging in nurseries and young plants, though older plants can be affected without obvious symptoms. Management involves harvesting softwood cuttings and rhizomes only from plants free of the crown gall bacterium, as no special disease management strategies are typically needed for crown gall in established plants.

- *Hop Shoot Proliferation*, This disease was first observed in Poland in early spring 2003 (Solarska et al., 2004). The disease, characterised by severe shoot proliferation, was confirmed through polymerase chain reaction (PCR), identifying the presence of a phytoplasma belonging to the 16SrI-B subgroup. Given the presence of phytoplasma in both symptomatic and asymptomatic plants, it is crucial to further investigate its etiological role and determine effective control measures to prevent potential spread.
- *Leaf Spot by Bipolaris sorokiniana*, A leaf spot and blight on common hop (*Humulus lupulus*), commercially grown for brewing, were observed in South Korea, caused by the fungus *Bipolaris sorokiniana*. Symptoms included dark-brown necrotic lesions with yellow halos on the leaves. In this case, control was required and fungicides such as fluxapyroxad, pyraclostrobin, and hexaconazole were effective in managing the disease (Liu et al., 2023).
- *Leaf Spot by Cercospora apii*, Reported within Brazil the symptoms include small, subcircular to irregular spots with a whitish to greyish centre and dark brown necrotic margins, which can coalesce into larger lesions causing extensive necrosis and stem dieback. The conidia of *Cercospora apii* sensu lato 'sp. Q' are acicular to filiform, measuring 47 to 210 μm in length and 2.5 to 5 μm in width. Given the severe damage observed, control measures may be essential to prevent this disease from becoming a significant threat to hop in Brazil (Pereira et al., 2023).
- *Rhizoctonia solani* causes sunken, brick-red to black lesions on young shoots, leading to stunted growth and potential collapse, particularly observed on the cultivar Brewer's Gold in British Columbia and North Carolina. This disease has been recently reported as causing a girdling of newly emerging shoots of cultivar Eureka in southern Idaho (Woodhall et al., 2020). The pathogen survives in soil and plant debris, infecting plants when conditions favour its growth, often exacerbated by agricultural practices like hilling soil immediately after spring crowning. Management includes avoiding such practices, managing soil moisture, and improving drainage to reduce infection risk.
- *Septoria Leaf Spot* has been noted to occur on hop in tropical or subtropical regions (e.g. India) but it is not prevalent in other production regions (Puttoo et al., 1975). It is caused by *Septoria humuli* but there has been some nomenclature confusion since it was initially described (Braun & Bensch, 2017). Lesions are initially grey and circular to irregular with dark brown margins and small black pycnidia will eventually become apparent within the lesion. Conidia are slight filiform, hyaline and septate measuring 37–53 × 1.3–2.5 μm. Control is rarely required for this disease as powdery mildew treatments likely provide some level of control in production fields.

5.3 Virus and Viroid Diseases

5.3.1 Apple Mosaic Virus

Apple mosaic virus (ApMV) (*Bromoviridae; Ilarvirus*) has a broad host range, from apples (*Malus* spp.), stone fruits (*Prunus* spp.), roses (*Rosa* spp.), and hop. It was first reported in the latter in the 1960s by Bock (1966, 1967) who identified a virus serologically related to ApMV and to prunus necrotic ringspot virus (PNRSV); subsequent studies have shown that while the hop infecting Ilarviruses are diverse, they are more closely related to ApMV than PNRSV (Sano et al., 1985; Smith & Skotland, 1986; Crowle et al., 2003). ApMV is found in all hop producing areas of the world, from Europe to Asia, Oceania, and the Americas.

Foliar symptoms of ApMV infection in hop are expressed as chlorotic ringspots, which may become necrotic as the growing season progresses, or oak-leaf patterns (Fig. 5.26). These symptoms are often scatted around the bine, with either localised clusters of symptomatic leaves, or randomly based on virus accumulation; foliar symptom expression and severity is cultivar specific, with many hop lines maintaining asymptomatic infections through their lives, or symptoms may only appear several seasons after infection or under ideal environmental conditions (Pethybridge et al., 2008).

Fig. 5.26 Chlorotic rings on hop leaves infected by apple mosaic virus. (Photo: Sebastjan Radišek)

ApMV also impacts the growth and productivity of infected bines, particularly when infected early in their lifespan (Fig. 5.27), reducing bine height by 10–20%, and cone mass and yield by up to 36% (Pethybridge et al., 2002a; Matsui et al., 2017). Finally, ApMV infection reduces alpha-acids content in some cultivars by approximately 10%, as well as decreasing monoterpenes and increasing sesquiterpene content (Pethybridge et al., 2002a; Matsui et al., 2017).

Propagation of ApMV infected plants is the primary means of introduction into hop plantings, and while this virus is pollen transmitted in other plant hosts, this does not appear to be the case in *Humulus* sp. (Pethybridge & Madden, 2003; Pethybridge et al., 2008). In-field spread is by mechanical means, either transferring virus-containing sap from plant to plant on cutting tools, and through plant-to-plant contact by hop shoot intertwining, or root grafting between neighbouring plants in a hop yard (Pethybridge et al., 2002b). As such, sanitary best practices in a hop yard (described in Sect. 5.3.9) combined with the propagation of ApMV-free hop plants (Nukari et al., 2013) can effectively manage this pathogen.

Fig. 5.27 Comparison of the growth of a young healthy 'Comet' bine (left) versus bines of the same cultivar infected with apple mosaic virus (right). (Photo: S. Harper)

5.3.2 *Arabis Mosaic Virus*

Arabis mosaic virus (ArMV) (*Secoviridae; Nepovirus*) has been identified as the causal agent of hop split leaf blotch, nettlehead, spidery hop and hop chlorotic diseases (Bock, 1966; Thresh et al., 1972; Pethybridge et al., 2008). These diseases are cultivar specific but share commonality in the symptoms presented, with shortened internodes and stunting, weak or curled bine growth that may fall away from the string (nettlehead, spidery hop and barebine). Foliar symptoms are more distinguishable, with upward leaf curling (nettlehead), small dark green leaves (barebine), chlorotic interveinal blotches and splitting of the leaf blades towards the leaf base (split leaf blotch), and chlorosis and distortion primarily on the lower leaves of a bine (Legg, 1959; Bock, 1966; Thresh et al., 1972; Pethybridge et al., 2008; Hay & Barbara, 2009). What role the hop cultivar versus the infecting ArMV strain plays in determining disease expression remains unknown, as in the U.K. infecting ArMV isolates were found to be serologically indistinguishable regardless of the symptoms presented (Adams et al., 1987). While Davies and Clark (1983) identified a putative satellite RNA that may be associated with disease expression and severity, to date there have been no molecular studies to reexamine this question in hops.

In hop, ArMV is primarily found in Europe, South Africa, and New Zealand, but rare or absent in Australia, Japan, and the United States (Hay et al., 1992; Pethybridge et al., 2008). Once introduced into a region through trade or plant movement, local spread relies on the presence of the dagger nematode (*Xiphinema diversicaudatum*) (Valdez et al., 1974). In an infected hop yard, spread from infected bines to their neighbors is slow as nematode movement is limited within a year, but it is nonetheless difficult to control as ArMV has a wide host range and *X. diversicaudatum* can remain infective for between 5–8 months (Harrison & Winslow, 1961; Pethybridge et al., 2008; Hay & Barbara, 2009). Rogueing, soil fumigation, and a minimum of two years of leaving the field fallow before replanting with ArMV free hops present the best option for control of this virus (Pitcher & McNamara, 1973). ArMV is mechanically transmissible, though it has been found to be an inefficient means of transmission in hop and can be controlled by basic virus sanitary practices (Pethybridge et al., 2008). It is also seed transmissible (Eppler, 1992), requiring rogueing of volunteer or male plants.

5.3.3 *Hop-Infecting Carlaviruses*

There are three viruses from the genus *Carlavirus* (Fam. *Betaflexiviridae*) that infect hop, these being hop mosaic virus (HpMV), hop latent virus (HpLV), and American hop latent virus (AHLV). The first two are present in most, if not all major hop growing regions of the world, while AHLV is rarer; while it is common in the U.S., it has been imported into and in some cases, become established in other countries (Adams & Barbara, 1982; Hay et al., 1992; Pethybridge et al., 2008; Dann et al.,

2014). HpMV and HpLV are closely genetically and serologically related to one another, while AHLV is more distant (Probasco & Skotland, 1978; Adams & Barbara, 1982; Hataya et al., 2000; Eastwell & Druffel, 2012).

Of the three viruses, HpMV is the most likely to cause observable symptoms and crop damage. On sensitive cultivars, chlorotic mosaic can develop between major leaf veins. Severely affected plants may establish poorly when planted, have weak bine growth, and often fail to attach to the string, followed by premature plant decline and death. Cultivars with Golding-type parentage are most susceptible to this virus, although this is suspected to be virus-strain specific. Cone yields and weights can be reduced by up to ~50% in cultivars sensitive to this virus as well as reductions in alpha- and beta-acids of up to ~20% (Probasco & Murphey, 1996; Pethybridge et al., 2000, 2002a, 2008).

HpLV is largely latent on most cultivars, though has been reported to cause chlorotic flecking on Cluster and Hersbrucker Spat, but not on cultivars from the Golding or Wye lineages which remain asymptomatic (Probasco & Skotland, 1978; Adams & Barbara, 1982; Pethybridge et al., 2008). HpLV has also been associated with reduced growth, shorter internodes and laterals, and fewer leaves on some cultivars, and yield reductions of between 10–70% (Probasco & Murphey, 1996; Pethybridge et al., 2002a), with corresponding reductions in alpha-acids. No diagnosable symptoms have been reported for AHLV on any cultivars, though it, along with HpLV may synergistically increase loss when in combination with more pathogenic viruses (Pethybridge et al., 2002a).

The primary means of spread of these viruses, other than movement of infected planting material, is via aphid transmission. All three viruses are transmitted by the hop aphid (*Phorodon humuli*) (Adams & Barbara, 1982), while HpMV and HpLV are also transmitted in the field by *Macrosiphum euphorbiae,* the potato aphid, and *Myzus persicae*, the green peach aphid, a species that is prolific and highly mobile (Crowle et al., 2006). In all cases, transmission is non-persistent, which means that the aphid can rapidly acquire and transmit the virus, but as the virus is not retained for long in the aphid's stylet, subsequent transmission to other plants it feeds on is inefficient or does not occur at all, unless it reacquires the virus from another infected plant. All three viruses have limited host ranges (Adams & Barbara, 1982), although incidental infection of weeds, garden plants, and other crops has been observed for HpLV due to feeding by virus-carrying aphid vectors, but alternative hosts are not considered to contribute to the spread of this virus (Pethybridge et al., 2008). There is currently no evidence of seed or pollen transmission for any of these viruses, but in field-spread may occur through mechanical transfer of virus-containing sap on cutting or pruning tools, through injury and contact between neighbouring plants, or by root grafting (Pethybridge et al., 2008). Standard phytosanitary practices of rouging, vector control, and use of clean propagative materials are applied for the control of these viruses (detailed in Sect. 5.3.8).

5.3.4 Hop Stunt Viroid (HSVd)

Hop stunt disease was first viroid observed in hop plants in Japan in the mid-twentieth century, characterised by dwarfing and low alpha-acids content in cones. Studies were helped by infected cucumber plants which exhibited symptoms much earlier than hop plants, suggesting a viroid as the causative agent (Sasaki & Shikata, 1977). The complete 297 bp sequence of the viroid was determined in 1983 (Ohno et al., 1983).

HSVd belongs to the genus *Hostuviroid*, family Pospiviroidae. It is one of the most variable viroid species in terms of host range and sequence variability. Currently, the National Center for Biotechnology Information (NCBI) database holds 1455 HSVd records, second only to *Potato spindle tuber viroid* (PSTVd). Shortly after the discovery of the hop variant, HSVd isolates were found in cucumber, grapevine, and citrus. HSVd has a broad host range, primarily affecting fruits (for a full list see (Hataya et al., 2017)) and is distributed worldwide. While its presence is often symptomless in many hosts, HSVd variants can cause additional diseases in sensitive hosts, such as cachexia disease in citrus, dapple fruit disease in plum and peach, and cucumber pale fruit (Marquez-Molins et al., 2021).

The epidemiology of hop stunt disease was extensively studied in Japan on two main Japanese hop cultivars. Symptoms in hop plants typically develop several years after infection and include stunted growth, curled leaves, bent leaves, and smaller cones. HSVd was confirmed in Korea in 1988 in a Japanese hop cultivar, likely introduced through the import of infected material (Lee et al., 1988). A 2004 survey in Washington State, USA, detected HSVd in eight out of 13 sampled hop cultivars, suggesting that its presence had gone unnoticed for an extended period. Symptoms in the cultivar 'Glacier' included bright yellow leaves, moderate epinasty, and delayed climbing. Older leaves developed pale yellow speckling along major veins (Eastwell & Nelson, 2007). In 2017, HSVd was confirmed in Ohio in plants of the Cascade variety, showing characteristic symptoms (Han et al., 2019). In 2008, HSVd was detected in the primary hop production area in China across three hop cultivars, with a high incidence in one cultivar (Guo et al., 2008). A survey of a severe stunting phenomena in Slovenia in 2012 confirmed its presence, though it was not linked to the etiology of the disease (Radisek et al., 2012). The latest report comes from Australia, where HSVd was found in a hop cultivar showing symptoms of chlorosis, downward-curled leaves, and distorted leaf growth (Chambers et al., 2021).

5.3.5 Hop Latent Viroid (HLVd)

Hop latent viroid (HLVd) was the second viroid discovered in hop. Evidence that hop could host an additional viroid emerged in 1987 when viroid-like RNA was detected in two hop cultivars grown in Spain. Nucleic acid dot-blot analysis did not reveal hybridizations to four different viroids, including HSVd (Pallas et al., 1987). A year later, its 256 bp sequence was deduced and named hop latent viroid due to its presumed latent nature, exhibiting no visible symptoms (Puchta et al., 1988). Later, HLVd was detected in all hop-growing regions, infecting the majority of hop plants (Faggioli et al., 2017).

HLVd belongs to the genus *Cocadviroid*, family Pospiviroidae. While HSVd viroid is highly diverse in terms of sequence variability, HLVd sequences are less variable (Atallah et al., 2024). For a long period, HLVd was considered a viroid only associated with hops. However, it was also reported in Japanese hop (*Humulus japonicus*) and stinging nettle (*Urtica dioica*), both species taxonomically related to hops (Knabel et al., 1999). Recent confirmation of HLVd in *Cannabis* greenhouse production in the US and Canada (Warren et al., 2019; Punja et al., 2024) describes it as a major threat to the cannabis industry, causing dudding disease (Atallah et al., 2024; Adkar-Purushothama et al., 2023).

HLVd in hops is not transmissible through seeds, or it has a very low level of possible transmission (Matoušek & Patzak, 2000). However, in *Cannabis*, seed transmission rates of up to 80% have been described, with new experimental hosts including tomato, cucumber, chrysanthemum, *Nicotiana benthamiana*, and *Arabidopsis thaliana* (Atallah et al., 2024).

Since HLVd is so widely distributed in commercial hop production and is considered latent in hop, not many precautions were taken in studying its epidemiology. However, it has been reported to cause symptoms in specific hop cultivars, contributing to lower cone yield, reduction of alpha-acids, increase of beta-acids, and changes in the essential oil composition (Barbara et al., 1990; Patzak et al., 2021). Global transcriptome analysis revealed its ability to change the expression of certain hop transcripts (Štajner et al., 2019).

5.3.6 *Citrus Bark Cracking Viroid (CBCVd)*

Citrus bark cracking viroid (CBCVd) is the latest viroid entity discovered to infect hop plants, causing devastating symptoms. Originally named Citrus viroid IV (CVd-IV) in 1988, it was identified as the fourth viroid species in citron (*Citrus medica*). CBCVd exhibits distinct electrophoretic mobility and no crossover hybridization with three other citrus viroid species or Citrus exocortis viroid (CEVd) (Duran-Vila et al., 1988). Its 284 bp sequence was first identified in 1991 from a dwarf grapefruit plant (*Citrus paradisi*) (Puchta et al., 1991), where its chimeric nature of CEVd and HSVd was proposed.

In hop, CBCVd was confirmed in Slovenia in 2015 through high-throughput sequencing, although symptoms were first reported as early as 2007 (Jakse et al., 2015). Subsequent confirmations of CBCVd occurred in Germany in 2019 (EPPO, 2019) and in Brazil in 2023 (Eiras et al., 2023). According to the latest taxonomy, CBCVd belongs to the same genus and family as HLVd, *Cocadviroid*, Pospiviroidae. Despite being less well-studied in citrus, as of May 2023, there are 124 CBCVd sequences deposited in the NCBI nucleotide database, with 16 of them originating from hop. The viroid's name reflects the bark cracking symptoms it induces in trifoliate oranges (Vernière et al., 2004, 2006).

Until recently, known natural hosts of CBCVd were various species and interspecies of the genus *Citrus*. In 2015, CBCVd was confirmed in hop and in 2018 in pistachio (*Pistacia vera*). There are also experimental hosts, both symptomatic and asymptomatic, including ornamentals, vegetables, and weeds (for a full list see EPPO, 2024).

CBCVd causes the most severe symptoms in hop of all viroids, including plant stunting, leaf yellowing, leaf down-curling, smaller cones, dry root rot, and longitudinal bine cracking (Jakse et al., 2015) (Fig. 5.28). Some symptoms resemble those of HSVd infection but are more intense and progress more rapidly. Typically, the first year of infection is symptomless, with symptoms developing after the dormancy phase, usually observed at the end of May or June in the northern hemisphere. Symptom severity is cultivar-dependent and more pronounced in hot weather conditions, with plants typically dying 3 to 5 years after infection (Radišek et al., 2017).

Fig. 5.28 Symptoms of severe hop stunt disease, caused by Citrus bark cracking viroid (CBCVd), on the susceptible hop cultivar Celeia: (**a**) yellowing leaves and stunted growth of lateral bines; (**b**) stunted plants before harvesting; (**c**) extensive bine cracking; (**d**) dry root rot; (**e**) smaller cones from infected plants (right) compared to healthy ones (left). (Photo: Sebastjan Radišek)

The transmission of CBCVd to hop likely occurred through biological citrus waste, as confirmed in experimental conditions (Hagemann et al., 2023). Therefore, the use of citrus biological waste or introducing citrus fruit into hop fields by any means should be avoided. Currently, single CBCVd-infected hop plants are not found in field conditions; CBCVd infection is always accompanied by HLVd, with the severity of the disease attributed to the synergistic effect of the two viroids (Štajner et al., 2019).

5.3.7 *Apple Fruit Crinkle Viroid (AFCVd)*

Apple fruit crinkle disease was first discovered in Japan in 1976 on apple trees for which a graft transmission agent was confirmed and also etiology of the disease suggested a viroid agent. Viroid-like RNA was confirmed and isolated in 1993 (Ito et al., 1993). In 2004, the AFCVd isolates from hop plants with 93–98% sequence homology with apple isolates were detected in Japan, and a survey suggested the viroid presence in hop as early as in 1980s (Sano et al., 2004).

The latest criteria for viroid taxonomy, based on pairwise identity scores, place apple fruit crinkle viroid in an unclassified viroid group due to its similarity with Australian grapevine viroid (AGVd) (Chiumenti et al., 2021). AFCVd has a narrow host range and has been found in apple, hop, and persimmon. Artificial infection studies have confirmed that it can also infect tomato and cucumber (Suzuki et al., 2017).

AFCVd is primarily restricted and reported in Japan. In hop, the symptoms are similar to those caused by hop stunt viroid but are slightly more severe (Fig. 5.29). Infected plants exhibit abnormal dwarfing and severe leaf curling in the upper part of the vines (Sano et al., 2004).

5.3.8 *Viruses and Virus-Like Agents of Minor Importance*

Over the past fifty years, other viruses beyond those described above have been detected and reported as being present in hop. Some are pathogenic, while others are not and many of these are localised or isolated reports of infection, likely representing opportunistic or vector-mediated transmissions from other crops or free-living plants in the region (Pethybridge et al., 2008). Also, given the age of some of these reports and the methodology in use at the time, caution is advised with both virus identification and causal association with any symptoms reported.

For example, alfalfa mosaic virus has been reported from both China and the former Czechoslovakia, causing systemic chlorotic spots on leaves (Novák & Lanzová, 1976; Yu & Liu, 1987). Cucumber mosaic virus was observed across

Fig. 5.29 Symptoms associated with AFCVd infection: (**a**) wild hop clone (*H. lupulus* var. *cordifolius*, Souma 1) inoculated with AFCVd and HSVd; (**b**) severe leaf curling of infected plant. (Photo: Teruo Sano)

much of Europe, in association with foliar distortion, mottling and necrosis, while tobacco necrosis virus and tobacco ringspot virus have been associated with local necrotic lesions and foliar ringspot symptoms in commercial hop (Chod et al., 1979; Macovei, 1980; Pethybridge et al., 2008). Members of the Tombusviridae, specifically tomato bushy stunt virus and petunia asteroid mosaic virus have been reported in hop in Europe. These have been associated with a broad range of foliar symptoms, including chlorosis, ring- and -line patterns, deformation and necrosis (Novák & Lanzová, 1976; Pethybridge et al., 2008).

Humulus japonicus latent virus was found in both common and Japanese hop in China and imported into the UK but have been reported to be symptomless (Adams et al., 1989). Finally, cherry leafroll virus, raspberry bushy dwarf virus, strawberry latent virus, and tobacco mosaic virus all exist only as isolated reports from the UK, the former Czechoslovakia, and China respectively, with no clear disease associations presented (Polák & Svoboda, 1989; Pethybridge et al., 2008) and are likely incidental and of no management or quarantine significance.

5.3.9 *Management of Viruses and Viroids*

Management of viruses and viroids infecting commercially grown hop follows the basic principles of virus management: prevention, identification and removal, and mitigation. One of the primary routes of hop viruses and viroids entering a hop yard is through infected propagative material, therefore for over 80 years countries across Europe, the Americas and Australasia have developed and operated clean stock programs. The programs acquire important hop germplasm, test the plants for viruses, viroids, and other systemically infecting pathogens, and eliminate pathogens found by tissue culture, thermotherapy, cryotherapy, or chemotherapy approaches. The clean stock material is then held under protected conditions and distributed to propagators for multiplication and sale to hop growers (Pethybridge et al., 2008; Davis et al., 2021; Lavagi-Craddock et al., 2024). At the local level, avoiding propagation of bines from established hop yards known to be infected, or from those of unknown infections status is recommended; it should not be assumed that propagules from infected bines will grow slower than healthy ones or exhibit obvious symptoms at a young age (Pethybridge et al., 2008). Where possible, bines should be spaced and propagated to prevent plant-to-plant contact, and using contact herbicides rather than mechanical pruning to control basal growth may reduce mechanical transmission of viruses between adjacent plants.

Control of vectors can reduce the risk of introduction and spread. However, application of insecticides to control aphids that transmit the three hop-infecting carlaviruses is of limited efficacy because the virus will be transmitted before the viruliferous aphids are by systemic insecticides, and as these aphids have a wide host range, targeted sprays are ineffective. However, reducing local aphid populations using contact or systemic insecticides can reduce the rate of secondary transmission within a hop yard. Nematode-transmitted viruses present a particular challenge for management. Sites known to possess a nepovirus such as ArMV or TRSV, and their respective nematode vector must be herbicide treated to eliminate infected bines and alternative hosts such as broadleaf weeds and fumigated to reduce nematode numbers (Harrison & Winslow, 1961; Hay & Barbara, 2009). Fallow periods with host plant control are employed to purge the nematode population of the virus, although this has not been effective in all crops (Demangeat et al., 2005; Hay & Barbara, 2009).

Bines suspected to be infected with a virus or viroid should be identified and removed, with herbicide applications to ensure death of infected rhizome material to reduce the risk of volunteer plants in subsequent seasons. Finally, mechanical transmission of hop-infecting viruses and various may be managed by spacing bines to reduce contact and rubbing injury or root grafting between neighbours. Tools and equipment should be sanitised to remove viruliferous sap between use, ideally between individual bines, and specifically between hop yards to reduce the risk of spreading or inducing infections.

5.4 Nematode Diseases

Up to 12 genera of plant-parasitic nematodes have been reported in hop, including *Criconemella, Helicotylenchus, Heterodera humuli, Meloidogyne, Paratylenchus, Pratylenchus, Paratrichodorus,* and *Tylenchorhynchus* (Malan et al., 1991; Hay & Pethybridge, 2003; Watson et al., 2020; Núñez et al., 2022). Of these, the most frequently reported plant-parasitic nematode across all hop-producing countries is *Heterodera humuli*, also known as the hop cyst nematode (Darling et al., 2023).

5.4.1 *Hop Cyst Nematode*

5.4.1.1 Taxonomy

Cysts of *Heterodera* spp. are commonly lemon-shaped with a vulval cone, which makes them different from *Globodera*, *Dolichodera*, and *Paradolichodera*. The genus *Heterodera* has been divided into nine groups: *Afenestrata*, *Avenae*, *Bifenestra*, *Cardiolata*, *Cyperi*, *Goettingiana*, *Humili*, *Sacchari* and *Schachtii*. *Heterodera humuli* belongs to the *Humuli* group, which is characterised by having a bifenestrated vulval cone and the presence of a weak vulval bridge (Handoo & Subbotin, 2018).

Second-stage juvenile measurements can be used to distinguish between species within the *Humuli* group. For example, J2 stylet length and body length of *H. humuli* are smaller when compared to J2 of *H. litoralis,* 22.5–24 μm vs 29.5 μm and 364–425 μm vs 520 μm, respectively (Subbotin et al., 2010). Molecular techniques have also been used to separate *H. humuli* from other cyst nematode species. Two ribosomal genes, the *28S* and the *ITS-5.8S-ITS2*, and a mitochondrial gene, the *cox1*, have been used to identify *H. humuli* (Madani et al., 2004; Mundo-Ocampo et al., 2008; Sekimoto et al., 2017).

5.4.1.2 Disease Symptoms

In general, there are no specific symptoms caused by plant-parasitic nematodes. However, symptoms such as wilting, stunting, and chlorosis can be observed when plant-parasitic nematodes feed on roots. Only a few studies have been conducted to evaluate the impact of *H. humuli* on the growth and yield of hop. A reduction of up to 68% in root weight of hop inoculated with *H. humuli* was observed relative to non-inoculated plants in a greenhouse study (Sen & Jensen, 1969). In another greenhouse experiment, bine height was reduced by 40% in plant inoculated with *H. humuli* compared with control plants (Darling et al., 2021). Hafez et al. (1999) observed a severe nutrient deficiency in greenhouse-grown hop inoculated with *H. humuli*. In a hop yard in Tasmania, dry hop per bine was reduced 38% in the presence of *H. humuli* (Hay & Pethybridge, 2003).

5.4.1.3 Life Cycle

Heterodera humuli is a sedentary endoparasitic nematode. Starting with the egg (Figs. 5.30 and 5.31) in the spring, the hop plant produces root diffusates necessary for *H. humuli* egg hatch (De Grisse & Gillard, 1963); *H. humuli* has been demonstrated to be highly dependent on root diffusates to stimulate hatch (Turner & Rowe, 2006). The second-stage juvenile (J2) hatches from an egg and penetrates the hop root and moves intracellularly to find a cell near the vascular tissue to establish a feeding site known as a syncytium where the nematode will obtain the required nutrients to complete its life cycle (Fig. 5.30; Palomares-Rius et al., 2017). As the nematode develops through two more juvenile stages and then to an adult female, it swells and ruptures through the root tissue (Fig. 5.31). The male remains wormlike and leaves the root to mate with a female. The female eventually dies and eggs are protected in the dead female body known as a cyst (Fig. 5.31). For some cyst nematodes, encysted eggs can remain viable for more than 15 years (Folkertsma et al., 1997), however viability not been quantified for *H. humuli.* The number of generations that *H. humuli* can complete in a year has been reported to be one or two (Von Mende & McNamara, 1995; Mikhailpukov, 1980).

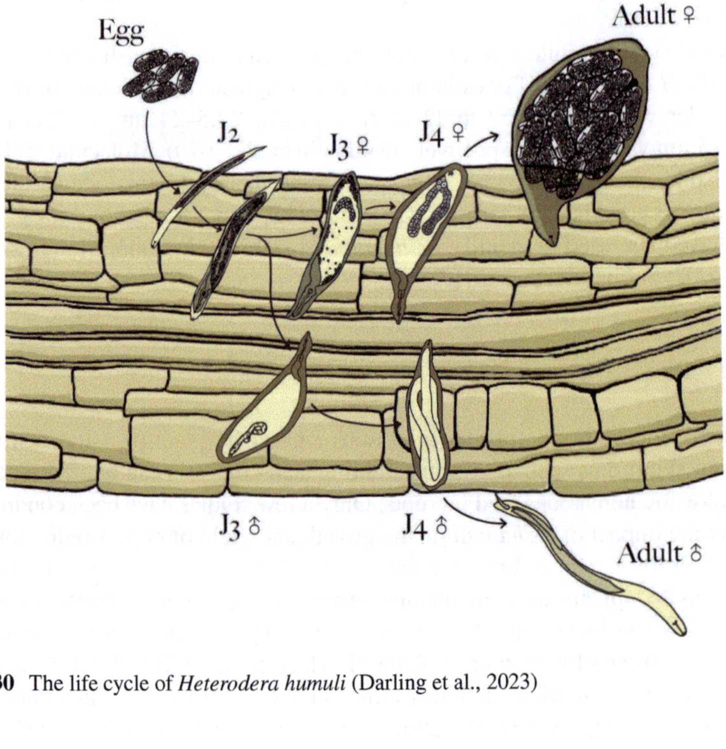

Fig. 5.30 The life cycle of *Heterodera humuli* (Darling et al., 2023)

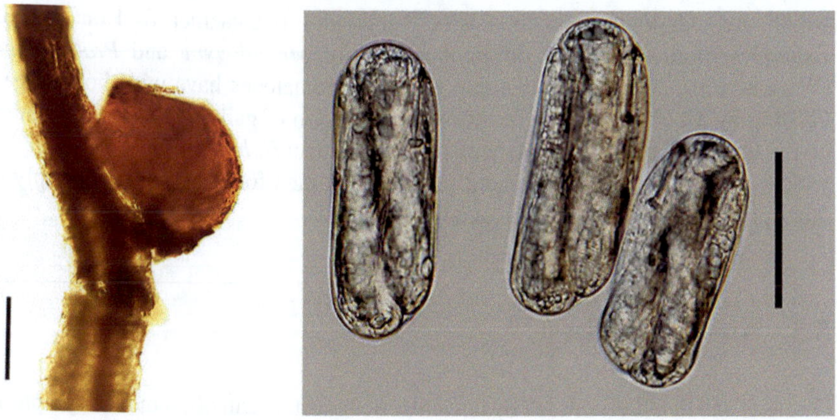

Fig. 5.31 Cyst containing eggs of *H. humuli* on the root surface (scale 200 μm) (left); Eggs of *Heterodera humuli* containing coiled second-stage juveniles (scale 50 μm) (right) (Photo: Lester Núñez-Rodríguez)

5.4.1.4 Disease Management

There is little information available on the control of *H. humuli* in hop yards. Fumigation is a practice that growers use to combat problems caused by soilborne pathogens, including plant-parasitic nematodes. In the USA, the estimated cost for fumigating is ~ US$1,000 per ha (Galinato, 2020). Increasing grower awareness of using safer molecules and discontinuation of hazardous products have resulted in the ban of some of the most effective nematicides (Desaeger et al., 2020), and in the future may impact the use of the molecules currently used for fumigating hop yards. It is highly encouraged that tissue culture plants be used to establish new hop yards to avoid introducing cysts of *H. humuli* to non-infested yards.

5.4.2 *Other Plant-Parasitic Nematodes in Hop*

Several genera of plant-parasitic nematodes can be found in hop yards. Ectoparasitic nematodes, nematodes that feed and develop in soil outside the hop root system (Dracaemer & Hunt, 2006), have been reported in hop including *Belonolaimus longicaudatus*, *Bitylenchus* sp., *Criconemella* sp., *Helicotylenchus dihystera*, *Mesocriconema xenoplax*, *Paratylenchus* spp., *Paratrichodorus minor*, *Tylenchorhynchus clarus*, *Xiphinema* spp., (Malan et al., 1991; Hay & Pethybridge, 2003; Watson et al., 2020; Núñez et al., 2022). Ectoparasitic nematodes are not only important for their potential direct impact when feeding on plant roots, but they can also play an important role in transmitting viruses. *Xiphinema diversicaudatum* is a species that can vector arabis mosaic virus to hop (Valdez et al., 1974). There are also nematodes that penetrate and develop most of their life cycle inside hop roots, which are classified as endoparasitic nematodes (Dracaemer & Hunt, 2006). Examples of these include different species of *Meloidogyne* and *Pratylenchus* (Watson et al., 2020). Both of these groups of nematodes have wide host ranges. Feeding by *Meloidogyne* can result in the formation of galls on roots. Information on the impact of plant-parasitic nematodes other than *H. humuli* on hop growth and yield is not available. Management practices outlined for *H. humuli* also apply to other plant-parasitic nematodes found in hop yards.

5.5 Abiotic Diseases

Diseases and disorders caused by non-infectious factors can often induce symptoms similar to those caused by biotic agents (Lucas et al., 1992). The most common abiotic causes encountered in hop growing are nutrient disorders, pesticide and fertilisers injuries, and unfavourable environmental conditions such as water stress, high temperatures, frost, wind, and hail. Additionally, animals such as voles or other rodents and deer can also cause significant damage.

5.5.1 Nutritional Imbalances

The occurrence of disease-like symptoms can be a result of inadequate or excessive levels of nutrients in the soil. Often, the availability of individual nutrients is influenced by soil properties such as pH, texture, and soil moisture content (Bergmann, 1992). Hop, like other plants, react to nutrient imbalances by expressing specific symptoms (OMAFRA, 2023), but soil and plant tissue analyses are needed to confirm the cause. Hop production requires intensive use of macronutrient fertilisers, as the plants develop abundant biomass in a very short time. However, unbalanced fertilization can lead to deficiencies of certain micronutrients and overstocking, resulting in antagonism and blockage of nutrient uptake. In springtime, deficiencies of boron and zinc are often observed, with some varieties showing a higher degree of sensitivity. A severe boron deficiency is manifested by poor shoot growth and stunting (Fig. 5.32). The leaves are poorly developed, small, and crinkled. Boron deficiency occurs mainly on sandy soils with excessive levels of potassium and calcium. Hop plants are also sensitive to zinc deficiency, which occurs mainly in soils overloaded with phosphorus or in alkaline soils. Plants develop narrow, elongated, curled chlorotic leaves. Shoots develop less vigorously and turn yellow, and overall plant growth is impaired (Fig. 5.33).

Fig. 5.32 Boron deficiency in hop cv. Aurora: (**a**) stunted shoots with small, distorted leaves; (**b**) comparison of normal and affected plants showing poor growth. (Photo: Sebastjan Radišek)

Fig. 5.33 Curled chlorotic leaves and yellow shoots caused by zinc deficiency in hop cv. Aurora. (Photo: Sebastjan Radišek)

5.5.2 Environmental Stress

In spring, hop can be affected by frost, causing damage that is visible on young shoots, which look water-soaked and may decline. Leaves may develop malformations that can resemble downy mildew (Fig. 5.34). On heavy soils, frequent and excessive water can cause waterlogging, affecting the roots and leading to leaf yellowing and wilting. The following year, plants often exhibit poor growth due to root damage. Hail and wind can cause mechanical damage to plants, and in severe cases, this can result in the breakdown of entire trellises. Wind can also damage cones by rubbing them against leaves and stems, causing browning on the exposed side. Similar browning can occur during harvesting if cones are suddenly exposed to direct summer sunlight for several hours, leading to sunburn on the exposed side.

5.5.3 Chemical Damages

Improper use of pesticides and fertilisers can cause phytotoxic responses in plants, which in some cases result in plant death, severe deformations, leaf spots, and leaf burns (Fig. 5.35). Phytotoxicity can result from using products at excessive doses, incompatible mixtures of pesticides and foliar fertilisers, applying them in unfavourable environmental conditions, or drift from other crops. For this reason,

Fig. 5.34 (**a**) Leaf malformations that develop after cold injury; (**b**) affected young sprouts after a spring freeze. (Photo: Sebastjan Radišek)

Fig. 5.35 Pesticide injuries on hop: (**a**) Stunted growth and leaf distortion caused by glyphosate drift from a neighbouring field; (**b**) affected growth of shoots and leaf yellowing after spraying with too high a concentration of metalaxyl-M; (**c**) leaf yellowing and affected growth caused by drift of a corn herbicide containing isoxaflutole and thiencarbazone-methyl; (**d**) phytotoxic reaction on leaves caused by incompatible mixtures of pesticides and foliar fertiliser. (Photo: Sebastjan Radišek)

diagnosing phytotoxicity requires analysis of the grower's spray and fertilizing records, weather conditions, growing stage of plants and examination of neighbouring crops.

Acknowledgment The section on Verticillium wilt and viroids was financially supported by the Slovenian Research Agency, grant no. P4-0077. The authors like to express their gratitude to Prof. Dr. Teruo Sano, Faculty of Agriculture and Life Science, Gene Research Center, Hirosaki University, Japan for figures of viroids AFCVd and HSVd.

References

Adams, A. N., & Barbara, D. J. (1982). Host range, purification and some properties of two carlaviruses from hop (*Humulus lupulus*): Hop latent and American hop latent. *Annals of Applied Biology, 101*(3), 483–494.

Adams, A. N., Barbara, D. J., & Vies, D. D. (1987). The etiology of hop chlorotic disease. *Annals of Applied Biology, 111*(2), 365–371.

Adams, A. N., Clark, M. F., & Barbara, D. J. (1989). Host range, purification and some properties of a new ilarvirus from *Humulus japonicus*. *Annals of Applied Biology, 114*(3), 497–508.

Adkar-Purushothama, C. R., Sano, T., & Perreault, J. P. (2023). Hop latent viroid: A hidden threat to the cannabis industry. *Viruses, 15*(3), 681. https://doi.org/10.3390/v15030681

Allan-Perkins, E., Li, D. W., Schultes, N., Yavuz, S., & LaMondia, J. (2020). The identification of a new species, *Diaporthe humulicola*, a pathogen causing diaporthe leaf spot on common hop. *Plant Disease, 104*(9), 2377–2390.

Arens, K. (1929). Untersuchungen über *Pseudoperonospora humuli* (Miyabe u. Takah.), den Erreger der neuen Hopfenkrankheit. *Phytopathologische Zeitschrift, 1*, 169–193.

Atallah, O. O., Yassin, S. M., & Verchot, J. (2024). New insights into hop latent viroid detection, infectivity, host range, and transmission. *Viruses, 16*(*1*), 30. https://doi.org/10.3390/v16010030

Augusto Moretti Ferreira Pinto, F., Araujo, L., de Andrade, C. C. L., Mendes Fagherazzi, M., Fontanella Brighenti, A., Schlichting de Martin, M., Gomes, L. B., Fernandes, J., Duarte, V., João Arioli, C., & Sommer, V. B. (2022). First report of *Fusarium meridionale* causing canker in hop plants. *Australasian Plant Disease Notes, 17*(1), 13.

Barbara, D. J., Morton, A., Adams, A. N., & Green, C. P. (1990). Some effects of hop latent viroid on two cultivars of hop (*Humulus lupulus*) in the UK. *Annals of Applied Biology, 117*(2), 359–366.

Beal, E. J., Waghorn, I. A. G., Perry, J. N., Clover, G. R. G., & Cromey, M. G. (2021). Susceptibility of garden plants to Phytophthora root rot. *Plant Disease, 105*(6), 1610–1620.

Bello, J. C., Hausbeck, M. K., & Sakalidis, M. L. (2021). Application of target enrichment sequencing for population genetic analyses of the obligate plant pathogens *Pseudoperonospora cubensis* and *P. humuli* in Michigan. *Molecular Plant-Microbe Interactions, 34*(10), 1103–1118.

Bergmann, W. (1992). *Nutritional disorders of plants: developments, visual and analytical diagnosis.* .

Bezuidenhout, C. M., Denman, S., Kirk, S. A., Botha, W. J., Mostert, L., & McLeod, A. (2010). Phytophthora taxa associated with cultivated *Agathosma*, with emphasis on the *P. citricola* complex and *P. capensis* sp. nov. *Persoonia, 25*(1), 32–49.

Bienapfl, J. C. (2003). *Hop cone tip blight: A new disease in the Pacific Northwest* [Master's Thesis, Oregon State University]. Oregon State University ScholarsArchive. https://ir.library.oregonstate.edu/concern/graduate_thesis_or_dissertations/vh53wz908

Bienapfl, J. C., Ocamb, C. M., Klein, R., & Nelson, M. (2005). Fusarium cone tip blight of Humulus lupulus. *Acta Horticulturae, 668*, 123–128.

Bock, K. R. (1966). Arabis mosaic and Prunus necrotic ringspot viruses in hop (*Humulus lupulus* L.). *Annals of Applied Biology, 57*(1), 131–140.

Bock, K. R. (1967). Strains of Prunus necrotic ringspot virus in hop (*Humulus lupulus* L.). *Annals of Applied Biology, 59*(3), 437–446.

Boerema, G. H., de Gruyter, J., Noordeloos, M. E., & Hamers, M. E. C. (2004). *Phoma identification manual. Differentiation of specific and infra-specific taxa in culture*. CABI.

Bradshaw, M. J., Boufford, D., Braun, U., Moparthi, S., Jellings, K., Maust, A., Pandey, B., Slack, S., & Pfister, D. H. (2024). An in-depth evaluation of powdery mildew hosts reveals one of the world's most common and widespread groups of fungal plant pathogens. *Plant Disease, 108*(3), 576–581.

Braun, U., & Bensch, K. (2017). The nomenclature of *Septoria humuli* – A case of confusion and misinterpretation. *Schlechtendalia, 32*, 79–80.

Braun, U., & Takamatsu, S. (2000). Phylogeny of Erysiphe, Microsphaera, Uncinula (Erysipheae) and Cystotheca, Podosphaera, Sphaerotheca (Cystotheceae) inferred from rDNA ITS sequences - Some taxonomic consequences. *Schlechtendalia, 4*, 1–33.

Brazee, N. J., Yang, X., & Hong, C. X. (2017). *Phytophthora caryae* sp. nov., a new species recovered from streams and rivers in the eastern United States. *Plant Pathology, 66*(5), 805–817.

Bressman, E. N., & Nichols, R. A. (1933). Germination of the oospores of *Pseudoperonospora humuli*. *Phytopathology, 23*(5), 485–487.

Chambers, G. A., Dodds, K., & Donovan, N. J. (2021). Hop stunt viroid detection in hops (*Humulus lupulus*) in Australia. *Australasian Plant Disease Notes, 16*(1), 3. https://doi.org/10.1007/s13314-021-00419-x

Chee, H., & Klein, R. (1998). Laboratory production of oospores in *Pseudoperonospora humuli*. *Korean Journal of Plant Pathology, 14*(6), 618–621.

Chen, X. Y., & Xiang, W. N. (1987). Studies on *Agrobacterium tumefaciens* isolated from hops in China. *Acta Microbiologica Sinica, 27*(1), 10–16.

Chen, Q., Jiang, J. R., Zhang, G. Z., Cai, L., & Crous, P. W. (2015). Resolving the *Phoma enigma*. *Studies in Mycology, 82*, 137–217.

Chiumenti, M., Navarro, B., Candresse, T., Flores, R., & Di Serio, F. (2021). Reassessing species demarcation criteria in viroid taxonomy by pairwise identity matrices. *Virus Evolution, 7*(1), Article veab001. https://doi.org/10.1093/ve/veab001

Chod, J., Jokeš, M., & Novak, M. (1979). The electron-microscopic proof of tobacco necrosis virus in hop plant. *Biologia Plantarum, 21*(2), 152–153.

Choi, Y. J., Hong, S. B., & Shin, H. D. (2005). A re-consideration of *Pseudoperonospora cubensis* and *P. humuli* based on molecular and morphological data. *Mycological Research, 109*(7), 841–848.

Christie, T. (1956). Black root-rot, *Phytophthora cactorum* (Leb. and Cohn) Schroet, of the hop plant. *New Zealand Journal of Science and Technology, 38*, 277–284.

Coley-Smith, J. R. (1962). Overwintering of hop downy mildew *Pseudoperonospora humuli* (Miy. and Tak.) Wilson. *Annals of Applied Biology, 50*(2), 235–243.

Coley-Smith, J. R. (1964). Persistence and identification of downy mildew *Pseudoperonospora humuli* (Miy. and Tak.) Wilson in hop rootstocks. *Annals of Applied Biology, 53*(1), 129–132.

Coley-Smith, J. R. (1966). Early-season control of hop downy mildew, *Pseudoperonospora humuli* (Miy. and Tak.) Wilson, with streptomycin and protectant fungicides in severely infected plantings. *Annals of Applied Biology, 57*(2), 183–191.

Crandall, S. G., Ramon, M. L., Burkhardt, A. K., Bello Rodriguez, J. C., Adair, N., Gent, D. H., Hausbeck, M. K., Quesada-Ocampo, L. M., & Martin, F. N. (2021). A multiplex TaqMan qPCR assay for detection and quantification of Clade 1 and Clade 2 isolates of *Pseudoperonospora cubensis* and *Pseudoperonospora humuli*. *Plant Disease, 105*(10), 3154–3161.

Cregeen, S., Radisek, S., Mandelc, S., Turk, B., Stajner, N., Jakse, J., & Javornik, B. (2015). Different gene expressions of resistant and susceptible hop cultivars in response to infection with a highly aggressive strain of *Verticillium albo-atrum*. *Plant Molecular Biology Reporter, 33*(3), 689–704. https://doi.org/10.1007/s11105-014-0767-4

Crous, P. W., Braun, U., Hunter, G. C., Wingfield, M. J., Verkley, G. J. M., Shin, H. D., Nakashima, C., & Groenewald, J. Z. (2013). Phylogenetic lineages in Pseudocercospora. *Studies in Mycology, 75*(1), 37–114. https://doi.org/10.3114/sim0005

Crowle, D. R., Pethybridge, S. J., Leggett, G. W., Sherriff, L. J., & Wilson, C. R. (2003). Diversity of the coat protein-coding region among Ilarvirus isolates infecting hop in Australia. *Plant Pathology, 52*(5), 655–662.

Crowle, D. R., Pethybridge, S. J., & Wilson, C. R. (2006). Transmission of hop latent and hop mosaic carlaviruses by *Macrosiphum euphorbiae* and *Myzus persicae*. *Journal of Phytopathology, 154*(11–12), 745–747.

Dann, A. L., Hössel, S. E., Cross, P. A., & Whittock, S. P. (2014). Phylogenetic analysis of American hop latent virus in Humulus lupulus surveyed in Tasmania and Victoria, Australia. *Australasian Plant Disease Notes, 9*, 1–4.

Darby, P. (1988). *Alternaria alternata* infection of hop (*Humulus lupulus*) cones. *Transactions of the British Mycological Society, 90*(4), 650–653.

Darby, P. (2001). Single gene traits in hop breeding. In E. Seigner (Ed.), *Proceedings of the Scientific Commissions* (pp. 76–80). International Hop Growers Convention IHGC.

Darby, P. (2013). Could there be a dominant gene for susceptibility? In E. Seigner (Ed.), *Proceedings of the Scientific Commissions* (pp. 17–20). International Hop Growers Convention IHGC.

Darby, P., Mansfield, J. W., & Godwin, J. R. (1989). The assessment of partial resistance to powdery mildew disease in hops. *Plant Pathology, 38*(2), 219–225.

Darling, E., Pu, J., Cole, E., Christian, R., Warner, F. W., Zasada, I., Chung, H., & Quintanilla, M. (2021). First report of the hop cyst nematode, *Heterodera humuli*, in two counties of the Yakima Valley Region, WA, USA. *Plant Disease, 105*(4), 1228. https://doi.org/10.1094/PDIS-08-20-1769-PDN

Darling, E., Núñez-Rodríguez, L., Chung, H., Zasada, I., & Quintanilla Tornel, M. (2023). The hop cyst nematode, *Heterodera humuli*: History, distribution, and impact on global hop production. *Phytopathololog*y, *113*(2), 142–149. https://doi.org/10.1094/PHYTO-04-22-0121-RVW

Davies, D. L. G. (1938). A rootstock canker of hops caused by *Gibberella pulicaris*. *The Annual Report of the Long Ashton Research Station for 1938, 15*, 115–123.

Davies, D. L., & Clark, M. F. (1983). A satellite-like nucleic acid of arabis mosaic virus associated with hop nettlehead disease. *Annals of Applied Biology, 103*(3), 439–448.

Davis, T. J., Gómez, M. I., Harper, S. J., & Twomey, M. (2021). The economic impact of hop stunt viroid and certified clean planting materials. *HortScience, 56*(12), 1471–1475.

De Grisse, A., & Gillard, A. (1963). Morphology and biology of hop cyst eelworm (*Heterodera humuli* Filipjev 1934). *Nematologica, 9*(1), 41–48.

Demangeat, G., Voisin, R., Minot, J. C., Bosselut, N., Fuchs, M., & Esmenjaud, D. (2005). Survival of Xiphinema index in vineyard soil and retention of Grapevine fanleaf virus over extended time in the absence of host plants. *Phytopathology, 95*(10), 1151–1156.

Desaeger, J., Wram, C., & Zasada, I. (2020). New reduced-risk agricultural nematicides-rationale and review. *Journal of Nematology, 52*(1), e2020. https://doi.org/10.21307/jofnem-2020-091

Dodlek Šarkanj, I., Vahčić, N., Markov, K., Haramija, J., Uršulin-Trstenjak, N., Hajdek, K., Sulyok, M., Krska, R., & Šarkanj, B. (2024). First report on mycotoxin contamination of hops (*Humulus lupulus* L.). *Toxins, 16*(7), 293. https://doi.org/10.3390/toxins16070293

Dracaemer, W., & Hunt, D. (2006). Structure and classification. In R. Perry & M. Moens (Eds.), *Plant nematology* (pp. 3–32). CABI.

Duran-Vila, N., Roistacher, C. N., Rivera-Bustamante, R., & Semancik, J. S. (1988). A definition of citrus viroid groups and their relationship to the exocortis disease. *Journal of General Virology, 69*(12), 3069–3080. https://doi.org/10.1099/0022-1317-69-12-3069

Eastwell, K. C., & Druffel, K. L. (2012). Complete genome organization of American hop latent virus and its relationship to carlaviruses. *Archives of Virology, 157*(7), 1403–1406.

Eastwell, K. C., & Nelson, M. E. (2007). Occurrence of viroids in commercial hop (*Humulus lupulus* L.) production areas of Washington State. *Plant Health Progress, 8*(1), 1. https://doi.org/10.1094/PHP-2007-1127-01-RS

Eiras, M., de Oliveira, A. M., de Fátima Ramos, A., Harakava, R., & Daròs, J. A. (2023). First report of citrus bark cracking viroid and hop latent viroid infecting hop in commercial yards in Brazil. *Journal of Plant Pathology, 105*(2), 603–603. https://doi.org/10.1007/s42161-023-01313-4

Engelhard, B. (2006). Plant protection in hop cultivation. In *Annual report 2005, special crop hops* (pp. 52–61). Bavarian State Research Center for Agriculture - Institute for Crop Science and Plant Breeding and the Society of Hop Research.

Eppler, A. (1992). Arabis mosaic virus in German hops. *Acta Horticulturae, 308*, 81–86.

EPPO. (2014). *Current situation of Verticillium albo-atrum hop strains in Slovenia*. EPPO Reporting Service no. 07-2014 Num. article: 2014/134. https://gd.eppo.int/reporting/article-3227. Accesses 27 April 2024.

EPPO. (2019). *First finding of Citrus bark cracking viroid (CBCVd) in Germany*. Reporting Service no. 08-2019 Num. article: 2019/165.

EPPO. (2024). *Citrus bark cracking viroid*. EPPO datasheets on pests recommended for regulation. https://gd.eppo.int. Accessed 16 Jul 2024.

Faggioli, F., Duran-Vila, N., Tsagris, M., & Pallás, V. (2017). Geographical distribution of viroids in Europe. In A. Hadidi, R. Flores, J. W. Randles, & P. Palukaitis (Eds.), *Viroids and satellites* (pp. 473–484). Academic Press. https://doi.org/10.1016/B978-0-12-801498-1.00044-9

Feiner, A., Pitra, N., Matthews, P., Pillen, K., Wessjohann, L. A., & Riewe, D. (2021). Downy mildew resistance is genetically mediated by prophylactic production of phenylpropanoids in hop. *Plant, Cell & Environment, 44*(1), 323–338.

Feltrin, D. M., Jadoski, S. O., Giacomin, R. M., Horst, M. V., Moccellin, R., & Da-Silva, P. R. (2024). First report of *Pseudoperonospora humuli* (Miy. and Tak.) Wilson causing downy mildew on hop plants in Brazil. *Crop Protection, 179*, 106609.

Folkertsma, R. T., Helder, J., Gommers, F., & Bakker, J. (1997). Storage of potato cyst nematode at −80°C. *Fundamental and Applied Nematology, 20*(3), 299–302.

French, N., Ludlam, F. A. B., & Wardlow, L. R. (1973). Biology, damage and control of rosy rnstic moth, *Hydraecia micacea* (esp.), on hops. *Plant Pathology, 22*(2), 58–64.

Galinato, S. P. (2020). *2020 estimated cost of establishing and producing conventional and organic hops in the Pacific Northwest*. Washington State University Extension. Retrieved February 19, 2024, from https://www.usahops.org/cabinet/data/Hop%20Enterprise%20Budget%202020_Draft.pdf

Gent, D. H., Nelson, M. E., George, A. E., Grove, G. G., Mahaffee, W. F., Ocamb, C. M., Barbour, J. D., Peetz, A., & Turechek, W. W. (2008). A decade of hop powdery mildew in the Pacific Northwest. *Plant Health Progress, 9*(1), 33.

Gent, D. H., Nelson, M. E., Farnsworth, J. L., & Grove, G. G. (2009). PCR detection of *Pseudoperonospora humuli* in air samples from hop yards. *Plant Pathology, 58*(6), 1081–1091.

Gent, D. H., Ocamb, C. M., & Farnsworth, J. L. (2010). Forecasting and management of hop downy mildew. *Plant Disease, 94*(4), 425–431.

Gent, D. H., Nelson, M. E., Grove, G. G., Mahaffee, W. F., Turechek, W. W., & Woods, J. L. (2012). Association of spring pruning practices with severity of powdery mildew and downy mildew on hop. *Plant Disease, 96*(9), 1343–1351.

Gent, D. H., Grove, G. G., Nelson, M. E., Wolfenbarger, S. N., & Woods, J. L. (2014). Crop damage caused by powdery mildew on hop and its relationship to late season management. *Plant Pathology, 63*(3), 625–639.

Gent, D. H., Probst, C., Nelson, M. E., Grove, G. G., Massie, S. T., & Twomey, M. C. (2016). Interaction of basal foliage removal and late-season fungicide applications in management of hop powdery mildew. *Plant Disease, 100*(6), 1153–1160.

Gent, D. H., Cohen, Y., & Runge, F. (2017a). Homothallism in *Pseudoperonospora humuli*. *Plant Pathology, 66*(9), 1508–1516.

Gent, D. H., Massie, S. T., Twomey, M. C., & Wolfenbarger, S. N. (2017b). Adaptation to partial resistance to powdery mildew in the hop cultivar Cascade by *Podosphaera macularis*. *Plant Disease, 101*(6), 874–881.

Gent, D. H., Claassen, B. J., Twomey, M. C., & Wolfenbarger, S. N. (2018a). Identification of resistance to powdery mildew in publicly available male hop germplasm. *Plant Health Progress, 19*(3), 258–264.

Gent, D. H., Claassen, B. J., Twomey, M. C., Wolfenbarger, S. N., & Woods, J. L. (2018b). Susceptibility of hop crown buds to powdery mildew and its relation to perennation of *Podosphaera macularis*. *Plant Disease, 102*(7), 1316–1325.

Gent, D. H., Adair, N., Knaus, B. J., & Grünwald, N. J. (2019a). Genotyping-by-sequencing reveals fine-scale differentiation in populations of *Pseudoperonospora humuli*. *Phytopathology, 109*(10), 1801–1810.

Gent, D. H., Bhattacharyya, S., & Ruiz, T. (2019b). Prediction of spread and regional development of hop powdery mildew: A network analysis. *Phytopathology, 109*(8), 1392–1403.

Gent, D. H., Mahaffee, W. F., Turechek, W. W., Ocamb, C. M., Twomey, M. C., Woods, J. L., & Probst, C. (2019c). Risk factors for bud perennation of *Podosphaera macularis* on hop. *Phytopathology, 109*(1), 74–83.

Gent, D. H., Block, M., & Claassen, B. J. (2020a). High levels of insensitivity to phosphonate fungicides in *Pseudoperonospora humuli*. *Plant Disease, 104*(5), 1400–1406.

Gent, D. H., Claassen, B. J., Gadoury, D. M., Grünwald, N. J., Knaus, B. J., Radišek, S., Weldon, W., & Wiseman, M. S. (2020b). Population diversity and structure of *Podosphaera macularis* in the Pacific Northwestern United States and other populations. *Phytopathology, 110*(5), 1105–1116.

Ginetti, B., Moricca, S., Squires, J. N., Cooke, D. E. L., Ragazzi, A., & Jung, T. (2014). *Phytophthora acerina* sp. *nov*., a new species causing bleeding cankers and dieback of *Acer pseudoplatanus* trees in planted forests in northern Italy. *Plant Pathology, 63*(4), 858–876.

Giraldo, A., & Crous, P. W. (2019). Inside Plectosphaerellaceae. *Studies in Mycology, 92*(1), 227–286. https://doi.org/10.1016/j.simyco.2018.10.005

Guo, L., Liu, S., Wu, Z., Mu, L., Xiang, B., & Li, S. (2008). Hop stunt viroid (HSVd) newly reported from hop in Xinjiang, China. *Plant Pathology, 57*(4), 764–764. https://doi.org/10.1111/j.1365-3059.2008.01875.x

Hafez, S., Sundararaj, P., & Barbour, J. (1999). Impact of Heterodera humuli on growth and mineral nutrition composition of hops, *Humulus lupulus* cv. Cascade. *International Journal of Nematology, 9*(1), 23–26.

Hagemann, M. H., Treiber, C., Born, U., Schrader, G., Stampfl, J., Jakše, J., & Radišek, S. (2023). Risk potential of international fruit trade for viroid spreading - Case study on hop viroids in Europe. *Journal of Plant Pathology, 105*(4), 1335–1346. https://doi.org/10.1007/s42161-023-01449-3

Han, J., Yao, X. L., Qu, F., Kaufman, R. M., & Lewis Ivey, M. L. (2019). First report of hop stunt viroid infecting hop in Ohio. *Plant Disease, 103*(7), 1802–1802. https://doi.org/10.1094/PDIS-02-19-0402-PDN

Handoo, Z. A., & Subbotin, S. A. (2018). Taxonomy, identification and principal species. In P. N. Roland, M. Moens, & J. T. Jones (Eds.), *Cyst nematodes* (pp. 365–398). CABI.

Harris, R. V. (1927). A wilt disease of hops. *East Malling Research Station Annual Report for, 1925*, 92–93.

Harrison, B. D., & Winslow, R. D. (1961). Laboratory and field studies on the relation of arabis mosaic virus to its nematode vector *Xiphinema diversicaudatum* (Micoletzky). *Annals of Applied Biology, 49*(4), 621–633.

Hataya, T., Uchino, K., Arimoto, R., Suda, N., Sano, T., Shikata, E., & Uyeda, I. (2000). Molecular characterization of hop latent virus and phylogenetic relationships among viruses closely related to carlaviruses. *Archives of Virology, 145*(12), 2503–2524.

Hataya, T., Tsushima, T., & Sano, T. (2017). Hop Stunt Viroid. In A. Hadidi, R. Flores, J. W. Randles, & P. Palukaitis (Eds.), *Viroids and satellites* (pp. 199–210). Academic Press. https://doi.org/10.1016/B978-0-12-801498-1.00019-X

Hatlen, R. J., Higgins, D. S., Venne, J., Rojas, J. A., Hausbeck, M. K., & Miles, T. D. (2022). First report of halo blight of hop (*Humulus lupulus*) caused by *Diaporthe humulicola* in Quebec, Canada. *Plant Disease, 106*(6), 1750.

Hatlen, R. J., Swift, R. M., Miles, L. A., Byrne, J. M., & Miles, T. D. (2023). First report of *Colletotrichum fioriniae* infecting hop (*Humulus lupulus*) in Michigan. *Plant Disease, 107*(10), 3280.

Hatlen, R. J., Lizotte, E., Sirrine, R., & Miles, T. D. (2024). *Managing Halo blight of hop*. Michigan State University. Retrieved August 5, 2024, from https://www.canr.msu.edu/resources/managing-halo-blight-of-hop

Havill, J. S., Wiseman, M. S., Henning, J. A., Gent, D. H., & Muehlbauer, G. J. (2023). Registration of seven powdery mildew-resistant wild hop germplasm lines. *Journal of Plant Registrations, 17*(1), 171–179.

Hay, F. S., & Barbara, D. J. (2009). Arabis mosaic. In W. F. Mahaffee, S. J. Pethybridge, & D. H. Gent (Eds.), *Compendium of hop diseases and pests* (pp. 41–44). APS Press.

Hay, F., & Pethybridge, S. (2003). Plant-parasitic nematodes associated with hop production in Tasmia, Australia. *Journal of Phytopathology, 151*(7–8), 369–375. https://doi.org/10.1046/j.1439-0434.2003.00733.x

Hay, F. S., Close, R. C., Hetcher, J. D., & Ashby, J. W. (1992). Incidence and spread of viruses in hop (*Humulus lupulus* L.) in New Zealand. *New Zealand Journal of Crop and Horticultural Science, 20*(3), 319–327.

Hellwig, K., Kremheller, H. T., & Agerer, R. (1991). Investigations on the resistance of *Pseudoperonospora humuli* (Mik. et Tak.) Wilson to metalaxyl. *Gesunde Pflanzen, 43*(12), 400–404.

Henning, J. A., Townsend, M. S., Gent, D. H., Bassil, N., Matthews, P., Buck, E., & Beatson, R. (2011). QTL mapping of powdery mildew susceptibility in hop (*Humulus lupulus* L.). *Euphytica, 180*(3), 411–420.

Henning, J. A., Gent, D. H., Twomey, M. C., Townsend, M. S., Pitra, N. J., & Matthews, P. D. (2015). Precision QTL mapping of downy mildew resistance in hop (*Humulus lupulus* L.). *Euphytica, 202*(3), 487–498.

Henning, J. A., Gent, D. H., Twomey, M. C., Townsend, M. S., Pitra, N. J., & Matthews, P. D. (2016). Genotyping-by-sequencing of a bi-parental mapping population segregating for downy mildew resistance in hop (*Humulus lupulus* L.). *Euphytica, 208*(3), 545–559.

Henricot, B., Pérez Sierra, A., & Jung, T. (2014). *Phytophthora pachypleura* sp. *nov*., a new species causing root rot of *Aucuba japonica* and other ornamentals in the United Kingdom. *Plant Pathology, 63*(5), 1095–1109.

Heungens, K., Debode, J., Van Poucke, K., Radišek, S., Willaert, L., & Haegeman, A. (2021). Population structure of *Verticillium* isolates from Belgian hops. In *72nd International Symposium on Crop Protection (ISCP)*.

Higgins, D. S., Hatlen, R. J., Byrne, J. M., Sakalidis, M. L., Miles, T. D., & Hausbeck, M. K. (2021). Etiology of halo blight in Michigan hopyards. *Plant Disease, 105*(4), 859–872.

Higgins, D. S., Miles, T. D., Byrne, J. M., & Hausbeck, M. K. (2022). Optimizing molecular detection for the hop downy mildew pathogen *Pseudoperonospora humuli* in plant tissue. *Phytopathology, 112*(11), 2426–2439.

Hoerner, G. R. (1940). The infection capabilities of hop downy mildew. *Journal of Agricultural Research, 61*(5), 331–334.

Hoerner, G. R. (1945). Crown gall of hops. *Plant Disease Reporter, 29*(4), 98–110.

Hoerner, G. R. (1949). Hop diseases in the United States. *Brewers Digest, 24*(4), 45–51.

Inderbitzin, P., & Subbarao, K. V. (2014). Verticillium systematics and evolution: How confusion impedes *Verticillium* wilt management and how to resolve it. *Phytopathology, 104*(6), 564–574. https://doi.org/10.1094/phyto-11-13-0315-ia

Inderbitzin, P., Bostock, R. M., Davis, R. M., Usami, T., Platt, H. W., & Subbarao, K. V. (2011). Phylogenetics and taxonomy of the fungal vascular wilt pathogen *Verticillium*, with the descriptions of five new species. *PLoS ONE, 6*(12), e28341. https://doi.org/10.1371/journal.pone.0028341

Inderbitzin, P., Davis, R. M., Bostock, R. M., & Subbarao, K. V. (2013). Identification and differentiation of *Verticillium* species and *V. longisporum* lineages by simplex and multiplex PCR Assays. *PLoS ONE, 8*(6), e65990. https://doi.org/10.1371/journal.pone.0065990

Isaac, I., & Keyworth, G. W. (1948). Verticillium wilt of the hop (*Humulus lupulus*) III. A study of the pathogenicity of the isolates from fluctuating and from progressive outbreaks. *Annals of Applied Biology, 35*(2), 243–249.

Ito, T., Kanematsu, S., Koganezawa, H., Tsuchizaki, T., & Yoshida, K. (1993). Detection of a viroid associated with apple fruit crinkle disease. *Annals of the Phytopathological Society of Japan, 59*(5), 520–527.

Jakše, J., Cerenak, A., Radišek, S., Satovic, Z., Luthar, Z., & Javornik, B. (2013). Identification of quantitative trait loci for resistance to Verticillium wilt and yield parameters in hop (Humulus lupulus L.). *Theoretical and Applied Genetics, 126*(6), 1431–1443. https://doi.org/10.1007/s00122-013-2062-4

Jakše, J., Radišek, S., Pokorn, T., Matousek, J., & Javornik, B. (2015). Deep-sequencing revealed Citrus bark cracking viroid (CBCVd) as a highly aggressive pathogen on hop. *Plant Pathology, 64*(4), 831–842.

Jakše, J., Jelen, V., Radišek, S., de Jonge, R., Mandelc, S., Majer, A., Curk, T., Zupan, B., Thomma, B. P. H. J., & Javornik, B. (2018). Genome sequence of a lethal strain of xylem-Invading Verticillium nonalfalfae. *Genome Announcements, 6*(2), e01458–17. https://doi.org/10.1128/genomeA.01458-17

Johnson, D. A. (1991). Two degree-day models for predicting initial emergence of hop shoots systemically infected with *Pseudoperonospora humuli*. *Plant Disease, 75*(3), 285–287.

Johnson, D. A., & Anliker, W. L. (1985). Effect of downy mildew epidemics on the seasonal carryover of initial inoculum in hop yards. *Plant Disease, 69*(2), 140–142.

Johnson, D. A., Alldredge, J. R., & Allen, J. R. (1994). Weather and downy mildew epidemics of hop in Washington state. *Phytopathology, 84*(5), 524–527.

Jones, W. (1932a). A new technique for obtaining oospores of the hop downy mildew by inoculating cotyledons. *Science, 75*(1934), 108.

Jones, W. (1932b). Downy mildew infection of hop and nettle seedlings in British Columbia. *Journal of the Institute of Brewing, 29*(4), 194–196.

Jung, T., & Burgess, T. I. (2009). Re-evaluation of *Phytophthora citricola* isolates from multiple woody hosts in Europe and North America reveals a new species, *Phytophthora plurivora* sp. *nov*. *Persoonia, 22*(1), 95–110.

Kitner, M., Runge, F., Lebeda, A., Vaculná, L., Sedláková, B., & Thines, M. (2021). *Pseudoperonospora humuli* might be an introduced species in Central Europe with low genetic diversity but high distribution potential. *European Journal of Plant Pathology, 159*(4), 903–915.

Klein, R. E. (1994). Occurrence and incidence of Metalaxyl resistance in *Pseudoperonospora humuli* in the Pacific Northwest. *Plant Disease, 78*(2), 161–163.

Klimes, A., Dobinson, K. F., Thomma, B. P., & Klosterman, S. J. (2015). Genomics spurs rapid advances in our understanding of the biology of vascular wilt pathogens in the genus Verticillium. *Annual Review of Phytopathology, 53*(1), 181–198. https://doi.org/10.1146/annurev-phyto-080614-120224

Knabel, S., Seigner, L., & Wallnöfer, P. R. (1999). Nachweis des Hop latent Viroids (HLVd) mit der Polymerase-Kettenreaktion (PCR). *Gesunde Pflanzen, 51*(7), 234–239.

Kopp, P. A. (2016). *Hoptopia: A world of agriculture and beer in Oregon's Willamette Valley* (1st ed.). University of California Press.

Kropf, S. M., Putnam, M. L., Serdani, M., Twomey, M. C., Woods, J. L., & Gent, D. H. (2012). Sclerotinia wilt of hop (*Humulus lupulus*) caused by *Sclerotinia sclerotiorum* in the Pacific Northwest United States. *Plant Disease, 96*(4), 583.

Kusch, S., & Panstruga, R. (2017). *mlo*-based resistance: An apparently universal "weapon" to defeat powdery mildew disease. *Molecular Plant-Microbe Interactions, 30*(3), 179–189.

Laurie, R. B., Claassen, B., Ross, C. J., & Gent, D. H. (2022). Yard age, cultivar susceptibility, and spring pruning practices as risk factors for overwintering of *Podosphaera macularis* on hop. *PhytoFrontiers, 3*(2), 390–398.

Lavagi-Craddock, I., Harper, S., Krueger, R., Quijia-Lamiña, P., & Vidalakis, G. (2024). Policies, regulations, and production of viroid-free propagative plant materials for sustainable agriculture. In *Fundamentals of viroid biology* (pp. 337–361). Academic Press.

Lee, J. Y., Puchta, H., Ramm, K., & Sänger, H. L. (1988). Nucleotide sequence of the Korean strain of hop stunt viroid (HSV). *Nucleic Acids Research, 16*(17), 8708. https://doi.org/10.1093/nar/16.17.8708

Legg, J. T. (1959). The effect of split leaf blotch and nettlehead virus diseases on the yield of Fuggle hops. *Journal of Horticultural Science, 34*(2), 122–125.

Liu, H., Park, S., & Sang, H. (2023). Identification and fungicide control of *Bipolaris sorokiniana* causing leaf spot and blight on common hop (*Humulus lupulus*) in Korea. *Plant Disease, 107*(10), 2939–2943.

Lucas, G. B., Campbell, C. L., & Lucas, L. T. (1992). Abiotic agents. In *Introduction to plant diseases* (pp. 316–331). Springer. https://doi.org/10.1007/978-1-4615-7294-7_20

Macovei, A. (1980). Identification and determination of the main characteristics of hop viruses in Romania IV. Data on the presence of cucumber mosaic virus in the hop cultivar Huller Bitterer. *Analele Institutului de Cercetari pentru Protectia Plantelor, 16*, 47–51.

Madani, M., Vovlas, N., Castillo, P., Subbotin, S. A., & Moens, M. (2004). Molecular characterization of cyst nematode species (*Heterodera* spp.) from the Mediterranean Basin using RFLPs and sequences of ITS-rDNA. *Journal of Phytopathology, 152*(4), 229–234.

Mahaffee, W. F., Thomas, C. S., Turechek, W. W., Ocamb, C. M., Nelson, M. E., Fox, A., & Gubler, W. D. (2003). Responding to an introduced pathogen: *Podosphaera macularis* (hop powdery mildew) in the Pacific northwest. *Plant Health Progress, 4*(1), 21.

Mahaffee, W. F., Engelhard, B., Gent, D. H., & Grove, G. G. (2009). Powdery mildew. In W. F. Mahaffee, S. J. Pethybridge, & D. H. Gent (Eds.), *Compendium of hop diseases and pests* (pp. 25–31). APS Press.

Malan, A., Pieterse, W., & Britis, G. (1991). Plant-parasitic nematodes associated with hops (*Humulus lupulus*) in South Africa. *Phytophylactica, 23*, 173–175.

Marquez-Molins, J., Gomez, G., & Pallas, V. (2021). Hop stunt viroid: A polyphagous pathogenic RNA that has shed light on viroid–host interactions. *Molecular Plant Pathology, 22*(2), 153–162. https://doi.org/10.1111/mpp.13022

Matoušek, J., & Patzak, J. (2000). A low transmissibility of hop latent viroid through a generative phase of Humulus lupulus L. *Biologia Plantarum, 43*(1), 145–148. https://doi.org/10.1023/A:1026531819806

Matsui, H., Kroupa, F., Krofta, K., Snidlova, J., Inui, T., Oka, K., & Fukui, N. (2017). Influence of apple mosaic virus on the growth, yield and qualities of Saaz Hop. *Journal of Plant Sciences, 5*(5), 152–159. https://doi.org/10.11648/j.jps.20170505.14

Mendes Fagherazzi, M., Sartori Pereira, F., Regianini Nerbass, F., Ferreira Zacca, E., Bizolo Sommer, V., Gonçalves, M. J., Rufato, L., Trezzi Casa, R., Bogo, A., & Nascimento da Silva, F. (2021). First report of *Podosphaera macularis* causing hop (*Humulus lupulus*) powdery mildew in Brazil. *Plant Disease, 105*(8), 2243.

Mikhailpukov, V. S. (1980). The ontogenesis of *Heterodera humuli*. *Vestnik Zoologii, 5*, 56–62.

Mitchell, M. N., Ocamb, C. M., Grünwald, N. J., Mancino, L. E., & Gent, D. H. (2011). Genetic and pathogenic relatedness of *Pseudoperonospora cubensis* and *P. humuli*. *Phytopathology, 101*(7), 805–818.

Mori, Y. (1966). Studies on downy mildew of hop plants IV. Some observations on life cycle of the causal fungus (*Pseudoperonospora humuli* (Miy. et Tak.) Wilson), and especially on the primary source of infection in hop downy mildew. *Japanese Journal of Phytopathology, 32*(5), 275–284.

Munda, A., Radisek, S., Seigner, E., Jakše, J., & Štajner, N. (2020). PM 7/78 (2) *Verticillium nonalfalfae* and *V. dahliae*. *EPPO Bulletin, 50*(3), 462–476. https://doi.org/10.1111/epp.12701

Mundo-Ocampo, M., Troccoli, A., Subbotin, S., Del Cid, J., Baldwin, J. G., & Inserra, R. N. (2008). Synonymy of *Afenestrata* with *Heterodera* supported by phylogenetics with molecular and morphological characterization of *H. koreana* comb. n. (*Tylenchida: Heteroderidae*). *Nematology, 10*(5), 611–632.

Mundt, C. C. (2014). Durable resistance: A key to sustainable management of pathogens and pests. *Infection, Genetics and Evolution, 27*, 446–455.

Munkvold, G. P., Proctor, R. H., & Moretti, A. (2021). Mycotoxin production in Fusarium according to contemporary species concepts. *Annual Review of Phytopathology, 59*(1), 373–402.

Mycobank. (2024). *Mycobank database*. Retrieved August 5, 2024, from https://www.mycobank.org/

Nelson, M. E., Gent, D. H., & Grove, G. G. (2015). Meta-analysis reveals a critical period for management of powdery mildew on hop cones. *Plant Disease, 99*(5), 632–640.

Neve, R. A. (1991). *Hops*. Chapman and Hall.

Novák, J. B., & Lanzová, J. (1976). Identification of alfalfa mosaic virus and tomato bushy stunt virus in hop (*Humulus lupulus* L.) and grapevine (*Vitis vinifera* subsp. *sativa* (DC./HEGI) plants in Czechoslovakia). *Biologia Plantarum, 18*(2), 152–154.

Nukari, A., Laamanen, J., Uosukainen, M., & Lemmetty, A. (2013). Comparison of virus eradication of Apple mosaic virus from hop by encapsulation-dehydration cryotherapy and meristem culture methods. *Acta Horticulturae, 1039*, 113–119.

Núñez, L., Peetz, A., Gent, D., Wram, C., & Zasada, I. A. (2022). Plant-parasitic nematodes associated with hop (Humulus lupulus) in the Pacific Northwest. *Phytopathology, 112*(8), 24. https://doi.org/10.1094/PHYTO-112-8-S2.22

Ohno, T., Takamatsu, N., Meshi, T., & Okada, Y. (1983). Hop stunt viroid: Molecular cloning and nucleotide sequence of the complete cDNA copy. *Nucleic Acids Research, 11*(18), 6185–6197. https://doi.org/10.1093/nar/11.18.6185

Olatoye, M. O., Wiseman, M., Gent, D. H., Henning, J. A., & Altendorf, K. R. (2023). Genetic characterization of downy mildew resistance from the hop (*Humulus lupulus* L.) line USDA 64035M. *Crop Science, 63*(3), 1082–1091.

OMAFRA-Ontario Ministry of Agriculture, Food and Rural Affairs. (2023). *Nutrient Deficiency ID Guide for Hops*. King's Printer for Ontario. Retrieved August 5, 2024, from https://www.ontario.ca/page/nutrient-deficiency-identification-guide-hops

Pallas, V., Navarro, A., & Flores, R. (1987). Isolation of a viroid-like rna from hop different from hop stunt viroid. *Journal of General Virology, 68*(12), 3201–3205. https://doi.org/10.1099/0022-1317-68-12-3201

Palomares-Rius, J. E., Escobar, C., Cabrera, J., Vovlas, A., & Castillo, P. (2017). Anatomical alterations in plant tissues induced by plant-parasitic nematodes. *Frontiers in Plant Science, 8*, 1987. https://doi.org/10.3389/fpls.2017.01987

Parke, J. L., Knaus, B. J., Fieland, V. J., Lewis, C., & Grünwald, N. J. (2014). Phytophthora community structure analyses in Oregon nurseries inform systems approaches to disease management. *Phytopathology, 104*(10), 1052–1062.

Patzak, J., Henychová, A., Krofta, K., Svoboda, P., & Malířová, I. (2021). The influence of hop latent viroid (hlvd) infection on gene expression and secondary metabolite contents in hop (*Humulus lupulus* l.) glandular trichomes. *Plants, 10(11)*, 2297. https://doi.org/10.3390/plants10112297

Pépin, N., Hebert, F. O., & Joly, D. L. (2021). Genome-wide characterization of the MLO gene family in Cannabis sativa reveals two genes as strong candidates for powdery mildew susceptibility. *Frontiers in Plant Science, 12*, 729261. https://doi.org/10.3389/fpls.2021.729261

Pereira, C. M., Ferreira, B. W., & Barreto, R. W. (2023). First report of *Cercospora apii sensu lato* 'sp. Q' leaf spot on hop in Brazil. *Plant Disease, 107*(3), 947.

Pérez, B. A., Martínez, E., Noetinger, F., & Wright, E. R. (2009). Hop downy mildew caused by *Pseudoperonospora humuli* in Argentina. *Plant Disease, 93*(8), 839–839.

Pethybridge, S. J., & Madden, L. V. (2003). Analysis of spatiotemporal dynamics of virus spread in an Australian hop garden by stochastic modeling. *Plant Disease, 87*(1), 56–62.

Pethybridge, S. J., Wilson, C. R., Sherriff, L. J., Leggett, G. W., & Munro, D. (2000). Virus incidence in Australian hop (*Humulus lupulus* L.) gardens and cultivar differences in susceptibility to infection. *Australian Journal of Agricultural Research, 51*(6), 685–689.

Pethybridge, S. J., Hay, F. S., Wilson, C. R., Sherriff, L. J., & Leggett, G. W. (2001a). First report of *Fusarium crookwellense* causing tip blight on cones of hop. *Plant Disease, 85*(11), 1208–1208.

Pethybridge, S. J., Hay, F. S., Wilson, C. R., Sherriff, L. J., & Leggett, G. W. (2001b). First report of infection of hop cones by *Alternaria alternata* in Australia. *Plant Disease, 85*(7), 804.

Pethybridge, S. J., Wilson, C. R., Hay, F. S., Leggett, G. W., & Sherriff, L. J. (2002a). Effect of viruses on agronomic and brewing characteristics of four hop (Humulus lupulus) cultivars in Australia. *Annals of Applied Biology, 140*(1), 97–105.

Pethybridge, S. J., Wilson, C. R., Hay, F. S., Leggett, G. W., & Sherriff, L. J. (2002b). Mechanical transmission of apple mosaic virus in Australian hop (*Humulus lupulus*) gardens. *Annals of Applied Biology, 141*(1), 77–85.

Pethybridge, S. J., Hay, F. S., Barbara, D. J., Eastwell, K. C., & Wilson, C. R. (2008). Viruses and viroids infecting hop: Significance, epidemiology, and management. *Plant Disease, 92*(3), 324–338.

Pichlmaier, J., & Zinkernagel, V. (1992). *Phytophthora citricola* (Sawada) als Erreger der Stockfäule des Hopfens. *Zeitschrift für Pflanzenkrankheiten und Pflanzenschutz, 99*(5), 505–518.

Pidopličko, N. M. (1978). Gribi paraziti kulturnih rastenij. *Opridelitelj, 3*, 145–153.

Pitcher, R. S., & McNamara, D. G. (1973). The control of *Xiphinema diversicaudatum*, the vector of arabis mosaic virus in hops. *Annals of Applied Biology, 75*(3), 468–469.

Polák, J., & Svoboda, P. (1989). Preliminary results of screening hop varieties and clones for the presence of some hop viruses in Czechoslovakia using ELISA. In *Proceedings, International Workshop on hop virus diseases* (pp. 27–32).

Probasco, G., & Murphey, J. M. (1996). The effects of hop viruses on brewing and agronomic characteristics in the hop variety chinook. *Technical quarterly-Master Brewers Association of the Americas, 33*(3), 160–165.

Probasco, E. G., & Skotland, C. B. (1978). Host range, general properties, purification and electron microscopy of hop latent virus. *Phytopathology, 68*, 277–281.

Puchta, H., Ramm, K., & Sanger, H. L. (1988). The molecular-structure of hop latent viroid (HLV), a new viroid occurring worldwide in hops. *Nucleic Acids Research, 16*(10), 4197–4216. https://doi.org/10.1093/nar/16.10.4197

Puchta, H., Ramm, K., Luckinger, R., Hadas, R., Barjoseph, M., & Sanger, H. L. (1991). Primary and secondary structure of citrus viroid-iv (CVd-IV), a new chimeric viroid present in dwarfed grapefruit in Israel. *Nucleic Acids Research,* ***19*****(23)**, 6640.

Punja, Z. K., Wang, K., Lung, S., & Buirs, L. (2024). Symptomology, prevalence, and impact of Hop latent viroid on greenhouse-grown cannabis (*Cannabis sativa* L.) plants in Canada. *Canadian Journal of Plant Pathology, 46*(2), 174–197. https://doi.org/10.1080/07060661.2023.2279184

Purayannur, S., Gent, D. H., Miles, T. D., Radišek, S., & Quesada-Ocampo, L. M. (2021). The hop downy mildew pathogen *Pseudoperonospora humuli*. *Molecular Plant Pathology, 22*(7), 755–768.

Puttoo, B. L., Koul, M. L., & Saraf, R. L. (1975). Septoria leaf disease of hops (*Humulus lupulus*) in Kashmir. *Indian Phytopathology, 27*(2), 247–248.

Radišek, S. (2009). Verticillium wilt. In W. Mahaffee, S. Pethybridge, & D. H. Gent (Eds.), *Compendium of hop diseases and pests* (pp. 33–36). The American Phytopathological Society.

Radišek, S., Jakše, J., & Javornik, B. (2006). Genetic variability and virulence among *Verticillium albo-atrum* isolates from hop. *European Journal of Plant Pathology, 116*, 301–314. https://doi.org/10.1007/s10658-006-9061-0

Radišek, S., Jakše, J., Javornik, B., & de Gruyter, J. (2007). First report of Phoma exigua as a pathogen of hop in Slovenia. *Plant Pathology, 57*(2), 381.

Radišek, S., Leskošek, G., Jakše, J., & Javornik, B. (2008). *Cercospora cantauriensis* on hop in Austria and Slovenia. *Hmeljarski Bilten, 15*, 5–14.

Radišek, S., Leskosek, G., Jakše, J., & Javornik, B. (2009). Occurrence of *Cercospora cantuariensis* on hop in Austria and Slovenia. *Plant Pathology, 58*(2), 400. https://doi.org/10.1111/j.1365-3059.2008.01962.x

Radišek, S., Majer, A., Jakše, J., Javornik, B., & Matousek, J. (2012). First report of hop stunt viroid infecting hop in Slovenia. *Plant Disease, 96*(4), 592–593. https://doi.org/10.1094/pdis-08-11-0640-pdn

Radišek, S., Guček, T., Leskošek, G., Benko Beloglavec, A., Jakše, J., & Javornik, B. (2017). Huda viroidna zakrnelost hmelja/Severe hop stunt disease. Inštitut za hmeljarstvo in pivovarstvo Slovenije, 29 pp. ISBN 978-961-93322-2-1. http://www.ihps.si/wp-content/uploads/2016/08/Huda-viroidna-zakrnelost-hmelja2017.compressed.pdf

Radišek, S., Košir, I. J., Čeh, B., & Jakše, J. (2023). Management of *Verticillium* wilt of hop by using green manure crops. In F. Weihrauch (Ed.), *Proceedings of the Scientific-Technical Commission* (pp. 80). International Hop Growers Convention.

Rahman, A., Góngora-Castillo, E., Bowman, M. J., Childs, K. L., Gent, D. H., Martin, F. N., & Quesada-Ocampo, L. M. (2019). Genome sequencing and transcriptome analysis of the hop

downy mildew pathogen *Pseudoperonospora humuli* reveal species-specific genes for molecular detection. *Phytopathology, 109*(8), 1354–1366.
Richardson, B. J., & Gent, D. H. (2023). Impact of infection timing and autumnal fungicide applications on perennation of *Pseudoperonospora humuli* and severity of hop downy mildew. *Plant Disease, 107*(11), 3430–3436.
Rivedal, H. M., Wiseman, M., Richardson, B., Massie, S. T., Garfinkel, A., Ocamb, C., & Gent, D. H. (2023). Characterization of powdery mildew fungi affecting hemp in the Pacific Northwest. *PhytoFrontiers, 4*(2), 205–212.
Roborgh, R. H. J. (1961). New varieties of hops tolerant to black root rot (*Phytophthora* spp.). *New Zealand Journal of Agriculture, 103*, 169–171.
Royle, D. J. (1968). Hop black root-rot (*Phytophthra citricola* Sawada) in England. *Plant Pathology, 17*(1), 36–40.
Royle, D. J. (1970). Infection periods in relation to the natural development of hop downy mildew (*Pseudoperonospora humuli*). *Annals of Applied Biology, 66*(2), 281–291.
Royle, D. J. (1978). Powdery mildew of the hop. In D. Spencer (Ed.), *The powdery mildews* (pp. 381–409). Academic Press.
Royle, D. J., & Kremheller, H. T. (1981). Downy mildew of the hop. In D. M. Spencer (Ed.), *Downy Mildews* (pp. 395–419). Academic Press.
Royle, D. J., & Thomas, G. G. (1973). Factors affecting zoospore responses towards stomata in hop downy mildew (*Pseudoperonospora humuli*) including some comparisons with grapevine downy mildew (*Plasmopara viticola*). *Physiological Plant Pathology, 3*(3), 405–417.
Runge, F., & Thines, M. (2012). Reevaluation of host specificity of the closely related species *Pseudoperonospora humuli* and *P. cubensis*. *Plant Disease, 96*(1), 55–61.
Runge, F., Choi, Y. J., & Thines, M. (2011). Phylogenetic investigations in the genus *Pseudoperonospora* reveal overlooked species and cryptic diversity in the *P. cubensis* species cluster. *European Journal of Plant Pathology, 129*(2), 135.
Sakai, M. (1976). Studies on grey mould of hop (*Botrytis cinerea*). II. Occurrence of grey mould and rainfall in flower period. *Bulletin of Brewing Science, 22*, 1–7.
Salmon, E. S., & Ware, W. M. (1925). The downy mildew of the hop and its epidemic occurrence in 1924. *Annals of Applied Biology, 12*(2), 121–151.
Salmon, E. S., & Ware, W. M. (1936a). The Cladosporium disease of the hop. *Journal of the South-Eastern Agricultural College, 38*, 53–54.
Salmon, E. S., & Ware, W. M. (1936b). Sclerotinia "canker" of the hop. *Journal of the South-Eastern Agricultural College, 37*, 25–26.
Salmon, E. S., & Ware, W. M. (1937). The honey fungus (*Armillaria mellea*) attacking fruit trees and hops; with observations on Pholiota squarrosa in cherry orchards. *Journal of the South-Eastern Agricultural College, 40*, 18–26.
Salmon, E. S., & Wormald, H. (1922). Hop "canker" or "growing off". *Journal of the Ministry of Agriculture, 29*(4), 354–359.
Salmon, E. S., & Wormald, H. (1923). A new Cercospora on Humulus. *Journal of Botany, British and Foreign, 61*, 134–136.
Sano, T., Sasaki, M., & Shikata, E. (1985). Apple mosaic virus isolated from hop plants in Japan. *Annals of Applied Biology, 106*(2), 305–312.
Sano, T., Yoshida, H., Goshono, M., Monma, T., Kawasaki, H., & Ishizaki, K. (2004). Characterization of a new viroid strain from hops: Evidence for viroid speciation by isolation in different host species. *Journal of General Plant Pathology, 70*(3), 181–187.
Sasaki, M., & Shikata, E. (1977). On some properties of hop stunt disease agent, a viroid. *Proceedings of the Japan Academy, Series B, 53*(3), 109–112.
Sawada, K. (1927). Descriptive catalogue of the Formosan fungi. *Report of the Government Research Institute Department of Agriculture Formosa, 27*, 21–24.
Schlenzig, A., Campbell, R., & Chard, J. (2015). Phytophthora species infecting hardy ornamentals in nurseries and the managed environment in Scotland. *Journal of Phytopathology, 163*(7–8), 686–689.

Schnathorst, W. C. (1981). Life cycle and epidemiology of Verticillium. In E. M. Mace, A. A. Bell, & C. H. Beckman (Eds.), *Fungal wilt diseases of plants* (pp. 81–111). Academic Press.

Scott, P. M., Burgess, T. I., Barber, P. A., Shearer, B. L., Stukely, M. J. C., Hardy, G. S. J., & Jung, T. (2009). *Phytophthora multivora* sp. nov., a new species recovered from declining Eucalyptus, Banksia, Agonis and other plant species in Western Australia. *Persoonia, 22*(1), 1–13.

Seefelder, S., Seigner, E., Niedermeier, E., Radišek, S., & Javornik, B. (2009). Genotyping of Verticillium pathotypes in the Hallertau: Basic findings to assess the risk of Verticillium infections. In E. Seigner (Ed.), *Proceedings of the Scientific Commission of the Internation Hop Growers Convention* (pp. 67–69). International Hop Growers` Convention.

Seigner, E., Seefelder, S., Haugg, B., Hesse, H., Rosch, H., & Felsenstein, F. (2001). Investigations on the virulence spectrum of hop powdery mildew (*Sphaerotheca humuli*) and on the effectiveness of race-specific resistance genes. In E. Seigner (Ed.), *Proceedings of the Scientific Commission of the Internation Hop Growers Convention* (pp. 40–43). International Hop Growers` Convention.

Sekimoto, S., Uehara, T., & Mizukubo, T. (2017). Characterization of populations of Heterodera trifolii Goffsrt, 1932 (Nematoda: Heteroderidae) in Japan and their phylogenetic relationships with closely related species. *Nematology, 19*(5), 543–558.

Sen, A., & Jensen, H. (1969). Host-parasite relationships of various plants and hop cyst nematode. *Heterodera humuli. Plant Disease Reporter, 53*(1), 37–40.

Sewell, G. W. F., & Wilson, J. F. (1958). Weed hosts of the 'progressive' Hop strain of Verticillium albo-atrum Reinke and Berth. In *Report of East Malling Research Station* (pp. 126–128).

Sewell, G. W. F., & Wilson, J. F. (1966). Verticillium wilt of the hop: The survival of Verticillium albo-atrum in soil. *Annals of Applied Biology, 58*(2), 241–249.

Sewell, G. W. F., & Wilson, J. F. (1967). Verticillium wilt of the hop: Field studies on wilt in a resistant cultivar in relation to nitrogen fertilizer applications. *Annals of Applied Biology, 59*(2), 265–273. https://doi.org/10.1111/j.1744-7348.1967.tb04435.x

Sewell, G. W. F., & Wilson, J. F. (1974). Hop wilt, soil temperature and nitrogen. *East Malling Research Station Annual Report for 1973*, 203–204.

Sewell, G. W. F., & Wilson, J. F. (1984). The nature and distribution of Verticillium albo-atrum strains highly pathogenic to the hop. *Plant Pathology, 33*(1), 39–51.

Sewell, G. W. F., Wilson, J. F., & Martin, D. G. (1962). Machine picking in relation to progressive verticillium wilt of the hop. II. The effect of composting on the infectivity of machine-picked hop waste. *East Malling Research Station Annual Report for 1961., A45*, 102–106.

Sharma, S., Strickland, D. A., Hay, F., & Pethybrige, S. J. (2023). First report of halo blight on hop caused by *Diaporthe humulicola* in New York. *Plant Disease, 107*(1), 216.

Simmons, E. G. (2002). *Alternaria* themes and variations (305-309) Lewial Alternaria revised. *Mycotaxon, 83*, 127–145.

Singh, G., Singh, R. N., & Bhandari, T. P. S. (1985). Ascochyta leaf spots on fig and hop. New records for India. *Indian Journal of Mycology and Plant Pathology, 14*(1), 100.

Skotland, C. B., & Johnson, D. A. (1983). Control of downy mildew of hops. *Plant Disease, 67*(11), 1183–1185.

Smith, D. R., & Skotland, C. B. (1986). Host range and serology of Prunus necrotic ringspot virus serotypes isolated from hops (*Humulus lupulus*) in Washington. *Plant Disease, 70*(11), 1019–1023.

Solarska, E. (1981). Więdnięcie infekcyjne chmielu (*Humulus lupulus* L.) uprawianego na Lubelszczyźnie. *Pamietnik pulawski, 74*, 147–151.

Solarska, E. (2003). Control of Fusarium spp. on hop cones with grapefruit extract. In E. Seigner (Ed.), *Proceedings of the Scientific Commission of the International Hop Growers' Convention* (pp. 32–35). International Hop Growers` Convention.

Solarska, E., Kamińska, M., & Śliwa, H. (2004). First report of Phytoplasma infection in hop plants. *Plant Disease, 88*(8), 908.

Sonoda, R. M., & Ogawa, J. M. (1968). Phytophthora crown and root rot of hops in California. *Plant Disease Reporter, 52*(1), 76–78.

Stack, G. M., Cala, A. R., Quade, M. A., Toth, J. A., Monserrate, L. A., Wilkerson, D. G., Carlson, C. H., Mamerto, A., Michael, T. P., Crawford, S., Smart, C. D., & Smart, L. B. (2023). Genetic mapping, identification, and characterization of a candidate susceptibility gene for powdery mildew in *Cannabis sativa. Molecular Plant-Microbe Interactions, 37*(1), 51–61.

Štajner, N., Radišek, S., Mishra, A. K., Nath, V. S., Matoušek, J., & Jakše, J. (2019). Evaluation of disease severity and global transcriptome response induced by Citrus bark cracking viroid, hop latent viroid, and their co-infection in hop (*Humulus lupulus* L.). *International Journal of Molecular Sciences, 20*(13), 3154. https://doi.org/10.3390/ijms20133154

Subbotin, S. A., Mundo-Ocampo, M., & Baldwin, J. G. (2010). *Systematics of cyst nematodes (Nematoda: Heteroderinae)*. Brill.

Summers, C. F., Gulliford, C. M., Carlson, C. H., Lillis, J. A., Carlson, M. O., Cadle-Davidson, L., Gent, D. H., & Smart, C. D. (2015). Identification of genetic variation between obligate plant pathogens *Pseudoperonospora cubensis* and *P. humuli* using RNA sequencing and genotyping-by-sequencing. *PLoS One, 10*(11), e0143665.

Sutton, B. C. (1980). *The Coelomycetes: Fungi imperfecti with pycnidia, acervuli and stromata*. Commonwealth Mycological Institute.

Suzuki, T., Fujibayashi, M., Hataya, T., Taneda, A., He, Y. H., Tsushima, T., Duraisamy, G. S., Siglová, K., & Matoušek, J. (2017). Characterization of host-dependent mutations of apple fruit crinkle viroid replicating in newly identified experimental hosts suggests maintenance of stem-loop structures in the left-hand half of the molecule is important for replication. *Journal of General Virology, 98*(3), 506–516. https://doi.org/10.1099/jgv.0.000693.eppo

Svobodová, I. (2023, April 17–19). *Interesting diseases in the Czech Republic* [Annual meeting]. Meeting of European Mycological Network, 25th Annual meeting, NIVIP, NPPO Wageningen.

Talboys, P. W. (1958). Association of tylosis and hyperplasia of the xylem with vascular invasion of the hop by *Verticillium albo-atrum. Transactions of the British Mycological Society, 41*(2), 249–260.

Thomas, A., Carbone, I., Choe, K., Quesada-Ocampo, L. M., & Ojiambo, P. S. (2017). Resurgence of cucurbit downy mildew in the United States: Insights from comparative genomic analysis of *Pseudoperonospora cubensis. Ecology and Evolution, 7*(16), 6231–6246.

Thomas, W. J., Borland, T. G., Bergl, D. D., Claassen, B. J., Flodquist, T. A., Montgomery, A. S., Rivedal, H. M., Woodhall, J., Ocamb, C. M., & Gent, D. H. (2022). A quantitative PCR assay for detection and quantification of *Fusarium sambucinum. Plant Disease, 106*(10), 2601–2606.

Thresh, J. M., Pitcher, R. S., McNamara, D. G., & Ormerod, P. J. (1972). The spread and control of nettlehead and related diseases of hop. *Report of East Malling Research Station for 1971*, 155–162.

Tsykun, T., Prospero, S., Schoebel, C. N., Rea, A., & Burgess, T. I. (2022). Global invasion history of the emerging plant pathogen *Phytophthora multivora. BMC Genomics, 23*(1), 153.

Turechek, W. W., Mahaffee, W. F., & Ocamb, C. M. (2001). Development of management strategies for hop powdery mildew in the Pacific Northwest. *Plant Health Progress, 2*(1), 8.

Turner, S., & Rowe, J. (2006). Cyst nematodes. In R. Perry & M. Moens (Eds.), *Plant nematology* (pp. 91–122). CAB International.

Twomey, M. C., Wolfenbarger, S. N., Woods, J. L., & Gent, D. H. (2015). Development of partial ontogenic resistance to powdery mildew in hop cones and its management implications. *PLoS One, 10*(3), e0120987.

Twomey, M. C., Stone, J., & Gent, D. (2016). Black wilt of hop (*Humulus lupulus*) caused by *Diplodia seriata* in New York state. *Plant Disease, 100*(4), 861.

Valdez, R. B., McNamara, D. G., Ormerod, P. J., Pitcher, R. S., & Thresh, J. M. (1974). Transmission of the hop strain of arabis mosaic virus by *Xiphinema diversicaudatum. Annals of Applied Biology, 76*(1), 113–122.

Vernière, C., Perrier, X., Dubois, C., Dubois, A., Botella, L., Chabrier, C., Bové, J. M., & Duran Vila, N. (2004). Citrus viroids: Symptom expression and effect on vegetative growth and yield of clementine trees grafted on trifoliate orange. *Plant Disease, 88*(11), 1189–1197. https://doi.org/10.1094/pdis.2004.88.11.1189

Vernière, C., Perrier, X., Dubois, C., Dubois, A., Botella, L., Chabrier, C., Bové, J. M., & Duran Vila, N. (2006). Interactions between citrus viroids affect symptom expression and field performance of clementine trees grafted on trifoliate orange. *Phytopathology, 96*(4), 356–368. https://doi.org/10.1094/phyto-96-0356

Von Mende, N., & McNamara, D. G. (1995). Biology of the hop cyst nematode *Heterodera humuli*: I. Life cycle and survival. *Annals of Applied Biology, 126*(3), 505–516.

Wallace, E. C., & Quesada-Ocampo, L. M. (2017). Analysis of microsatellites from the transcriptome of downy mildew pathogens and their application for characterization of *Pseudoperonospora populations*. *PeerJ, 5*, e3266.

Ware, W. M. (1926). *Pseudoperonospora humuli* and its mycelial invasion of the host plant. *Transactions of the British Mycological Society, 11*(1–2), 91–107.

Ware, W. M. (1929). Experiments on the production of diseased shoots by the hop downy mildew, *Pseudoperonospora humuli* (Miy. et Takah.), Wils. *Annals of Botany, 43*(172), 683–710.

Warren, J. G., Mercado, J., & Grace, D. (2019). Occurrence of hop latent viroid causing disease in Cannabis sativa in California. *Plant Disease, 103*(10), 2699–2699. https://doi.org/10.1094/PDIS-03-19-0530-PDN

Waterhouse, G. M. (1957). *Phytophthora citricola* Sawada (Syn. *P. Cactorum* var. *aplanata* Chester). *Transactions of the British Mycological Society, 40*(3), 349–357.

Watson, T., Suarez, M., Deng, Z., & Desaeger, J. A. (2020). Plant-parasitic nematodes associated with the root zone of hop cultivars planted in Florida field soil. *Journal of Nematology, 52*(1), 1–10. https://doi.org/10.21307/jofnem-2020-040

Weiland, J. E., Nelson, A. H., & Hudler, G. W. (2010). Aggressiveness of *Phytophthora cactorum*, *P. citricola* I, and *P. plurivora* from European Beech. *Plant Disease, 94*(8), 1009–1014.

Weldon, W. A., Gent, D. H., & Gadoury, D. M. (2021a). Management of hop powdery mildew in the context of recent advances in pathogen ecology and population genetics. *Plant Health Progress, 22*(4), 450–458.

Weldon, W. A., Knaus, B. J., Grünwald, N. J., Havill, J. S., Block, M. H., Gent, D. H., Cadle-Davidson, L. E., & Gadoury, D. M. (2021b). Transcriptome-derived amplicon sequencing markers elucidate the U.S. *Podosphaera macularis* population structure across feral and commercial plantings of *Humulus lupulus*. *Phytopathology, 111*(1), 194–203.

Weldon, W. A., Marks, M. E., Gevens, A. J., D'Arcangelo, K. N., Quesada-Ocampo, L. M., Parry, S., Gent, D. H., Cadle-Davidson, L. E., & Gadoury, D. M. (2021c). A comprehensive characterization of ecological and epidemiological factors driving perennation of *Podosphaera macularis* chasmothecia on hop (*Humulus lupulus*). *Phytopathology, 111*(11), 1972–1982.

Wilcox, W. (2017, June). *Grape Disease Control: Taking Stock and Looking Forward*. Research Focus 2017-6, Appellation Cornell. Cornel University. Retrieved August 5, 2024, from https://cpb-us-e1.wpmucdn.com/blogs.cornell.edu/dist/3/7313/files/2016/12/Wilcox-Grape-Disease-Control-2017-1uzpoqi.pdf

Williamson, B., Tudzynski, B., Tudzynski, P., & Van Kan, J. A. L. (2007). *Botrytis cinerea*: The cause of grey mould disease. *Molecular Plant Pathology, 8*(5), 561–580. https://doi.org/10.1111/j.1364-3703.2007.00417.x

Wolfenbarger, S. N., Eck, E. B., & Gent, D. H. (2014). Characterization of resistance to powdery mildew in the hop cultivars Newport and Comet. *Plant Health Progress, 15*(2), 55–56.

Wolfenbarger, S. N., Twomey, M. C., Gadoury, D. M., Knaus, B. J., Grünwald, N. J., & Gent, D. H. (2015). Identification and distribution of mating-type idiomorphs in populations of *Podosphaera macularis* and development of chasmothecia of the fungus. *Plant Pathology, 64*(5), 1094–1102.

Wolfenbarger, S. N., Massie, S. T., Ocamb, C., Eck, E. B., Grove, G. G., Nelson, M. E., Probst, C., Twomey, M. C., & Gent, D. H. (2016). Distribution and characterization of *Podosphaera macularis* virulent on hop cultivars possessing *R6*-based resistance to powdery mildew. *Plant Disease, 100*(6), 1212.

Woodhall, J. W., Barbour, J. D., Fairchild, K., & Wharton, P. S. (2020). First report of *Rhizoctonia solani* anastomosis Group 2-1 affecting shoots of *Humulus lupulus* in Idaho. *Plant Disease, 104*(9), 2520.

Woods, J. L., & Gent, D. H. (2016). Susceptibility of hop cultivars to downy mildew: Associations with chemical characteristics and region of origin. *Plant Health Progress, 17*(1), 42–48.

Wormald, H. (1928). The parasitism of the hop leaf-spot fungus *Cercospora cantuariensis*. *Transactions of the British Mycological Society, 13*(1–2), 32–39. https://doi.org/10.1016/S0007-1536(28)80002-7

Wormald, H. (1946). *Diseases of fruits and hops*. Crosby Lockwood & Son.

Yu, J., & Liu, Y. (1987). The occurrence of three viruses in hop (*Humulus lupulus*) in China. *Plant Pathology, 36*(1), 38–44.

Zhang, M., Wu, H., & Li, D. (2009). A new species of *Alternaria* on *Humulus scandens* from China. *Mycotaxon, 108*(1), 49–52.

Chapter 6
Entomology of Hops (Pests)

Magda Rak Cizej, FrančFek Poličnik, and Octave Lacroix

6.1 Introduction

Hop, *Humulus lupulus* L., is an important plant in the brewing industry, contributing to the aroma, flavour, and stability of beer. However, hop cultivation is often threatened by various pests that can cause significant crop damage, reducing both yield and quality. Among the most common pests, which cause important economic damage every year in hop gardens are: the damson-hop aphid, *Phorodon humuli* (Schrank), the two-spotted spider mite, *Tetranychus urticae* Koch, the alfalfa snout beetle, *Otiorhynchus ligustici* (Linnaeus), the hop flea beetle, *Psylliodes attenuata* (Koch), and the European Corn Borer (ECB), *Ostrinia nubilalis* Hübner. These pests not only affect the yield but also the quality of hops, which in turn impacts the entire brewing industry.

Besides the aforementioned pests, hop damage can also be caused by the rosy rustic, *Hydraecia micacea* Esper, the cabbage moth, *Mamestra brassicae* (L.), hop snout weevil, *Neoplinthus tigratus* subsp. *porcatus* (Panzer), slugs, *Arion* spp., and other wireworms.

Understanding the biology and ecology of these pests is crucial for developing effective management strategies. In this chapter, we will explore the different types of pests that attack hops, their life cycles, the damage they cause, and methods for their management.

M. Rak Cizej (✉) · F. Poličnik · O. Lacroix
Plant Protection Department, Slovenian Institute of Hop Research and Brewing, Žalec, Slovenia
e-mail: magda.rak-cizej@ihps.si; franci.policnik@ihps.si; octave.lacroix@ihps.si

K. Carbone (ed.), *All About Hops: The Crop, its Cultivation, and its Uses in Brewing and Beyond*, https://doi.org/10.1007/978-3-031-96606-4_6

6.2 Damson-Hop Aphid, *Phorodon humuli*

The damson-hop aphid, *Phorodon humuli* (Schrank), is an important pest of hops that appears annually and causes both direct and indirect damage. Direct damage is caused by sucking plant sap, which inhibits the growth and development of the plant, while indirect damage is caused by transmitting pathogenic viruses.

The primary hosts of the hop aphid are plants from the genus *Prunus*, such as the prune plum, *Prunus domestica* L., while the secondary host is hop, *Humulus lupulus*. The hop aphid belongs to the family of true aphids (Aphididae) and can completely destroy the hop crop.

Aphids overwinter in the form of eggs on stone fruit trees, and in the spring, they hatch and reproduce asexually. The first winged aphids appear on hops at the end of April or the beginning of May, and their flight lasts until the beginning of July. On hops, wingless larvae quickly develop new generations of aphids, which can lead to a massive presence on the underside of leaves and in hop cones.

6.2.1 *Determination*

Hop aphid on primary host is light green with green stripes on the back (Fig. 6.1). On secondary host, aphid is light green to yellow, almost white. Its body length ranges from 1 to 1.9 mm. Hop aphid on primary and secondary host has brown/black head and dorsal part of thorax. Ventral part of body shows 3 lateral spots and a central scattered spot, with a body length of 2.1–2.8 mm.

Fig. 6.1 Damson-hop aphid winged, nymphs and adults. (Photo: Magda Rak Cizej)

6.2.2 Biology and Life History

P. humuli is a holocyclic aphid. Summer hosts are cultivated hops, wild hops, *Humulus lupulus* and *Humulus scandens* (Loureiro) Merrill, and nettle, *Urtica dioica* L. Winter hosts are *Prunus* such as *P. domestica* L., *P. spinosa* L., *P. padus* L., *P. cerasifera* Ehrh., *P. serotina* Ehrh. and *P. persicae* (L.) Stokes (Campbell et al., 1993).

On the primary host (winter host), larvae of the 1st generation of fundatrix emerge in March when *Prunus* starts to bud. We observe up to 6 parthenogenetic generations of aphids on primary host. The development time of the first generation ranges from 14 to 18 days. Each individual produces on average 21 descendants. From the 3rd generation on, *P. humuli* starts producing more and more winged aphids who starts emigrating to their secondary host (summer host) due to a quality loss of food substrate from *Prunus*. The peak migration to hops is reached at the end of May (Table 6.1 and Fig. 6.2). Aphids prioritise young yellow-green leaves of hops at their arrival. *P. humuli* multiplies until the migration back on primary host in autumn. First females are observed during the second half of September. After mating, in October, females lay eggs in sheltered places, between buds and leaves. On secondary hosts, 7 to 9 parthenogenetic generations of flightless aphids follow each

Table 6.1 The beginning and ending flight period of damson-hop aphid, *Phorodon humuli*, between 1996 and 2024

Monitored year	Beginning *flight* period (day/month)	Ending *flight* period (day/month)	Duration *flight* period (days)
1996	27/5	7/7	42
1997	12/5	26/6	46
1998	4/5	8/6	36
1999	10/5	2/7	54
2000	29/4	21/6	54
2001	5/5	1/7	58
2002	4/5	11/6	39
2003	4/5	15/6	43
2004	8/5	27/6	51
2005	10/5	29/6	51
2006	8/5	25/6	49
2007	23/4	30/5	38
2008	2/5	16/6	45
2009	4/5	17/6	44
2010	5/5	18/6	44
2011	2/5	20/6	50
2012	1/5	6/6	37
2013	3/5	21/6	50
2014	2/5	11/6	40
2015	6/5	10/6	36

Monitored year	Beginning *flight* period (day/month)	Ending *flight* period (day/month)	Duration *flight* period (days)
2016	2/5	13/6	42
2017	25/4	12/6	49
2018	30/4	8/6	40
2019	8/5	14/6	38
2020	27/4	4/6	38
2021	15/5	23/6	39,0
2022	3/5	8/6	36
2023	18/5	21/6	34
2024	29/4	3/6	35
Average duration flight period (days) in 28 years			43.4

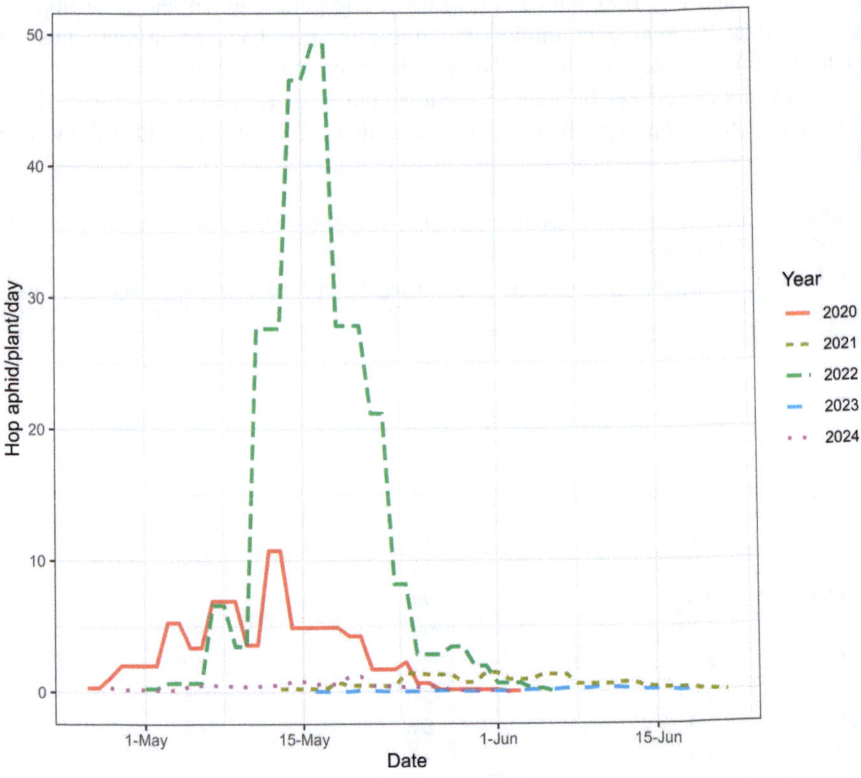

Fig. 6.2 Population of winged damson-hop aphid, *Phorodon humuli*, on hops in different years between April and June

other. Each individual produces between 30 and 80 descendants. Return on primary host is allowed thanks to winged gynopares (pre-sexual females) aphids and males (Loxdale et al., 1998). Aphids lay 7 to 15 larvae under the leaves. Then, on the second part of September, the first female aphids can be observed laying from 6 to 12 eggs that will overwinter on *Prunus* for a new cycle.

Sexual morph production in the damson-hop aphid is controlled by day length (Campbell & Tregidga, 2005, 2006). Wingless aphids that had matured in short-day conditions (Light Dark (LD) 12:12 h) produced gynopares and males, while those reared in long-day conditions (LD 18:6) produce mainly wingless aphids and gyropares in a smaller portion.

Host detection is allowed thanks to volatiles perceived by winged aphids (Campbell et al., 1993).

6.2.3 Damage

Damage of the damson-hop aphid is caused by direct feeding on leaves by insertion of stylet into the phloem strands and excretion of honeydew, which works as a growing substrate for sooty mould and as a vector of hop viruses (Hop mosaic virus (*Carlavirus humuli*) and American hop latent virus (*Carlavirus americanense*)) (Fig. 6.3). A threshold of 200 aphids per hop leaf was defined by Knan & Kremheller (1984) under which no foliar damage was observed before flowering, however this density differs according to productions and countries. Heavy feeding of the aphid may cause defoliation and reduce plant vigour. Once hop exhibits cones, aphids can turn into a large depreciator of the quality of the hop products. Indeed, cones usually

Fig. 6.3 Damson-hop aphid, *Phorodon humuli*, on hop leaf. (Photo: Magda Rak Cizej)

remain "sheltered" from insecticides and offer a refuge to aphids, which destroy the plant's organs. Excess honeydew and sooty mould also contribute to the downgrade of crop quality.

6.2.4 *Management/Control*

6.2.4.1 Biological Control

Hippodamia convergens Guérin-Méneville and *Coccinella transversoguttata* Falderman are the two main predators of the damson-hop aphid (Campbell & Cone, 1999). Observed an average larvae consumption of 318 and 413 adult aphids, for *H. convergens* and *C. transversoguttata* (female larvae here), respectively. Lacewings, hover flies and birds are also responsible for the regulation of the damson-hop aphid. Campbell (2018) found that meadow-mix of flowering plants as interrow in hop was a great means to ensure a higher aphid natural enemy density, where hop yield loss could easily be corrected with a heightened fertilization (Campbell & Cone, 1999; Campbell, 2018).

6.2.4.2 Insecticides

Resistances to conventional insecticides such as organophosphates, carbamates and pyrethroids were already observed more than 40 years ago in hop aphid populations (Lewis & Madge, 1984; Muir, 1979). Weichel & Nauen (2003) found no resistance of *P. humuli* to imidacloprid, amitraz and pymetrozine, and a moderate resistance to pyrethroids, xydemeton-methyl and pirimicarb in German hop. In the past, highly effective neonicotinoids were used to control aphids but have been banned in Europe since 2019. Today, only a few systemic insecticide compounds exist, and resistance can be expected to develop in the very near future.

6.2.4.3 Cultural and Mechanical Methods

Secondary host (hop) recognition by damson-hop aphid response to semiochemicals can be disrupted, however significant results have not been obtained yet (Lösel et al., 1996). Lösel et al. (1996) found out that semiochemicals produced by aphids and/or hosts can be used to increase trap efficiencies. Inter crop (Campbell & Ridout, 2001) or cover crop (Calderwood et al., 2017) did not have any effect on damson-hop aphid population.

6.2.4.4 Resistances

Certain cultivars were identified as less sensitive to damson-hop aphid, such as "Spalter Select" and "Boadicea", however these varieties cannot be named resistant to aphids as they still receive attacks of *P. humuli* and are suitable hosts for population survival (reasonable lifespans and cumulative offspring) (Weihrauch & Moreth, 2005; Weihrauch, 2009; Campbell, 2018). Paul et al. (1996) used electrical penetration graphs to assess the mechanisms of resistance of hops less susceptible to aphids and found that *P. humuli* had a reduced uptake of phloem and a 2-minute non-probing time, which could be explained by a mechanical factor blocking the stylet of aphids, antinutritional factors and an inadequate supply of nutrients of these hops. Metabolite profiling on hop led to interesting conclusion on the possible role of metabolites responsible for aroma and flavour that may play a role in resistance against aphids (Undas et al., 2018).

6.3 Two-Spotted Spider Mite, *Tetranychus urticae*

Two-spotted sider mite, *Tetranychus urticae* Koch, is an arthropod belonging to the class of Arachnida, responsible for sap sucking on more than 1000 species of plant (De Rouck et al., 2023), including hop (Wu et al., 2019). With numerous generations (5–9) per year, spider mite attacks can be extremely fast and damaging on many crops and ornamental plants.

The mites suck the contents of the cells from the leaves, causing small pale spots or scars, where the green epidermal cells have been destroyed. Although the individual lesions are very small, the attack of hundreds or thousands of mites can cause thousands of lesions, which can significantly reduce the photosynthetic ability of plants. The mite multiplies rapidly in hot and dry conditions, which can lead to an exponential increase in the population.

6.3.1 Determination

Eggs are round, about 0.14 mm of diameter, first transparent then opaque and finally they become yellow before hatching. The first eggs laid by overwintering females have pigments similar to the female. Larvae have 3 pairs of legs, and look pale at first but quickly turn greenish. Two red eye-spots and two dark spots in the middle of their body can be observed. Protonymphs are bigger, with 4 pairs of legs and theirs colour varies from pale to dark green. Their two spots on side of body become more visible. Deutonymphs are even bigger than protonymphs but show identical colour (Fig. 6.4). Sexual dimorphism can only be observed when mites reach adult

Fig. 6.4 Different stages of two-spotted spider mite, *Tetranychus urticae*, on underside hops leaf. (Photo: Magda Rak Cizej)

stage. Males are ¾ the size of females, which is more oval with a body length of 0.5–0.6 mm. Females can be orange, light yellow, light to dark green and red. Male is more active and thinner than females. Its colour can vary between light and dark yellow, orange and brown. Overwinter females are red-orange due to their high concentration of hydroxy-keto carotenoids. As they start feeding, they slowly return to their greenish colours.

6.3.2 *Biology and Life History*

T. urticae has more than 1000 host plants. During its life, the two-spotted spider mite goes through 5 stages: egg, larva, protonymph, deutonymph and adult; a cycle lasting between 7 days at 30 °C and 36 days at 15 °C. Diapause of the females enables the overwintering of the spider mite, induced by lowering of temperature, day length and food availability. Per year we observe from 5 to 9 generations during summer. The female lays her eggs mostly under leaves. Under low humidity and high temperatures, spider mite thrives. At 80% RH, mite stops feeding and at 90% RH, it stops reproducing. Males mature earlier than females, therefore they can be observed in resting rate next to female deutonymphs, ready to copulate as soon as the adult female emerges. Eggs are produced in a ratio 1:3 (Male: Female). Without fertilization of eggs, parthenogenesis takes over and produces haploid males, resulting in a potential outbreak from a single female. On average 6 eggs per day are laid by a female, ensuing a total of 100 to 150 eggs produced by one female in her lifetime.

6.3.3 Damage

Larvae, nymphs and adults are located mostly under leaves where they use their stylet-like mouthparts to suck the cells' content. Their stylet-like mouthparts can pierce and penetrate epidermal and parenchyma cells, particularly causing the removal of chlorophyll. Damaged cells stand out on the leaves, leaving yellow chloroses on the leaf surface, resulting in a reduce photosynthesis rate and transpiration rate for infested crops (Ferree & Hall, 1980; Hall & Ferree, 1975; Lacroix et al., 2024; Sances et al., 1981; Youngman et al., 1986). In the later stages of the growing season, when hop cones are present, different stages of the two-spotted spider mite can damage not only the leaves but also the hop cones (Fig. 6.5). This damage can be significant, as mites feed on the cones by piercing plant cells and sucking out their contents, leading to discolouration, decreased quality and reduced yield. Proper monitoring and management of spider mite populations during the cone development stage are crucial to prevent economic losses in hop production.

6.3.4 Management/Control

6.3.4.1 Biological Control

Beneficial organisms, such as *Feltiella acarisuga* (Vallot) (Mo & Liu, 2007; Topakci, 2022; Xiao et al., 2011) or *Macrolophus caliginosus* Wagner (Enkegaard et al., 2001; Foglar et al., 1990; Hansen et al., 1999) show great control of the

Fig. 6.5 Damage on hops: webbing of the two-spotted mite, *Tetranychus urticae*. (Photo: Magda Rak Cizej)

two-spotted spider mite. Predatory mites, *Phytoseiulus persimilis* Athias-Henriot and *Amblyseius andersoni* (Chant) (Chacon-Hernandez et al., 2016), can also be used for the control of *T. urticae*. Predatory mites tend to avoid infected mites (i.e., *T. urticae* that have been exposed to *Beauveria bassiana* (Bals.-Criv.) Vuill, used in the biological control of the European Corn Borer), meaning that several natural enemies can be used in biological control without negative interference (Seiedy et al., 2013).

6.3.4.2 Insecticides

Acaricides and insecticides can be used to control the two-spotted spider mite, however care must be taken to use a variety of substances to overrule or delay resistances to acaricide and insecticide (Piraneo et al., 2015; Wu et al., 2019). Barati and Hejazi (2015) observed significantly higher reproductive rate (i.e., oviposition rate and survival rate) of female mites in the use of neonicotinoids (i.e., acetamiprid, thiamethoxam and thiacloprid at recommended dose of 150, 125, and 120 mg ai/l). One of the mechanisms responsible for an increase in mite populations is the reduction of natural enemies; indeed, the use of acaricides can also be detrimental to naturally occurring enemies (Lilley & Campbell, 1999).

6.3.4.3 Cultural and Mechanical Methods

To prevent the spread of population of mites in the hop field, material, tools and irrigation system need to be disinfected. In the same direction, overtillage can favour the dust propagation and mites' dispersion (Strong et al., 1999). Additionally, to prevent outbreaks of mites, a high care must be taken regarding any excessive nitrogen fertilization which can lead to 4 times the amount of *T. urticae* compared to unfertilised crop (Hata et al., 2019; Ahmad et al., 2022) and irrigation due to the high resistant (76% of survival rate) of diapausing female submerged in water (Popov et al., 2021).

6.4 Alfalfa Snout Beetle, *Otiorhynchus ligustici*

Otiorhynchus ligustici (L.) is a wingless weevil feeding on leaves' margins, buds, shoots and flowers of hop. Larvae damage the roots of hop as well (Muška et al., 2020). *O. ligustici* is also observed on leguminous, especially alfalfa, *Medicago sativa* L., and clovers, *Trifolium* spp., grapevine, *Vitis vinifera* L., young beetroot, *Beta vulgaris* L., etc. (Harcourt & Bereza, 2017).

6.4.1 Determination

Adult is 10–14 mm long, in a convex shape, sheltered by grey-brown scales and light-coloured spurs with metallic reflections (Fig. 6.6). Its abdomen is oval and its elytra are finely grained. Rostrum is wide, finely cut on low part. Its superior face has a median hull (carinated shape). Femur has a short and blunt tooth on the lower side. Legless larva is 10–12 mm long, its body is white and slightly curved, and the head capsule is brown (Fig. 6.7). Description of 1st instar and mature larvae was deeply described by Gosik and Sprick (2012).

6.4.2 Biology and Life History

O. ligustici's main host is hop, however it can also strongly damage alfalfa and clover meadows. When these hosts are absents, other leguminous (pea, *Lathyrus oleraceus* Lam., beans, *Phaseolus vulgaris* Blanco), grapevine, young beetroots, asparagus. *Asparagus officinalis* L. and cereals can attract them.

Snout beetle has a 2-year life cycle. Eggs hatch in soil in autumn. The new larva will diapause and overwinter in soil at 30 to 40 cm soil depth. While remaining in soil the following year, the snout beetle completes its development to adult stage until July and overwinter without leaving the soil. Only as second spring comes do we observe adults emerging from soil (Schroeder et al., 1995). They start searching for hosts in April. They cannot fly as their elytra are joined; therefore, they usually remain on the same host plant. Later on, eggs are laid on the roots of host plant, either alone or in patch. Reproduction is mostly parthenogenetic, as males are rare. One female lay between 400 and 800 eggs. Egg development time is 1.5–3 weeks long, while larva stage lasts one year with the overwintering diapause. If spring conditions are not adequate (wet and cold climate), the larva can delay its development into an adult until after a second winter, in its third year of development.

6.4.3 Damage

The economic threshold level for *O. ligustici* is 1 adult per 40 m² on alfalfa, but on hops, thresholds are lower, with 1 individual per 2–3 plants as the limit. Soil sampling is the most efficient way to assess density of snout beetle (Harcourt & Binns, 2017).

Most damage occurs as larva stage on roots. They feed on roots from late summer to following spring. Bites in the roots can be identified as grooves and holes, leading to the withering, interruption of growth and death of the plant. Severe infestations can drastically reduce hop yard life. Damage on roots can also be used for

Fig. 6.6 Adult of alfalfa snout beetle, *Otiorhynchus ligustici*, damage on hops stems. (Photo: Magda Rak Cizej)

Fig. 6.7 Larvae of alfalfa snout beetle, *Otiorhynchus ligustici*, on hop roots. (Photo: Magda Rak Cizej)

secondary pathogens as an entrance pathway. Snout beetle is a nocturnal animal and is hardly observed in field but damage from adults can also be observed on leaves to assess the infestation level. Dispersion in the field is favoured by man activity.

6.4.4 *Management/Control*

6.4.4.1 Biological Control

Nematodes (e.g., *Heterorhabditis bacteriophora* Poinar) in controlled conditions showed great reduction in number of snout beetles, yet in field the control was not always as effective due to low dispersal and vertical movement of nematodes in soil (Ferguson et al., 1995; Neumann & Shields, 2008; Schroeder et al., 1994).

Beauveria bassiana used in the control of other hop pests can also be an effective fungal agent to control the snout beetle.

6.4.4.2 Insecticides

Snout beetle was controlled during their adult stage before oviposition with 1.12 kg/ha of methidathion, 0.56 kg/ha of carbofuran and 0.56 kg/ha of azinphosmethyl, offering an average control of 76.3, 83.2 and 79.8% of the snout beetle (York & Gyrisco, 1978). Currently, in Europe, there are no chemical compounds available that are effective against the larvae or adults of *Otiorhynchus* spp. in hops.

6.5 Hop Flea Beetle, *Psylliodes attenuata*

Psylliodes attenuata (Koch) is a hop flea beetle (Coleoptera, Chrysomelidae) damaging hop by feeding on its leaves and cones, which can affect the yield (Rak Cizej & Žolnir, 2003). Its name comes from the ability of adults to jump when disturbed thanks to the enlarged femur. Polyphagous, the hop flea beetle feeds on different plants: hops, hemp, *Cannabis sativa* L. and stinging nettle (Rak Cizej et al., 2007). In the last ten years, the hop flea beetle has been regularly present in hop gardens and causes great damage (Rak Cizej, 2018).

6.5.1 *Determination*

P. attenuata adult is metallic green to bronze colour and about 1.5 to 2.8 mm long, oval, convex and having pubescent body (Fig. 6.8). Their enlarged femur on hind legs allows them to jump up 60 cm high. Eggs are whitish-yellow, oval and around 0.4 mm in diameter. Larvae are 0.5 cm long and off-white with a brown head.

Fig. 6.8 Female adult of hop flea beetle, *Psylliodes attenuata*. (Photo: Magda Rak Cizej)

6.5.2 Biology and Life History

P. attenuata overwinters as an adult to later emerge in spring when the temperature rise above 5 °C (from March to May). The hop flea beetle can use many shelters to overwinter, such as plant debris, loose soil, bark, and others. From May to June, we observe copulation and oviposition. Eggs are laid in loosen soil. Adults then die around end of July. Eggs hatch after a development stage of 10 to 12 days. Larvae feed on the roots of the plant in the first 25 cm of soil then pupate and later on an adult emerge to start overwintering in late autumn.

6.5.3 Damage

Adult feeding causes shot holes on the leaves of young bines as soon as the first hop shoots emerge above ground in spring. Under heavy infestation, this can reduce photosynthesis in the first 1–2 m of the canopy, especially when the leaves begin to appear skeletonized (Figs. 6.9, 6.10, and 6.11) (Weihrauch et al., 2017). Larvae feed on fine roots of hops. Damage is usually not significantly affecting the crop thanks to the rapid growth of hop. The hop flea beetle causes little damage to hops, therefore no intervention threshold has been set. Any injury can be quickly outgrown by healthy hops, only young hops may be susceptible.

Fig. 6.9 Adults of hop flea beetle, *Psylliodes attenuata*, and damage on hop leaves. (Photo: Magda Rak Cizej)

Fig. 6.10 Adults of hop flea beetle, *Psylliodes attenuata*, on hop cones. (Photo: Magda Rak Cizej)

Fig. 6.11 Damage of hop flea beetles, *Psylliodes attenuata*, during summer, left on hop cones, right on hop leaves (photo: Magda Rak Cizej)

6.5.4 Management/Control

6.5.4.1 Biological Control

Entomopathogenic nematodes can be used to reduce the overwintering population of *P. attenuata*. Little is known on the symbiotes of this pest, Guo et al. (2022) described the microbial communities of *P. attenuata* on 9 sites in China and found 1960 known species of bacteria, which most common were *Rickettsiales* and 410 known species in the fungal communities. Most prevalent bacteria were protobacteria, known to perform many important functions such as nitrogen fixation, insecticide resistance, and protection against parasites and pathogen. This study could be a good lead in the identification of biocontrol agent against *P. attenuata*.

Preparation with sea algae (i.e., Algo-Plasmin in 3–4 kg/ha dose applied 3 times) and calcium extract of sea algae (i.e., Coralite KR+ in 1.5 kg/ha dose applied 2 times) showed different efficacy: only calcium extract of algae showed a significant effect on *P. attenuata* (Rak Cizej & Poličnik, 2022). Physical barrier such as kaolinite (i.e., CutiSan in 15 kg/ha dose applied 2 times) and talc-based solution (i.e., Invelop in 15 kg/ha dose applied 2 time) sprayed on hop leaves in year 2021 and 2022 did not show any significant effect on the control of *P. attenuata* (Rak Cizej & Poličnik, 2022).

6.5.4.2 Insecticides

Control of *P. attenuata* can be achieved with insecticides such as cyantraniliprole (i.e., Exirel 0.45 l/ha and 0.75 l/ha) and lambda-cyhalothrin (i.e., Karate Zeon 5 CS 50 g/l) with a reduction of *P. attenuata* of 56.5 and 61.6%, respectively (Rak Cizej & Poličnik, 2022) and pyrethrins. Treatments of hemp on early growth stage (i.e., 2–4 leaves) with imidacloprid (i.e., Gaucho 600 FS 6.0 l/t) and thiamethoxam (i.e., Cruiser 350 FS 10.0 l/t) showed a reduction from 27.6% of shoot attacked to only 2.8 and 3.1%, respectively (Trotuş et al., 2011).

6.5.4.3 Cultural and Mechanical Methods

To capture adult, trap crops such as radish, *Raphanus sativus* L. or mustard, *Sinapis alba* L., can be used, to monitor and manage pest before it reaches hops. Tilling and/or plowing of weeds and hop residue in autumn destroy the overwintering adults of *P. attenuata*. 5 volatile organic compounds (VOCs) of interest (i.e., (E)-alpha Bergamotene, Sesquiterpene, Sesquiphellandrene, Linalool, and 1-hexadecanol) were identified when *P. attenuata* feeds on hop bine, which may be exploited for trapping beetles in the hopyard either as monitoring or mass trapping (Mumm et al., 2017).

6.5.4.4 Resistances

Cultivars of hop present different levels of attractiveness to the hop flea beetle. “Magnum” and “Celeia” hop cultivars showed a high attractiveness in Slovenian hop garden, while “Bucket” hop cultivar remained relatively less visited by the hop flea beetle (Rak Cizej et al., 2007).

6.6 European Corn Borer, *Ostrinia nubilalis*

Ostrinia nubilalis Hübner is a lepidoptera most known for its damage on corn, *Zea mays* L. The larva is responsible for the boring of the stem of corn, leading to the decay of the plant. *O. nubilalis* has been observed on a variety of plants from corn to hop and different strains can be observed among populations (Bourguet et al., 2000; Pelozuelo et al., 2004; Gargani et al., 2018).

6.6.1 Determination

O. nubilalis adult size ranges from 2 to 3 cm wingspan. Female wings are yellow pale with thin brown, serrated transverse streaks (Fig. 6.12). Male wings are darker, with ochre to dark brown shades. Males when resting can be seen exposing their last abdominal segments under folded wings. Eggs are observed under leaves, especially on corn, in masses of about 20 eggs. Fully developed larva body length is around 2 cm. Larva is grey to yellow colours and exhibits on each segment 6 plates, carrying spurs.

Fig. 6.12 Left: Female adult of European Corn Borer (ECB), *Ostrinia nubilalis*; middle: male adult of ECB; right: eggs of ECB just before hatching. (Photo: Franček Poličnik)

6.6.2 Biology and Life History

ECB has up to 200 host plants, major ones are corn, hops, raspberry bush, *Rubus idaeus* L., bell pepper, *Capsicum annuum* L., gladiolus, *Gladiolus* spp., etc. *O. nubilalis* has been observed to be genetically different depending on its host crop and not their geographical distribution (Pelozuelo et al., 2004), exhuming a different pheromone to attract a partner (according to their E or Z strain). However, other countries observed E-strain ECB both on corn and hop (e.g., Slovenia).

One or two generations can be observed, regarding the climate in spring and summer. *O. nubilalis* overwinters in the larval stage and pupates in spring with the increase of temperature. With a base temperature of 10 °C, first flight of adult is expected at 288 Degree-Days (DD), when mating happens and the first eggs are laid at 321DD. Egg development lasts between 5 and 15 days and is favoured by high hygrometry. On the first days of its development (up to its third stage), the larva disperses to surrounding plants, with a high mortality of the ECB. Diapause of *O. nubilalis* is done as larva (5th stage) in the crop debris. Peak of flight is at 482 and 1199DD for the first and second generation, respectively.

6.6.3 Damage

During the first week of its development, the larva mainly feeds on the leaves before boring in the stem (Fig. 6.13). Boring of larvae leads to the disruption of vascular tissues and the flow of nutrients, which can lead to the decay of the bine above the feeding site. Boring of larvae is also known to allow secondary infection by opening a pathway for pathogens into vascular system.

Fig. 6.13 Larvae of European corn borer, *Ostrinia nubilalis*, left in hops cone and right into the stem of hop. (Photo: Magda Rak Cizej)

6.6.4 Management/Control

6.6.4.1 Biological Control

ECB is very efficiently controlled by *Trichogramma brassicae* Bezdenko (Hoffmann et al., 2002, 2006; Razinger et al., 2016; Russell & Bessin, 2009) with inundative release in field. Parasitoid wasps are easily stored in UV-killed eggs of *Ephestia kuehniella* Zeller that they parasitised and the release pattern is designed in a way that covers the whole egg development of *O. nubilalis* in field.

The use of *Bacillus thuringiensis* (Bt) Berliner, is highly effective against *O. nubilalis* (Tan et al., 2011) and genes of resistance to Bt toxin are located in more than one locus on *O. nubilalis* and are not common in its founding population, suggesting a possible delay in the resistance of the ECB to Bt (Alves et al., 2006; Smith & Farhan, 2023). Entomopathogenic fungus *Beauvaria bassiana* showed great reduction of *O. nubilalis* number (Bing & Lewis, 1991; Lewis et al., 2002; Rahman et al., 2010). Zimmermann et al. (2016) and Lewis et al. (2009) also found a high frequence of entomopathogenic microsporidium *Nosema pyrausta* (A. Paillot) J. Weiser in the control of *O. nubilalis*, causing a chronic disease in larvae and adults.

Occurrences of nematodes (e.g., Pristionchus lheritieri (Maupas) Paramonov) (Maupas) Paramonov on the control of *O. nubilalis* was also found (Zimmermann et al., 2016).

6.6.4.2 Cultural and Mechanical Methods

Crop residue removal or destruction by tillage reduce the overwintering population of the ECB (Bing & Lewis, 1993).

6.6.4.3 Resistances

Genetically modified corn, bearing Cry genes (Bt toxin) showed good efficiency in the resistance against *O. nubilalis* (Alves et al., 2006; Labatte et al., 1996; Yu et al., 2018).

6.7 Rosy Rustic Moth, *Hydraecia micacea*

Hydraecia micacea Esper is a moth lepidopteran (Noctuidae: Lepidoptera) responsible for serious damage to a number of crops, including potato, *Solanum tuberosum* L., corn, and hops. *H. micacea* is responsible for destroying the main stalk of corn and hops, and can lead to reduced growth and plant death. *H. micacea* is also described as a gateway to secondary pests such as *Fusarium* ssp.

6.7.1 Determination

The rosy rustic moth has a wingspan of 28–45 mm. Its eggs are observed on stem of weeds as egg mass. The larva is reddish flesh colour, with a deeper dorsal line and yellowish sides. Its head is shiny red brown, while thoracic and anal plates are yellowish. The last instar larva is 25–31 mm long (Godfrey, 1981). The adult's forewings are pale greyish ochre, or dark grey, or brownish grey with a strong russet tinge. The median zone is darker, with olive-brown shading, except along the central and inner margins. The paler ground colour only appears in the terminal third, with brown inner and outer lines. The hindwings are lutei form, generally tinged with grey, with a dark outer line and a submarginal cloud. The typical form is brownish-grey tinged with pink, but paler, more ochre-coloured forms with very little red also exist (Seitz, 1900).

6.7.2 Biology and Life History

H. micacea has many hosts from vegetable, fruit and field crops to a wide variety of weeds. Eggs are laid on couchgrass (*Elymus repens* (L.) Gould) stem mainly in late-Summer to Autumn where larva can overwinter after hatching (French et al., 1973). Eggs were also observed to overwinter, to then hatch mid-May of the following year (Deedat et al., 1983; Giebink et al., 1999). Larvae first develop on weeds where they can feed and later on colonise plants with larger diameter stem such as corn, potato or hop, due to their internal feeding behavior. Larvae enter the hop bine within the first 30 cm from soil in mid-May (at their 4th instar), and exit the bine around the same height after 4–5 weeks of feeding in June (French et al., 1973), with a body length of 2–2.5 cm. Larvae continue feeding on the same hop on the base of the bine, focusing either on stem base, crown or roots. First flight of adults is observed in end of June until October (Weihrauch, 2020). *H. micacea* only has one generation per year, supported with the help of weeds.

6.7.3 Damage

H. micacea is responsible for the boring of the bine and roots of hop, leading to a reduce growth, leaf yellowing in July and August, or even the death of the plant in extreme infestation conditions. Larval holes in the bine can be easily identified. Damage is more detrimental on newly planted hop, where death of the plant can be observed. By the type of damage of *H. micacea*, secondary pests (i.e., *Fusarium* spp.) can be observed on moth infested hop bine (Gryndler et al., 2008).

Wet climate is reported to be beneficial to the settlement of the moth (French et al., 1973; Šedivý et al., 2018).

6.7.4 Management/Control

6.7.4.1 Biological Control

Bt is highly effective against lepidopterans (Tan et al., 2011) and could be used as biological treatment of *Hydraecia micacea*. Occurrence of the entomopathogenic fungus *B. bassiana*, the common parasitoid *Ichneumon sarcitorius* L. and another less occurring parasitoid *Lydella thompsoni* Herting were observed by Šedivý (2018) on larvae and pupae of *H. micacea*, resulting in a 17% of naturally parasited *H. micacea*. Deedat et al. (1983) reported the emergence of 32 trichogrammatids *Trichogramma retorridum* (Girault) on parasitised eggs of *H. micacea*.

6.7.4.2 Insecticides

It's difficult to assess the efficacy of an insecticide in field, as rosy rustic moth infestations are sporadic and the control situation is often free of the moth. However, Weihrauch (2020) observed excellent control of *H. micacea* with the application of chlorantraniliprole with plot of hop 100% free of the moth.

6.7.4.3 Cultural and Mechanical Methods

As weeds are the main hosts of *H. micacea*, their management (i.e., removal of weed) can provide good control of the moth (Šedivý et al., 2018; Weihrauch, 2020), and drastically reduce infestation over a few years by reducing potential oviposition sites. Targeted weeds should primarily be Poaceaes such as *Elymus repens* and *Echinochloa crus-galli* (L.) P. Beauv, but it seems that *Avena sativa* L. and numerous broad-leaved weeds can also receive *H. micacea* eggs (Giebink et al., 1992, 1999).

6.8 Hop Snout Weevil, *Neoplinthus tigratus* subsp. *porcatus*

The hop snout weevil, *Neoplinthus tigratus* subsp. *porcatus* (Panzer), is a monophagous pest that feeds exclusively on hops. In addition to Slovenia, the species is present in Austria, Croatia, the Czech Republic, Germany, Hungary, Italy, Poland, Romania, Slovakia and Ukraine ("Neoplinthus tigratus subsp. porcatus (Panzer 1798)," Fauna Europaea, 2024). In the literature, it is stated that the hop snout weevil migrated from wild hops to cultivated hops. In the Savinja Valley, hop snout weevil was first observed in large numbers in 1893, when, due to their high abundance, damage to hop gardens was first visible (Janežič, 1951).

6.8.1 Determination

The adult specimen of the hop snout weevil is dark brown to black, covered with grey scales, measuring 12 to 14 mm in length (Fig. 6.14). The larvae are white with a brown head and no legs (Fig. 6.15).

Fig. 6.14 Adult of hop snout weevil, *Neoplinthus tigratus* subsp. *porcatus*. (Photo: Magda Rak Cizej)

Fig. 6.15 Larvae of hop snout weevil, *Neoplinthus tigratus* subsp. *porcatus*, on underground root. (Photo: Magda Rak Cizej)

Table 6.2 Sum of effective temperature required to complete the individual larval development stages of the hop snout weevil, *Neoplinthus tigratus* subsp. *Porcatus* at a temperature threshold of 8.0 °C (Šparl, 2015)

Developmental stages (body length/ description)	Sum of effective temperatures above the threshold of 8 °C
L1 (0.1–0.4 cm)	182.6
L2 (0.5–0.8 cm)	169.4
L3 (0.9–1.2 cm)	165.7
L4 (1.3–1.7 cm)	154.3
L5 (pupae)	124.3
Total development from 1st to 5th instar	796.4

6.8.2 *Biology and Life History*

Adults hop snout weevil appear from early May to June. At this time, females lay eggs above the ground of the hop plant. The hatched larvae spin into the underground part of the stem, where they move down the hop plant by feeding (Janežič, 1951). The larvae of the hop moth overwinter in the underground parts of the stem and rhizome of hops, where they pupate during Spring (Table 6.2) (Rak Cizej & Oset Luskar, 2012; Šparl, 2015).

6.8.3 *Damage*

Damage is caused by larvae, eating out the underground parts of hops (underground stems and rhizomes). Damaged hop rhizomes and underground parts of the stems result in limited uptake and transport of water and nutrients throughout the plant (Čerenak et al., 2002). There can be between 8 and 10 larvae of the hop snout weevil in one plant, up to 20 in the case of a stronger attack (Janežič, 1951). Damaged plants grow worse, and as a result, plants achieve a lower yield that is of lower quality. With a stronger attack of hop snout weevil, individual hop plants may completely perish (Čerenak et al., 2002).

6.8.4 *Management/Control*

As of today, we do not have any plant protection product available that has been registered to reduce the population of hop snout weevil. The most effective measure to reduce the pest population is the careful cutting of underground parts of the hops, followed by their removal and appropriate destruction.

6.9 Cabbage Moth, *Mamestra brassicae*

Mamestra brassicae (L.) is a lepidopteran (Lepidoptera Noctuidae) that mainly attacks cabbage leaves (i.e., white cabbage (*Brassica oleracea* var. *capitata* f. *alba* L.), rutabaga (*B. napus* L.), collard (*B. oleracea* var. *acephala* L.) and red cabbage (*B. oleracea* var. *capitata* f. *rubra* L.)) (Metspalu et al., 2004), however, the moth is polyphagous and can feed on many species, including hops (Badenes-Pérez, 2022).

6.9.1 *Determination*

Cabbage moth has a wingspan of 38–49 mm with a body length of 15–22 mm. Females are on average slightly larger than males. Adult forewings are blackish-grey with a conspicuous reniform stigma, surrounded by white points. Its foretibia has a big black slightly curved and pointed spine, which can help identify the species. The eggs of the moth are subspherical with a flat base, 0.60–0.65 mm wide, white-pale yellow when laid, then becoming pale-yellow with an above-equatorial rust ring line as the embryo develops. The first instar larva is white-pale yellow with a black head, while mature larva ranges in colour from pale green to olive green-black and greyish with pink shades. A characteristic sub-circular yellow spot can be seen in the ocellar position (Sannino & Espinosa, 1999).

6.9.2 *Biology and Life History*

M. brassicae has numerous hosts as described by Badenes-Prérez (2022), mainly in cultivated and agricultural aeras. Pupae of the moth diapauses in the soil during overwintering (or during the summer months in case of bivoltine when climate conditions are right), and the adult emerges in late June – early July (Metspalu et al., 2004). Females oviposit at night in single layer cluster of 14–37 eggs, on the underside of host plant leaves. Around 500–1500 eggs are laid by one female per year, and larvae can be found for several months due to its long oviposition period. A few hours after hatching larvae spread over the host plant, where first- to fourth- instar larvae develop and mainly target the host's outer leaves. From the fifth instar to pupa, the larvae move into the central part of the plant, between the young leaves where they complete their larval development. In warm climates, two generations can be observed per year.

6.9.3 Damage

On cabbage (*Brassicae* spp.), larvae feed on old leaves between first and third instars and then move to young leaves in the fourth and fifth instars, leaving holes in the leaves while respecting the leaf nerves. From fifth instar onwards, the larvae begin to enter the cabbage head.

On hops, cones are mainly damaged by the moth's feeding. *Mamestra brassicae* causes superficial damage to the edges of petals and bracts, but severs almost 30% of the strands, causing the distal part of undamaged cones to fall off (Campbell, 2019), reducing yield both qualitatively and quantitively.

6.9.4 Management/Control

6.9.4.1 Biological Control

Turnock and Carl (1995) listed 12 species of larval parasitoids in Europe between 1974 and 1987, the most constant and abundant being the braconid *Microplitis mediator* (Hal.), the ichneumonid *Exetastes atrator* (Forster) and the two tachinids *Siphona flavifrons* Staeger and *Panzeria consobrina* (Meigen). Natural enemies such as the egg parasitoid *Trichogramma* are also widely used to control the population of the cabbage moth (Badenes-Pérez, 2022), with the example of *Trichogramma semblidis* (Aurivillius).

As a lepidopteran, Bt could be used against *M. brassicae*, but this specific moth seems to be less susceptible to Bt toxins (Lightwood et al., 2000).

Entomopathogenic nematodes such as *Steinernema carpocapsae* (Weiser, 1955) Wouts, Mráček, Gerdin & Bedding show good control of the cabbage moth (Beck et al., 2014; Philipsen & Nielsen, 2003).

Polyphagous predators can reduce population of the cabbage moth by feeding on eggs, such as *Philonthus atratus* (Gravenhorst) (Staphylinidae) and *Bembidion tetracolum* Say (Carabidae), or by focusing on a broader spectrum of *M. brassicae* life stages with predation of both larvae and eggs, such as *Pterostichus melanarius* (Illiger) and *Harpalus rufipes* (DeGeer) (Carabidae) (Johansen, 1997).

6.9.4.2 Insecticides

As with other lepidopteran, good control can be achieved with chlorantraniliprole (i.e., Coragen 20% concentration) (Median lethal concentration (LC50) = 0.35 mg/L) and indoxacarb (i.e., Avaunt 15% concentration) (LC50 = 1.71 mg/L); the cabbage moth shows high sensitivity to these active substances, especially on second instar larva (Moustafa et al., 2023).

6.9.4.3 Cultural and Mechanical Methods

Increasing in-field plant diversity can enhance the biological regulation of the cabbage moth by natural enemies, but incorporating specific plant can also directly harm the cabbage moth, such as *L. oleraceus*, leading to a reduction in cabbage moth survival (Metspalu et al., 2013) and depleting the population by providing a less favourable host.

6.10 Slugs, *Arion* spp.

Arion spp. is a large group of slugs with similar physical characteristics causing damage in gardens and vegetable production mainly. Main species met in Europe are *Arion hortensis* Ferussac, *A. distinctus* Mabille, *A. vulgaris* Moquin-Tandon, *A. rufus* (L.) and *A. subfuscus* (Draparnaud). These slugs are responsible for tissues consumption on crop, located on leaves, stems, fruits, tubers and bulbs.

6.10.1 Determination

Arion slug eggs are 2 mm of diameter, clear and transparent, clustered in masses of 10–50 eggs. The adult is 30–40 mm long, grey to dark brown-black in colour. Its body is stocky and its back is round. Related to the snail, the *Arion* slug's shell is reduced and has a cluster of calcareous granules under the mantle, where its pneumostome (respiratory pore of molluscs) is located. The slug's movement is made possible by the contraction of its foot, leaving a characteristic yellow mucus in its wake. The mucus is secreted by numerous glands under the skin. The slug's head harbours 2 pairs of antennae or horns as sensory organs: one serves as a light receptor (eyes), the other as an odour detector (Rueda et al., 2023).

6.10.2 Biology and Life History

Arion spp. is found in a very large number of host plants and is extremely polyphagous. Its repartition is mostly directed by relative humidity, badly surviving drought periods. In general, slugs are amorphs beings, with a relatively reduced dispersion. Adults dig a hole in soil or choose a hollow rock to lay their eggs under. Hatching of eggs is expected 3 to 4 weeks after oviposition, although in warm conditions hatching happens already after 2 weeks. After 3 weeks, slugs reach sexual maturity, as hermaphrodites. After mating, slugs can lay from 300 to 500 eggs in several waves, from May to Autumn. Slugs' life span rarely extends over a year (Briner & Frank, 1998).

6.10.3 Damage

Arion slugs cause most of their damage at night, when they leave their shelters (damp areas under rocks, logs and other objects with high humidity and low heat) to feed. In a single day, they can ingest up to 40% of their mass. Slugs focus on the vegetative part of plants, using their rasping tongue-like organs to scrape or cut food before it enters their oesophagus. Their tongue, called a radula, is composed of a tiny, toothed chitinous ribbon. The radula is unique to molluscs, making it easier to identify slug damage on leaves. On leaves, slug damage follows the nerves of the leaf, with holes in the middle and at the edges. On bulbs, beetroots and cabbage heads, slugs leave holes of varying depths. On the tips of plants, the entire flower head can be cut off. On hop, they feed on leaves, stems and flowers and can lead to a reduce growth of plants and lower quantity and quality of cones (Fig. 6.16). In extreme infestation, they can cause the decay of the hop plant.

Arion slugs' infestation generally starts at the edges of crops, causing the most damage in persistent wet weather. Young plants are mainly targeted and may be severely affected. Secondary infection may also occur with pathogenic fungi.

6.10.4 Management/Control

6.10.4.1 Biological Control

Biological control of *Arion* slugs can be obtained by directly harming the slug or by setting a barrier to prevent its infestation in the crop (repellent). A large panel of essential oils showed great result as lethal control of slug or as repellent, even essential oil of hops can be of use with main constituents of myrcene, humulene and β-caryophyllene which was found to be toxic with LC50 values of 118.653 µl/l and

Fig. 6.16 Slugs damage hop leaves. (Photo: Magda Rak Cizej)

essential oils of *Cannabis sativa* with an LC50 of 35.370 μL/L on the freshwater pan-pulmonate snail *Physella acuta* (Draparnaud) (Bedini et al., 2016; Radwan & Gad, 2021; Perilla Rueda et al., 2023). Plant extract such as garlic concentrate showed a great barrier efficacy, ranging from 65% to 100% while causing low mortality (Schüder et al., 2003). Schüder et al. (2003) also showed that natural solutions bring great barrier efficacy (72.5% and 80%, respectively) with the use of urea-formaldehyde (derived production of urea and acids) or copper-based product. Salt in 10% concentration offer a 20% reduction of slugs (Laznik et al., 2010) and citric acid in 5% concentration reduced slug population by 64% when sprayed on leaves of lettuce, yet did not increase mortally when ingested by *Arion* sp. (Steele et al., 2021).

Natural predators of *Arion* slugs are scarce but can be enhanced in the proximity of the crop by adding diverse beneficial habitat, and can target either adult slugs (e.g., birds) or eggs (e.g., carabid beetles (Pianezzola et al., 2013)).

The nematode *Pellioditis hermaphrodita* (A. Schneider) can be used on *Arion* slugs, however due to the behaviour and life cycles of slugs, it might be necessary to resort to at least 2 applications for an effective control (Speiser et al., 2001; Wilson et al., 1993, 1999).

6.10.4.2 Insecticides

Molluscicides such as metaldehyde (e.g., Terminator) and methiocarb (e.g., Mesurol) can be used on field crops against all types of slugs, including *Arion* slugs with a 100% control of *Arion* spp. in 2 days (Laznik et al. 2010), but they offer a broad spectrum of action and can be very damaging to non-target species and they remain active for only a short period of time (9 days for metaldehyde) (Kozłowski et al., 2010). Organic molluscicide is also available as iron phosphate, and can even be used by hobbyists.

6.10.4.3 Cultural and Mechanical Methods

Mechanical barriers are hardly effective on their own and need to be combined to some repellent such as waterglass or copper (Watz & Nyqvist, 2021).

Mulch or plants, especially invasive species such as staghorn sumac *Rhus hirta* (L.) Südw., giant goldenrod *Solidago gigantea* Aiton, and Japanese knotweed *Reynoutria japonica* Houtt. can as well be a mean to repel *Arion* slugs (Laznik et al., 2020; Schüder et al., 2003).

References

Ahmad, S., Javed, H., Ahmad, M., & Khan, M. A. (2022). Utilization of fertilizers application methods for the management of *Tetranychus urticae* population in Brinjal crop. *Fresenius Environmental Bulletin, 31*(4), 3910–3917.

Alves, A. P., Spencer, T. A., Tabashnik, B. E., & Siegfried, B. D. (2006). Inheritance of resistance to the Cry1Ab Bacillus thuringiensis toxin in Ostrinia nubilalis (Lepidoptera: Crambidae). *Journal of Economic Entomology, 99*(2), 494–501. https://doi.org/10.1603/0022-0493-99.2.494

Badenes-Pérez, F. (2022). *Mamestra brassicae (cabbage moth)*. CABI Compendium. https://doi.org/10.1079/cabicompendium.8491

Barati, R., & Hejazi, M. J. (2015) Reproductive parameters of Tetranychus urticae (Acari: Tetranychidae) affected by neonicotinoid insecticides. *Experimental and Applied Acarology, 66*(4), 481–489. https://doi.org/10.1007/s10493-015-9910-7

Beck, B., Brusselman, E., Nuyttens, D., Moens, M., Temmerman, F., Pollet, S., Van Weyenberg, S., & Spanoghe, P. (2014). Improving the biocontrol potential of entomopathogenic nematodes against Mamestra brassicae: Effect of spray application technique, adjuvants and an attractant. *Pest Management Science, 70*(1), 103–112. https://doi.org/10.1002/ps.3533

Bedini, S., Flamini, G., Cosci, F., Ascrizzi, R., Benelli, G., & Conti, B. (2016). Cannabis sativa and Humulus lupulus essential oils as novel control tools against the invasive mosquito Aedes albopictus and fresh water snail Physella acuta. *Industrial Crops and Products, 85*, 318–323. https://doi.org/10.1016/j.indcrop.2016.03.008

Bing, L., & Lewis, L. (1991). Suppression of Ostrinia-Nubilalis (hubner) (lepidoptera, Pyralidae) by Endophytic Beauveria-Bassiana (balsamo) Vuillemin. *Environmental Entomology, 20*(4), 1207–1211. https://doi.org/10.1093/ee/20.4.1207

Bing, L., & Lewis, L. (1993). Occurrence of the entomopathogen Beauveria bassiana (Balsamo) Vuillemin in different tillage regimes and in Zea mays L. and virulence towards Ostrinia nubilalis (Hübner). *Agriculture Ecosystems & Environment, 45*(1 2), 147 156. https://doi.org/10.1016/0167-8809(93)90065-W

Bourguet, D., Bethenod, M. T., Trouvé, C., & Viard, F. (2000). Host-plant diversity of the European corn borer Ostrinia nubilalis: What value for sustainable transgenic insecticidal Bt maize? *Proceedings of the Royal Society of London. Series B: Biological Sciences, 267*(1449), 1177–1184. https://doi.org/10.1098/rspb.2000.1126

Briner, T., & Frank, T. (1998). The palatability of 78 wildflower strip plants to the slug Arion lusitanicus. *Annals of Applied Biology, 133*(1), 123–133. https://doi.org/10.1111/j.1744-7348.1998.tb05808.x

Calderwood, L., Cubins, J., Vesty, D., & Darby, H. (2017). Effect of drive row ground covers on hop (Rosales: Cannabaceae) yard arthropod pests in Vermont, USA. *Environmental Entomology, 46*(2), 183–190. https://doi.org/10.1093/ee/nvw069

Campbell, C. A. M. (2018). Influence of companion planting on damson hop aphid Phorodon humuli, two spotted spider mite Tetranychus urticae, and their antagonists in low trellis hops. *Crop Protection, 114*(August), 23–31. https://doi.org/10.1016/j.cropro.2018.08.014

Campbell, C. A. M. (2019). New and resurgent insect pests on low trellis hops. *Agricultural and Forest Entomology, 21*(2), 209–218. https://doi.org/10.1111/afe.12322

Campbell, C. A. M., & Cone, W. W. (1999). Consumption of damson-hop aphids (Phorodon humuli) by larvae of Coccinella transversoguttata and Hippodamia convergens (Coleoptera: Coccinellidae). *Biocontrol Science and Technology, 9*(1), 75–78. https://doi.org/10.1080/09583159929929

Campbell, C. A. M., & Ridout, M. S. (2001). Effects of plant spacing and interplanting with oilseed rape on colonisation of dwarf hops by the damson-hop aphid, Phorodon humuli. *Entomologia Experimentalis et Applicata, 99*, 211–216.

Campbell, C. A. M., & Tregidga, E. L. (2005). Photoperiodic determination of gynoparae and males of damson-hop aphid Phorodon humuli. *Physiological Entomology, 30*(2), 189–194. https://doi.org/10.1111/j.1365-3032.2005.00447.x

Campbell, C. A. M., & Tregidga, E. L. (2006). A transgenerational interval timer inhibits unseasonal sexual morph production in damson-hop aphid, Phorodon humuli. *Physiological Entomology, 31*(4), 394–397. https://doi.org/10.1111/j.1365-3032.2006.00531.x

Campbell, C. A. M., Pettersson, J., Pickett, J. A., Wadhams, L. J., & Woodcock, C. M. (1993). Spring migration of damson-hop aphid, *Phorodon humuli* (Homoptera, Aphididae), and summer host plant-derived semiochemicals released on feeding. *Journal of Chemical Ecology, 19*(7), 1569–1576.

Čerenak, A., Dolinar, M., Ferant, N., Friškovec, I., Knapič, M., Knapič, V., Košir, I., Kovačeveič, M., Pavlovič, M., Rode, J., Simončič, A., Šuštar-Vozlič, J., Virant, M., Zmrzlak, M., & Žolnir, M. (2002). *Priročnik za hmeljarje* (D. Majer (Ed.)). Inštitut za hmeljarstvo in pivovarstvo Žalec.

Chacon-Hernandez, J., Cerna-Chavez, E., Ochoa-Fuentes, Y., Hernandez-Juarez, A., Palacios-Bazaldua, L., & Landeros-Flores, J. (2016). Spatial distribution and population fluctuation of *Tetranychus urticae* (Koch) and *Phytoseiulus persimilis* (Athias-Henriot) in rose crops. *Phyton-International Journal of Experimental Botany, 85*, 236–245. https://www.webofscience.com/wos/woscc/full-record/WOS:000392213300031

De Rouck, S., İnak, E., Dermauw, W., & Van Leeuwen, T. (2023). A review of the molecular mechanisms of acaricide resistance in mites and ticks. *Insect Biochemistry and Molecular Biology, 159*(June), 103981. https://doi.org/10.1016/j.ibmb.2023.103981

Deedat, Y. D., Ellis, C. R., & West, R. J. (1983). Life history of the potato stem borer (Lepidoptera: Noctuidae) in Ontario 1. *J. Econ.Entomol, 76*(5), 1033–1037. https://doi.org/10.1093/jee/76.5.1033

Enkegaard, A., Brødsgaard, H. F., & Hansen, D. L. (2001). Macrolophus caliginosus: Functional response to whiteflies and preference and switching capacity between whiteflies and spider mites. *Entomologia Experimentalis et Applicata, 101*, 81–88.

Ferguson, C. S., Schroeder, P. C., & Shields, E. J. (1995). Vertical distribution, persistence, and activity of entomopathogenic nematodes (Nematoda: Heterorhabditidae and Steinernematidae) in Alfalfa Snout Beetle-(Coleoptera: Curculionidae) infested fields. *Environmental Entomology, 24*(1), 149–158.

Ferree, D. C., & Hall, F. R. (1980). Effects of soil water stress and twospotted spider mites on net photosynthesis and transpiration of apple leaves. *Photosynthesis Research, 1*(3), 189–197. https://doi.org/10.1007/BF00020598

Foglar, H., Malausa, C., & Wajnberg, E. (1990). The functional-response and preference of Macrolophus-Caliginosus [heteroptera, Miridae] for 2 of its prey - Myzus-Persicae and Tetranychus-Urticae. *Entomophaga, 35*(3), 465–474.

French, N., Ludlam, F. A. B., & Wardlow, L. R. (1973). Damage and control of Rosy Rnstic Moth, Hydraecia micacea. *Plant Pathology, 22*(2), 58–64. https://doi.org/10.1111/j.1365-3059.1973.tb01773.x

Gargani, E., Faggioli, F., & Haegi, A. (2018). A survey on pests and diseases of Italian hop crops. *Italus Hortus, 24*, 1–17. https://doi.org/10.26353/j.itahort/2017.2.117

Giebink, B. L., Scriber, J. M., & Wedberg, J. L. (1992). Suitability of selected broad-leaved weeds for survival and growth of two stalk-boring Hydraecia species (Lepidoptera: Noctuidae). *The Great Lakes Entomologist, 25*(4). https://doi.org/10.22543/0090-0222.1790

Giebink, B. L., Scriber, J. M., & Wedberg, J. (1999). Survival and growth of two *Hydraecia* species (Noctuidae: Lepidoptera) on eight midwest grass species. *The Great Lakes Entomologist, 32*(3), 247–256. https://doi.org/10.22543/0090-0222.1999

Godfrey, G. L. (1981). Identification and descriptions of the ultimate instar larvae of Hydraecia immanis (hop vine borer) and H. micacea (potato stem borer) (Lepidoptera, Noctuidae). *Illinois Natural History Survey, 114*. https://agris.fao.org/search/en/providers/122376/records/647472edd02e8fbe07eeb937

Gosik, R., & Sprick, P. (2012). Larval morphology of Otiorhynchus ligustici, O. porcatus and O. salicicola (Coleoptera, Curculionidae, Otiorhynchini). *Deutsche Entomologische Zeitschrift, 59*(2), 301–316. https://doi.org/10.1002/mmnd.201200025

Gryndler, M., Krofta, K., Gryndlerová, H., Soukupová, L., Hršelová, H., & Gabriel, J. (2008). Potentially dangerous fusarioid microorganisms associated with rot of hops (Humulus lupulus L.) plants in field culture. *Plant, Soil and Environment, 54*(4), 149–154. https://doi.org/10.17221/387-PSE

Guo, L., Tang, C., Gao, C., Li, Z., Cheng, Y., Chen, J., Wang, T., & Xu, J. (2022). Bacterial and fungal communities within and among geographic samples of the hemp pest Psylliodes attenuata from China. *Frontiers in Microbiology, 13*(September). https://doi.org/10.3389/fmicb.2022.964735

Hall, F. R., & Ferree, D. C. (1975). Influence of twospotted spider mite populations on photosynthesis of apple leaves. *Journal of Economic Entomology, 68*(4), 517–520. https://doi.org/10.1093/jee/68.4.517

Hansen, D. L., Brødsgaard, H. F., & Enkegaard, A. (1999). Life table characteristics of Macrolophus caliginosus preying upon Tetranychus urticae. *Entomologia Experimentalis et Applicata, 93*, 269–275.

Harcourt, D. G., & Bereza, K. (2017). Survival and longevity of *Otiorhynchus Ligustici* (Coleoptera: Curculionidae) in Alfalfa Hay Bales in Eastern Ontario. *The Great Lakes Entomologist, 23*(1). https://doi.org/10.22543/0090-0222.1687s

Harcourt, D. G., & Binns, M. R. (2017). Sampling technique for larvae of the Alfalfa Snout Beetle, *Otiorhynchus Ligustici* (Coleoptera: Curculionidae). *The Great Lakes Entomologist, 22*(3). https://doi.org/10.22543/0090-0222.1678

Hata, F. T., Ventura, M. U., de Souza, M. S. J., de Sousa, N. V., Oliveira, B. G., & da Silva, J. B. (2019). Mineral and organic fertilization affects Tetranychus urticae, pseudofruit production and leaf nutrient content in strawberry. *Phytoparasitica, 47*(4), 513–521. https://doi.org/10.1007/s12600-019-00744-z

Hoffmann, M. P., Wright, M. G., Pitcher, S. A., & Gardner, J. (2002). Inoculative releases of *Trichogramma ostriniae* for suppression of *Ostrinia nubilalis* (European corn borer) in sweet corn: Field biology and population dynamics. *Biological Control, 25*(3), 249–258. https://doi.org/10.1016/S1049-9644(02)00105-6

Hoffmann, M. P., Pitcher, S. A., Cheever, S. A., Gardner, J., Losey, J. E., Kuhar, T. P., Laub, C. A., & Youngman, R. R. (2006). Efficacy of inoculative releases of Trichogramma ostriniae (Hymenoptera: Trichogrammatidae) against European corn borer Ostrinia nubilalis (Lepidoptera: Crambidae) in field corn. *Biological Control, 36*(3), 345–349. https://doi.org/10.1016/j.biocontrol.2005.10.008

Janežič, F. (1951). *Varstvo rastlin pred boleznimi in škodljivci* (Vol. 86). Državna založba Slovenije. https://plus.cobiss.net/cobiss/si/sl/bib/gimms/185345

Johansen, N. S. (1997). Mortality of eggs, larvae and pupae and larval dispersal of the cabbage moth, Mamestra brassicae, in white cabbage in South-Eastern Norway. In *Entomologia Experimentalis et Applicata* (Vol. 83). Kluwer Academic Publishers.

Knan, A., & Kremheller, T. (1984). Festlegung von Schadensschwellen bei der Bekämpfung der Hopferblattlaus, Phorodon humuli. *Bulletin SROP, 6*, 51–55. http://pascal-francis.inist.fr/vibad/index.php?action=getRecordDetail&idt=8957074

Kozłowski, J., Kałuski, T., Jaskulska, M., & Kozłowska, M. (2010). Initial evaluation of the effectiveness of selected active substances in reducing damage to rape plants caused by Arion lusitanicus (Gastropoda, Pulmonata, Arionidae). *Journal of Plant Protection Research, 50*(4), 520–526. https://doi.org/10.2478/v10045-010-0086-1

Labatte, J.-M., Meusnier, S., Migeon, A., Chaufaux, J., Couteaudier, Y., Riba, G., & Got, A. (1996). Field evaluation of and modeling the impact of three control methods on the larval dynamics of Ostrinia nubilalis (Lepidoptera: Pyralidae). In *Journal of Economic Entomology* (Vol. 89, Issue 4).

Lacroix, O., Lescourret, F., Génard, M., Memah, M. M., Vercambre, G., Valsesia, P., Bevacqua, D., & Grechi, I. (2024). Modeling the effect of multiple pests on ecosystem services provided by fruit crops: Application to apple. *Agricultural Systems, 213*, 103808. https://doi.org/10.1016/j.agsy.2023.103808

Laznik, Ž., Mihičinac, M., Rupnik, J., Vidrih, M., Prša, I., & Trdan, S. (2010). Preizkušanje učinkovitosti različnih snovi za zatiranje lazarjev (arion spp., Arionidae) v laboratorijskih razmerah. *Acta Agriculturae Slovenica, 95*(2), 129–140. https://doi.org/10.2478/v10014-010-0009-x

Laznik, Z., Bohinc, T., Franin, K., Majić, I., & Trdan, S. (2020). Efficacy of invasive alien plants in controlling arionidae slugs. *Spanish Journal of Agricultural Research, 18*(1), e1001. https://doi.org/10.5424/sjar/2020181-15542

Lewis, G. A., & Madge, D. S. (1984). Esterase activity and associated insecticide resistance in the damson-hop aphid, Phorodon humuli (Schrank) (Hemiptera: Aphididae). *Bulletin of Entomological Research, 74*(2), 227–238.

Lewis, L. C., Bruck, D. J., & Gunnarson, R. D. (2002). On-farm evaluation of *Beauveria bassiana* for control of *Ostrinia nubilalis* in Iowa, USA. *BioControl, 47*(2), 167–176. https://doi.org/10.1023/A:1014574526071

Lewis, L. C., Bruck, D. J., Prasifka, J. R., & Raun, E. S. (2009). Nosema pyrausta: Its biology, history, and potential role in a landscape of transgenic insecticidal crops. *Biological Control, 48*(3), 223–231. https://doi.org/10.1016/j.biocontrol.2008.10.009

Lightwood, D. J., Ellar, D. J., & Jarrett, P. (2000). Role of proteolysis in determining potency of Bacillus thuringiensis Cry1Ac-endotoxin. In *Applied and Environmental Microbiology* (Vol. 66, Issue 12).

Neoplinthus tigratus subsp. porcatus (Panzer 1798). (2024). In *Fauna Europaea*. https://fauna-eu.org/cdm_dataportal/taxon/7ea16a8d-4226-40a0-abdd-84cf8cefc2db

Lilley, R., & Campbell, C. A. M. (1999). Biological, chemical and integrated control of two-spotted spider mite Tetranychus urticae on dwarf hops. *Biocontrol Science and Technology, 9*(4), 467–473. https://doi.org/10.1080/09583159929433

Lösel, P. M., Lindemann, M., Scherkenbeck, J., Maier, J., Engelhard, B., Campbell, C. A. M., Hardie, J., Pickett, J. A., Wadhams, L. J., Elbert, A., & Thielking, G. (1996). The potential of semiochemicals for control of Phorodon humuli (Homoptera: Aphididae). *Pesticide Science, 48*(4), 293–303. https://doi.org/10.1002/(SICI)1096-9063(199612)48:4<293::AID-PS479>3.0.CO;2-5

Loxdale, H. D., Brookes, C. P., Wynne, R., & Clark, S. J. (1998). Genetic variability within and between English populations of the damson-hop aphid, Phorodon humuli (Hemiptera:

Aphididae), with special reference to esterases associated with insecticide resistance. *Bulletin of Entomological Research, 88*, 513–526.

Metspalu, L., Jõgar, K., Hiiesaar, K., & Grishakova, M. (2004). Food plant preference of the cabbage moth Mamestra brassicae (L.). *Latvian Journal of Agronomy, 7*, 15–19. https://llufb.llu.lv/conference/agrvestis/content/n7/AgrVestis-Nr7-15-19.pdf

Metspalu, L., Kruus, E., Jogar, K., Kuusik, A., Williams, I. H., Veromann, E., Luik, A., Ploomi, A., Hiiesaar, K., & Kivimaegi, I. (2013). Larval food plants can regulate the cabbage moth, Mamestra brassicae population. *Bulletin of Insectology, 66*(1), 93–101. https://www.researchgate.net/profile/Luule-Metspalu/publication/235914331_Larval_food_plants_can_regulate_the_cabbage_moth_Mamestra_brassicae_population/links/0c96053bbb935148e8000000/Larval-food-plants-can-regulate-the-cabbage-moth-Mamestra-brassicae-population.pdf

Mo, T. L., & Liu, T. X. (2007). Predation and life table of Feltiella acarisuga (Diptera: Cecidomyiidae) preying on eggs of Tetranychus urticae (Acari: Tetranychidae). *Environmental Entomology, 36*(2), 369–375. https://doi.org/10.1603/0046-225X(2007)36[369:PALTOF]2.0.CO;2

Moustafa, M. A. M., Fouad, E. A., Ibrahim, E., Erdei, A. L., Kárpáti, Z., & Fónagy, A. (2023). The comparative toxicity, biochemical and physiological impacts of chlorantraniliprole and indoxacarb on Mamestra brassicae (Lepidoptera: Noctuidae). *Toxics, 11*(3). https://doi.org/10.3390/toxics11030212

Muir, R. C. (1979). Insecticide resistance in damson-hop aphid, Phorodon humuli in commercial hop gardens in Kent. *The Annals of Applied Biology, 92*(1), 1–9.

Mumm, R., van Tol, R. W. H. M., & Weihrauch, F. (2017). Elucidation of the role of volatile compounds in the chemical communication of the hop flea beetle Psylliodes attenuatus. *Proceedings of the Scientific-Technical Commission*, 60–64.

Muška, F., Kazda, J., Rožnovský, J., & Muška, A. (2020). Škodlivé výskyty lalokonosce libečkového na řepě cukrové a krmné na území Česka. *Listy cukrovarnické a řepařské, 136*(12), 415–420.

Neumann, G., & Shields, E. J. (2008). Turnbull and chant. *Journal of Economic Entomology, 101*(5), 1533–1539. https://academic.oup.com/jee/article/101/5/1533/2198903

Paul, T. A., Darby, P., Green, C. P., Hodgson, C. J., Rossiter, J. T., College, E., & Tn, K. (1996). Electrical penetration graphs of the damson-hop aphid, Phorodon humuli on resistant and susceptible hops (Humulus lupulus). *Entomologia Experimentalis et Applicata, 80*, 335–342.

Pelozuelo, L., Malosse, C., Genestier, G., Guenego, H., & Frerot, B. (2004). Host-plant specialization in pheromone strains of the European Corn Borer Ostrinia nubilalis in France. *Journal of Chemical Ecology, 30*(2), 335–352. https://doi.org/10.1023/B:JOEC.0000017981.03762.ed

Perilla Rueda, K. Y., Castellanos González, L., & Chaparro, A. L. (2023). Evaluación del efecto de formulados de aceites esenciales de especies Eucalyptus contra Arion spp. *Inge CuC, 19*(2), 131–142. https://doi.org/10.17981/ingecuc.19.2.2023.11

Philipsen, H., & Nielsen, O. (2003). Host potential of insects from cruciferous crops to entomopathogenic nematodes and augmentation of nematodes through oil seed rape growing. *IOBC Bulletin, 26*, 141–146. 10/1/rapsproceeding.doc.

Pianezzola, E., Roth, S., & Hatteland, B. A. (2013). Predation by carabid beetles on the invasive slug Arion vulgaris in an agricultural semi-field experiment. *Bulletin of Entomological Research, 103*(2), 225–232. https://doi.org/10.1017/S0007485312000569

Piraneo, T. G., Bull, J., Morales, M. A., Lavine, L. C., Walsh, D. B., & Zhu, F. (2015). Molecular mechanisms of Tetranychus urticae chemical adaptation in hop fields. *Scientific Reports, 5*. https://doi.org/10.1038/srep17090

Popov, S. Y., Alyokhin, A. V., & Kiselyov, E. D. (2021). Survival of diapausing females of two-spotted spider mite Tetranychus urticae Koch (Acari: Tetranychidae) submerged in water. *International Journal of Acarology, 47*(7), 564–567. https://doi.org/10.1080/01647954.2021.1976834

Radwan, M. A., & Gad, A. F. (2021). Essential oils and their components as promising approach for gastropod mollusc control: A review. In Journal of Plant Diseases and Protection (Vol. 128, 4, pp. 923–949). Springer Science and Business Media Deutschland GmbH. https://doi.org/10.1007/s41348-021-00484-5.

Rahman, K. M. A., Barta, M., & Cagáň, Ľ. (2010). Effects of combining Beauveria bassiana and Nosema pyrausta on the mortality of Ostrinia nubilalis. *Central European Journal of Biology, 5*(4), 472–480. https://doi.org/10.2478/s11535-010-0035-z

Rak Cizej, M., & Oset Luskar, M. (2012). Bolezni in škodljivci hmelja. *V: Hmelj Od Sadike Do Storžkov. Čeh B.(Ur.), Žalec*, 21–37.

Rak Cizej, M. (2018). *Hmeljev bolhač*. Integrirano Varstvo Rastlin. https://www.ivr.si/skodljivec/hmeljev-bolhac/.

Rak Cizej, M., Milevoj, L., & Kastelec, D. (2007). Ugotavljanje preference hroščev hmeljevega bolhača (Psylliodes attenuatus Koch) v hmeljišču, na različnih sortah hmelja. *Hmeljarski Bilten*. https://repozitorij.uni-lj.si/IzpisGradiva.php?id=59332.

Rak Cizej, M., & Poličinik, F. (2022). Control of hop flea beetle (Psylliodes attenuatus Koch) on hops (Humulus Lupulus L.) using alternative preparations. *Hmeljarski Bilten Hop Bulletin, 29*, 97–108.

Rak Cizej, M., & Žolnir, M. (2003). Hmeljev bolhač (Psylliodes attenuatus Koch) vse pogostejši škodljivec hmelja v Sloveniji. In *Zbornik Predavanj in Referatov 6. Slovenskega Posvetovanja o Varstvu Rastlin*, 233–238.

Razinger, J., Vasileiadis, V. P., Giraud, M., Van Dijk, W., Modic, Š., Sattin, M., & Urek, G. (2016). On-farm evaluation of inundative biological control of Ostrinia nubilalis (Lepidoptera: Crambidae) by Trichogramma brassicae (Hymenoptera: Trichogrammatidae) in three European maize-producing regions. *Pest Management Science, 72*(2), 246–254. https://doi.org/10.1002/ps.4054

Russell, K., & Bessin, R. (2009). Integration of Trichogramma ostriniae releases and habitat modification for suppression of European corn borer (Ostrinia nubilalis Hübner) in bell peppers. *Renewable Agriculture and Food Systems, 24*(1), 19–24. https://doi.org/10.1017/S174217050800238X

Sances, F. V., Wyman, J. A., Ting, I. P., Van Steenwyk, R. A., & Oatman, E. R. (1981). Spider mite interactions with photosynthesis, transpiration and productivity of strawberry. *Environmental Entomology, 10*(4), 442–448. https://doi.org/10.1093/ee/10.4.442

Sannino, L., & Espinosa, B. (1999). Morphological notes on Mamestra brassicae (Lepidoptera Noctuidae). *Il Tabacco, 7*, 13–24. https://citeseerx.ist.psu.edu/document?repid=rep1&type=pdf&doi=e0ee30db72c1490b6a0dc25fb63af1b719afcc5b

Schroeder, P. C., Ferguson, C. S., Elson, A., & Shields, J. (1995). Spring activity of adult alfalfa snout beetles (Coleoptera: Curculionidae): Upward movement from overwintering sites, dispersal, and oviposition. *Journal of Economic Entomology, 88*(4), 859–863. http://jee.oxfordjournals.org/

Schroeder, P. C., Ferguson, C. S., Shields, E. J., & Villanii, M. G. (1994). Pathogenicity of Rhabditid Nematodes (Nematoda: Heterorhabditidae and Steinernematidae) to Alfalfa Snout Beetle (Coleoptera: Curculionidae) Larvae. *Journal of Economic Entomology, 87*(4), 917–922. http://jee.oxfordjournals.org/

Schüder, I., Port, G., & Bennison, J. (2003). Barriers, repellents and antifeedants for slug and snail control. *Crop Protection, 22*(8), 1033–1038. https://doi.org/10.1016/S0261-2194(03)00120-0

Šedivý, J., Born, P., & Vostřel, J. (2018). Harmful occurrence of rosy rustic moth (Hydraecia micacea) (Noctuidae: Lepidoptera) on hop in The Czech Republic. *Plant Protection Science, 41*, 150–157. https://doi.org/10.17221/2748-PPS

Seiedy, M., Saboori, A., & Zahedi-Golpayegani, A. (2013). Olfactory response of Phytoseiulus persimilis (Acari: Phytoseiidae) to untreated and Beauveria bassiana-treated Tetranychus urticae (Acari: Tetranychidae). *Experimental and Applied Acarology, 60*(2), 219–227. https://doi.org/10.1007/s10493-012-9652-8

Seitz, A. (1900). *Die Gross-Schmetterlinge der Erde : eine systematische Bearbeitung der bis jetzt bekannten Gross-Schmetterlinge*. Alfred Kernen. http://archive.org/details/diegrossschmette01seit

Smith, J. L., & Farhan, Y. (2023). Monitoring resistance of *Ostrinia nubilalis* (Lepidoptera: Crambidae) in Canada to Cry toxins produced by Bt corn. *Journal of Economic Entomology, 116*(3), 916–926. https://doi.org/10.1093/jee/toad046

Šparl, M. (2015). *Preučevanje razvojnega kroga hmeljevega rilčkarja (Neoplinthus tigratus porcatus (Panzer), Coleoptera, Curculionidae) v laboratorijskih razmerah in na prostem*. Univerza v Ljubljani, Biotehniška fakulteta oddlelek za agronomijo.

Speiser, B., Zaller, J. G., & Neudecker, A. (2001). Size-specific susceptibility of the pest slugs Deroceras reticulatum and Arion lusitanicus to the nematode biocontrol agent Phasmarhabditis hermaphrodita. *BioControl, 46*, 311–320.

Steele, K. A., Peters, B. J., Price-Christenson, G. J., & Blair, A. C. (2021). Citric acid as a potential molluscicide and deterrent for the invasive slug Arion fasciatus. *American Malacological Bulletin, 38*(2), 85–90. https://doi.org/10.4003/006.038.0213

Strong, W. B., Slone, D. H., & Croft, B. A. (1999). Hops as a metapopulation landscape for tetranychid-phytoseiid interactions: Perspectives of intra- and interplant dispersal. *Experimental and Applied Acarology, 23*(7), 581–597. https://doi.org/10.1023/A:1006208218771

Tan, S. Y., Cayabyab, B. F., Alcantara, E. P., Ibrahim, Y. B., Huang, F., Blankenship, E. E., & Siegfried, B. D. (2011). Comparative susceptibility of *Ostrinia furnacalis*, *Ostrinia nubilalis* and *Diatraea saccharalis* (Lepidoptera: Crambidae) to *bacillus thuringiensis* Cry1 toxins. *Crop Protection, 30*(9), 1184–1189. https://doi.org/10.1016/j.cropro.2011.05.009

Topakci, N. (2022). Spider mite predator *Feltiella acarisuga* (Vallot, 1827) (Diptera: Cecidomyiidae) in greenhouse strawberry cultivation in Antalya Province: Recognition, population dynamics and parasitization by *Aphanogmus* sp. *Turkiye Entomoloji Dergisi, 46*(1), 25–36. https://doi.org/10.16970/entoted.1019437

Trotuş, E., Naie, M., & Buburuz, A. A. (2011). Protecţia culturilor de cânepă pentru sămânţă împotriva agenţilor patogeni şi dăunătorilor. *Protectia Plantelor, LXXIX*(l), 161–170.

Turnock, W. J., & Carl, K. P. (1995). Evaluation of the Palaearctic Eurithia consobrina (Diptera: Tachinidae) as a potential biocontrol agent for Mamestra configurata (Lepidoptera: Noctuidae) in Canada. *Biocontrol Science and Technology, 5*(1), 55–68. https://doi.org/10.1080/09583159550040015

Undas, A. K., Weihrauch, F., Lutz, A., van Tol, R., Delatte, T., Verstappen, F., & Bouwmeester, H. (2018). The use of metabolomics to elucidate resistance markers against damson-hop aphid. *Journal of Chemical Ecology, 44*(7–8), 711–726. https://doi.org/10.1007/s10886-018-0980-y

Watz, J., & Nyqvist, D. (2021). Artificial barriers against arionid slug movement. *Crop Protection, 142*, 105525. https://doi.org/10.1016/j.cropro.2020.105525

Weichel, L., & Nauen, R. (2003). Monitoring of insecticide resistance in damson hop aphid, Phorodon humuli Schrank (Hemiptera: Aphididae) from German hop gardens. *Pest Management Science, 59*(9), 991–998. https://doi.org/10.1002/ps.711

Weihrauch, F. (2009). First steps towards a revised control threshold for the damson-hop aphid Phorodon humuli. *Proceedings of the Scientific Commission*, 89–92.

Weihrauch, F., Baumgartner, A., Eisenbraun, D., Wolf, S., van Tol, R. W. H. M., & Mumm, R. (2017). Flea-beetle control in organic hops: Are there options? *Proceedings of the Scientific-Technical Commission*, 59.

Weihrauch, F. (2020). Rosy rustic moth as a hop pest in the Hallertau: History and current situation. *Mitteilungen Der Deutschen Gesellschaft Für Allgemeine Und Angewandte Entomologie, 22*, 125–128.

Weihrauch, F., & Moreth, L. (2005). Behavior and population development of Phorodon humuli (Schrank) (Homoptera: Aphididae) on two hop cultivars of different susceptibility. *Journal of Insect Behavior, 18*(5), 693–705. https://doi.org/10.1007/s10905-005-7020-9

Wilson, M. J., Hughes, L. A., Jefferies, D., & Glen, D. M. (1999). Slugs (Deroceras reticulatum and Arion ater agg.) avoid soil treated with the Rhabditid Nematode Phasmarhabditis hermaphrodita. http://www.idealibrary.com

Wilson, M. J., Glen, D. M., & George, S. K. (1993). The Rhabditid Nematode Phasmarhabditis hermaphrodita as a potential biological control agent for slugs. *Biocontrol Science and Technology, 3*(4), 503–511. https://doi.org/10.1080/09583159309355306

Wu, M., Adesanya, A. W., Morales, M. A., Walsh, D. B., Lavine, L. C., Lavine, M. D., & Zhu, F. (2019). Multiple acaricide resistance and underlying mechanisms in Tetranychus urticae on hops. *Journal of Pest Science, 92*(2), 543–555. https://doi.org/10.1007/s10340-018-1050-5

Xiao, Y., Osborne, L. S., Chen, J., Mckenzie, C., Houben, K., & Irizarry, F. (2011). Evaluation of corn plant as potential banker plant for supporting predatory gall midge, Feltiella acarisuga (Diptera: Cecidomyiidae) against Tetranychus urticae (Acari: Tetranychidae) in greenhouse vegetable production. *Crop Protection, 30*(12), 1635–1642. https://doi.org/10.1016/j.cropro.2011.09.004

York, A. C., & Gyrisco, G. G. (1978). Dosage-mortality response and field control of adult alfalfa snout beetle. *Journal of Economic Entomology, 71*(5), 783–784. https://doi.org/10.1093/jee/71.5.783

Youngman, R. R., Jones, V. P., Welter, S. C., & Barnes, M. M. (1986). Comparison of feeding damage caused by four tetranychid mite species on gas-exchange rates of almond leaves. *Environmental Entomology, 15*, 190–193. http://ee.oxfordjournals.org/

Yu, T., Li, X., Coates, B.-S., Zhang, Q., Siegfried, B.-D., & Zhou, X. (2018). microRNA profiling between *bacillus thuringiensis* Cry1Ab-susceptible and -resistant European corn borer, *Ostrinia nubilalis* (Hubner). *Insect Molecular Biology, 27*(3), 279–294. https://doi.org/10.1111/imb.12376

Zimmermann, G., Huger, A. M., Langenbruch, G. A., & Kleespies, R. G. (2016). Pathogens of the European corn borer, *Ostrinia nubilalis*, with special regard to the microsporidium *Nosema pyrausta*. *Journal of Pest Science, 89*(2), 329–346. https://doi.org/10.1007/s10340-016-0749-4

Part III
Hops: Sustainable Approach to Hop Waste Management

Chapter 7
Post-harvest Hop Waste Management

Barbara Čeh, Lucija Luskar, Julija Polanšek, and Ana Karničnik Klančnik

7.1 Background

Each year an average of 15 tonnes of waste plant biomass (fresh matter), which consists of leaves and stems, arises from each harvested hectare of hop (*Humulus lupulus* L.) (Čeh et al., 2019), so large volumes of waste hop biomass are generated annually near harvest facilities. For instance, in the Hallertau region, the world's largest hop-growing area covering around 17,000 hectares, approximately 230,000 tons of hop bines are generated each year as a by-product of the harvest (Institute for Crop Science, 2022); in Slovenia approximately 23,000 tons from around 1500 hectares (Čeh et al., 2022a). Unfortunately, this biomass often poses a challenge, particularly when plastic twines (solely or in combination with iron) are used as support structures for hop plants in the field. These twines remain tightly entangled within the waste biomass, making it impossible to separate them, which complicates its effective utilisation. Furthermore, if this biomass is left in the environment, it not only contributes to environmental issues but also leads to the loss of valuable nutrients and organic matter from the farm.

When iron strings are used to support hop plants, the resulting waste biomass can be utilised for biogas production in biogas plants (Renewable Carbon News, 2021), typically in conjunction with maize silage. However, it is essential to first remove the iron wires in the initial step of the biogas production process, using magnets. Once the wires are eliminated, the chopped biomass may be processed through anaerobic digestion to produce biogas (Biomethane plant, 2025). The input material

B. Čeh (✉) · L. Luskar · A. Karničnik Klančnik
Department of Plants, Soil and Environment, Slovenian Institute of Hop Research and Brewing, Žalec, Slovenia
e-mail: barbara.ceh@ihps.si

J. Polanšek
Agricultural Institute of Slovenia, Ljubljana, Slovenia

K. Carbone (ed.), *All About Hops: The Crop, its Cultivation, and its Uses in Brewing and Beyond*, https://doi.org/10.1007/978-3-031-96606-4_7

consists of hop silage (65%) and maize / grass silage (35%), which is mechanically pretreated before entering the process (Trošt & Čeh, 2022). Biogas is used for heat generation for households, while digestate from biogas plant is sent back to farmers for use as a fertiliser (European Biogas Association, 2016). As reported by Renewable Carbon News (2021), researchers have tested the hop waste biogas digestate also to produce a composite material that can be used to make furniture. The presented production process begins with the environmentally friendly cleaning of plant-based residues, refined material is then transformed into a wet-laid nonwoven, which is subsequently combined with a bio-based resin system and pressed into a durable composite material.

On the other side, there are numerous other opportunities for the utilisation of this waste plant material; some possibilities with the specific aim of increasing sustainability in the hop industry through circular economy principles are presented in Fig. 7.1. However, also other possibilities have been investigated. For example, this

Fig. 7.1 Circular economy principle in hop production: the BioTHOP model. (Sketch: LIFE BioTHOP, 2022)

biomass can be considered as a valuable source of functional molecules and nutrients that are still underexploited (Astray et al., 2020).

7.2 Post-harvest Waste Hop Biomass as a Raw Material

7.2.1 Composting

On-farm composting emerges as an efficient, cost-effective, and environmentally friendly biological process for recycling residual biomass, resulting in stabile organic manure (Maniadakis et al., 2004). This is a process during which, if done properly, sufficient high temperatures during the thermophilic phase eliminate pathogens and weed seeds, resulting in sanitised biomass and good quality organic fertiliser. This bioconversion of organic waste into value-added product can turn the problem of environmental pollution into an opportunity by reducing the wastes as well as minimizing the dependency on fossil fuels (Awasthi et al., 2020). It offers a sustainable solution for managing agricultural biomass waste while simultaneously providing farmers with a self-sufficient source of high-quality compost to enhance soil health and fertility (Scotti, 2016).

Although there are various possible options for utilizing hop biomass, composting remains the most promising approach for managing it (Luskar & Čeh, 2021). As calculated by Eblagon et al. (2022), on site composting presents an effective and sustainable solution for waste hop biomass, achieving a net reduction in greenhouse gas (GHG) emissions of 20% compared to open burning while enabling farmers to utilise the produced compost on their fields. This process can be implemented directly on-site by growers, offering a scalable and adaptable approach to waste management. Additionally, it enhances waste treatment capacity by integrating other farm residues, such as manure and biomass from various crops, further supporting sustainable agricultural practices.

Post-harvest waste hop biomass contains organic matter and lots of nutrients (Table 7.1). The carbon-to-nitrogen (C:N) ratio varies depending on the plant parts

Table 7.1 Chemical composition of waste hop biomass before composting (Čeh et al., 2019)

Component	Percentage in hop biomass	Amount per 1 ha (≈15 tons fresh matter, equivalent to waste hop biomass from 1 ha)
Organic matter	18%	
Nitrogen (N)	0.8%	88–99 kg
Potassium (K_2O)	0.3%	50–71 kg
Phosphorus (P_2O_5)	0.1%	21–22 kg
Magnesium (MgO)		32–41 kg
Sulphur (S)		5.5 kg

being composted, with ratio of 13:1 when composting stems and leaves together, and 23:1 when composting only stems (Čeh et al., 2019; Luskar & Čeh, 2021). Given the optimal C:N ratio, stems are more suitable for composting than leaves, however Čeh et al. (2022a) suggest to compost the entire waste biomass, stems and leaves together, to ensure that the process proceeds appropriately and efficiently. If only the stems are composted, the pile will dry out too quickly due to excessive empty spaces, which will negatively affect the process, namely. Trials have indicated that also the temperature in compost pile during the sole composting of hop stems is not reached high enough for complete sanitisation (Čeh et al., 2022a). This may be due to a high percentage of woody material and the coarse structure of chopped hop stems (Lohr et al., 2021). A mixture of leaves and stems has proven to be a better option for the input material in terms of composting (Čeh et al., 2022a).

In addition to the C:N ratio, particle size also has a crucial role. If harvest machines are set to cut hop stems into longer pieces (approximately 15 cm), this allows sufficient oxygen circulation within the compost pile but may also result in excessive moisture loss (Čeh et al., 2019). To optimise the process, it is recommended to cut biomass into smaller pieces (around 3 cm), mix the compost pile periodically based on temperature monitoring, and maintain temperatures between 50 °C and 65 °C for a minimum of 5 weeks. The moisture content of the compost pile must be monitored throughout the process as well. A simple method, such as the fist squeeze test, can be used to assess the moisture level. If the pile becomes too dry, additional water should be added during mixing to maintain optimal composting conditions. If the pile becomes excessively wet, it is left opened and uncovered for a period to allow moisture reduction (Čeh et al., 2022a).

The degradation of organic components in composting occurs during the thermophilic phase when microorganisms break down the organic material. The addition of biochar can accelerate the composting process by enhancing aeration and improve pile structure and serve as a habitat for microorganisms, resulting in higher microbial activity and thus a higher composting temperature. For instance, already Talboys (1961) was investigating the temperature requirements for killing *Verticillium albo-atrum* (now *Verticillium nonalfalfae*) in hop bines, therefore showing that composting might be beneficial for pathogens reduction (Lutz et al., 2022; Talboys, 1961). The addition of biochar may help to achieve complete hygienisation of the pile by raising the composting temperature, as tested by Luskar et al. (2022), however the findings of Čeh et al. (2025) suggest that further testing is necessary to optimise the process when incorporating biochar.

Given the substantial quantities produced, this biomass presents significant potential for replenishing nutrients and organic matter in agricultural fields. Compared to composting, the common practice of putting back chopped stems in hop gardens right after the harvest season bears phyto-pathological risks, particularly related to the spread of the hop verticillium wilt (Talboys, 1987). However, one challenge that composting does not solve is the problem of plastic twine that remains undegraded in the mature compost. To address this issue, potential solutions include (a) using biodegradable twine as a support for hop plants on the field or (b) sieving the plastic twines out of the final compost before spreading it on the agricultural fields.

Although composting is considered a natural biodegradation process that transforms organic waste into valuable products, certain factors must be carefully managed to ensure its efficiency and effectiveness. Key recommendations include immediately establishing compost pile after harvest, using both leaves and stems together and regularly mixing the material during the thermophilic phase, related to temperature measurements, which is crucial for temperature control and aeration. It is recommended to establish a compost pile from at least one hectare of hop field, which is approximately 15 tons of hop waste. The ideal shape of the pile is triangular, with a slope that is not too steep to ensure proper drainage of precipitation. The pile should be formed with the height of approximately 3 meters and a maximum settled height of 1.5 meters, constructed in a way that allows easy drainage, maintaining moisture content of 60–70% (Čeh et al., 2022a) (Fig. 7.2).

Regular mixing and moisturisation are essential to maintain the desired temperature range. Temperature measurements should be taken daily at the same time in all four cardinal directions. The mixing should be performed when we measure 3 days temperature above 60 °C, or 2 days temperature above 65 °C. Once temperatures drop below 45 °C, mixing should occur once a week. Around 2 months after setup, as the cooling phase begins, the pile should be covered with a semi-permeable membrane to retain heat and moisture, prevent leaching of nutrients, and facilitate optimal microbial activity for proper maturation (Čeh et al., 2022a). Covering hop waste composting piles with a semipermeable membrane following the thermophilic phase and maintaining coverage until spring, when the compost became ready for use, significantly reduced leachate volume and nutrient leaching in the experiment by Čeh et al. (2025). During winter and spring, all treatments exhibited negligible leachate production and minimal nutrient loss. The compost pile without

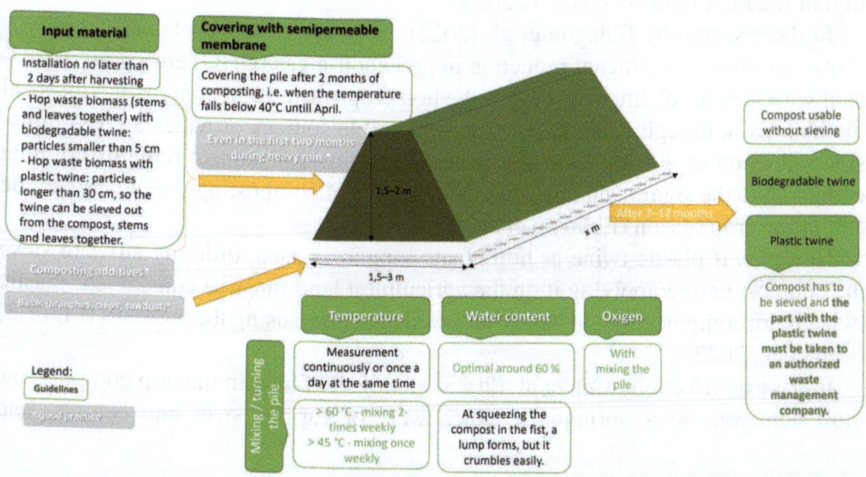

Fig. 7.2 Guidelines for on-farm waste hop biomass composting. (Sketch: L. Luskar and B. Čeh)

additives, consisting of waste hop biomass particles ranging from 2 to 10 cm and properly mixed based on regular temperature measurements during the thermophilic phase, achieved the most favourable results. This treatment produced the lowest leachate volume and nutrient loss, while the final compost exhibited significantly higher nutrient content and the highest germination index. However, all treatments, differing in particles size and composting additives) produced compost without a bad smell, had a total nitrogen content greater than 2%, could be considered as mature (the C: N ratio was below 20) and stable, and the biomass hygienisation threshold was reached in all of them.

The stabilisation phase begins when the temperature drops below 40 °C while the compost is ready for use when the core of the pile reaches the same temperature as the surroundings and emits the smell of soil, but not before 7 months after the start of composting process; this is usually in April next year (Čeh et al., 2022a).

Final compost (dry matter; DM) contains about 3–4% nitrogen, 0.3–0.4% phosphorus, 1.0–2.5% potassium, 35–43% total organic carbon (Luskar et al., 2022); in 100 g DM, 2.7, 0.38 and 1.08 g of N, P and K respectively. Considering the radish germination index, which averaged 89%, the hop waste compost prepared according to composting guidelines for hop stems and leaves together exhibited non-phytotoxic characteristics, indicating stability and suitability for plant cultivation. According to the cress germination tests, it is a nutrient-rich and stimulating substrate for plant growth, as evidenced by a germination index ranging from 125% to 213%, with an average of 196%.

When using biodegradable BioTHOP twine (made from polylactic-acid (PLA); having OK Compost Certification—www.compliancegate.com/ok-compost) in the hop fields for hop plants support, we get pure compost with no artificial leftovers, no sieving is needed; this twine at proper on-site composting completely degrades to water, CO_2 and organic matter in this time period namely, as the same conditions as are required for the sanitisation of waste biomass as are needed for the degradation of the PLA twine (Čeh et al., 2023).

In the research by Eblagon et al. (2022) the usage of compostable twines was shown to allow a significant reduction in greenhouse gas (GHG) emissions associated with the end-of-life disposal for the hop crop waste via a composting disposal method. Even though using twine from renewable sources provides a GHG emission reduction of 400 kg CO_2-eq per hectare, its overall impact remains minimal compared to the significant environmental benefit of composting green waste at the end of the crop season (Eblagon et al., 2022).

However, if plastic twine as hop plants support is used, ordering sifter to sieve the compost before applying it on the agricultural land must be considered; plastic twine is not compostable and has to be sieved out before using the compost (Fig. 7.3) (Čeh et al., 2022a).

In the experiment by Čeh et al. (2023) among the fauna in the hop compost, the most numerous were springtails, which were of various sizes and colours, and

Fig. 7.3 Machine sieving plastic twine from compost pile before use. (Photo: Ana Karničnik Klančnik)

mites; amoebas, earthworms, centipedes, spiders, larvae, beetles and insects were also detected (Fig. 7.4). Microbial respiration in the hop waste biomass compost was observed to be between 0.1 and 0.2 C-CO_2/g of compost per day (Luskar & Čeh, 2021). The microbial world of composted hop biomass was reported to be dominated by bacteria (Luskar & Čeh, 2021).

Čeh et al. (2022c) conducted a pot experiment with Chinese cabbage (*Brassica rapa* L. subsp. *chinensis* (L.) Hanelt), mixing in the substrate:

- Fresh waste hop biomass,
- Mature compost prepared from hop waste biomass,
- No additives were incorporated into the substrate (control pots).

After 47 days from sowing, the above-ground biomass of the Chinese cabbage was statistically significant the most abundant in the pots with mature composts (Fig. 7.5). Leaves of plants in these pots were the most intensively green, the content of nitrate in the leaves was significantly higher (130 mg/L) compared to other two treatments (36 mg/L and 19 mg/L) and the density and branching of the root system were the highest (Fig. 7.5).

Fig. 7.4 Some of the meso- and macrofauna, detected in the mature waste hop biomass compost (after 7 months of composting). (Photo: Ana Karničnik Klančnik, originally published in Barbara Čeh, 'Increasing the Value of Waste Hop Biomass by Composting: Closing the Nutrient Cycle on Hop Farms,' in *Organic Fertilizers—New Advances and Applications*, IntechOpen, Year. Licensed under CC BY 3.0)

The results show the effectiveness of mature compost in increasing soil fertility. By reintroducing waste hop biomass material back into the soil in the form of high-quality organic fertiliser, farmers can enrich the nutrient content and organic matter of the soil, promoting more resilient and sustainable agricultural practices; the closed-loop system within the hop farm is effectively established.

Fig. 7.5 (**a**) Fertilisation pot trial with Chinese cabbage (*Brassica rapa* L. subsp. *chinensis* (L.) Hanelt) after 47 days from sowing; top photo, left: fresh waste hop biomass in the substrate, in the middle: mature hop compost in the substrate, right: control—just substrate; (**b**) left: fresh waste hop biomass in the substrate, middle: control—just substrate, right: mature hop compost in the substrate. (Photos: Ana Karničnik Klančnik)

7.2.2 Waste Leaves as a Source of Polyphenols

Antioxidants are inherent components of plants and include a variety of compounds such as vitamins C and E, carotenoids, flavonoids, and phenolic acids. These compounds have the remarkable ability to attenuate or decelerate oxidative damage caused by free radicals. Among these naturally occurring antioxidants, phenolic compounds are particularly abundant, and those containing phenolic groups are among the most widely utilised (Narayana Saibaba, 2023). Hop leaves are attracting attention as potential sources of antioxidants, antibacterial, and antiviral agents (Macchioni et al., 2022; Sabbatini et al., 2024).

As confirmed by Čeh et al. (2007), the content of phenolic compounds in hop leaf extracts corresponds to their antioxidative potential. The study of Abram et al. (2015) encompassed two hop varieties (Hallertauer Magnum and Aurora) grown in four different hop-growing regions (Žalec in Slovenia; Hüll in Germany; Leutschach (Kranach) in Austria; Žatec in Czech Republic) in the same investigated years. The total phenolic concentration in the extracts varied significantly, with leaf extracts ranging from 0.099 to 0.542 mg CAE/mL, while cone extracts ranged from 0.738 to 1.734 mg CAE/mL (CAE—chlorogenic acid equivalents in mg per gram of dry sample; a higher value indicates stronger antioxidant activity). So, the leaves contain 3- to 30-fold less total phenolics than cones, based on ethanol-extracted supernatants, however, considering that leaves are typically treated as waste material, their phenolic content presents a valuable opportunity for utilisation.

Choi et al. (2018) reported a total polyphenols of 15.45 mg/g dry weight (assuming 20% dry matter) in *Humulus japonicus* leaves, whereas Proestos et al. (2006) quantified the total polyphenol content in hop leaves at 2.9 mg GA/g (gallic acid equivalents) of dry mass. Important factor responsible for differences in polyphenolic content in hop leaves appears to be different climatic conditions in particular years, as noted by Muzykiewicz et al. (2019) as well as different measurement methods (Siebert & Lynn, 2005). Their study documented total polyphenol content in alcoholic extracts of hop leaves ranging from 0.09 to 6.22 mg GA per gram of raw material, with variations depending on the solvent type and the harvest year. Additionally, Morcol et al. (2021) identified several bioactive compounds, including bitter acids, prenylchalcones, and phenolic acids, in hydroalcoholic extracts of hop leaves from various commercial varieties emphasizing the impact of genotype on their metabolic profile.

Macchioni et al. (2022) worked on evaluating the antioxidant potential and nutraceutical profile of hop leaves from different hop varieties. Their results indicate the presence of soluble polyphenolic compounds and antioxidant pigments in the non-phenolic fraction from hop leaves, in quantities comparable to or even exceeding those reported in literature for dried herbs, such as green tea.

7.2.3 *Waste Hop Stems as a Source of Fibres*

Natural fibres can be utilised in the production of biodegradable and compostable composites, serving as an alternative to synthetic plastics. This approach contributes to reducing plastic waste accumulation on the Earth's surface, supporting more sustainable waste management solutions (Harder et al., 2023). Natural fillers and fibres serve as reinforcement materials in such plastic alternatives (Mohanty et al., 2022).

Waste hop stems are rich in lignocellulosic content and can serve as a source of short fibres (Haunreiter et al., 2021). Research on the production and use of hop-derived fibres is limited, with most studies focusing on chemical extraction or isolation techniques. However, recently reports can be found also related to the production of fibres from hop stems using physical methods (Harder et al., 2023). Fibres produced through physical methods offer a cost-effective alternative to chemical extraction and can easily be incorporated into composites, so they are preferred as reinforcements for fibre-plastic composites (Zou et al., 2010; Harder et al., 2023). The findings by Harder et al. (2023) indicate that hop fibres obtained through attrition milling of hop stems have potential for biocomposite applications, offering a sustainable alternative for biocomposite development. They also found that the fibre size distribution is a critical factor influencing the overall performance of biocomposites.

The research by Harder et al. (2023) is looking to expand the applications of natural fibres in composites. One of the methods to get the fibre from the stems employs hop steam explosion, a degumming process shown also by Limousin et al. (2023) to be particularly efficient for defibrating hop stems, yielding fibres with excellent tensile strength. Similarly, the study by Harder et al. (2023) investigated natural fibres from hop residues for biocomposites, and Haunreiter et al. (2021) explored alternative extraction methods for hop-derived fibres. These studies provide valuable insights into the potential of hop fibres in composite applications.

The bio-composites produced with hop fibres can demonstrate enhanced mechanical and thermal properties. For instance, in a similar approach using hop fibres, composites made with poly butylene succinate-co-butylene adipate (PBSA) showed significant improvements in tensile strength, flexural moduli, and heat deflection temperature (HDT) when compared to the neat matrix, with the optimal performance achieved by using fibres of specific sizes (1 mm). These results indicate the crucial role of fibre size and distribution in achieving superior material properties for sustainable applications (Harder et al., 2023).

Within the EU LIFE project BioTHOP (www.life-biothop.eu), two innovative procedures for extracting fibres from waste hop stems were developed (Figs. 7.6 and 7.7), transforming waste hop stems into a valuable raw material and facilitating their utilisation in diverse applications (LIFE BioTHOP, 2022). In the first procedure pulverised fibre is produced, which is then dried to specific dry matter content before undergoing the compounding process. The dried fibres are mixed with a biobased polymeric matrix derived from starch, a biodegradable material. The

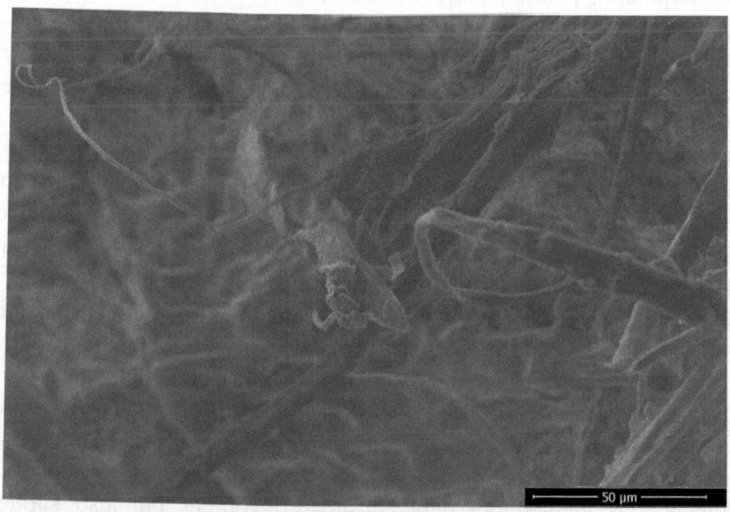

Fig. 7.6 Processed defibrillated/engineered ligno-cellulose fibre from hop waste. (Photo: LIFE BioTHOP, 2022)

Fig. 7.7 Modified fibres derived from hop waste biomass. (Photo: LIFE BioTHOP, 2022)

extruder machine mixes materials by heating and propelling them forward until the mixture comes through a die in a cylinder cross-section shape (a thread). A long water bath, positioned in front of the machine die, decreases the temperature of the obtained thread. At the end of the water bath, a pelletiser chops the thread into small granules. These granules are then dried, mixed with polylactic acid (PLA, which is a well-known bioplastic – biobased and biodegradable under industrial composting conditions), and in this form used for production of injection moulded products (LIFE BioTHOP, 2022). The bio-composite (Fig. 7.8) is designed for manufacturing a variety of biodegradable and compostable products through injection moulding, such as planting pots and playing bricks (Fig. 7.9). At the end of their life cycle, these products can be industrially composted, breaking down into water, carbon dioxide (CO_2), and organic matter. The resulting compost restores valuable nutrients and organic matter to agricultural soils, closing a sustainable loop of resource utilisation and waste reduction. This process completes the cycle within the farm's ecosystems, albeit in a slightly longer manner.

The second application of hop fibres developed within the EU LIFE BioTHOP project involves incorporating hop fibres into biodegradable and compostable packaging for wine bottle transport via pulp moulding (Figs. 7.10 and 7.11). This solution offers a sustainable and environmentally friendly alternative to conventional packaging materials, significantly reducing packaging waste and its environmental impact (LIFE BioTHOP, 2022). Market analysis indicated that the BioTHOP wine bottle packaging represented a pioneering innovation on a global scale; the companies involved were recognised as key innovators by the European Commission's Innovation Radar, which identified "Hop waste fibre sheets reused for wine bottle packaging" as a Market Creating Innovation in 2021 (European Commission, 2021). Additionally, this method not only conserves resources but also supports the

Fig. 7.8 Biocomposite pellets from hop fibres. (Photo: LIFE BioTHOP, 2022)

Fig. 7.9 Biodegradable playing bricks from hop fibres as example of injected BioTHOP products. (Photo: LIFE BioTHOP, 2022)

principles of a circular economy, thereby enhancing sustainability in agricultural practices. The product is biodegradable and compostable, and the resulting compost can be returned to agricultural fields.

Panels were also developed from hop fibres got from waste stems representing a novel application area for waste plant biomass and are presented on the Fig. 7.12.

Fig. 7.10 Biodegradable bottle transport packaging production from mix of hop fibres and old cartoon in Tridas company, Czech Republic. (Photo: LIFE BioTHOP, 2022)

Fig. 7.11 Biodegradable and on-site compostable bottle transport packaging as produced by pulp moulding. (Photo: Barbara Čeh)

Fig. 7.12 Panels from hop fibres, developed within EU LIFE project BioTHOP by Zelfo Techology GmbH. (Photo: Barbara Čeh)

7.2.4 Other Possible Applications for Waste Hop Biomass

Open burning is a waste management solution that can have very high negative impact well beyond its greenhouse gas emissions while also landfilling is an unsustainable waste management practice and should be minimised to the greatest extent possible. Whenever feasible, it should be replaced with more environmentally responsible and safer alternatives (Eblagon et al., 2022). Environmentally controversial method for managing post-harvest waste is also decomposition in static piles. This is leading to lack of agitation, poor porosity and inadequate air distribution within the pile (Onwosi et al., 2017). Insufficient aeration can create anaerobic conditions (Gao et al., 2011), while higher summer temperatures further reduce oxygen solubility. The anaerobic activity of bacteria results in elevated odour emissions (Wiśniewska et al., 2020). In addition to odor, static piles also produce leachate, leading to the loss of valuable nutrients. This not only compromises the quality of the final biomass, as many nutrients are lost, but also poses a risk to groundwater contamination (Onwosi et al., 2017). Furthermore, methane can form in the piles under anaerobic conditions (van Calcar & Eblagon, 2019). Combustion is possible in controlled environment (while not open burning which generates various pollutants) for energy production, however fresh hop biomass has moisture of approx. 70%, while moisture content of biomass should not exceed 60% for this process, otherwise high heating values (HHV) will be lower than the one required for evaporation of moisture (González et al., 2012). Therefore, this is a viable option, but the cost of drying should be considered to determine if it is financially feasible (Trošt & Čeh, 2022).

From emissions perspective, the most effective solution is incineration or incineration with co-generation. However, both technologies are highly capital-intensive and require waste feedstocks with low moisture content to operate efficiently. For optimal combustion, waste streams used in co-generation should contain no more than 50% organic fractions, making it unfeasible to rely solely on hop waste. Additionally, the disposal of ash generated during the process must be considered, as it represents a residual waste that requires landfill management, which also brings doubts about the sustainability of this solution in the long term (Eblagon et al., 2022).

Rutto et al. (2020) highlighted the potential of waste hop biomass as a feed resource to enhance feed efficiency in ruminants. Iglesias et al. (2021) proposed the use of alcoholic extracts from hop leaves for the development of an environmentally friendly bee pesticide, emphasizing the role of genotype in influencing the content of secondary metabolites. Dziedzinski et al. (2020) reported the potential for producing functional teas incorporating hop leaves and low concentrations of hop cones, resulting in a product with both desirable flavour and health-promoting properties. They highlighted the impact of genotype in determining these functional attributes.

The agricultural by-products of the hop plant were investigated to determine their potential for use in the removal of heavy lead(II) ions from contaminated aqueous solutions by Gardea-Torresdey et al. (2022). Their findings show that they may be a viable alternative for the removal and recovery of aqueous lead(II) ions from contaminated waters. The beneficial effects of hop waste biomass as a source of potential nutritional and pharmacological compounds that could be used as functional and cosmetic ingredients was also investigated (del Río et al., 2013); researches suggest its potential for use in skin care products as anti-acne, anti-odour, and whitening agents, though further in vivo testing is needed (Pereira et al., 2022). The study by Mitton et al. (2025) investigated hop leaves as a natural antioxidant in rough scad patties during storage conditions. Their research suggests that hop leaves could be a valuable natural source of antioxidants to mitigate lipid oxidation and preserve polyunsaturated fatty acids (PUFAs) in fish-based products.

There are also other opportunities. As we see on the Fig. 7.13, paper, beer coasters, panels, mulching films, buttons, cosmetic pots, shoe horns, phone holders, and thermoformed trays for automotive industry as well as harvesting machine fingers, can be produced from hop waste, as presented by the EU LIFE project BioTHOP.

7.3 Conclusions

In terms of circular economy in agriculture, the most useful application of waste hop biomass would be on-site (on-farm) composting, as it returns nutrients back to farming lands in the narrowest circle. It facilitates the recycling of nutrients and

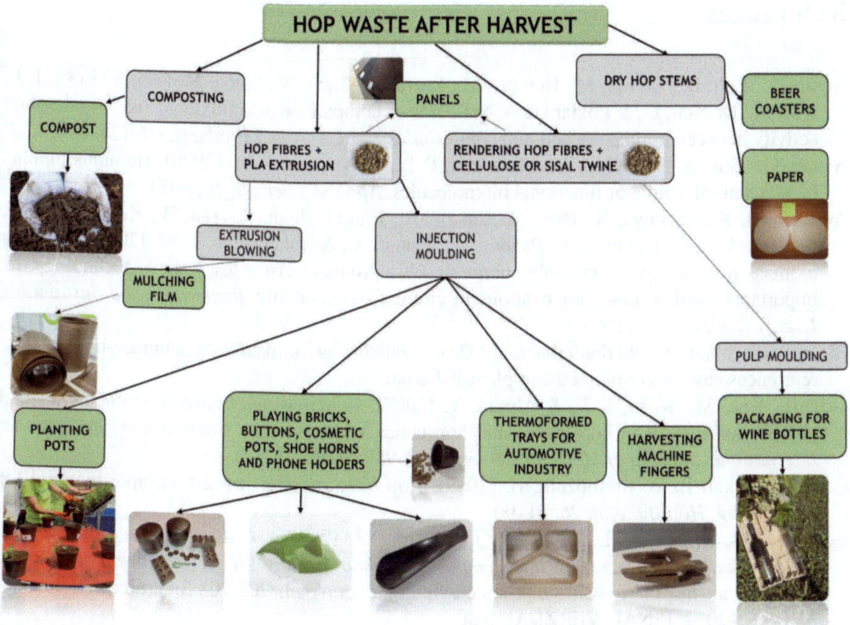

Fig. 7.13 Products from waste hop biomass, developed within EU LIFE project BioTHOP. (Figure: B. Čeh)

organic matter back into agricultural fields, effectively minimizing the environmental footprint and enhancing sustainability within the tightest possible loop. This method not only conserves resources but also contributes significantly to the maintenance of soil health and fertility, aligning with sustainable agricultural practices. Composting as an alternative disposal technique was shown to reduce GHG emissions from 18% up to 42% compared to the available end-of-life alternatives.

On the other hand, biogas production stands out as the most prevalent strategy for waste management and resource recovery. However, hop waste biomass can be repurposed also through a variety of other sustainable and practical applications. These include, for instance, polyphenol extraction, use in cosmetics, and the production of fibres for biodegradable and compostable packaging materials or biocomposites, offering an environmentally friendly alternative to conventional plastics. Once these materials reach the end of their lifecycle, they can be composted and reintegrated into agricultural soils, thereby supporting closed-loop systems within the on-farm circular economy.

Funding Acknowledgement Statement This paper was prepared as a part of After LIFE programme of the EU LIFE project BioTHOP and under the auspices of the professional task Hop Production and Processing Technologies, financed by the Ministry of Agriculture, Forestry and Food of the Republic of Slovenia.

References

Abram, V., Čeh, B., Vidmar, M., Hercezi, M., Lazić, N., Bucik, V., Smole Možina, S., Košir, I. J., Kač, M., Demšar, L., & Poklar Ulrih, N. (2015). A comparison of antioxidant and antimicrobial activity between hop leaves and hop cones. *Industrial Crops and Products, 64*, 124–134.

Astray, G., Gullón, P., Gullón, B., Munekata, P. E., & Lorenzo, J. M. (2020). Humulus lupulus L. as a natural source of functional biomolecules. *Applied Sciences, 10*, 5074.

Awasthi, M. K., Sarsaiya, S., Patel, A., Juneja, A., Prasad Singh, R., Yan, B., Kumar Awasthi, S., Jain, A., Liu, T., Duan, Y., Pandey, A., Zhang, Z., & Taherzadeh, J. M. (2020). Refining biomass residues for sustainable energy and bio-products: An assessment of technology, its importance, and strategic applications in circular bio-economy. *Renewable and Sustainable Energy Reviews, 127*, 109876.

Biomethane plant in Oberlauterbach. (2025). Available at: https://www.schmack-biogas.it/en/references/case-study-biomethane-plant-hallertau/

Čeh, B., Kač, M., Košir, I. J., & Abram, V. (2007). Relationships between xanthohumol and polyphenol content in hop leaves and hop cones with regard to water supply and cultivar. *International Journal of Molecular Sciences, 8*, 989–1000.

Čeh, B., Luskar, L., & Čremožnik, B. (2019). Hop biomass after harvest as input material for composting. *Hop Bulletin, 26*, 81–90.

Čeh, B., Flis, J., Luskar, L., Polanšek, J., & Trošt, Ž. (2022a). *Guidelines for handling hop residues and their processing into compost on a hop farming estate (in Slovene language).* Available at: https://www.life-biothop.eu/wp-content/uploads/2022/08/Smernice_hmeljevina-AVGUST-2022_FINAL-VERZIJA-1.pdf

Čeh, B., Luskar, L., Hladnik, A., Trošt, Ž., Polanšek, J., & Naglič, B. (2022b). The quantity and composition of leachate from hop plant biomass during the composting process. *Applied Sciences, 12*, 2375.

Čeh, B., Trošt, Ž., & Karničnik Klančnik, A. (2022c). Comparison of fertilization with fresh hop biomass versus mature compost from hop biomass – pot experiment. *Hop Bulletin, 29*, 65–72. ISSN 0350-0756.

Čeh, B., Luskar, L., Polanšek, J., Karničnik Klančnik, A., & Trošt, Ž. (2023). Increasing the value of waste hop biomass by composting – closing the nutrient cycle on hop farms. In *Organic fertilizers – new advances and applications*. Available at: https://www.intechopen.com/chapters/1131594

Čeh, B., Polanšek, J., Trošt, Ž., & Karničnik Klančnik, A. (2025). On-site composting of waste hop biomass: the impact of covering piles on leachate quantity and compost quality. *Plant, Soil and Environment, 71*. https://doi.org/10.17221/197/2024-PSE

Choi, J. Y., Desta, K. T., Lee, S. J., Kim, Y. H., Shin, S. C., & Kim, G. S. (2018). LC-MS/MS profiling of polyphenol-enriched leaf, stems and root extracts of Korean Humulus japonicus Siebold & Zucc and determination of their antioxidant effects. *Biomedical Chromatography, 32*, 1–8.

del Río, J. C., Prinsen, P., & Gutiérrez, A. (2013). Chemical composition of lipids in brewer's spent grain: A promising source of valuable phytochemicals. *Journal of Cereal Science, 58*, 248–254. https://doi.org/10.1016/j.jcs.2013.07.001

Dziedzinski, M., Szczepaniak, O., Telichowska, A., & Kobus-Cisowska, J. (2020). Antioxidant capacity and cholinesterase inhibiting properties of dietary infusions with Humulus lupulus. *Journal of Elementology, 25*.

Eblagon, F., Trošt, Ž., & Čeh, B. (2022). End of life biomass waste management: impact of guiding material on supported crops. In *SUM2022, Symposium on circular economy and urban mining: 6th symposium on circular economy and urban mining, 18–20 May 2022, Capri, Italy*. IWWG. 16 p. Available at: https://members.sumsymposium.it/search-in-papers.php?c=sum2022

European Biogas Association. (2016). *SUCCESS STORIES Of the Members of the European Biogas Association Good practices and innovations in the Biogas Industry. [Brochure]*. https://

docplayer.net/47507179-Success-stories-of-the-members-of-the-european-biogas-association-goodpractices-and-innovations-in-the-biogas-industry.htm

European Commission. (2021). *Discover great EU-funded Innovations; BioTHOP: Hop waste fibre sheets reused for wine bottle packaging*. Available at: https://innovation-radar.ec.europa.eu/innovation/39125?fbclid=IwAR1ZJ5AzbZE-Dl9JOQyCTGJzBLhhu4S5_d_LUHDPEkQh6RWS_Oqrct42k6w

Gardea-Torresdey, J., Hejazi, M., Tiemann, K., Parsons, J. G., Duarte-Gardea, M., & Henning, J. (2022). *Journal of Hazardous Materials, 91*(1–3), 95–112.

Gao, D. W., Tao, Y., An, R., Fu, Y., & Ren, N. Q. (2011). Fate of organic carbon in UAFB treating raw sewage: Impact of moderate to low temperature. *Bioresource Technology, 102*(3), 2248–2254.

González, Z., Feria, M. J., Vargas, F., & Rodríguez, A. (2012). Comparison of the heating values of various types of fuel from non-wood raw materials. *American Journal of Environmental Engineering, 4*, 91–96. https://doi.org/j.ajee.20120204.04

Harder, N., Rodriguez-Uribe, A., Snowdon, M. R., Misra, M., & Mohanty, A. K. (2023). Hop natural fiber-reinforced poly(butylene succinate-*co*-butylene adipate) (PBSA) biodegradable plastics: Effect of fiber length on the performance of biocomposites. *Materials Advances, 4*, 1502. Available at: https://pubs.rsc.org/en/content/articlelanding/2023/ma/d2ma00831a

Haunreiter, K. J., Dichiara, A., & Gustafson, R. (2021). Structural and chemical characterization of hop stem fibers and their applications in the paper industry. *Industrial Crops and Products, 174*, 114217. Available at: https://www.sciencedirect.com/science/article/abs/pii/S0926669021009821

Iglesias, A., Martinez, P. G., Ramirez, C., Mitton, G., Arcerito, F. M., Fangio, M. F., Churio, M. S., Fuselli, S., Fanovich, A., & Eguaras, M. (2021). Valorization of hop leaves for development of eco-friendly bee pesticides. *Apidologie, 52*, 186–198.

Institute for Crop Science and Plant Breeding, Hops Department. (2022). *Annual Report 2022*. Special Crop: Hops; Bavarian State Research Center for Agriculture.

LIFE BioTHOP, Final report, Covering the project activities from 1st July 20191 to 31st Dec 2022 (2022). Available at: https://www.life-biothop.eu/wp-content/uploads/2019/12/final-report-biothop_june-2023-2.pdf

Limousin, R., Guendoul, F., Vibert, C., Antoun, K., Ségovia, C., Ziegler-Devin, I., & Brosse, N. (2023). Hop stems, a new fibrous bioresource. *Industrial Crops and Products, 201*, 116966. Available at: https://www.sciencedirect.com/science/article/abs/pii/S0926669023007318?via%3Dihub

Lohr, D., Görl, J., & Meinken, E. (2021). Nitrogen dynamics of chopped hop stems-effect of leaf: Stem ratio. *Acta Horticulturae, 1328*, 127–134.

Luskar, L., & Čeh, B. (2021). Hop biomass composting approach impact on compost microbiological properties. *Agricultural Research & Technology, 26*(2), 1–6. ISSN 2471-6774.

Luskar, L., Polanšek, J., Hladnik, A., & Ceh, B. (2022). On-farm composting of hop plant green waste—chemical and biological value of compost. *Applied Sciences, 12*, 4190.

Lutz, K., Euringer, S., & Schlagenhaufer, A. (2022). Verticillium: Thermal treatment of hop waste—bioassay using the indicator plant eggplant. In *Proceedings of the Scientific Commission/International Hop Growers' Convention*, Lugo, Spain, 3–7 July 2022; Volume 59.

Macchioni V, Picchi V, Carbone K (2022) Hop leaves as an alternative source of health-active compounds: Effect of genotype and drying conditions. *Plants, 11, 99*. https://doi.org/10.3390/plants11010099

Maniadakis, K., Lasaridi, K., Manios, Y., Kyriacou, M., & Manios, T. (2004). Integrated waste management through producers and consumers education: Composting of vegetable crop residues for reuse in cultivation. *Journal of Environmental Science and Health, Part B, 39*, 169–183.

Mitton, F. M., Turina, Y., Kulisz, N., Vittone, M., & Massa, A. (2025). Hop (*Humulus lupulus*) leaves as a functional ingredient to mitigate lipid oxidation and preserve PUFAs in rough scad patties. *Food Chemistry Advances, 6*, 100865. https://doi.org/10.1016/j.focha.2024.100865

Mohanty, A. K., Wu, F., Mincheva, R., Hakkarainen, M., Raquez, J. M., Mielewski, D. F., Narayan, R., Netravali, A. N., & Misra, M. (2022). Sustainable polymers. *Nature Reviews Methods Primers, 2*, 46.

Morcol, T. B., Matthews, P. D., & Kennelly, E. J. (2021). Differences in leaf chemistry and glandular trichome density between Wild Southwestern American Hop (Humulus neomexicanus) and Commercial Hop Cultivars. *Journal of Agricultural and Food Chemistry, 69*, 7798–7814.

Muzykiewicz, A., Nowak, A., Zielonka-Brzezicka, J., Florkowska, K., Duchnik, W., & Klimowicz, A. (2019). Comparison of antioxidant activity of extracts of hop leaves harvested in different years. *Herba Polonica, 65*(3), 1–9.

Narayana Saibaba, K. V. (2023). Next generation biosurfactants and their practical application in the food sector. In *Applications of next generation biosurfactants in the food sector* (pp. 349–360).

Onwosi, C. O., Igbokwe, V. C., Odimba, J. N., Eke, I. E., Nwankwoala, M. O., Iroh, I. N., & Ezeogu, L. I. (2017). Composting technology in waste stabilization: On the methods, challenges and future prospects. *Journal of Environmental Management, 190*, 140–157.

Pereira, O. R., Santos, G., & Sousa, M. J. (2022). Hop by-products: pharmacological activities and potential application as cosmetics. *Cosmetics, 9*, 139. https://doi.org/10.3390/cosmetics9060139

Proestos, C., Boziaris, I. S., Nychas, G. J. E., & Komaitis, M. (2006). Analysis of flavonoids and phenolic acid in Greek aromatic plants: Investigation of their antioxidant capacity and antimicrobial activity. *Food Chemistry, 95*(4):664–671. https://doi.org/10.1016/j.foodchem.2005.01.049

Renewable Carbon News. (2021). *Furniture from the biogas plant*. Available at: https://renewable-carbon.eu/news/furniture-from-the-biogas-plant/

Rutto, L. K., Temu, V. W., Ferreira, G., & Kering, M. K. (2020). Nutritive value and in vitro digestibility of non-cone hop (humulus lupulus l.) biomass. *Journal of Animal & Plant Sciences, 30*, 794–802.

Sabbatini, G., Mari, E., Ortore, M. G., Di Gregorio, A., Fattorini, D., Di Carlo, M., et al. (2024). Hop leaves: From waste to a valuable source of bioactive compounds–A multidisciplinary approach to investigating potential applications. *Heliyon, 10*(18).

Siebert, K. J., & Lynn, P. Y. (2005). Comparison of methods for measuring polyphenols in beer. *Journal of the American Society of Brewing Chemists, 64*(3), 127–134.

Scotti, R., Pane, C., Spaccini, R., Palese, A. M., Piccolo, A., Celano, G., & Zaccardelli, M. (2016). On-farm compost: A useful tool to improve soil quality under intensive farming systems. *Applied Soil Ecology, 107, 13–23.* https://doi.org/10.1016/j.apsoil.2016.05.004

Talboys, P. W. (1961). Time-temperature requirements for killing Verticillium albo-atrum in hop stem. *Plant Pathology, 10*, 162–163.

Talboys, P. W. (1987). Verticillium wilt in English hops: retrospect and prospect. *Canadian Journal of Plant Pathology, 9*, 68–77.

Trošt, Ž., & Čeh, B. (2022). Waste hop biomass management options. In *SUM2022, Symposium on circular economy and urban mining: 6th symposium on circular economy and urban mining, 18–20 May 2022, Capri, Italy*. IWWG. 8 p. Available at: https://members.sumsymposium.it/search-in-papers.php?c=sum2022

van Calcar, J., & Eblagon, F. (2019). Sustainable hop crop support using compostable PLA twines. *Hop Bulletin, 26*, 69–80.

Wiśniewska, M., Kulig, A., & Lelicińska-Serafin, K. (2020). Odour Emissions of municipal waste biogas plants— Impact of technological factors, air temperature and humidity. *Applied Sciences, 10*(3), 1093.

Zou, Y., Reddy, N., & Yang, Y. (2010). *Journal of Applied Polymer Science, 116*, 2366–2373.

Chapter 8
Spent Hops: Utilisation, Impact, and Future Prospects

Katya Carbone and Erika Celi

8.1 Spent Hops in the Brewing Industry: Environmental Impact and Challenges

Beer is the most widely consumed alcoholic beverage globally, ranking third in overall popularity after water and tea. It is primarily made from water, malt, hops, and yeast. The top beer-producing nations are China, the United States, and Brazil (Statista, 2025). Nevertheless, the brewing process is considered to be one of the most environmentally impactful agro-industrial processes. The information provided by the literature indicates that the total output of the by-product(s) of the top three beer producers can be estimated at around 19,6 million tons/year, based on average production rates (Thiago et al., 2014).

The solid waste products of the brewing industry include brewer's spent grain (BSG), hot trub (HT; often referred to as spent hops SH), residual yeast and diatomaceous earth (Fig. 8.1) (Hcjna et al., 2024).

BSG is the most significant solid waste generated in the brewery, accounting for approximately 85% of the total amount of by-product(s). It represents the insoluble fraction of the barley grain that remains after lautering (Chattaraj et al., 2024) and is estimated to amount to about 100–250 g/L of beer (Pasquet et al., 2024), whereas spent yeast, obtained by filtration at the end of the fermentation process, constitutes approximately 10–15% of the total by-product(s) of beer (Olivares-Galván et al., 2022).

HT and SH represent the remaining by-product(s) after the boiling and dry-hopping processes, respectively. HT is the precipitate formed during the wort boiling process (Claudia Salanță et al., 2023). It is primarily composed of water-insoluble tannins, protein conglomerates, and hop residue (Stachnik et al., 2021), and,

K. Carbone (✉) · E. Celi
CREA – Research Centre for Olive, Fruit and Citrus Crops, Rome, Italy
e-mail: katya.carbone@crea.gov.it

K. Carbone (ed.), *All About Hops: The Crop, its Cultivation, and its Uses in Brewing and Beyond*, https://doi.org/10.1007/978-3-031-96606-4_8

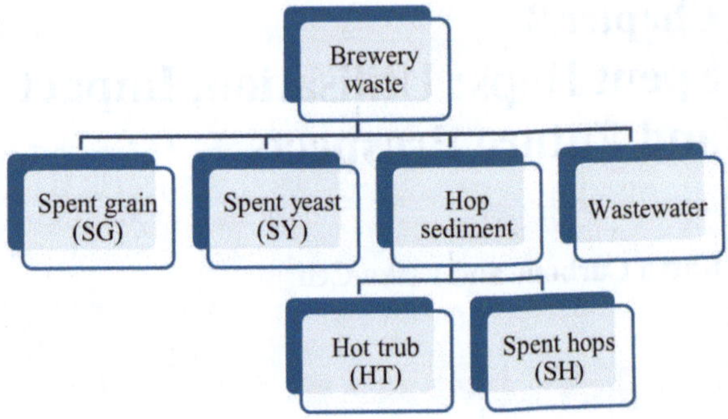

Fig. 8.1 General outline of brewery by-product(s)

currently, it is removed in a clarifying tank (also known as a whirlpool) and discarded. However, due to its chemical composition, rich in valuable molecules, this waste is well suited to recovery and valorisation. SH is the residual hop material collected after dry hopping from the maturation tank (Kopeć et al., 2022). According to Karlovic′et al. (2020), the production of 1 hL of beer produces 0.3 kg of spent hops, which are currently reused mainly as fertiliser or compost. As for SH, it has been reported that only 15% of hop components are incorporated into beer, with the remaining 85% becoming waste material, which is rich in polyphenols, essential oils, proteins and fibres (Karlović et al., 2020).

In light of previous evidence, transitioning towards circular economy models represents a promising opportunity for the reduction of waste and the recovery of resources in the beer processing industry.

8.2 Overview of the Brewing Process

In the current market, consumers are showing a growing interest in high-quality beverages that provide a unique sensory experience. This has resulted in an increased demand for innovative beers with distinctive aromas, leading to significant growth in the craft brewery sector (Statista, 2024).

Since hop is one of the ingredients that most influence the sensory characteristics of beer, this demand has been translated into an increased focus on hoppy beers (Gomes et al., 2021). To craft such beers, brewers often employ techniques such as late- or dry-hopping. The former refers to the addition of hops at the end of the boiling phase or in the whirlpool, whereas in dry hopping they are added during the fermentation or maturation. Both methods require the separation of hop sediment from the final beer product, a critical step in ensuring clarity and stability.

The timing of hop addition in these processes is crucial to allow the extraction of aromatic compounds, mainly essential oils, without increasing the perception of bitterness in the beer. Essential oils, which contain a complex array of volatile compounds, contribute significantly to the aroma profile of beer, imparting several floral and herbal notes. As a result, dry-hopping has become one of the most widely used techniques to enrich the sensory profile of beer, especially for styles that focus on robust, hop-forward characteristics, such as India Pale Ales (IPAs) and other types of craft beer. This method has gained popularity due to its ability to enhance aroma without compromising the balance of flavour. This goal is achieved by fine-tuning several variables, including contact time, extraction mode (static or dynamic), hop varieties and dosage (Lafontaine et al., 2019). Over the years, the quantity of hops added during dry hopping has risen significantly. In the United States, recent reports have indicated a range from 5–8 g/L up to as much as 22 g/L (Lafontaine & Shellhammer, 2018). Moreover, the growing popularity of non- and low- alcoholic beers (NABs and LABs respectively) has further fuelled the use of dry-hopping. These beers, which are produced through physical or biological processes that reduce or eliminate alcohol, often suffer from diminished aroma. The use of dry hopping has been identified as an effective method to compensate for this loss and restore the aromatic profile of the beer, making it more appealing to consumers seeking flavourful options without the adverse effects of alcohol (Brendel et al., 2020). Consequently, a substantial quantity of waste in the form of spent hops is produced.

Analysing the beer production process is therefore essential for understanding how and when the different types of waste are generated (Fig. 8.2) (Goldammer, 2022; Stewart & Priest, 2006; Eßlinger, 2009).

8.2.1 *Malting*

It is the process of transforming barley into malt, with the main purpose of developing the enzyme content and begin the conversion of starches in simpler sugars. It involves three stages as described below.

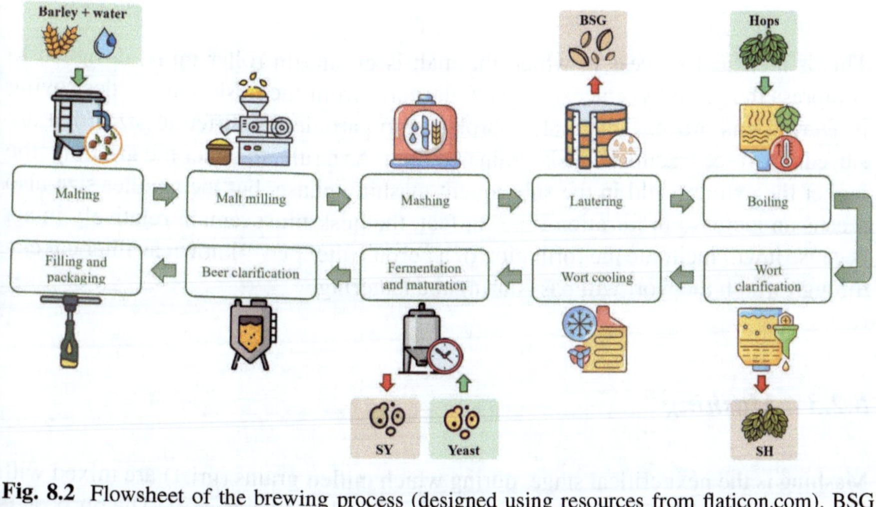

Fig. 8.2 Flowsheet of the brewing process (designed using resources from flaticon.com). BSG brewery spent grain; BSH brewery spent hop; BSY brewery spent yeast

- *Steeping:* barley grains are usually soaked in water until their weight reaches an average moisture content of 42-45%. (Nobis et al., 2019). This percentage indicates the degree of steeping, i.e. the amount of water required to rehydrate the endosperm and trigger the germination process.
- *Germination*: the soaked barley is germinated under controlled conditions. The absorbed water activates enzymes such as amylases and proteases, which are essential for breaking down the proteins and the carbohydrate matrix that encloses the starch reserves of the seeds.
- *Drying or Kilning*: during this stage, the heating process stops the sprouts from growing more and reduces the water activity. The temperature is maintained just high enough to stop all biological activities without damaging the desired enzymes (typically below 50 °C). This results in a stable product known as barley malt, which retains a significant quantity of enzymes and reserve materials for subsequent wort production. Variations in time and temperature during this step affect the colour, flavour, and mouthfeel (often referred to as "body") of the final product.

Malting is often excluded from the overall brewing process because it typically occurs at a separate location from the brewery. However, it is important to mention it in this context, as 3–5% (w/w) of the produced malt consists of by-product(s), mostly barley malt rootlets (also known as malt culms, coombes, or sprouts). Other malting by-product(s) include barley and malt dust, small or broken grains, embryonic shoots, and husk fractions. These materials can be repurposed for animal feed, composting, or biofuel production, yet they still represent a significant source of waste within the malting stage.

8.2.2 *Malt Milling*

This is a critical process in which the malt is ground in roller mills, designed to compress the grain by squeezing the endosperm from the husk without destroying it. During this process, the malt is broken into particles of different sizes that are sieved into three fractions: husk, grain and flour. As a rule, the finer the grinding, the higher the extract yield in the subsequent mashing phase, but the smaller size also means an increase in filtration time. In fact, the husk must remain relatively intact because it will facilitate the formation of a porous filter bed—known as filter cake—through which the wort will pass during the lautering.

8.2.3 *Mashing*

Mashing is the next critical stage, during which milled grains (grist) are mixed with hot water (about 50–78 °C) to create a sugary liquid known as wort. The purpose is to produce the carbohydrates necessary for the yeast to ferment and thus produce alcohol. Since yeast readily consumes simple sugars such as maltose and glucose, during mashing the long-chain carbohydrate molecules are broken down into monomers that can be fermented. Several enzymes are involved in the malting phase: α-amylase (which converts starch to dextrins and maltotriose at temperatures between 65 °C and 75 °C), β-amylase (which produces maltose at temperatures between 60 and 65 °C), β-glucanase (which reduces viscosity at temperatures between 35 °C and 45 °C), protease (which degrades proteins into amino acids for yeast nutrition at temperatures between 45 °C and 55 °C).

8.2.4 *Lautering*

Lautering is the process of separating the wort from the spent grains. The two most widely used devices to achieve this aim are the "mash lauter tan" and the "mash filter". In the lauter tan separation is driven by gravity, typically aided by a false bottom. Instead, in the mash filter a transversal pressure is applied to pump the mash into a chamber containing a series of filter plates.

Lautering is essential for obtaining a clear wort that will undergo the next stages of brewing. The obtained BSG is composed of barley shells, remaining endosperm starch granules, and other materials. It is characterised by high moisture content (63–80%) and considerable fibre content (up to 70%) (Olivares-Galván et al., 2022). The quantity of produced spent grains varies depending on the size of the brewery and the mashing recipe, but on average, industrial breweries generate between 10 and 25 kg of spent grains per hectolitre of beer produced (Pasquet et al., 2024).

Currently, BSGs are increasingly being used in the formulation of animal feed, human food fortifications and bioenergy production, or as a substrate for fermentation in other industries (Chattaraj et al., 2024).

8.2.5 *Wort Boiling and Hopping*

At this point, wort is boiled (at about 100 °C, for about 60–90 min) and hops are added (usually in the form of pellets, extract and, in some cases, cones) at different times depending on the desired effect in the final product, the beer. As boiling proceeds, malt enzymes are deactivated by high temperature, wort is concentrated and sterilised, and proteins (from both malt and hops) coagulate. The dense residue obtained at this stage is known as hot trub.

8.2.6 Wort Clarification

After boiling, the wort undergoes clarification to separate any hop residues, proteins and other solids from the liquid. This process is typically carried out in a whirlpool separator, where the wort is swirled at high speed, causing the heavier particles to settle. The by-product(s) obtained during this stage, the hot trub, includes spent hops, protein agglomerates and other impurities that must be removed before fermentation (Fărcaş et al., 2017).

8.2.7 Wort Cooling and Aeration

Once clarified, the wort is adjusted, making it ready for the subsequent phase of fermentation. It is quickly cooled (to reduce the risk of contamination) to an optimal temperature for fermentation, typically between 18–25 °C for ales and 7–15 °C for lagers (Eßlinger, 2009). It is diluted with water, if necessary, and aerated to achieve a sufficient level of dissolved oxygen, ensuring conditions for an adequate fermentation process.

While wort cooling and aeration do not generate significant solid waste, any water used for cooling or dilution represents a water consumption by-product(s). Water use efficiency is a critical factor in the sustainability of the brewing process.

8.2.8 Fermentation and Maturation

The wort is used as a fermentation medium, into which yeast (belonging to the genus *Saccharomyces*) is inoculated, converting fermentable sugars into alcohol, carbon dioxide and various other metabolic by-product(s). These elements significantly impact the flavour profile and taste of the final product. Fermentation is a process that can be subdivided into two distinct stages. Primary fermentation typically lasts 5–9 days, during which the microorganisms multiply rapidly, converting 60–70% of the total sugars present in the wort (Goyal et al., 2023).

Depending on the yeast they use (e.g. *Saccharomyces cerevisiae* and *Saccharomyces pastorianus*), brewers can produce top- or bottom- fermented beer. As the name suggests, top-fermented beer has the yeast floating on top of the tank, while bottom-fermented beer has the yeast sinking to the bottom.

Following primary fermentation, the young beer is transferred from the fermenter to a maturation tank, where secondary fermentation (beer maturation) is conducted under controlled conditions and during which the remaining sugar is converted into alcohol and carbon dioxide. In contrast to primary fermentation, secondary fermentation occurs at lower temperatures and higher pressure values and lasts longer (3 to 8 weeks). During this time, the proteins, polyphenols and hops settle at the bottom of the tank.

Usually, yeast is harvested at the end of fermentation, when it still has a high cellular vitality. Right after, it is washed and sieved to remove large impurities. Finally, it is stored under gentle agitation at 0 °C, ready to be re-used for four to six times in future brewing cycles. Nevertheless, a portion of this yeast is discarded, and it is the principal by-product(s) of the whole fermentation process along with carbon dioxide. The average amount of surplus yeast produced at this stage ranges from 2 to 4 kg/hL of beer (Karlović et al., 2020).

8.2.9 Beer Clarification and Stabilisation

Beer clarification and stabilisation are essential steps in ensuring the final product's clarity, microbiological stability, and consistent quality. These processes aim to remove suspended solids, such as residual yeast, protein-polyphenol complexes and other insoluble material, which coagulate due to the low pH and the cold temperature during aging. If left in the beer, these particles can cause haziness or negatively affect sensory characteristics, so they must be removed prior to packaging. Four main clarification techniques are commonly used, either separately or in combination:

- gravity sedimentation: the simplest clarification method, where fermented beer is chilled to about 0 °C for extended periods, allowing particles to settle naturally;
- use of fining agents: these substances ease the remotion of unwanted particles by binding to them or causing them to settle. Common fining agents include Irish moss (a seaweed), isinglass (a fish-derived gelatine), and silica gel. The choice of agent depends on factors such as beer style, personal preference, and dietary restrictions;
- centrifugation: this technique exploits centrifugal force to separate solid particles from the beer depending on their density;
- filtration: the beer is forced to pass through a filter medium that captures suspended solids.

Consumers often perceive haziness as a flaw and a potential health risk, so beer must remain bright and clear throughout its shelf life. To address this issue, beer undergoes stabilisation processes to ensure physical stability and prevent haze formation. Since turbidity is typically caused by proteins and polyphenols, most stabilisation treatments and agents aim to remove them or reduce their concentration. The most effective method is cold storage of the beer at −1 to −2 °C for about seven days. By combining these clarification and stabilisation techniques, brewers ensure that the final beer meets the desired clarity, stability, and consistent quality standards for commercial production.

8.2.10 Filling and Packaging

The final stage of the brewing process is bottle filling and packaging. If the finished beer is to be kept in glass bottles, it is first bottled and then batch-pasteurised to prolong its shelf life. On the contrary, when using cans or kegs beer undergoes flash-pasteurisation and is then packed aseptically. Whether the beer is bottled, canned, or kegged, the packaging process generates significant waste in the form of containers, labels, and packaging materials. Efficient recycling systems and the increasing use of sustainable packaging materials are critical for reducing the environmental footprint of this final stage.

8.3 Spent Hops: Chemical Composition and Applications

SH and HT are by-product(s) of brewing industry characterised by a complex chemical composition that is influenced by several factors, including hop variety, hop formulation (cones, pellet or extract), brewing process, and the stage at which hops are added. Based on these factors and the available literature, SH can be classified as spent hops associated with HT (SHHT), spent hops from dry hopping (SHDH), spent hops from supercritical CO_2 hop extraction or other extraction processes.

8.3.1 Chemical Composition of SH

In their study, Karlović et al. (2020) reported the composition of spent hops from different sources as rich in soluble and insoluble carbohydrate (known as nitrogen free extract and crude fibre respectively), proteins, lipids, inorganic elements and essential oils (Karlović et al., 2020) (Fig. 8.3).

With regard to the latter class of compounds, it is interesting to note that although the overall concentration of essential oils represents only 0.11% of the by-product(s)

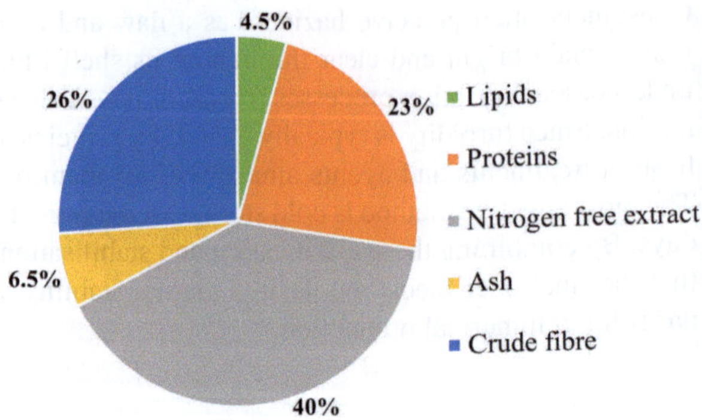

Fig. 8.3 Overall composition of spent hops from different sources (adapted from Karlović et al., 2020)

dry matter, this value is comparable to, or even higher than, that obtained from many aromatic and/or officinal plants, i.e. *Salvia officinalis* L. (0.2–2.4%), *Rosmarinus officinalis* L. (0.9–1.9%), or *Daucus carota* L. (0.5–0.8%). Moreover, the composition of essential oils in spent hops is not significantly different from that reported for fresh hops, with major constituents including sesquiterpene hydrocarbons (37%), monoterpene hydrocarbons (27%), non-terpene derivatives (18%), oxygenated sesquiterpenes (8%), and oxygenated monoterpenes (4%) (Karlović et al., 2020) (Fig. 8.4).

More in detail, the authors identified 31 constituents of spent hop essential oils, extracted by hydrodistillation. The most concentrated were myrcene (24.2%), α-humulene (16.2%) and β-caryophyllene (6.6%) (Fig. 8.5).

Several studies have also investigated the polyphenolic content of spent hops SHHT obtaining slightly differing results. For instance, Gandolpho et al. (2021) reported an extraction of 7.23 milligrams of gallic acid per gram of HT using ultrasound-assisted extraction with ethanol (Gandolpho et al., 2021), while a similar method was applied by Senna Ferreira Costa et al. (2021), yielding 8.94 mg of gallic acid per g of HT, along with 1.31 mg of quercetin per g of HT. Additionally, Bravi et al. (2021) obtained 10.90 mg and 17.58 mg of gallic acid per g of spent hop

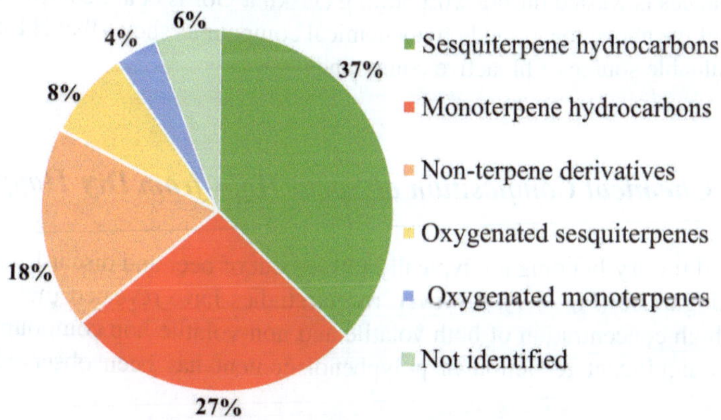

Fig. 8.4 Composition of spent hop essential oils (adapted from Karlović et al., 2020)

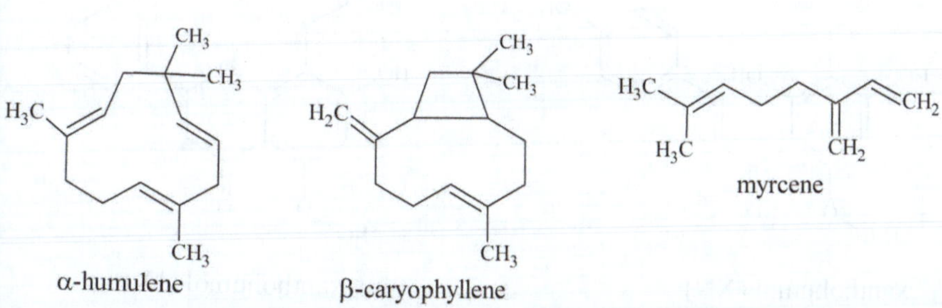

Fig. 8.5 Chemical structure of α-humulene, β-caryophyllene and myrcene

(depending on the beer style) by employing ultrasound-assisted extraction in a methanol, water, acetic acid (70/29.5/0.5) solvent mixture (Bravi et al., 2021).

Among hop polyphenols that we can find in SHHT xanthohumol (XN) (Fig. 8.6) deserves a special mention for its high value as bioactive compounds. In fact, it has an anti-inflammatory action, a sterol regulatory ability, and also the capacity to prevent liver and heart disease (de Andrade Silva et al., 2023).

XN is a prenylated chalcone, representing from 0.1% to 1% of hop cone dry weight (Stevens & Page, 2004). Due to its low solubility in water, this molecule is not significantly present in beer, with a concentration ranging from 1.99 to 81.22 μg/L (Wannenmacher et al., 2018). More in detail, during the wort boiling a fraction of XN is thermally isomerised to iso-xanthohumol (IX) (Fig. 8.6) that is more soluble in water, while the non-isomerised molecules precipitate and, as consequence, are removed in the HT (Stevens, 2020).

HT also contains hop bitter acids (such as hulupone, cohulupone, adhulupone, humulone, cohumulone, isocohumulone, isohumulone, etc.), which are prenylated derivatives of phloroglucinol (Spetsig, 1968) (Fig. 8.7).

This class of compounds undergoes isomerisation at high temperatures during wort boiling process (Fig. 8.8).

According to Jaskula-Goiris et al. (2018) approximately 50% of the mass of bitter substances is wasted during wort boiling (Jaskula-Goiris et al., 2018). The presence of all the above mentioned phytochemical compounds shows that HT could be a high valuable source of bioactive compounds.

8.3.2 *Chemical Composition of Spent Hops from Dry-Hopping*

Hops used for dry-hopping are typically filtered out of beer and discarded as waste after a single use (Fig. 8.9); however, recent studies have revealed that they still contain high concentration of both volatile and non-volatile hop compounds, even though a significant reduction in polyphenol content has been observed (Alfeo

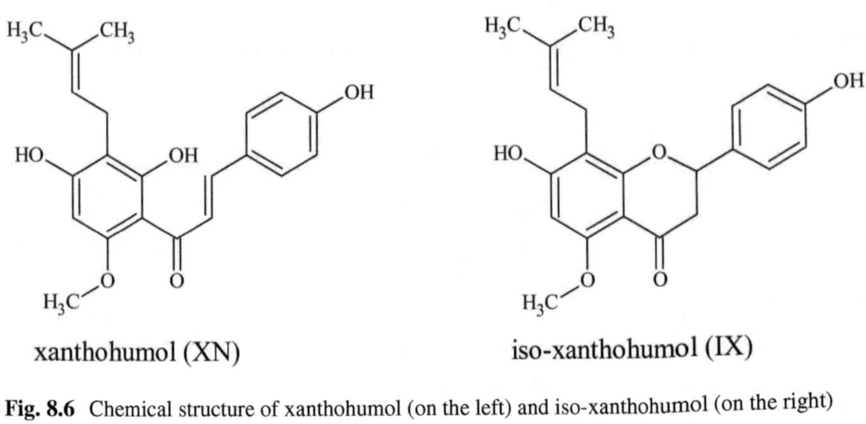

Fig. 8.6 Chemical structure of xanthohumol (on the left) and iso-xanthohumol (on the right)

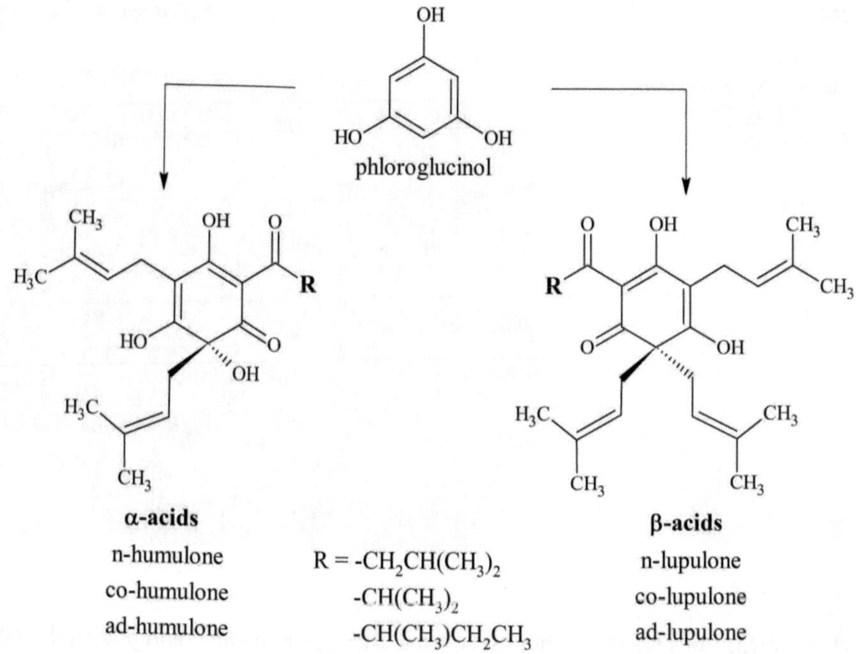

Fig. 8.7 Chemical structure of phloroglucinol and its derivatives, α-acids (on the left) and β-acids (on the right)

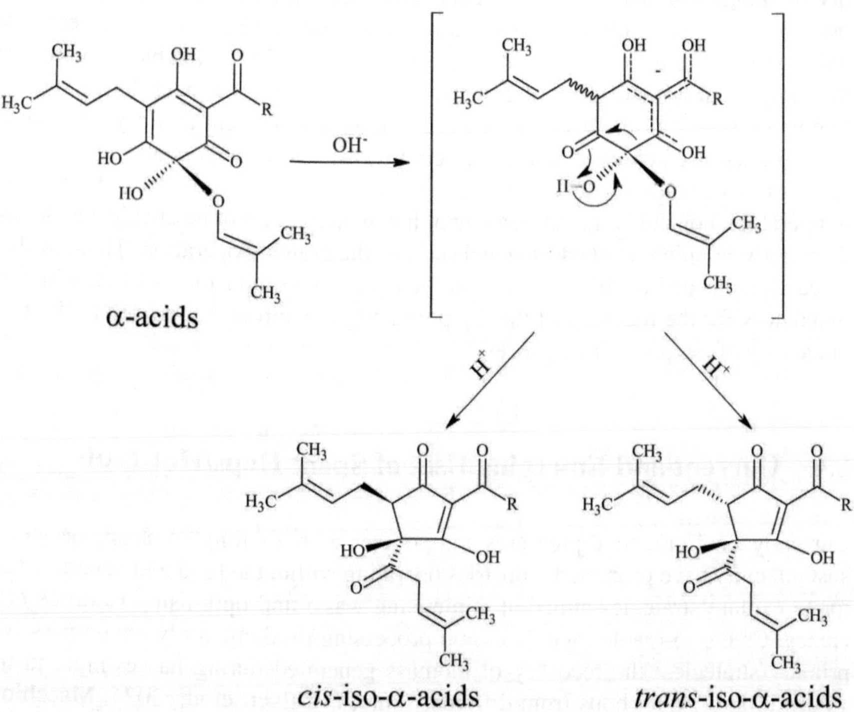

Fig. 8.8 Isomerisation mechanism of α-acids

Fig. 8.9 Spent hop from dry hopping. (Photo: Allendale Brew Co)

et al., 2023). As a result, the hops recovered after dry-hopping could potentially be recycled to kettle-hop beer for bittering purposes, reducing resource waste and lowering beer production costs (Hauser et al., 2019).

With regard to the transfer of hop volatiles into beer (the main objective of the dry-hopping technique), it is important to consider that the polarity of terpene alcohols increases their solubility, while non-polar compounds, such as myrcene, are only extracted to a limited extent (Alfeo et al., 2023). The available data suggest that only an average of 50% of essential oils is transferred to beer during the dry hopping (Hauser et al., 2019). It is worth taking into account that dry-hopping is not a standardised practice, and the extent to which hop components are extracted or retained depends on several system parameters, including but not limited to: temperature, hop cultivars, dosage rate of hop material, use of mechanical agitation during dry-hopping, contact time and scale of the brewing operation. Thus, further research is needed to fully understand the chemical composition of SH, which is mandatory for the recovery of this by-product(s) in a circular approach to the sustainability of the production process.

8.4 Current and Emerging Uses of Spent Hops/Hot Trub

Currently, environmental pressures and policies with a strong emphasis on global sustainability have prompted industries operating within the food and beverage sector to explore strategies aimed at minimising waste and optimising resource efficiency. Cradle-to-cradle models in hop processing predominantly encompass two primary strategies: the recovery of biomass generated during harvesting and the valorisation of spent hops from different sources (Calvert et al., 2024; Macchioni et al., 2021b; Silva et al., 2024).

The following paragraphs outline some examples, derived from the literature, concerning the most studied approaches for the recovery and enhancement of SH and HT (considering it as a source of spent hops).

8.4.1 Food

The potential for HT to be reused in the food industry is significant due to its composition and nutritional value. Nevertheless, its utilisation in food and feed is less frequent than that of other brewery waste, primarily due to its bitterness (Karlović et al., 2020).

Saraiva et al. (2019) investigated the possibility of using HT as an alternative source of vegetable protein for the food sector (Saraiva et al., 2019). In order to remove the compounds responsible for bitterness, the authors proposed a debittering protocol involving extraction of HT using a five-step aqueous extraction process at a high temperature of 100 °C for one hour in each step. They checked the trub before (TBP) and after (TAP) the process to see how its chemical composition, amino acid profile, and antioxidant activity had changed. As a result, a high protein concentration was obtained following this protocol, ranging from 24.91% to 70.26%. Furthermore, TAP demonstrated an oil absorption capacity of approximately 73%, indicating its potential application in fat-rich products. The extraction process led to a transformation of the trub structure from a compact, surface-coated configuration to a "clean" branched network. Although antioxidant activity and total phenolic compounds decreased in the TAP, it retained substantial concentrations of essential amino acids. The findings emphasise the promising potential of TAP for applications in the food industry, particularly in food fortification and as a novel source of vegetable protein, especially for fat-rich products.

As already mentioned, the availability of spent hops from dry-hopping has prompted scientists to explore the feasibility of reusing this by-product(s) in kettle additions for bittering purposes (alone or in combination with fresh hops), as well as in late- or dry-hopping (Alfeo et al., 2023; Hauser et al., 2019).

In their study, Hauser et al. (2019) investigated on the preservation of essential hop components in spent hops and their subsequent transfer into the beer (Hauser et al., 2019). The research, carried out at both pilot and industrial scale, showed that regardless of the hop variety (8 varieties were considered in the study: Amarillo®, Cascade and Centennial for the pilot scale trials, and Citra®, Amarillo®, Centennial, Mosaic®, and CTZ® for the industrial trials), about a third of the dry matter content of the hops is lost during dry-hopping, making it a relatively inefficient process. A significant proportion of α-acids (77% in the pilot scale, 52% in the industrial scale) and hop essential oil (51% in the pilot scale, 33% in the industrial scale) remained in the SHDH. Interestingly, the by-product(s) retained mainly terpenic and sesquiterpenic hydrocarbons rather than oxygenated compounds. In this regard, the authors also evaluated the extraction rates of specific hop compounds in beer and found that terpene alcohols, such as linalool and geraniol, had higher extraction rates (43–51%) compared to terpene hydrocarbons, such as β-myrcene (0.05%). However, the latter is an important terpene because it can undergo autooxidation to form potent odourants, such as linalool and geraniol. The findings of the research indicate that SHDH possesses significant brewing value and could be utilised directly in the brewery, thereby enhancing the economic and environmental sustainability of the production process. However, further research is necessary to investigate the various technical aspects of such practices, primarily the effects they may have on the sensory profile of beers. In this regard, Alfeo et al. (2023) also conducted a study on the potential reuse of this by-product(s) in the brewing process, carried out at both laboratory and pilot scales (Alfeo et al., 2023). The study reported that despite the differences in volatile compound composition between fresh hop and spent hop beers, sensory evaluation and triangle tests revealed no significant perceptible differences between the two. This finding suggests that SHDH can be utilised effectively in brewing applications without compromising the sensory attributes of the final product.

In a separate study, Gasínski et al. (2022) investigated the potential of utilising recycled hops (Hallertauer Mittelfrüh T90 hop pellets that had previously been employed for the process of dry hopping) in the production of beer. The study involved the brewing of test batches of beer with varying dosages of recycled hop material and the subsequent comparison of these batches with beer brewed with fresh hops (Gasiński et al., 2022). Despite a significant reduction in bitterness and a decrease in the concentration of hop-derived volatiles, beers brewed with recycled hops maintained similar levels of antioxidant activities, phenolic compounds, and many yeast-derived volatile metabolites as beers brewed with fresh hops. These findings suggest that recycled hops can still contribute significantly to beer quality. In fact, the sensory analysis revealed that the volatile profile of this beer exhibited notable distinctions from that of beer brewed with 100% fresh hops, whereas the sensory evaluation indicated that the beer brewed with recycled hops achieved higher scores in terms of taste, suggesting that SHDH can contribute significantly to the quality of the final product. The reuse of hops has the potential to engender substantial cost reductions for the brewing industry, especially for breweries that produce hoppy beers and currently dispose of large quantities of hops following the process of dry hopping. However, there are certain aspects that should be the focus of further investigation for the purpose of a practical extension of the study results. These include the potential long-term effects of the use of recycled hops on beer quality or stability, as well as how specific factors in the dry-hopping process (such as hop dosage, duration, variety and α-acid concentration) influence the properties of spent hops.

In addition to all these studies, Schrickel et al. (2024) set out an approach to the utilisation of SHDH in beer production (Schrickel et al., 2024). The authors reported that samples of SHDH from commercial breweries showed humulone retention rates of between 28% and 94%, indicating significant potential for its reuse. In the study, they brewed experimental beers using SHDH as the sole source of bitterness, which proved to be up to the target set. Sensory analysis of the experimental beers revealed that they were comparable to the reference beers brewed with hop pellets, with no significant negative impacts on taste or bitterness quality. However, the high hop aroma intensity achieved could potentially compromise the desired neutrality for lagers or pilsners. Before utilising SH/HT in the dry hopping process, it is imperative to consider the potential limitations or disadvantages inherent in the study. For instance, the study recognised considerable variability in the composition of spent hops utilised, which has the potential to result in inconsistent outcomes when employing them for bittering purposes. This heterogeneity poses a significant challenge for achieving uniform beer quality in different batches. In order to achieve this, it is also necessary to consider the microbiological stability of SH, which could pose a significant problem for food safety and the quality of the final product. Finally, while the study indicates the possibility of cost savings, a comprehensive economic analysis is necessary to ascertain the financial viability of implementing this approach on a commercial scale.

Kühbeck et al. (2007) investigated the effect of HT and particle addition on the fermentation performance of *Saccharomyces cerevisiae* in beer production. Results showed that the addition of HT enhanced suspended cell concentrations for all yeast vitalities examined, with more significant effects observed as the amount of trub increases. Pilot-scale fermentations showed significantly lower pH values and accelerated extract degradation with HT addition, advancing fermentation by approximately one day compared to extremely bright wort. The addition of model particles (PVPP and kieselguhr) also improved fermentation performance, although not to the same extent as HT. The study suggests that particulate material may improve the development of CO_2 from the fermenting medium, reducing its concentration and thereby its inhibitory effect on yeast metabolism (Kühbeck et al., 2007).

Spent hops could also be introduced in non-alcoholic drinks market by dry hopping the beverages, acting both as flavouring agents and sources of mild sedative effect, due to the presence of a molecule known as 2-methyl-3-buten-2-ol (Fig. 8.10), which is a by-product(s) of the oxidation of specific bitter compounds found in hops (Carbone & Gervasi, 2022; Claudia Salanță et al., 2023).

In addition, SH has a range of applications in the beverages industry, including high protein drinks, hop-infused honey, and hop alcoholic drinks (Claudia Salanță et al., 2023).

The applications of brewing hop waste in the food sector are limited by its bitterness and astringency. However, recent studies by numerous authors have identified several promising opportunities for the use of spent hops in food sectors other than beverages. One such approach involves the use of hot trub in food matrices such as pasta and ice cream, with a view to enhancing product quality and acceptability. Lomuscio et al. (2022) investigated the effect of fortifying fresh pasta with debittered trub powder (DTP) on the technological and physical-chemical characteristics,

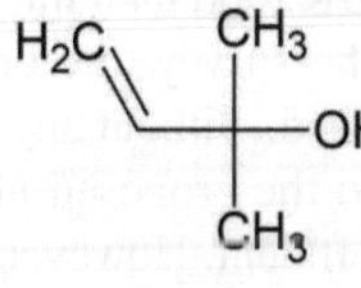

Fig. 8.10 Chemical structure of 2-methyl-3-buten-2-ol

in vitro digestibility, and sensory attributes of the final product. The results showed that DTP fortification significantly altered the properties of the pasta, with cooking loss increasing with DTP addition, but remaining within acceptable limits for fresh pasta. The fully cooked time, swelling index and textural properties showed distinctive features, particularly in PT15 samples, whose fibres may form a barrier to water uptake during cooking, delaying core pasta starch gelatinization, indicating complex interactions between the gluten network, starch and DTP components. In vitro digestion experiments showed reduced glucose release in fortified samples compared to the control at extended periods, while protein hydrolysis increased only in PT15 samples. This study found that DTP is a suitable ingredient for fortifying fresh pasta at up to 10 g/100 g, improving nutrition without affecting the quality and taste of the final product. The study demonstrates how the circular economy can be applied in food production by repurposing a brewing by-product(s) as a high-value pasta ingredient (Lomuscio et al., 2022).

In a separate study, Saraiva et al. (2020) investigated the use of debittered trub (DT) as a protein source for enriching liquid nitrogen ice cream (ICT), comparing it with ice cream formulations enriched with whey protein (ICW). The addition of trub proteins altered the chemical composition of the final product, which had higher fat, fibre, ash, and calorie content compared to ICW and control (standard ice cream formulation, ICS). Furthermore, ICT demonstrated higher viscosity, consistency index, and hysteresis area, suggesting a more robust three-dimensional network and enhanced resistance to flow, along with increased hardness but reduced adhesiveness. Notably, while ICT was well-received by consumers, it garnered lower scores compared to ICS and ICW in most attributes (e.g. flavour and texture). Indeed, despite the debittering process generally proving effective, it is likely that residual bitterness was a contributing factor to the moderate appreciation expressed by the ICT panel. DT can be utilised as a protein ingredient in ice cream, enhancing specific functional properties; however, further studies are necessary to enhance its sensory acceptability (Saraiva et al., 2020).

8.4.2 Feed

One of the first strategies proposed for the use of spent hops is their conversion into animal feed. Given the high nutritional content, particularly the fibre, protein, and mineral ones, they are considered a promising alternative feed source for livestock, mainly for ruminants such as cattle and sheep (Singh Nee' Nigam & Pandey, 2009).

In their study on the development of a methodology for sustainable management of food waste, Garcia-Garcia et al. (2017) considered a case study from the UK food manufacturing sector, including a brewery (Molson Coors). According to the survey, the brewery produced approximately 700 t/year of trub, which was used, without any pretreatment, as animal feed, mixed with spent grain. According to the proposed methodology, this is the most effective way to manage this solid effluent. However, the authors did not report on the percentage of HT added to the ration, nor on the effects on animal nutrition or on the palatability of the resulting diet (Garcia-Garcia et al., 2017). However, it should be noted that, even though HT are the richest in crude fibre among brewing waste, their fibre is of low quality and the net energy that cattle and sheep can derive from it is less than 50% of that obtainable from BSG. Furthermore, although it contains about 24% crude protein on a dry basis, the protein efficiency index is very low. As a result, its addition to the brewing grain compromises the quality of the feed mix (Huige, 2006). Nevertheless, it remains one of the most convenient and most popular methods of disposing of SH at the brewery site, after direct disposal into the sewage system (Bonato et al., 2022), which, however, generates a biological oxygen demand (BOD) load of just under 46 kg and a suspended solids load of just over 90 kg per 1600 hL of product. Consequently, careful consideration must be given to the necessity of pre-treating effluents prior to their entry into the sewerage system (Huige, 2006).

Recently, Mattioli et al. (2020) evaluated the impact of supplementing rabbit diets with beer trub (2% in freeze-dried form) on their in vivo oxidative state and the quality of their meat, focusing on its fatty acid profile and cholesterol levels (Mattioli et al., 2020). Moreover, the authors also explored the antioxidant properties of trub used in conjunction with dietary linseed supplementation (as 3% extruded linseed), which is known for its high n-3 polyunsaturated fatty acids (n-3 PUFA) content and susceptibility to oxidation. Productive performance was found to be similar across all groups, while supplementation with trub resulted in increased plasma retinol levels, without causing a significant change to the oxidative state in vivo. Conversely, the combination of linseed and trub exerted an adverse effect on the oxidative stability of the meat, leading to an augmented oxidation of the lipid and protein fraction in the supplemented groups. This outcome demonstrates the ineffectiveness of trub in counteracting the oxidative instability induced by the elevated PUFA content of linseed. Notably, the supplementation of both trub and linseed resulted in a reduction in the cholesterol content of the meat. However, the findings indicate that trub exerts no effect on the overall quality of meat, and its utilisation as a supplementary feed necessitates further research. Moreover, the study is not without its weaknesses, which include the following: a limited sample size, a lack of dose-response relationship, and above all the lack of sensory evaluation. Notwithstanding, the study under discussion is of interest to the feed industry, as rabbits in general show a good level of tolerance to bitterness, which allows for the bypassing of a major critical issue in using trub as a feed supplement.

However, direct incorporation of SH as a feed supplement is not recommended due to the presence of bitter compounds in this by-product(s), as animals generally tend to be reluctant to eat bitter feed. In this regard, Huszcza and Bartmańska (2008) proposed an interesting debittering protocol for SH by the yeast *Candida parapsilosis* under various conditions, allowing this protein-rich industrial by-product(s) to be used as animal feed (Huszcza & Bartmańska, 2008). During their comparative studies, the authors observed that *C. parapsilosis* degraded 100% of bitter acids within a period of four days, at a relatively high pH level, in light, and at 40 °C. The study also revealed a close correlation between the efficiency of degradation and the concentration of riboflavin in the *C. parapsilosis* culture. However, the study indicated that despite the lower bitter acid content of spent hops (33.6 mg/L) compared to hop extract (440 mg/L), the degradation process by the yeast was slower (4 days and 24 h, respectively), suggesting that the matrix of the substrate plays a key role in its efficiency and dynamics.

8.4.3 Bioenergy Production

Converting spent hops into biogas and bioethanol is another promising economic opportunity. Indeed, the bioenergy sector has grown significantly in recent years, and there is considerable untapped potential in the use of organic waste, such as spent hops, to produce biofuels. Consequently, the utilisation of biogas produced from SH/HT in a circular approach holds considerable potential as a direct source of fuel for brewing operations. This approach would be a valuable step in reducing brewery's dependence on external energy suppliers and lowering costs. In this regard, Ziemiński et al. (2012) demonstrated that the pretreatment of SH with enzymes can enhance the biogas yield and accelerate the process of organic matter degradation during anaerobic fermentation. The study showed that the fermentation of methane from enzymatically pretreated SH yielded more than 120 mL/d of biogas from 1 g of chemical oxigen demand (COD), at an organic matter loading of 6 g of COD d/L, pointing out that the biogas yield from the pretreated biomass was 13% greater than that from untreated SH. However, there are several issues that require further in-depth investigation. For instance, a cost-benefit analysis of the economic feasibility of enzyme pretreatment would provide valuable insights for the practical implementation of the research findings. Further research is necessary to ascertain the replicability of the results on an industrial scale, to explore alternative enzyme combinations or ratios that may enhance the outcomes, or to investigate further the impact of enzymatic pretreatment on the microbial communities involved in anaerobic fermentation. However, given the absence of prior research on anaerobic fermentation of SH, this study establishes an exciting new area for further investigation and potential valorisation of this brewing industry by-product(s) (Ziemiński et al., 2012).

Breweries are high-energy users, and there is an opportunity for them to adopt more circular practices. Specifically, breweries can generate energy from waste streams, a strategy that is particularly beneficial for small craft breweries, as it can help contain costs and enhance profit margins (Sperandio et al., 2017).

Andrić et al. (2017) investigated the potential for energy recovery from brewery waste through anaerobic digestion (AD), with a focus on a micro-brewery located in the North East of the UK (Andrić et al., 2017). The research examined waste generation, management practices, and energy auditing at the brewery, which produced approximately 1100 tonnes of solid waste annually, composed of 90% residual grains and 10% residual hops and yeast. Interestingly, the study employs several approaches and methods to investigate energy recovery from brewery waste: experimental setup, simulation model using Aspen Plus software, and boiler simulation.

The results obtained demonstrated the potential for biogas production under both mesophilic (35 °C) and thermophilic (55 °C) conditions, with thermophilic digestion yielding 2.6 litres *per* day of biogas with 65% methane content, and mesophilic digestion producing 3.0 litres *per* day with 55% methane content. The proposal of a predictive model for process performance when scaling up is a concept of great interest. The model predicted that for 1500 kg/day spent grain, 452.62 m^3/day of biogas will be produced at thermophilic conditions and 213.33 m^3/day at mesophilic conditions. It is recommended that the thermophilic process be employed due to the higher energy content and easier digestate handling. Furthermore, the simulation of a steam boiler fuelled by the produced biogas indicated that the thermophilic process could power a 126.01 kW boiler, while the mesophilic process could fuel a 76.48 kW boiler. A key finding was the potential for integrating waste heat recovery into the AD process. Typically, brewery waste is disposed of at approximately 60 °C; however, it was found that this waste could support the thermophilic AD process, thereby minimising the additional energy input required. This approach enhances waste management practices and provides a sustainable solution for energy generation within the brewery. Notwithstanding the intriguing approach proposed by the authors, certain limitations necessitate further investigation in future studies, including the composition of the biogas produced. For instance, the Aspen Plus model exhibited substantial discrepancies in predicting its composition, underestimating the methane content by 28% for the thermophilic condition and overestimating it by 16.36% for the mesophilic condition. Moreover, while the study addressed technical feasibility, it did not provide a comprehensive economic analysis of the implementation of the proposed system, which is crucial for the exploitation of the results achieved.

8.4.4 Agriculture

The integration of SH/HT into agricultural applications has been explored as a means to enhance soil quality and promote sustainable farming practices, given their high content of nitrogen and other nutrients.

Composting hop sediments has been identified as a viable method for the management of this brewing waste. However, it is imperative that careful consideration be given to the type of sediment utilised, and its effects on the composting process and the quality of the resulting compost. In this regard, Kopeć et al. (2022) investigated the biological activity of composts obtained from hop waste generated during beer brewing, focusing on hot trub from wet hopping (SHHT) and spent hops from dry hopping (SHDH), with the latter showing a significant total nitrogen content, affecting the composting process and requiring substrates with a wide C:N ratio. Moreover, composted SHDH resulted in a greater number of microorganisms compared to SHHT, which showed inhibitory properties for plant development, likely due to its organic compound content. While this might seem negative, the researchers suggest it could be beneficial in certain agricultural scenarios, potentially as a natural growth retardant. As a consequence, the findings underscore the importance of considering the type of hop waste in developing composting strategies for breweries. Finally, the study highlighted that the hop residues used were free of potentially pathogenic microorganisms and that they can be used as soil amendments without any concerns for safety (Kopeć et al., 2022).

Recently, Toninato et al. (2024) investigated the potential of spent hop extract in controlling *Meloidogyne javanica* (root-knot nematode) in tomato plants, in two greenhouse experiments using a 5% aqueous extract of SH, and its impact on soil biological activity. The results obtained from this study are very promising. Indeed, the extract significantly reduced the reproduction of *M. javanica* in tomato plants, with a reduction of the total number of nematodes at 30 days after inoculation of about 70%. Furthermore, the extract demonstrated effective persistence, with reductions of approximately 80% in the total number of nematodes and 73% in nematode population density at 60 days after nematode inoculation. Furthermore, application of the extract generally increased basal soil respiration and microbial carbon biomass, even in some cases up to more than 90% compared to controls. The research suggests that the nematicidal effect may be due to compounds present in hops, such as prenylated chalcones and bitter acids, known for their antimicrobial properties (Toninato et al., 2024). However, field trials would be needed to validate the efficacy of the extract under various environmental conditions and to analyse any potential impacts on beneficial soil microorganisms and long-term changes in soil ecology.

Moreover, SH essential oil (SHEO) was evaluated as a source of repellents against stored food insect pests (*Rhyzopertha dominica* and *Sitophilus granaries*), showing strong repellent activity against both insect species. Among bioactive compounds identified in the essential oil, myrcene was the most effective against *R. dominica* and limonene against *S. granarius*. It is noteworthy that while certain individual terpenes exhibited notable repellent properties, SHEO demonstrated a more pronounced effect, likely attributable to the synergistic interaction of multiple compounds present in the oil (Bedini et al., 2015).

In a separate study, the antifeedant activity of XN and supercritical carbon dioxide extract of SH against stored product pests, such as *Sitophilus granarius* L. (Granary Weevil), *Tribolium confusum* Duv. (Confused Flour Beetle), and *Trogoderma granarium* Everts (Khapra Beetle), was investigated (Jackowski et al., 2015). The results demonstrated that crude SH extract exhibited stronger deterrent activity against *T. confusum* adults in comparison to pure xanthohumol. This finding suggests potential applications for the crude spent hops extract also in stored product protection.

Yoon and Tak (2023) also demonstrated that SH oil and its primary component, β-myrcene, enhanced the efficacy of deltamethrin, emamectin benzoate and spinetoram against *Spodoptera frugiperda* (gunworm). This pest has the potential to damage and destroy a wide range of crops, primarily cereals and legumes. The study demonstrated that the observed synergistic effect between the tested insecticides and SH oil and β-myrcene may not be directly related to enzyme inhibition, as no inhibitory activities on the main detoxifying enzymes were observed. Moreover, neither spent hop oil nor β-myrcene was toxic to the insect itself. The study also revealed that the neural response of these conventional insecticides in the larval central nervous system was enhanced. The authors also reported that β-myrcene displayed spatial repellency against *S. frugiperda* larvae, suggesting that SH can be considered a novel candidate for developing insecticidal synergists and repellents (Yoon & Tak, 2023). However, further research is required to ascertain the full implications of the observation that the repellent effect of β-myrcene diminished significantly after 36 hours. This limited efficacy may, in fact, limit its practical application as a repellent.

A substantial body of evidence has also indicated that the utilisation of waste-derived extracts in agriculture could facilitate more sustainable pest management practices, thereby reducing reliance on synthetic fungicides. Within this context, HT and SH may also be used as natural potential crop-protection agent, thanks to their antimicrobial properties. It has been shown, in fact, that the ethyl acetate, acetone, and methanol crude extracts from spent hops, as well as the extracts formed after the removal of XN, demonstrated significant antifungal activity against various *Fusarium spp*., whereas the methylene chloride extract from SH demonstrated effectiveness against *Botrytis cinerea*, a serious plant pathogen (Bartmańska et al., 2018).

In their study, Giorni et al. (2023) explored the prospect of utilising extracts (1 mg/L) from agri-food by-product(s) to inhibit the in vitro growth of mycotoxigenic fungi and their associated mycotoxin production. In addition to a range of other by-product(s) commonly associated with the agri-food industry, the study also incorporated SH (Giorni et al., 2023). In the study, SH were subjected to hydrodistillation in order to obtain the essential oil, the main components of which were identified as myrcene, α-caryophyllene, β-caryophyllene, cadinene, and humulene epoxide. Overall, the SH essential oil and extracts exhibited the most promising results in terms of reducing ochratoxin A, aflatoxin B1 by *Aspergillus flavus* (about 58%), zearalenone, and tenuazonic acid (the most abundant *Alternaria* toxin in cereals) production. However, it is important to note that the effectiveness of the essential oil and extracts varied depending on the specific mycotoxin and the extraction method used. Moreover, the extracts were tested at a single concentration and the study does not provide information on dose-dependent effects or minimum effective concentrations.

8.4.5 Functional Extracts

A high-added value strategy for the economic valorisation of spent hops/hot trub involves the extraction of valuable bioactive compounds, including polyphenols, flavonoids, and hop acids, which have applications in nutraceuticals, pharmaceuticals, and cosmetics (Karlović et al., 2020).

It is estimated that the global nutraceuticals market size was around USD 591.1 billion in 2024, and it is thought that it could grow at a compound annual growth rate (CAGR) of 7.6% from 2025 to 2030 (Grand View Research, 2025). The growing demand for natural, plant-based products represents a significant market opportunity for compounds extracted from spent hops. Hop-derived compounds, especially polyphenols, have shown antioxidant, anti-inflammatory, and anti-cancer properties, making them suitable for incorporation into dietary supplements and cosmetics (Carbone & Gervasi, 2022). In addition, hops contain unique phytoconstituents, such as prenylated flavonoids, which have a high oestrogenic potential, related to the presence of 8-prenylnaringenin (8PN), among others, and may be useful in the treatment of menopausal-related symptoms. In this regard, Chadwick et al. (2004) conducted a pioneering investigation into the chemical composition of a unique form of spent hops that are the remains after CO_2 extraction of hop pellets. The CO_2 extraction process has been shown to result in the retention of chalcones and flavanones in spent hops, with a concomitant loss of the bitter acids and volatile oils that were originally present in hop pellets. Consequently, spent hops emerge as a cost-effective, pre-purified starting material for the subsequent extraction of estrogenic-like compounds (Chadwick et al., 2004).

A range of techniques has been used to extract valuable compounds from these byproduct(s), with the main compounds extracted including phenolic compounds and essential oils. Conventional solid-liquid extraction (CSLE) is the most common method, but other techniques such as ultrasound-assisted extraction (UAE), supercritical fluid extraction (SFE) and extraction with deep eutectic solvents (DES) have also been used. The selection of solvent and extraction method has been demonstrated to significantly influence the yield and type of compounds extracted. For instance, the use of ethanol and methanol has been shown to be effective in the extraction of bitter compounds such as iso-humulones, while hexane has been identified as a more efficient extraction solvent for humulones and lupulones. Among the phenolic compounds, particular attention has been given to xanthohumol due to its noteworthy biological properties. A range of extraction methods, including CSLE, UAE, SFE, and DES, have been employed for the purpose of XN extraction.

In recent times, the introduction of natural eutectic solvents has engendered a novel sustainable approach to the extraction of encapsulated extracts from ready-to-use plant matrices (González-Laredo et al., 2023; Macchioni et al., 2021a).

Grudniewska & Pastyrczyk (2023) studied the utilisation of SH, which are derived from the supercritical CO_2 (SC-CO_2) hop cone extraction process, for the simultaneous extraction of proteins and XN using deep eutectic solvents (Grudniewska & Pastyrczyk, 2023). Four different DESs were prepared by mixing choline chloride (hydrogen bond acceptor) with various hydrogen bond donors (glycerol, ethylene glycol, propylene glycol, or lactic acid) in a 1:2 molar ratio. In the study, the XN extraction efficiency varied between 0.52–1.92 mg/g SH depending on the DES used. However, the authors observe that an increase in the scale of the process (from 0.5 g to 4 g of SH) resulted in disparities in extraction efficiency for certain DESs, particularly DES-3 (Hydrogen Bond Acceptor: Choline chloride; Hydrogen Bond Donor: Glycerol). This finding suggests the potential for inconsistencies in results when the process is scaled up, emphasising the necessity for further research in this area.

In a separate study, Gandolpho et al. (2021) presented several significant insights into the extraction of phenolic compounds from trub using ethanol at different percentages as solvent. The findings of the study indicated that the solvent concentration, with an optimal range of 55–66%, had the most substantial impact on the extraction of total phenolic content (TPC) compared to the other process parameters tested (extraction temperature and solid-liquid ratio).The study demonstrated that trub contains a substantial quantity of valuable phenolic compounds that are comparable to, or higher than, those found in other brewing by-product(s), such as spent grain. Furthermore, the study emphasised the efficacy of UAE as a method for extracting phenolic compounds from trub, providing a cost-effective and efficient alternative to conventional extraction techniques (Gandolpho et al., 2021). However, it should be noted that the study concentrated primarily on three variables. Other potentially significant factors, such as extraction time, ultrasound frequency and the polyphenolic profile of the extract obtained, were not examined.

Pectins are complex polysaccharides that are found in the cell walls of plants and are widely used in the food industry as gelling and thickening agents, but not only (Vanitha & Khan, 2020). In this regard, research has shown that spent hops can be a cost-effective source of pectins with a relatively high molecular weight and intrinsic viscosity that is comparable to commercial apple and citrus pectins. Oosterveld (2002) successfully extracted SH pectins using acid extraction conditions and analysed their chemical composition and structure (Oosterveld, 2002). His findings demonstrated that these pectins exhibited a low degree of methyl esterification, which limits their use as gelling agents in acid and sugar gels but makes them suitable for calcium gels. Furthermore, the SH pectins exhibited a high degree of acetylation and a high neutral sugar content. Consequently, future research should focus on removing part of the acetyl groups and neutral sugars to improve the properties of pectins obtained from spent hops.

SH extracts have demonstrated a variety of bioactivities, rendering them potentially valuable for a range of applications (Olivares-Galván et al., 2022). The main bioactivities reported include the following:

- *Antioxidant and Antimicrobial Activities*

  SH extracts have shown significant antioxidant properties and exhibit antimicrobial activity, particularly against Gram-positive bacteria. Intriguingly, these activities are not solely dependent on the presence of XN, as extracts devoid of it still exhibited considerable antimicrobial properties (Rozalski et al., 2013). Furthermore, several compounds derived from spent hops have been shown to possess significant activity against both methicillin-sensitive and methicillin-resistant strains of *Staphylococcus aureus* and *Staphylococcus epidermidis*. In their study, Bartmańska et al. (2018) reported indications of structure-activity relationships (Bartmańska et al., 2018). The antistaphylococcal activity of the tested compounds was found to be related to several structural features, including the absence of a double bond in the prenyl group, a hydrogenated α,β double bond in chalcones, a free C-5 hydroxyl group in flavanones, an aurone skeleton, and finally the presence of a hydrophobic prenyl group, which generally increased antimicrobial activity, as observed in 8-prenylnaringenin (Fig. 8.11) compared to naringenin without the prenyl group.

It is also worthy of note that the presence of methoxy groups, particularly at the C-5 position, was found to have a tendency to decrease antibacterial activity. This finding underscores the potential for these compounds to be utilised in the treatment of staphylococcal infections, including those caused by antibiotic-resistant strains;

- *Antiproliferative Activity*

  SH extracts, particularly those containing XN and its derivatives, have shown antiproliferative activity that is comparable to that of resveratrol (Boncler et al., 2014);

- *Anticoagulant Activity*

  The anticoagulant properties of SH extracts are attributed to various compounds, including flavanols, hydroxycinnamic acids, and proanthocyanidin oligomers.

  This activity has been observed in both the phenolic and lipid fractions of the extracts (Luzak et al., 2016);

- *Anti-inflammatory Activity*

  The anti-inflammatory properties of extracts rich in flavonols and proanthocyanidins from SH are evidenced by their ability to inhibit the expression of genes encoding cytokines and other molecules involved in inflammation. These included interleukin-1β, interleukin-6, tumour necrosis factor-α, prostaglandin-endoperoxide synthase 2, nitric oxide synthase, nitric oxide, and nuclear factor-enhancing kappa light chains of activated B cells (Caban et al., 2020).

As evidenced by the studies analysed, there is considerable potential for the application of SH extracts in a variety of fields, including food preservation, pharmaceuticals and nutraceuticals. However, it is important to note the limitations to the industrial exploitation of these results, which primarily pertain to the high operating

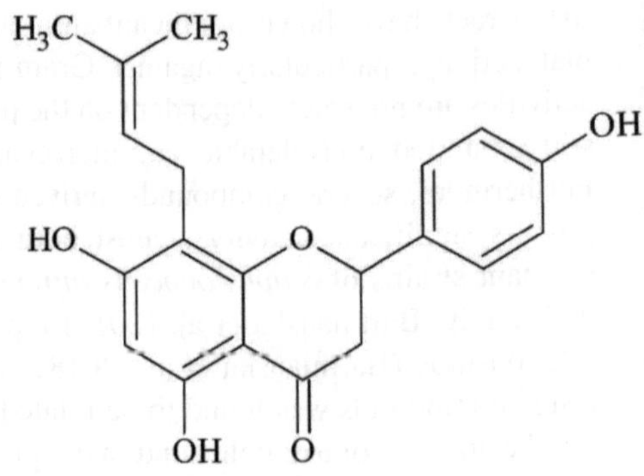

Fig. 8.11 Chemical structure of 8-prenylnaringenin

costs of the extraction and purification processes. In addition, the market for hop-derived compounds is relatively niche when compared to other natural products, and the creation of a reliable consumer base for these compounds may require a significant investment of time. Despite the significant economic potential, the scalability of the scientific outcomes discussed in this section is contingent on the availability of industrial-scale processes that can reduce extraction costs and enhance the efficiency of recovering the compounds of interest.

8.4.6 Phytotherapeutic Applications

It is well established that hops have been utilised for their therapeutic properties since ancient times (Carbone & Gervasi, 2022). As previously mentioned, spent hops have been shown to retain significant biological activity, and numerous studies have investigated their potential in the therapeutic field.

In the study by Luzak et al. (2016), for instance, the effect of spent hop extract (SHE), derived from hops extraction by supercritical CO_2, on blood platelet aggregation and endothelial cell function was analysed (Luzak et al., 2016). The results demonstrated that SHE, a rich source of flavanols, hydroxycinnamic acids, proanthocyanidin oligomers, flavan-3-ol monomers and flavonol glycosides (e.g. quercetin-3-O-hexoside), exhibited concentration-dependent inhibition of ADP-induced platelet aggregation. Furthermore, the extract enhanced the anticoagulant activity of human umbilical vein endothelial cells (HUVECs), a phenomenon that may have arisen from increased expression and activity of ecto-ADPase (CD39) and increased activation of nitric oxide synthase (eNOS). These effects were observed at concentrations that did not significantly reduce cell viability. Interestingly, results pointed out synergistic effects with aspirin. In fact, in patients with coronary artery disease treated with aspirin, SHE demonstrated enhanced inhibition of platelet aggregation, suggesting that it may operate through pathways that are complementary to cyclooxygenase-1 inhibition. Finally, the authors also analysed the effect of the extract in vivo by feeding diabetic rats with SHE doses of 2.5 and 5 mg/kg body weight per day, which showed significantly longer survival times than the placebo group. This enhanced survival was not accompanied by improvements in glycaemic control, suggesting that the extract may have ameliorated secondary effects of hyperglycaemia rather than directly affecting blood glucose levels. The results of the study also highlighted the potential for reusing spent hops as a dietary supplement that could be beneficial for cardiovascular health.

Hajirahimkhan et al. (2013) conducted a series of experiments to evaluate the oestrogenic potency of spent hops. In this case, however, the authors did not test the brewery by-product(s) directly; rather, they prepared an extract of spent hops from hop pellets cv Nugget (Hajirahimkhan et al., 2013). The extract contained 5.4% XN, 0.084% 8PN, 0.076% 6-prenylnaringenin (6PN), and 0.65% IX, as determined by weight percentage of the spent hops extract. The results of the study demonstrated that the spent hops extract exhibited stronger estrogenic properties than the licorice extracts in several assays, such as alkaline phosphatase induction in Ishikawa cells and estrogen responsive element (ERE)-luciferase induction in breast cancer (MCF-7) cells. The research indicates that the stronger estrogenic activities of hops are attributable to the presence of 8PN and its precursors. In contrast to the compounds found in licorice, 8PN from hops exhibited high affinity for both estrogen receptor (ER) subtypes (ERα and ERβ) and demonstrated full agonist activity similar to oestradiol in various assays. The study suggests that while hop has strong estrogenic properties, its effects might be less predictable compared to licorice. This is because the conversion of xanthohumol to the estrogenic 8PN in hops depends on cytochromeP450 metabolism and gut microbiota, which could vary among individuals.

More recently, Caban et al. (2022) investigated the chemopreventive properties of SHE against colorectal cancer (CRC) (Caban et al., 2022). The research focused on SHE's effects on angiogenesis, invasion, and migration of CRC cells, utilising two human colorectal adenocarcinoma cell lines, SW-480 and HT-29, to evaluate SHE's impact on various cellular mechanisms associated with cancer progression. The study pointed out that SHE significantly inhibited the invasion and migration of both SW-480 and HT-29 cells, accompanied by a decrease in the expression and activity of matrix metalloproteinases (MMP-2 and MMP-9), which are enzymes involved in the degradation of extracellular matrix components. Interestingly, SHE showed greater anti-invasive and anti-migratory effects than epigallocatechin gallate (EGCG), a known chemopreventive agent. Furthermore, SHE treatment led to a reduction in the expression of vascular endothelial growth factor (VEGF) and hypoxia-inducible factor 1α (HIF-1α), suggesting its ability to inhibit angiogenesis. This evidence, when taken together with the previously demonstrated anti-inflammatory and anti-proliferative properties, suggests that SHE may have significant potential as a natural chemopreventive agent for colorectal cancer. Nevertheless, it is important to acknowledge the study's limitations and to recognise the necessity for further research to address these issues. For instance, the need for further in vivo studies as the research is entirely based on in vitro experiments. Moreover, whilst the study under discussion provides evidence of effects on various markers, it does not provide detailed mechanistic insights into the causative factors of these changes. It is suggested that further detailed pathway analyses could have strengthened the findings.

Recently, silver nanoparticles (AgNPs, average hydrodynamic size: 92 ± 2 nm) have been proposed for use in the medical industry. They are synthesised using an extract of SH (40% ethanol; spent Hallertauer Perle hop waste in the form of crushed and dried powder) and have been found to be non-genotoxic and non-haemolytic (Das et al., 2022). The biosynthesised AgNPs demonstrated significant antibacterial activity against both *E. coli* and *S. aureus*, with MIC-50 values of about 202 μg/mL and 213 μg/mL, respectively. In addition, they exhibited anticancer effects against MCF-7 breast cancer cells with an IC-50 of 147.175 μg/mL, showing low toxicity and genotoxicity against normal cells and displayed superior hemocompatibility, which are crucial criteria for drug selection.

8.4.7 Cosmetics

One of the most promising sectors for the valorisation of SH, due to its high added value, is certainly that of natural cosmetics or cosmeceuticals (Pereira et al., 2022). It has been demonstrated that hops possess antioxidant, anti-inflammatory and anti-microbial compounds, which are essential for skin care formulations (Carbone & Gervasi, 2022; Hoang et al., 2021). To the best of our knowledge, there is currently only one study that reports on the use of SH in cosmetic products. In their study, Paredes-Ramos et al. (2022) proposed the valorisation of SH for the production of high value cosmetic ingredients due to the presence of cis-iso-α-cohumulone and 8PN, which can improve skin integrity by inhibiting elastase activity (Paredes-Ramos et al., 2022).

The authors conducted a blind docking analysis to detect the region of interaction of the elastase receptor against hop α and β-acids and prenylflavonoids. The results obtained demonstrated that the key residue binding energies were most favourable for cis- and trans-tetrahydroisohumulone, cis-iso-α-cohumulone and trans-isohumulone, followed by 8PN and lupulone. The high affinity of these compounds appears to be attributable to the hydroxy groups, which are optimally positioned to function as acceptor groups and bind the key residues His57 and Ser195. After the computational analysis, the optimisation of various SH extracts was undertaken, employing propanediol-water and water as solvents. The study demonstrated that propanediol-water extracts exhibited higher anti-elastase activity compared to water-only extracts, which, despite their high phenolic content and antioxidant capacity, did not demonstrate anti-elastase potential. This observation was attributed to the higher concentration of cis-iso-α-cohumulone and 8PN molecules in the propanediol-water extracts.

Although studies in the literature have shown that hop extracts can be applied in the cosmetic field by virtue of their rich array of secondary metabolites, e.g. as antioxidants in skin care products and as antimicrobials for anti-acne formulations (Weber et al., 2019), and deodorants (Dumas et al., 2009), as whitening agents due to tyrosinase inhibition (Liu et al., 2022) and finally in oral care products (Shinada et al., 2007). To date there are no studies on the recovery of SH for such purposes, with the exception of the study mentioned above, leaving room for future research in a high value-added cosmetic sector.

8.4.8 Other Uses

More recently, the potential of spent hops for unconventional applications has been explored, including in the field of materials.

In their study, Hejna et al. (2024) investigated the potential use of brewing industry by-product(s) as functional fillers for polymer composites (Hejna et al., 2024). These materials were incorporated into poly(ε-caprolactone) (PCL) based composites at a 50% w/w loading. As far as SH was concerned, SEM analysis revealed that SH showed a bimodal particle size distribution, which was attributed to the residual content of hop cones' fibrous parts in the hop pellets. The bimodal distribution, with a large share of particles up to 200 μm, contributed to higher complex viscosity values in SH-rich composites, thus affecting the material's flow characteristics and processability. Furthermore, composites with high SH content exhibited a substantial decrease in water contact angle, indicating increased hydrophilicity, the highest values of complex viscosity, and noticeable voids between filler particles and the polymer matrix, indicating poor interfacial adhesion. Despite the evidence that SH has a higher antioxidant activity than BSG and thus has the potential to improve the oxidative stability of composites, it is evident that there are challenges associated with its use that need to be addressed to achieve optimal performance in polymer composites.

Another interesting application is that reported by Moens et al. (2020), who explore the bioconversion of IX from SH into 8PN using resting cells of *Eubacterium limosum* (Moens et al., 2020). The researchers evaluated the XN content in spent hop pellets from different varieties, finding values ranging from 0.36 to 2.14 m% (mass percentage on the samples as such), for Golding and Hallertauer exp. variety, respectively. Nevertheless, it is important to acknowledge that the values have not been standardised with regard to dry matter content.

A key finding of this research is the successful conversion of spent hop-IX using stationary phase biomass. The researchers discovered that non-growing metabolically active cells of *E. limosum* could effectively convert IX to 8PN, even in the presence of antibacterial compounds from SH. This finding is of particular significance as it overcomes the challenge of bacterial growth inhibition by hop compounds. Furthermore, the study emphasised the pivotal role of pH in the conversion process, with optimal conversion attained at pH 7.8–8. These results provide a foundation for the development of a cost-effective process for producing 8PN from spent hops, with the potential to utilise an industrial by-product(s) in a value-added application. Further research is necessary to validate these preliminary findings, particularly the scalability of the process.

8.5 Final Considerations and Future Perspectives

Hop residues are a valuable component of waste from beer production, as they contain bitter acids, aromatic oils, polyphenols, proteins, fibres, and minerals. They therefore represent both a challenge and an opportunity for the brewing industry as a sustainable resource, offering opportunities in various sectors such as nutraceuticals, functional foods, cosmetics, and biofuel production. However, there are several areas that require further exploration to fully unlock SH value, including improving techniques for extracting bioactive compounds, increasing the efficiency and cost-effectiveness of processing methods, and optimising fermentation processes for biofuel production. In addition, issues related to availability, consumer acceptance, regulatory compliance, and supply chain logistics need to be addressed. Variability in the composition of spent hops and the need for consistent quality across different brewing operations pose additional challenges for large-scale applications. To overcome these challenges, interdisciplinary collaboration between brewers, biotechnology companies, and regulatory bodies is needed to develop standardised guidelines, optimise processing techniques, and conduct comprehensive life cycle assessments. By addressing these challenges, the brewing industry can transition towards a more circular economy model that integrates spent hops as valuable by-product(s) and promotes sustainability. Future research should focus on developing innovative technologies and business models that can effectively exploit the potential of SH, taking into account the associated economic and environmental trade-offs.

References

Alfeo, V., Sileoni, V., Bravi, E., Belardi, I., De Francesco, G., & Marconi, O. (2023). A sustainable valorisation of spent hops from dry-hopping. *LWT, 186*, 115248. https://doi.org/10.1016/j.lwt.2023.115248

Andrić, I., Pina, A., Ferrão, P., Fournier, J., Lacarrière, B., & Corre, O. L. (2017). Assessing the feasibility of using the heat demand-outdoor temperature function for a long-term district heat demand forecast. *Energy Procedia, 116*, 460–469. https://doi.org/10.1016/j.egypro.2017.05.093

Bartmańska, A., Wałecka-Zacharska, E., Tronina, T., Popłoński, J., Sordon, S., Brzezowska, E., Bania, J., & Huszcza, E. (2018). Antimicrobial properties of spent hops extracts, flavonoids isolated therefrom, and their derivatives. *Molecules, 23*(8), Article 8. https://doi.org/10.3390/molecules23082059

Bedini, S., Flamini, G., Girardi, J., Cosci, F., & Conti, B. (2015). Not just for beer: Evaluation of spent hops (Humulus lupulus L.) as a source of eco-friendly repellents for insect pests of stored foods. *Journal of Pest Science, 88*(3), 583–592. https://doi.org/10.1007/s10340-015-0647-1

Bonato, S. V., Augusto De Jesus Pacheco, D., Schwengber Ten Caten, C., & Caro, D. (2022). The missing link of circularity in small breweries' value chains: Unveiling strategies for waste management and biomass valorization. *Journal of Cleaner Production, 336*, 130275. https://doi.org/10.1016/j.jclepro.2021.130275

Boncler, M., Różalski, M., Krajewska, U., Podsędek, A., & Watala, C. (2014). Comparison of PrestoBlue and MTT assays of cellular viability in the assessment of anti-proliferative effects of plant extracts on human endothelial cells. *Journal of Pharmacological and Toxicological Methods, 69*(1), 9–16. https://doi.org/10.1016/j.vascn.2013.09.003

Bravi, E., Francesco, G. D., Sileoni, V., Perretti, G., Galgano, F., & Marconi, O. (2021). Brewing by-product upcycling potential: Nutritionally valuable compounds and antioxidant activity evaluation. *Antioxidants, 10*(2), 165. https://doi.org/10.3390/antiox10020165

Brendel, S., Hofmann, T., & Granvogl, M. (2020). Dry-hopping to modify the aroma of alcohol-free beer on a molecular level—Loss and transfer of odor-active compounds. *Journal of Agricultural and Food Chemistry, 68*(32), 8602–8612. https://doi.org/10.1021/acs.jafc.0c01907

Caban, M., et al. (2020). Spent hops (Humulus Lupulus L.) extract as modulator of the inflammatory response in lipopolysaccharide stimulated raw 264.7 macrophages. *Journal of Physiology and Pharmacology, 71*(1), 67–78. https://doi.org/10.26402/jpp.2020.1.05

Caban, M., et al. (2022). Chemopreventive properties of spent hops (Humulus Lupulus L.) extract against angiogenesis, invasion and migration of colorectal cancer cells. *Journal of Physiology and Pharmacology*. https://doi.org/10.26402/jpp.2022.3.11

Calvert, D., Dew, T., Gadon, A., Gros, J., & Cook, D. (2024). Valorisation of hop leaves for their bioactive compounds: Identification and quantification of phenolics across different varieties, crop years and stages of development. *Food Chemistry*, 142005. https://doi.org/10.1016/j.foodchem.2024.142005

Carbone, K., & Gervasi, F. (2022). An updated review of the genus Humulus: A valuable source of bioactive compounds for health and disease prevention. *Plants, 11*(24), 3434. https://doi.org/10.3390/plants11243434

Chadwick, L. R., Nikolic, D., Burdette, J. E., Overk, C. R., Bolton, J. L., Van Breemen, R. B., Fröhlich, R., Fong, H. H. S., Farnsworth, N. R., & Pauli, G. F. (2004). Estrogens and congeners from spent hops (*Humulus l upulus*). *Journal of Natural Products, 67*(12), 2024–2032. https://doi.org/10.1021/np049783i

Chattaraj, S., Mitra, D., Ganguly, A., Thatoi, H., & Das Mohapatra, P. K. (2024). A critical review on the biotechnological potential of brewers' waste: Challenges and future alternatives. *Current Research in Microbial Sciences, 6*, 100228. https://doi.org/10.1016/j.crmicr.2024.100228

Claudia Salanţă, L., Corina Fărcaş, A., Borşa, A., & Rodica Pop, C. (2023). Current strategies for the management of valuable compounds from hops waste for a circular economy. *Food Chemistry: X, 19*, 100876. https://doi.org/10.1016/j.fochx.2023.100876

Senna Ferreira Costa, F., Roquete Amparo, T., Brandão Seibert, J., Silveira, B. M., Gomes da Silva, R., Inocêncio Pereira, D., Gontijo Garcia Barbosa, R., dos Santos, O. D. H., Brandão, G. C., de Medeiros Teixeira, L. F., Melo de Abreu Vieira, P., & Bianco de Souza, G. H. (2021). Reuse of hot trub as an active ingredient with antioxidant and antimicrobial potential. *Waste and Biomass Valorization, 12*(4), 2037–2047. https://doi.org/10.1007/s12649-020-01163-6

Das, P., Dutta, T., Manna, S., Loganathan, S., & Basak, P. (2022). Facile green synthesis of non-genotoxic, non-hemolytic organometallic silver nanoparticles using extract of crushed, wasted, and spent Humulus lupulus (hops): Characterization, anti-bacterial, and anti-cancer studies. *Environmental Research, 204*, 111962. https://doi.org/10.1016/j.envres.2021.111962

de Andrade Silva, V., Demaman Arend, G., Antonio Ferreira Zielinski, A., Di Luccio, M., & Ambrosi, A. (2023). Xanthohumol properties and strategies for extraction from hops and brewery residues: A review. *Food Chemistry, 404*, 134629. https://doi.org/10.1016/j.foodchem.2022.134629

de Gomes, F. O., Guimarães, B. P., Ceola, D., & Ghesti, G. F. (2021). Advances in dry hopping for industrial brewing: A review. *Food Science and Technology, 42*, ctaAR60620. https://doi.org/10.1590/fst.60620

Dumas, E. R., Michaud, A. E., Bergeron, C., Lafrance, J. L., Mortillo, S., & Gafner, S. (2009). Deodorant effects of a supercritical hops extract: Antibacterial activity against Corynebacterium xerosis and Staphylococcus epidermidis and efficacy testing of a hops/zinc ricinoleate stick in humans through the sensory evaluation of axillary deodorancy. *Journal of Cosmetic Dermatology, 8*(3), 197–204. https://doi.org/10.1111/j.1473-2165.2009.00449.x

Eßlinger, H. M. (2009). *Handbook of brewing*. Wiley-VCH.

Fărcaş, A. C., Socaci, S. A., Mudura, E., Dulf, F. V., Vodnar, D. C., Tofană, M., Salanţă, L. C., Fărcaş, A. C., Socaci, S. A., Mudura, E., Dulf, F. V., Vodnar, D. C., Tofană, M., & Salanţă, L. C. (2017). Exploitation of brewing industry wastes to produce functional ingredients. In *Brewing technology*. IntechOpen. https://doi.org/10.5772/intechopen.69231

Gandolpho, Almeida, A. R., Gandolpho, G. M., Freitas, D. Z., Gasparini, O. C., MacHado, M. H., & Barreto, P. L. M. (2021). Optimization of brewing waste's (TRUB) phenolic compounds extraction by ultrasound assisted using response surface methodology. *Quimica Nova, 44*(4), 478–483. https://doi.org/10.21577/0100-4042.20170688

Garcia-Garcia, G., Woolley, E., Rahimifard, S., Colwill, J., White, R., & Needham, L. (2017). A methodology for sustainable management of food waste. *Waste and Biomass Valorization, 8*(6), 2209–2227. https://doi.org/10.1007/s12649-016-9720-0

Gasiński, A., Kawa-Rygielska, J., Paszkot, J., Pietrzak, W., Śniegowska, J., & Szumny, A. (2022). Second life of hops: Analysis of beer hopped with hop pellets previously used to dry-hop a beer. *LWT, 159*, 113186. https://doi.org/10.1016/j.lwt.2022.113186

Giorni, P., Bulla, G., Leni, G., Soldano, M., Tacchini, M., Guerrini, A., Sacchetti, G., & Bertuzzi, T. (2023). Enhancement of agri-food by-products: Green extractions of bioactive molecules with fungicidal action against mycotoxigenic fungi and their mycotoxins. *Frontiers in Nutrition, 10*, 1196812. https://doi.org/10.3389/fnut.2023.1196812

Goldammer, T. (2022). *The Brewer's handbook: A complete book to brewing beer*. Apex Publishers. https://www.beer-brewing.com/

González-Laredo, R. F., Sayago-Monreal, V. I., Moreno-Jiménez, M. R., Rocha-Guzmán, N. E., Gallegos-Infante, J. A., Landeros-Macías, L. F., & Rosales-Castro, M. (2023). Natural deep eutectic solvents (NADES) as an emerging technology for the valorisation of natural products and agro-food residues: A review. *International Journal of Food Science & Technology, 58*(12), 6660–6673. https://doi.org/10.1111/ijfs.16641

Goyal, A., Shukla, G., Mishra, S., Mallik, S., Singh, A., & Dubey, M. (2023). Beer production by fermentation process: A review. *Journal of Microbiology, Biotechnology and Food Sciences, e9532*. https://doi.org/10.55251/jmbfs.9532

Grand View Research. (2025). *Nutraceuticals market size & trends* (GVR-2-68038-059-0). https://www.grandviewresearch.com/industry-analysis/nutraceuticals-market.

Grudniewska, A., & Pastyrczyk, N. (2023). New insight for spent hops utilization: Simultaneous extraction of protein and xanthohumol using deep eutectic solvents. *Biomass Conversion and Biorefinery, 13*(16), 14975–14986. https://doi.org/10.1007/s13399-022-03462-5

Hajirahimkhan, A., Simmler, C., Yuan, Y., Anderson, J. R., Chen, S.-N., Nikolić, D., Dietz, B. M., Pauli, G. F., Van Breemen, R. B., & Bolton, J. L. (2013). Evaluation of estrogenic activity of licorice species in comparison with hops used in botanicals for menopausal symptoms. *PLoS One, 8*(7), e67947. https://doi.org/10.1371/journal.pone.0067947

Hauser, D. G., Lafontaine, S. R., & Shellhammer, T. H. (2019). Extraction efficiency of dry-hopping. *Journal of the American Society of Brewing Chemists, 77*(3), 188–198. https://doi.org/10.1080/03610470.2019.1617622

Hejna, A., Barczewski, M., Kosmela, P., Aniśko, J., Szulc, J., Skórczewska, K., Piasecki, A., & Kuang, T. (2024). More than just a beer—Brewers' spent grain, spent hops, and spent yeast as potential functional fillers for polymer composites. *Waste Management, 180*, 23–35. https://doi.org/10.1016/j.wasman.2024.03.023

Hoang, H. T., Moon, J.-Y., & Lee, Y.-C. (2021). Natural antioxidants from plant extracts in skincare cosmetics: Recent applications, challenges and perspectives. *Cosmetics, 8*(4), 106. https://doi.org/10.3390/cosmetics8040106

Huige, N. J. (2006). Brewery by-products and effluents. In *Handbook of brewing* (2nd ed.). CRC Press.

Huszcza, E., & Bartmańska, A. (2008). The implication of yeast in debittering of spent hops. *Enzyme and Microbial Technology, 42*(5), 421–425. https://doi.org/10.1016/j.enzmictec.2008.01.004

Jackowski, J., Hurej, M., Rój, E., Popłoński, J., Kośny, L., & Huszcza, E. (2015). Antifeedant activity of xanthohumol and supercritical carbon dioxide extract of spent hops against stored product pests. *Bulletin of Entomological Research, 105*(4), 456–461. https://doi.org/10.1017/S0007485315000255

Jaskula-Goiris, B., De Cooman, L., & Goiris, K. (2018). Humulus Lupulus: Hop alpha-acids isomerization—A review. *BrewingScience, 71*, 85–95. https://doi.org/10.23763/BrSc18-12jaskula

Karlović, A., Jurić, A., Ćorić, N., Habschied, K., Krstanović, V., & Mastanjević, K. (2020). By-products in the malting and brewing industries—Re-usage possibilities. *Fermentation, 6*(3), 82. https://doi.org/10.3390/fermentation6030082

Kopeć, M., Mierzwa-Hersztek, M., Gondek, K., Wolny-Koładka, K., Zdaniewicz, M., & Jarosz, R. (2022). Biological activity of composts obtained from hop waste generated during the brewing. *Biomass Conversion and Biorefinery, 12*(4), 1271–1279. https://doi.org/10.1007/s13399-020-00746-6

Kühbeck, F., Müller, M., Back, W., Kurz, T., & Krottenthaler, M. (2007). Effect of hot trub and particle addition on fermentation performance of Saccharomyces cerevisiae. *Enzyme and Microbial Technology, 41*(6–7), 711–720. https://doi.org/10.1016/j.enzmictec.2007.06.007

Lafontaine, & Shellhammer. (2018). Impact of static dry-hopping rate on the sensory and analytical profiles of beer. *Journal of the Institute of Brewing, 124*(4), 434–442. https://doi.org/10.1002/jib.517

Lafontaine, S., Varnum, S., Roland, A., Delpech, S., Dagan, L., Vollmer, D., Kishimoto, T., & Shellhammer, T. (2019). Impact of harvest maturity on the aroma characteristics and chemistry of Cascade hops used for dry-hopping. *Food Chemistry, 278*, 228–239. https://doi.org/10.1016/j.foodchem.2018.10.148

Liu, J., Chen, Y., Zhang, X., Zheng, J., Hu, W., & Teng, B. (2022). Hop tannins as multifunctional tyrosinase inhibitor: Structure characterization, inhibition activity, and mechanism. *Antioxidants, 11*(4), 772. https://doi.org/10.3390/antiox11040772

Lomuscio, E., Bianchi, F., Cervini, M., Giuberti, G., Simonato, B., & Rizzi, C. (2022). Durum wheat fresh pasta fortification with Trub, a beer industry by-product. *Food, 11*(16), 2496. https://doi.org/10.3390/foods11162496

Luzak, B., Golanski, J., Przygodzki, T., Boncler, M., Sosnowska, D., Oszmianski, J., Watala, C., & Rozalski, M. (2016). Extract from spent hop (Humulus lupulus L.) reduces blood platelet aggregation and improves anticoagulant activity of human endothelial cells in vitro. *Journal of Functional Foods, 22*, 257–269. https://doi.org/10.1016/j.jff.2016.01.029

Macchioni, V., Carbone, K., Cataldo, A., Fraschini, R., & Bellucci, S. (2021a). Lactic acid-based deep natural eutectic solvents for the extraction of bioactive metabolites of Humulus lupulus L.: Supramolecular organization, phytochemical profiling and biological activity. *Separation and Purification Technology, 264*, 118039. https://doi.org/10.1016/j.seppur.2020.118039

Macchioni, V., Picchi, V., & Carbone, K. (2021b). Hop leaves as an alternative source of health-active compounds: Effect of genotype and drying conditions. *Plants, 11*(1), 99. https://doi.org/10.3390/plants11010099

Mattioli, S., Castellini, C., Mancini, S., Roscini, V., Cartoni Mancinelli, A., Cotozzolo, E., Pauselli, M., & Dal Bosco, A. (2020). Effect of trub and/or linseed dietary supplementation on *in vivo* oxidative status and some quality traits of rabbit meat. *Meat Science, 163*, 108061. https://doi.org/10.1016/j.meatsci.2020.108061

Moens, E., Bolca, S., Van De Wiele, T., Van Landschoot, A., Goeman, J. L., Possemiers, S., & Verstraete, W. (2020). Exploration of isoxanthohumol bioconversion from spent hops into 8-prenylnaringenin using resting cells of Eubacterium limosum. *AMB Express, 10*(1), 79. https://doi.org/10.1186/s13568-020-01015-5

Nobis, A., Röhrig, A., Hellwig, M., Henle, T., Becker, T., & Gastl, M. (2019). Formation of 3-deoxyglucosone in the malting process. *Food Chemistry, 290*, 187–195. https://doi.org/10.1016/j.foodchem.2019.03.144

Olivares-Galván, S., Marina, M. L., & García, M. C. (2022). Extraction of valuable compounds from brewing residues: Malt rootlets, spent hops, and spent yeast. *Trends in Food Science & Technology, 127*, 181–197. https://doi.org/10.1016/j.tifs.2022.06.002

Oosterveld, A. (2002). Characterization of hop pectins shows the presence of an arabinogalactan-protein. *Carbohydrate Polymers, 49*(4), 407–413. https://doi.org/10.1016/S0144-8617(01)00350-2

Paredes-Ramos, M., Conde Piñeiro, E., & Lopez Vilariño, J. M. (2022). Brewers' spent hop revalorization for the production of high added-value cosmetics ingredients with elastase inhibition capacity. *Scientific Reports, 12*(1), 22074. https://doi.org/10.1038/s41598-022-26149-3

Pasquet, P.-L., Villain-Gambier, M., & Trébouet, D. (2024). By-product valorization as a means for the brewing industry to move toward a circular bioeconomy. *Sustainability, 16*(8), 3472. https://doi.org/10.3390/su16083472

Pereira, O. R., Santos, G., & Sousa, M. J. (2022). Hop by-products: Pharmacological activities and potential application as cosmetics. *Cosmetics, 9*(6), 139. https://doi.org/10.3390/cosmetics9060139

Rozalski, M., Micota, B., Sadowska, B., Stochmal, A., Jedrejek, D., Wieckowska-Szakiel, M., & Rozalska, B. (2013). Antiadherent and antibiofilm activity of *Humulus lupulus* L. derived products: New pharmacological properties. *BioMed Research International, 2013*, 1–7. https://doi.org/10.1155/2013/101089

Saraiva, B. R., Anjo, F. A., Vital, A. C. P., da Silva, L. H. M., Ogawa, C. Y. L., Sato, F., Coimbra, L. B., & Matumoto-Pintro, P. T. (2019). Waste from brewing (trub) as a source of protein for the food industry. *International Journal of Food Science & Technology, 54*(4), 1247–1255. https://doi.org/10.1111/ijfs.14101

Saraiva, B. R., Da Silva, L. H. M., Anjo, F. A., Vital, A. C. P., Da Silva, J. B., Bruschi, M. L., & Matumoto Pintro, P. T. (2020). Technological and sensorial properties of liquid nitrogen ice cream enriched with protein from brewing waste (trub). *International Journal of Food Science & Technology, 55*(5), 1962–1970. https://doi.org/10.1111/ijfs.14484

Schrickel, F., Bilge, D., Pahl, R., & Rettberg, N. (2024). Valorization of spent hops from dry hopping for bittering of Pilsener style beer. *Journal of the American Society of Brewing Chemists*, 1–13. https://doi.org/10.1080/03610470.2024.2327123

Shinada, K., Tagashira, M., Watanabe, H., Sopapornamorn, P., Kanayama, A., Kanda, T., Ikeda, M., & Kawaguchi, Y. (2007). Hop bract polyphenols reduced three-day dental plaque regrowth. *Journal of Dental Research, 86*(9), 848–851. https://doi.org/10.1177/154405910708600908

Silva, K. F. C. E., Feltre, G., Zandonadi, F. S., Rabelo, R. S., Sussulini, A., & Hubinger, M. D. (2024). Unlocking hot trub's potential: A simple method for extracting bitter acids and xanthohumol. *Journal of the Science of Food and Agriculture, 104*(9), 5381–5390. https://doi.org/10.1002/jsfa.13371

Singh Nee' Nigam, P., & Pandey, A. (Eds.). (2009). *Biotechnology for agro-industrial residues utilisation*. Springer. https://doi.org/10.1007/978-1-4020-9942-7

Sperandio, T. A., Carbone, K., Fedrizzi, M., Monteleone, A., Tarangioli, S., & Pagano, M. (2017). Increasing the value of spent grains from craft microbreweries for energy purposes. *Chemical Engineering Transactions, 58*, 487–492. https://doi.org/10.3303/CET1758082

Spetsig, L. O. (1968). The bitter substances of spent hops, Trub, and yeast cover: A chromatographic study. *Journal of the Institute of Brewing, 74*(4), 346–351. https://doi.org/10.1002/j.2050-0416.1968.tb03138.x

Stachnik, M., Sterczyńska, M., Smarzewska, E., Ptaszek, A., Piepiórka-Stepuk, J., Ageev, O., & Jakubowski, M. (2021). Rheological properties of industrial hot trub. *Materials, 14*(23), 7162. https://doi.org/10.3390/ma14237162

Statista. (2024). *Alcoholic Drinks—Worldwide*. https://www.statista.com/outlook/cmo/alcoholic-drinks/worldwide

Statista. (2025). *Beer production worldwide from 1998 to 2023*. https://www.statista.com/statistics/270275/worldwide-beer-production/

Stevens, J. F. (2020). Xanthohumol and structurally related Prenylflavonoids for cancer chemoprevention and control. In J. M. Pezzuto & O. Vang (Eds.), *Natural products for cancer chemoprevention: Single compounds and combinations* (pp. 319–350). Springer International Publishing. https://doi.org/10.1007/978-3-030-39855-2_10

Stevens, J. F., & Page, J. E. (2004). Xanthohumol and related prenylflavonoids from hops and beer: To your good health! *Phytochemistry, 65*(10), 1317–1330. https://doi.org/10.1016/j.phytochem.2004.04.025

Stewart, G. G., & Priest, F. G. (Eds.). (2006). *Handbook of brewing*. CRC Press. https://doi.org/10.1201/9781420015171

Thiago, R. D. S. M., Pedro, P. M. D. M., & Eliana, F. C. S. (2014). Solid wastes in brewing process: A review. *Journal of Brewing and Distilling, 5*(1), 1–9. https://doi.org/10.5897/JBD2014.0043

Toninato, B. O., Calandrelli, A., Fontana, L. F., & Dias-Arieira, C. R. (2024). Aqueous extract of spent hops suppresses root-knot nematode and enhances soil biological activity. *Horticultura Brasileira, 42*, e282871. https://doi.org/10.1590/s0102-0536-2024-e282871

Vanitha, T., & Khan, M. (2020). Role of pectin in food processing and food packaging. In M. Masuelli (Ed.), *Pectins—Extraction, purification, characterization and applications*. IntechOpen. https://doi.org/10.5772/intechopen.83677

Wannenmacher, J., Gastl, M., & Becker, T. (2018). Phenolic substances in beer: Structural diversity, reactive potential and relevance for brewing process and beer quality. *Comprehensive Reviews in Food Science and Food Safety, 17*(4), 953–988. https://doi.org/10.1111/1541-4337.12352

Weber, N., Biehler, K., Schwabe, K., Haarhaus, B., Quirin, K.-W., Frank, U., Schempp, C. M., & Wölfle, U. (2019). Hop extract acts as an antioxidant with antimicrobial effects against Propionibacterium acnes and staphylococcus aureus. *Molecules, 24*(2), Article 2. https://doi.org/10.3390/molecules24020223

Yoon, J., & Tak, J.-H. (2023). Potential utilization of the brewery's hop wastes as an insecticidal synergist and repellent against Spodoptera frugiperda. *Journal of Pest Science, 96*(4), 1441–1454. https://doi.org/10.1007/s10340-023-01640-6

Ziemiński, K., Romanowska, I., & Kowalska, M. (2012). Enzymatic pretreatment of lignocellulosic wastes to improve biogas production. *Waste Management, 32*(6), 1131–1137. https://doi.org/10.1016/j.wasman.2012.01.016

Part IV
Hops: Chemistry, Functionality and Health-Promoting Properties

Chapter 9
Hop Chemistry and Biochemistry

Katya Carbone, Erika Celi, and Alessia D'Andrea

9.1 Introduction

Hops (*Humulus lupulus* L.) are perennial climbing plants belonging to the Cannabaceae family, genus *Humulus* (Neve, 1991). In addition to *H. lupulus*, two other species belong to this genus: *Humulus japonicus*, native to Japan, a deciduous annual plant that differs from *H. lupulus* for the scaling of the cones and for a smaller amount of lupulin, therefore it is not used in brewing; and *Humulus yunnanensis*, native to China, about which little information is available and which, unlike the previous species, is capable of growing at lower latitudes (Neve, 1991).

Hops are primarily cultivated for their unique contribution to the brewing industry. Female inflorescences, termed hop cones or strobiles, are abundant in secondary metabolites that contribute to the distinctive flavours, aromas, and bitterness of beer, in addition to their antimicrobial and preservative properties. Recent studies have also revealed the presence of a significant quantity of valuable molecules in other plant tissues (e.g. leaves, sprouts and seeds).

Hop cones are constituted by a central strig (rachis) bearing petal-like structures, called bracts and bracteols, at each node (Fig. 9.1). During flowering, the rachis elongates and the bracts grow. Glandular trichomes then develop on the underside of the bracts and secrete a yellow resin called lupulin. This resin contains a variety of specific secondary metabolites (Carbone & Gervasi, 2022; Hong et al., 2021).

Beyond their role in brewing, hops are increasingly recognised for their plethora of bioactive compounds, which exhibit pharmacological activities, including antioxidant, anti-inflammatory, and anticancer effects (Carbone & Gervasi, 2022).

The hop cone phytochemical profile includes three main classes of compounds: bitter acids, essential oils, and polyphenols (Fig. 9.2).

K. Carbone (✉) · E. Celi · A. D'Andrea
CREA – Research Centre for Olive, Fruit and Citrus Crops, Rome, Italy
e-mail: katya.carbone@crea.gov.it

K. Carbone (ed.), *All About Hops: The Crop, its Cultivation, and its Uses in Brewing and Beyond*, https://doi.org/10.1007/978-3-031-96606-4_9

Fig. 9.1 Hop cones: whole (left), cut in half (top right), detail of lupulin (bottom right). (Photos: Katya Carbone)

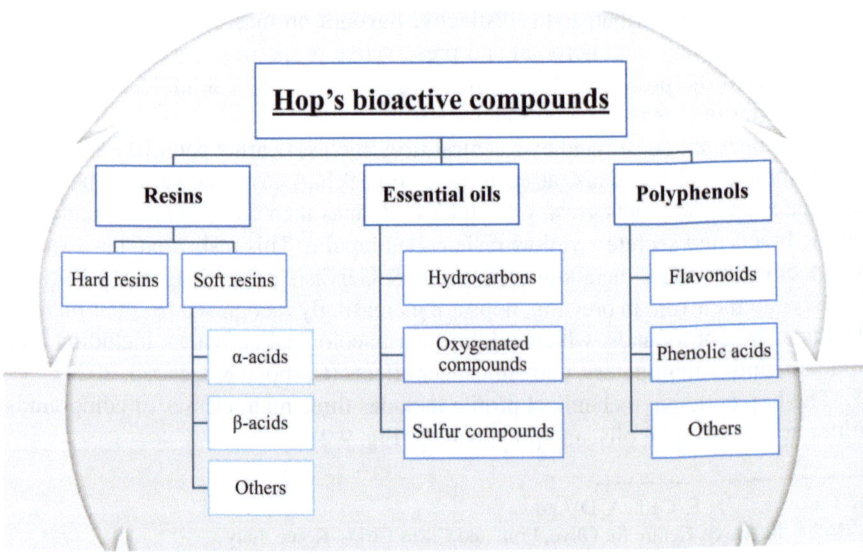

Fig. 9.2 Phytochemical composition of hop cones

Bitter acids are critical for the characteristic bitterness of beer and contribute to its stability. During the brewing process, α-acids are subject to thermal isomerisation, resulting in the conversion to iso-α-acids. These iso-α-acids contribute to the enhancement of the flavour profile and exhibit antimicrobial activity.

Hop essential oils are responsible for its characteristic aroma, which varies significantly between varieties. These volatile compounds are of increasing interest due to their contributions to sensory quality in beverages and their potential therapeutic applications.

Polyphenols, comprising flavonoids, tannins and proanthocyanidins, represent a substantial component of both hop cones and leaves. They contribute to beer stability, colour and mouthfeel, and exhibit potent antioxidant activity, which has been associated with health benefits including cardiovascular protection and a reduced risk of chronic diseases. The biosynthesis of these compounds is intricately regulated by genetic, environmental, and agronomic factors, which influence the chemical composition of hops. Advances in analytical techniques, such as high-performance liquid chromatography (HPLC) and mass spectrometry (MS), have facilitated the detailed characterisation of hop metabolites, enabling a deeper understanding of their chemistry and biological activities. Moreover, the exploration of underutilised hop varieties and breeding programs targeting specific traits are paving the way for the development of novel hop cultivars with tailored chemical profiles.

The present chapter aims to provide a comprehensive overview of the chemistry and biochemistry of hops, highlighting their significance in brewing, their health-related properties, and their potential applications in the food, pharmaceutical, and cosmetic industries. The emphasis of this study is twofold: firstly, to place recent advances in the understanding of hop phytochemistry and biosynthesis in the context of other, less investigated tissues, such as leaves; and secondly, to consider the implications of these advances for the future of the hops industry.

9.2 An Overview of the Phytochemical Composition of Hop Cones

Plants synthesise a range of essential compounds for their life processes, which are involved in various reactions such as glycolysis, the Krebs cycle, and oxidative phosphorylation, defining their primary metabolism. Primary metabolites are compounds involved in plant nutrition, growth, development, and reproduction. They include sugars, lipids, amino acids, peptides, and proteins, as well as their derivatives. Plants can also synthesise secondary metabolites through biosynthesis and degradation reactions, using intermediates from primary metabolism. These metabolites are not essential for plant growth, development, or reproduction, but they play a crucial role in the interaction between plants and their environment. Secondary metabolites often exhibit biological properties that are of significant interest to humans (Bocquet et al., 2018; Carbone & Gervasi, 2022). Secondary metabolism reactions occur in various plant tissues, such as flowers, leaves, and seeds. The resulting products may accumulate or be converted into simpler compounds. Secondary metabolites can arise due to deficiencies in plant cells, such as enzyme inhibition or biochemical imperfections, or in response to specific biological or climatic conditions.

To date, more than 1000 compounds have been identified in hops, including both primary and secondary metabolites (Chadwick et al., 2006) (Fig. 9.3).

9.2.1 *Primary Hop Metabolites*

Primary metabolites have a direct impact on plant growth, development, and structure. They are crucial in processes such as photosynthesis, the citric acid cycle, and electronic transport chains (Fang et al., 2019). These molecules include carbohydrates, proteins, lipids and nucleotides, which serve as fundamental building blocks for plant metabolism. Carbohydrates, such as glucose, sucrose and starch, provide energy and structural components, while proteins and amino acids play crucial roles in enzymatic activities and cellular signalling. Lipids, including phospholipids and fatty acids, are integral to membrane structure and energy storage, whereas nucleotides contribute to DNA and RNA synthesis, enabling genetic information transfer and cell division (Chinou, I, 2008).

The biosynthesis of primary metabolites is subject to regulation by key metabolic pathways, including glycolysis, the tricarboxylic acid (TCA) cycle, and the Calvin cycle. These interconnected processes ensure the production of energy and the assimilation of carbon, which are both fundamental to plant survival. In addition, primary metabolites also serve as precursors for secondary metabolites, which are responsible for plant defense, pigmentation, and interactions with the environment (Obata, 2019).

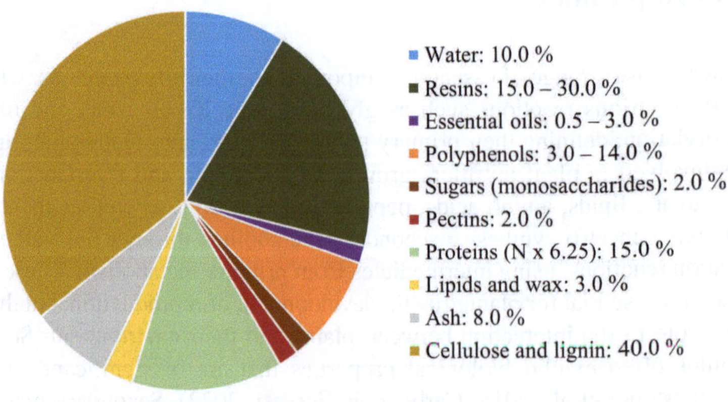

Fig. 9.3 Main classes of chemical compounds in hops (Carbone & Gervasi, 2022)

The primary metabolites of hops are discussed in detail below, highlighting that, to date, in-depth studies on these metabolites remain limited, indicating a need for further research to elucidate their roles and potential applications.

9.2.1.1 Carbohydrates

As previously stated, the literature provides limited information regarding the primary metabolites of hops. In his "*The chemistry of hop constituents*", Stevens reported the first information on the sugar profile of hops, dating back to 1913. The main sugars (% on a dry basis) found in Early Choice hop cones were fructose (0.51), glucose (0.4170), sucrose (0.2273), raffinose (0.09), stachyose (0.06) and an unknown sugar (0.06). In addition, the author reported a pectin content of 1–2%, and the presence of inositol found in hop leaves (Stevens, 1967).

More recently, Carbone et al. (2020) conducted a pioneering study, utilizing ^{1}H-NMR spectral deconvolution, to identify and quantify primary metabolites in wild hop cones, focusing on amino acids and sugars. Glucose and fructose were found to be the most abundant sugars, followed by maltose, galactose, xylose, and mannose (Fig. 9.4).

In relation to hop polysaccharides, Oosterveld (2002) reported the characterisation of hop pectins extracted from spent hops (Fig. 9.5). The researchers found that acid extraction of the cones yielded 2% of material, containing 59% polysaccharides. The hop pectins had a relatively high molecular weight and an intrinsic viscosity comparable to commercial apple and citrus pectins. However, their low degree of methyl esterification (DM) limits their use as gelling agents with acid and sugar. The degree of acetylation (DA) was 8.4%, comparable to acid-extracted apple pectins, while the neutral sugar content was relatively high compared to apple and citrus pectins, indicating the presence of substantial rhamnogalacturonan regions.

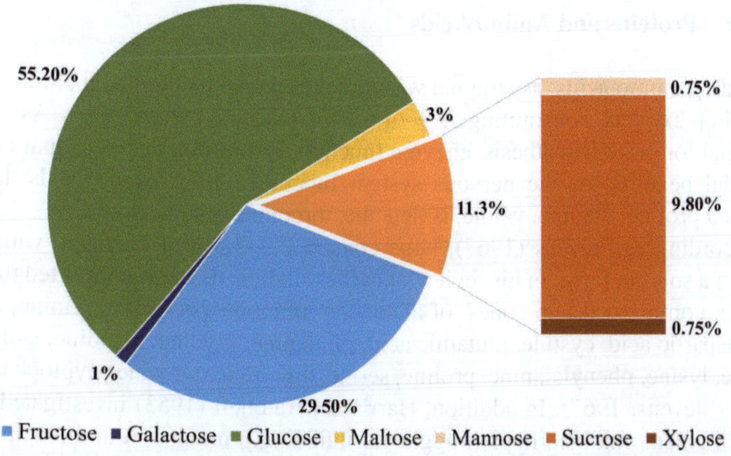

Fig. 9.4 Average hop cone sugar composition. (Adapted from Carbone et al. (2020))

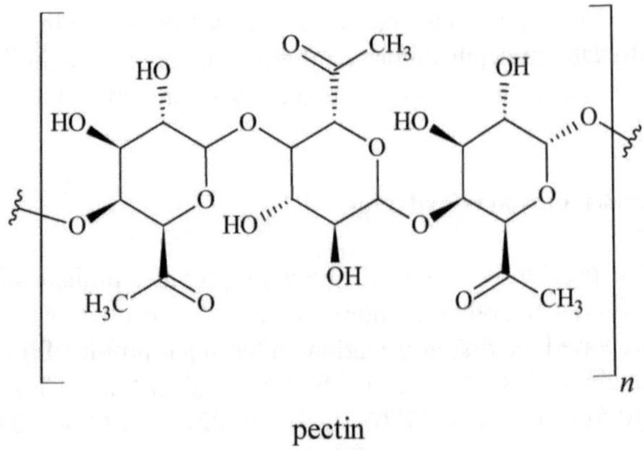

Fig. 9.5 Pectin structure motif

A significant finding of this study was the presence of arabinogalactan-proteins (AGPs) in the hop pectin extract. The AGPs were found in a high molecular weight fraction and consisted of (1 → 3)- and (1 → 3,6)-linked galactans highly branched with arabinose and galactose. The protein component of the AGP was found to be rich in cystein, threonin, serinin, alanin, and hydroxyprolin. Enzymatic degradation experiments indicated that the AGPs were linked to the pectin molecules and possessed a relatively low molecular weight themselves (Oosterveld, 2002). Furthermore, the authors reported the sugar composition (% w/w) of spent hops, which revealed the presence of significant quantities of glucose, uronic acid and xylose (15.3, 12.9 and 3.7%, respectively). Specifically, the pectic sugars (uronic acid, rhamnose, arabinose and galactose) accounted for 46% of the polysaccharides in spent hops. The arabinose and galactose content were comparatively low in relation to other sources, such as apple cell walls. Conversely, the high xylose content suggests the presence of xyloglucans in spent hops.

9.2.1.2 Proteins and Amino Acids

Regarding amino acids, asparagine was found to be the predominant amino acid in green hop extracts, constituting 75% of the total detected amino acids. Asparagine is crucial for protein synthesis, enzyme function, and muscle tissue formation, with potential benefits for the nervous system function. Other amino acids detected included proline, alanine, valine, isoleucine, threonine, and leucine.

According to Stevens (1967), hops contain 2.0–3.5% of nitrogen, with about 0.5% in a soluble form. In his review of earlier studies, the author reported the presence, in commercial hop cones, of alanine, γ-aminobutyric acid, arginine, asparagine, aspartic acid, cystine, glutamic acid, glutamine, glycine, histidine, isoleucine, leucine, lysine, phenylalanine, proline, serine, threonine, tyrosine, tryptophane, and valine (Stevens, 1967). In addition, Harris and Tatchell (1953) investigated amino acids in green, unkilned hops (Fuggle var.), finding *p*-alanine, α-aminoadipic acid, γ-methyleneglutamic acid, and γ-methylene glutamine in addition to those listed above. Notably, the last two compounds were found to be absent in kilned hops (Harris & Tatchell, 1953). Regarding the impact of hop processing on the amino acid profile, Bruner et al. (2023) analysed the amino acid composition of T90 hop pellets, variety Centennial. The analysis revealed that asparagine/aspartic acid was found in the highest relative concentration in the hops, followed by arginine (Bruner et al., 2023).

In a separate study, Oosterveld (2002) provided the amino acid composition of spent hops. According to the data, the most abundant amino acids were glutamic acid, aspartic acid, serine, alanine and glycine, whereas the amino acids present in lower amounts included valine, hydroxyproline, proline, isoleucine and tyrosine (Oosterveld, 2002).

Nezi et al. (2022) conducted a recent investigation onto the distribution of protein classes in hop cones from the Cascade and Chinook varieties cultivated in southern Italy (Nezi et al., 2022). The study revealed that proteins associated with photosystems were the most prevalent, followed by those linked to ATP synthase and flavonoid pathways. Conversely, proteins linked to the terpenoid pathway were the least abundant, comprising only 9% of the total. Notably, the research identified that differentially expressed proteins predominantly belonged to two main pathways: the bitter acid pathway and the xanthohumol pathway. In the former, the initial and final proteins involved were identified as mitochondrial branched chain aminotransferase 1 and humulone synthase 2, respectively. These proteins are particularly significant due to their contribution to the desirable bitter flavouring and overall microbial stabilisation in hops. In the xanthohumol pathway, CCL13 and naringenin-chalcone synthase were recognised as two proteins involved in xanthohumol synthesis. The proteomic analysis revealed several key differences between the Cascade and Chinook hop cultivars. Specifically, the Chinook cultivar was found to contain all the identified proteins, while the Cascade one lacked five specific proteins (humulone synthase 2, CCL13, 2-C-methyl-D-erythritol 2, mitochondrial branched-chain aminotransferase 1 and naringenin-chalcone synthase 1). These differences in protein composition between Cascade and Chinook could lead to variations in their chemical profiles, potentially affecting their use in brewing and their pharmacological properties.

9.2.1.3 Lipids

Understanding hop lipids is important for several aspects of brewing, including flavour stability, foam quality, and the potential development of valuable by-products from hop extraction (Bravi et al., 2009). To the best of our knowledge, the international literature has not yet paid sufficient attention to the lipid fraction of hops.

In their study on hop lipidomic, Rettberg et al. (2013) focused their efforts on the isolation and characterisation of non-polar, neutral, and polar hop lipids (Rettberg et al., 2013). The study analysed five US hop varieties (Nugget, Columbus, Chinook, Cascade, and Mount Hood) and the researchers found that hop pellets contain 30–50 mg lipids g^{-1} dry matter, which equals 3–5% total lipids. Non-polar lipids, such as triacyl glycerides and waxes, account for approximately 80% of total lipids, while polar lipids account for about 15% and neutral lipids for about 5%. The major fatty acids identified in hops were linoleic acid, α-linolenic acid, and palmitic acid. The study also revealed differences in fatty acid patterns between polar and non-polar lipids, with the polar fraction being rich in α-linolenic and linoleic acid (>70% in sum), while the non-polar fraction contained considerable amounts of palmitic acid (≤19%) and long-chain fatty acids, with carbon chains of C20–C22.

Hop variety has a significant impact on the lipidomic profile. For the Saaz variety, for example, Bravi et al. (2009) found a lipid content of 5.71%, with a hop total extract of 13.22 ± 0.03 g/100 g (dry weight basis), of which 44% was lipid. About the fatty acid profile, the main fatty acids found in Saaz hop pellets were linoleic, linolenic, lauric, palmitic, capric, and arachidonic acids (Bravi et al., 2009).

9.2.1.4 Other Compounds

Additionally, several other metabolites, including myo-inositol, lactate, choline, *o*-phosphocholine, ethanolamine, 4-aminobutyrate, and trigonelline (Fig. 9.6), were detected in hop cone extracts obtained through ultrasound- and microwave-assisted extractions (Carbone et al., 2020). Among them, trigonelline has been reported in literature studies to possess various medicinal properties such as antidiabetic, antibacterial, antiviral, sedative, anti-hemicranial, and antitumor activities (Zhou et al., 2022).

9.2.2 Secondary Hop Metabolites

Secondary metabolites are products derived from primary metabolic pathways (Srivastava, 2024). These metabolites can be traced back to three biosynthetic pathways connected to primary carbon metabolism:

1. The shikimate pathway, which leads to the formation of phenolic compounds;
2. The Mevalonic Acid Pathway and 2-C-Methyl-D-Erythritol 4-Phosphate/Deoxy-Xylulose-5-Phosphate (MEP/DOXP), which generate terpenoids;

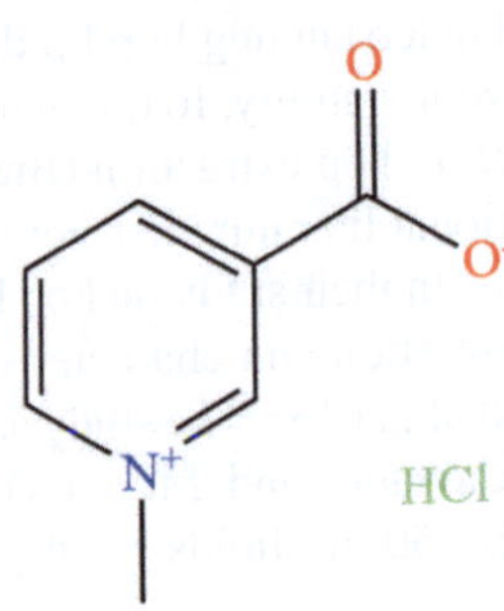

Fig. 9.6 Chemical structure of trigonelline

3. The acetate pathway, which leads to the formation of fatty acids and polyketides.

Secondary metabolites are natural organic compounds with molecular masses lower than 3000 Da (Dalton) that primarily serve three functions:

- Defense against pests and pathogens;
- Attracting pollinators;
- Protecting against environmental stresses like UV radiation.

The amount and composition of secondary metabolites in hops depend on several factors, such as genetic traits, soil and climatic conditions, harvest time, post-harvest treatments, etc. (Biendl et al., 2015; Hagemann et al., 2024; Sharp et al., 2014).

The following discussion will focus on the various secondary metabolites of hops, exploring them in-depth.

9.2.2.1 Bitter Acids

Bitter acids are prenylated polyketides synthesised in the lupulin glands of hops (Clark et al., 2013). They are part of lupulin's resins, which are soluble in a mixture of ether and methanol. These resins can be divided into categories depending on whether they dissolve in hexane: in particular, soft resins do, while hard resins do not (Astray et al., 2020; Przybyś & Skomra, 2020). These latter are divided into α-, β-, δ- and ε-hard resins. Soft resins are also known as bitter acids, as they are responsible for the bitter taste in beer, and have also been detected in male hop inflorescences, even though in considerably lower amounts (De Keukeleire et al., 2003).

Resins constitute 10–30% of the dry matter of the cone, with bitter acids accounting for 9–27.5% of the dry weight and hard resins accounting for the remaining 1.5–2.5% (Biendl et al., 2015; Sun et al., 2022). Other sources report slightly different figures, with hard resins constituting 3–5% of the cone's dry matter; anyhow, the resins' composition strongly depends on hop cultivar, climate and growing conditions (Almaguer et al., 2014; Biendl et al., 2015; Przybyś & Skomra, 2020).

Soft resins are divided into α-acids (3–17% w/w) and β-fraction, this latter composed of β-acids (2–7% w/w) and uncharacterised soft resins. The term "resin" is extensively used to describe these compounds, even though both α- and β-acids, the major representatives of the soft fraction, are crystalline, not resinous (Fig. 9.7) (Almaguer et al., 2014; Astray et al., 2020).

Bitter acids stem from the same precursor: they are prenylated derivatives of phloroglucinol (1,3,5-trihydroxybenzene) (Fig. 9.8) (Carbone & Gervasi, 2022).

Namely, α-acids are di-prenylated, while β-acids are tri-prenylated (Przybyś & Skomra, 2020; Zanoli & Zavatti, 2008). These two classes differ in their C6 substituent: -OH for α-acids and another prenyl unit for β-acids.

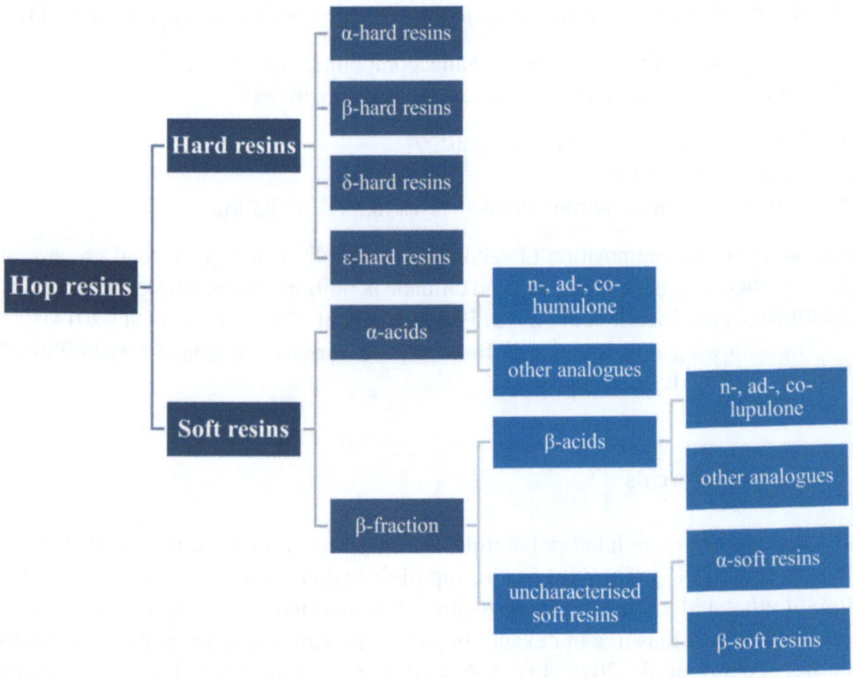

Fig. 9.7 Hop resins' classification

9.2.2.1.1 α-acids

The most representative α-acids are n-, co- and ad- humulone (Fig. 9.8). Humulone's structure was first proposed in 1925 (Wieland, 1925a, b) and then amended in 1950 (Cook & Harris, 1950); this latter option was later confirmed (Carson, 1951). This α-acid was separated for the first time in the crystalline state the following year (Carson, 1952). Its chiroptical structure, though, was only recently confirmed as (−)-humulone (Urban et al., 2013).

α-acids were long thought to be almost entirely composed of humulone, before Rigby and Bethune proved they are a complex mixture of homologues and analogues (Rigby & Bethune, 1952, 1955).

Depending on the acyl side chain, α-acids can be distinguished into n- (isovaleryl moiety), co- (isobutyryl moiety), ad- (2-methylbutyroyl moiety), pre- (isohexanoyl moiety), post- (propanoyl moiety) (Astray et al., 2020; Carbone & Gervasi, 2022), and ad-pre- (hexanoyl moiety) (Zanoli & Zavatti, 2008). They are distributed as follows: n-humulone (35–70% of the total α-acids), co-humulone (20–65%), ad-humulone (10–15%) and, in much smaller percentages, pre-humulone, post-humulone and ad-pre-humulone (Zanoli & Zavatti, 2008). Pre- and post-humulone were among the last compounds to be identified by partition chromatography (Velde & Verzele, 1986; Verzele, 1958). The relative contents of these homologues greatly

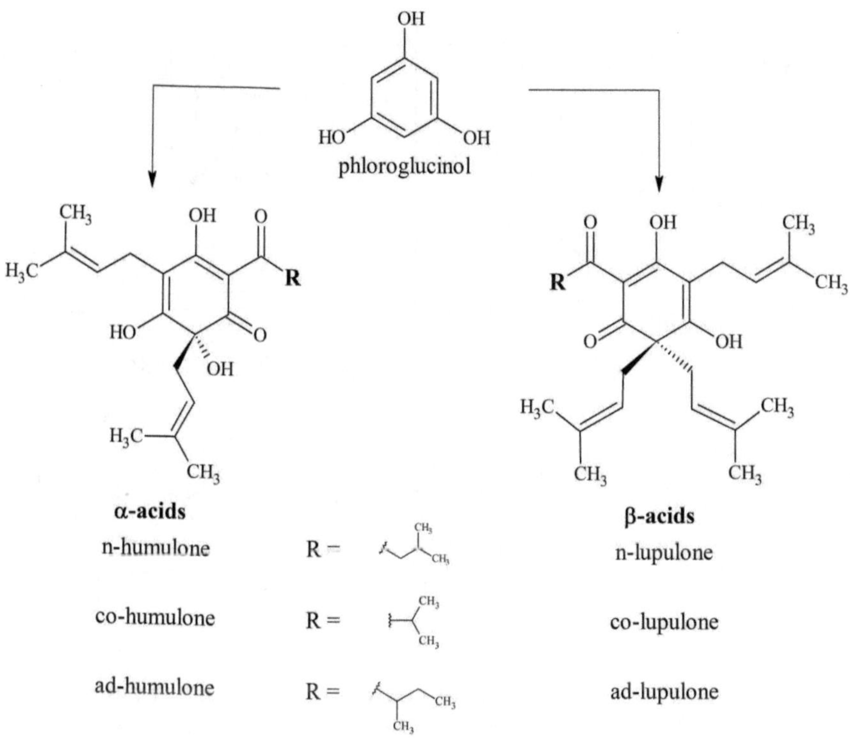

Fig. 9.8 Phloroglucinol and its derivatives, the α- and β-acids

vary depending on hop variety, cultivar conditions, harvesting time and climate (Astray et al., 2020; Zhang et al., 2021), but it was found that, regardless of the other conditions, pre- and post-humulone are much higher for late harvesting times (Almaguer et al., 2014), and that the amount of ad-humulone is relatively constant in different hops (Van Cleemput et al., 2009).

9.2.2.1.2 iso-α-acids

When exposed to high temperatures during brewing (100–130 °C) at alcaline pH (8–10), α-acids isomerise to iso-α-acids (Astray et al., 2020; Przybyś & Skomra, 2020).

Although the isomerisation process was known since the late 1920s, the exact configuration of the iso-α-acids has only been ascertained in 2013 through X-ray crystallography (Urban et al., 2013), proving that the affected stereo centre is C5 (Almaguer et al., 2014).

The isomerisation consists in an acyloin-type ring contraction (Fig. 9.9) (Jaskula-Goiris et al., 2010).

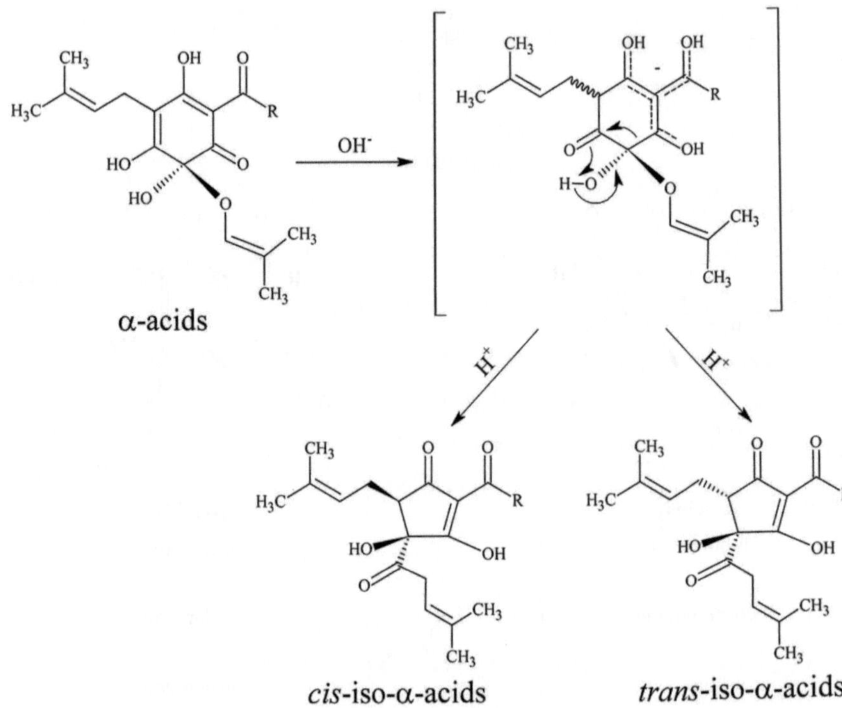

Fig. 9.9 Isomerisation process of α-acids

The isomers are six: *trans*-iso-humulone, *trans*-iso-co-humulone, *trans*-iso-ad-humulone, *cis*-iso-humulone, *cis*-iso-co-humulone, and *cis*-iso-ad-humulone (Zhang et al., 2021). In fact, each isomer occurs in diastereomer form.

These products are more bitter and soluble than their precursors (Carbone & Gervasi, 2022), and can be beneficial to the stability of beer (Van Cleemput et al., 2009).

The isomerisation yield reaches at most 50–60% of the total α-acids, depending on the different brewing processes. For example, scarce isomerisation levels are attributed to limited solubility of α-acids in wort, incomplete isomerisation during boiling, and depletion of the chemicals due to adsorption on the trub (Sun et al., 2022).

In general, hops with higher co-humulone levels contribute more to beer bitterness, since the isomerisation yield for this compound is higher than for the other homologues (Rutnik et al., 2023).

When iso-α-acids are exposed to light, they convert into three different "light-struck/sunstruck" derivatives, each with its solubility, bitterness and stabilising characteristics (Zhang et al., 2021).

9.2.2.1.3 β-acids

This fraction is overall less studied than the α counterpart. β-acids include n-lupulone (30–55% of total β-acids), co-lupulone (20–65%), ad-lupulone (10–15%), pre-lupulone and post-lupulone (Fig. 9.8) (Astray et al., 2020).

Lupulone was first isolated in crystalline form in the late 1800s. Its structure was proposed about 60 years later (Wöllmer, 1925), and corrected later on (Verzele & Govaert, 1949); its natural chiroptical structure was established as (−)-lupulone (Riedl, 1952).

During the twentieth century, homologues of lupulone were identified; their names bear the same prefixes as the homologues of lupulone (i.e. n-, co-, ad-), since they present the same prenyl moieties (Almaguer et al., 2014).

At first, they were thought to be lost during brewing and not to contribute to beer bitterness. In fact, the presence of a prenyl moiety *in lieu* of a hydroxyl group on the aromatic ring makes them less acidic and soluble than α-acids and prevents their isomerisation. Nevertheless, they contribute to beer's bitterness and, above all, to its biological properties, including its antimicrobial properties (Carbone & Gervasi, 2022; Rutnik et al., 2023). In general, β-acids in beer have also been associated with antidepressant activity due to activation of the Ca^{2+} channels (Schulz et al., 2018) and antiproliferative effect on human colon cancer cells (Cattoor et al., 2010).

9.2.2.1.4 Oxidation and Degradation Products of Bitter Acids

Hop bitter acids are very sensitive to oxidation and degradation (Van Cleemput et al., 2009). In fact, both α- and β-acids are subject to oxidation initiated by airborne oxygen during natural ageing, converting to hard resins (Almaguer et al., 2014; Sun et al., 2022).

In this regard, Taniguchi et al. (2013) studied the composition of bitter acids in hops upon forced ageing. Namely, they stored hop samples at different temperatures (20, 40 and 60 °C) to assess whether temperature impacted the formation of derivatives of soft resin components. The generated hydrophilic oxidation products at the three temperatures were almost identical, including minor ones (Taniguchi et al., 2013).

Derivatives of α-acids

Oxidative derivatives of hop α-acids, such as humulinones, are formed during the storage and production of beer. These compounds are formed via the oxidation of α-acids and contribute to the bitterness of the final product, though they are less bitter than iso-α-acids. The main oxidation products of alpha acids as reported in the literature are discussed below.

The oxidation of humulones on the 3-methyl-2-butenyl side chains yields humulinones (García-Villalba et al., 2006) (Fig. 9.10), more polar than the iso-α-isomers due to an additional hydroxyl moiety (Rutnik et al., 2023) and present in hop from 1 to 2% (Cook & Harris, 1950).

Humulinone was obtained in laboratory upon storing ethereal solutions of humulone in contact with a saturated solution of sodium hydrogen carbonate (Verzele & De Keukeleire, 1991). Its presence in aged hops was first determined by Taniguchi et al. (2013), while humulinones' contribution to beer bitterness has been demonstrated by another group (Taniguchi et al., 2013; Wang et al., 2024). However, to the best of our knowledge, no specific condition has been indicated to elucidate the formation mechanism of this class of compounds in beer.

α- and iso-α-acids derivates also include humulinic acids, which are their hydrolysis products and present two chiral centres (Fig. 9.10) (Przybyś & Skomra, 2020; Zhang et al., 2021).

In addition to the compounds mentioned above, which are the main and best known oxidation products of bitter acids, many others have been identified over time (Hao et al., 2020; Taniguchi et al., 2013).

Abeo-isohumulone (Fig. 9.10) is an oxidation product of α-acids presenting a five-membered structure, derived from humulinones via oxidation of their 3-methyl-2-butenyl side chain and cyclization, via either intramolecular

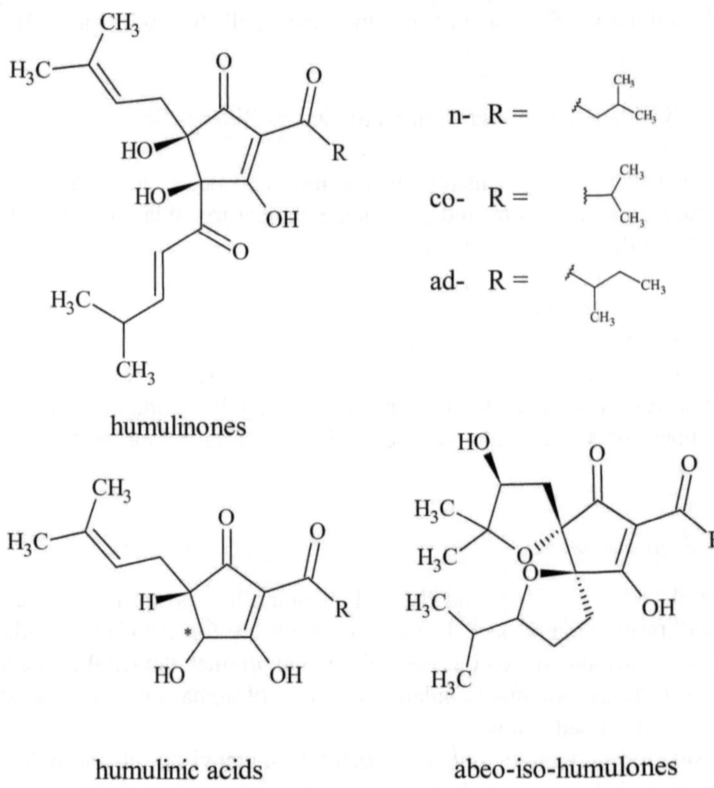

Fig. 9.10 Most common oxidation products of α-acids

dehydration or nucleophilic cleavage of the intermediate oxirane ring. It was found by García-Villalba et al. (2006) in forcibly oxidised hops (García-Villalba et al., 2006).

4′-hydroxyallohumulinones (Fig. 9.11) were identified by Taniguchi et al. (2013) as oxidation-derived products of humulinones, with a tentative pathway resembling that of iso-α-acids to hydroxyalloisohumulones (Taniguchi et al., 2013). These compounds were also labelled as the main constituents of the hard resins, whereas humulinones and hulupones are the main components of the soft resin fraction.

4′-hydroxyallohumulinones present a higher stability as compared to humulinones, hence they could be used to assess the degree of hop oxidation. Nevertheless, they can still undergo further modifications: isomerisation to 4′-hydroxyallo*cis*-humulinones and subsequent decomposition into *cis*-oxyhumulinic acids (Taniguchi et al., 2014b).

4′-hydroxyallohumulinones can also be subjected to cyclisation into scorpiohumulinols A/B and di-cyclohumulinols A/B (Fig. 9.11) during model solution storage (Taniguchi et al., 2015), with a proposed mechanism similar to that from hydroxyl*trans*-alloisohumulones to scorpiohumols.

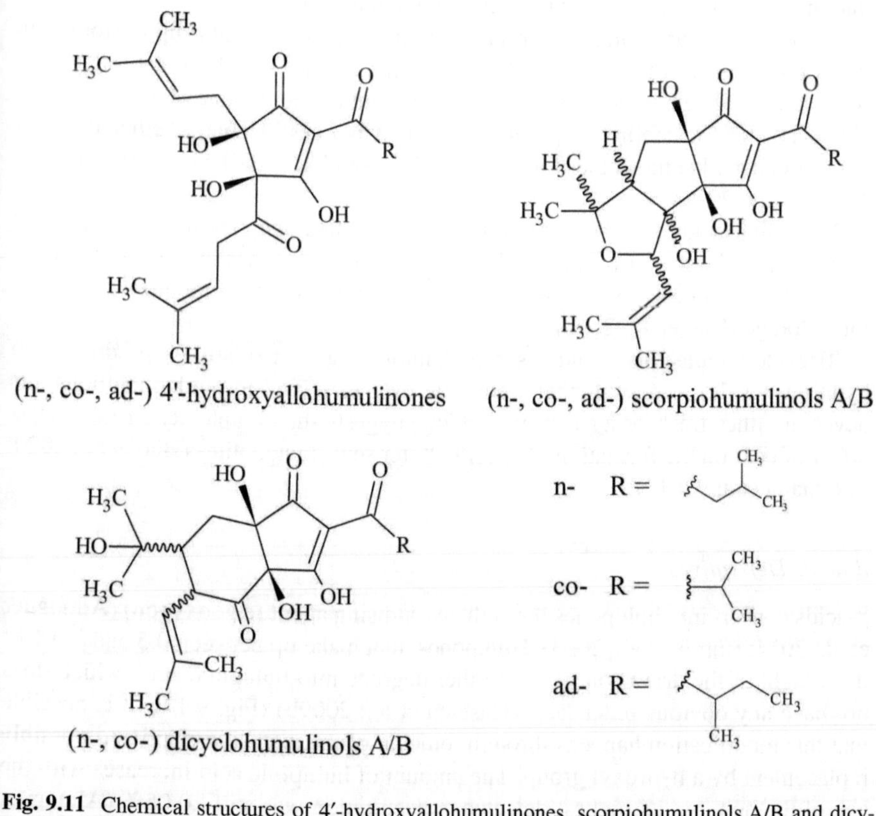

Fig. 9.11 Chemical structures of 4′-hydroxyallohumulinones, scorpiohumulinols A/B and dicyclohumulinols A/B

In another work, Taniguchi et al. (2014a) isolated some cyclisation products of humulone, among which tricycloisohumulones A and B, proposed their structures, and classified them as hard resins components (Taniguchi et al., 2014a).

Over time, many cyclised and oxidised products of iso-α-acids have also been identified, such as alloisohumulones compounds (hydroperoxy*cis*−/*trans*-alloisohumulones, hydroxyl*cis*−/*trans*-alloisohumulone, *cis*−/*trans*-alloisohumulones), humulinic acids (*cis*- and *trans*-), scorpiohumols, tricyclohumulactols, and more (Hao et al., 2020). Not all of them have been detected in hops, but rather in beers or model solutions deriving from hop extracts of some kind; still, they have been reported because they stem from α-acids found in hop.

Alloisohumulones form via the isomerisation of the double bond of the iso-3-hexenoyl side chain of iso-n-humulone and the consecutive hydration of the enone; the humulinic acids form by proton catalysing reaction or alkali catalysed formation by both *trans*- and *cis*-iso-α-acids. Traces of *cis*-alloisohumulones have been detected in pilsner beers of varying ages, while trans isomers have not been detected; however, their possible role in beer is still unclear (Intelmann et al., 2011). Hydroperoxy*cis*−/*trans*-alloisohumulones derive from a reaction cascade passing through hydroxylated and hydroperoxidated iso-α-acids. This kind of oxidation is common during beer brewing (Intelmann et al., 2011).

They are primary autoxidation products in beer, in particular upon storage in oxygen-permeable PET bottles. They are formed rather swiftly in aerobic conditions and could be used as a measure of the current oxidative state of a beer sample. However, their instability is pronounced, so it is questionable whether they are appropriate marker molecules for oxidative deterioration of the beer taste (Intelmann & Hofmann, 2010).

Humulinic acids, non-bitter derivatives of iso-α-acids, supposedly have a marked effect on the bitterness evolution of beers packaged in presence of oxygen, and have been detected in beers in increasing amounts over the course of months- and years-long storage (Hao et al., 2020).

Tricyclohumulactols and scorpiohumols are oxidative products of hydroperoxyalloiso-α-acids that were only ever identified in model solutions and never in either fresh or aged beers, which suggests the complexity of the matrix effect of beer on the formation of specific bitter sensometabolites (Hao et al., 2020; Intelmann et al., 2011).

β-acids Derivatives

β-acids convert into hulupones through an oxidising agent (e.g. oxygen) (Almaguer et al., 2014; Rutnik et al., 2023). Hulupones, that make up between 0.5 and 3.0% of the weight of the dried cones, can further degrade into hulupinic acid, which does not have any obvious bitter taste (Haseleu et al., 2009b) (Fig. 9.12). It is possible that this modification happens through removal of the acyl side chain and possibly replacement by a hydroxyl group. The amount of hulupinic acid increases with the age of the hop, but its highest detected concentration was still <0.05% (Almaguer et al., 2014).

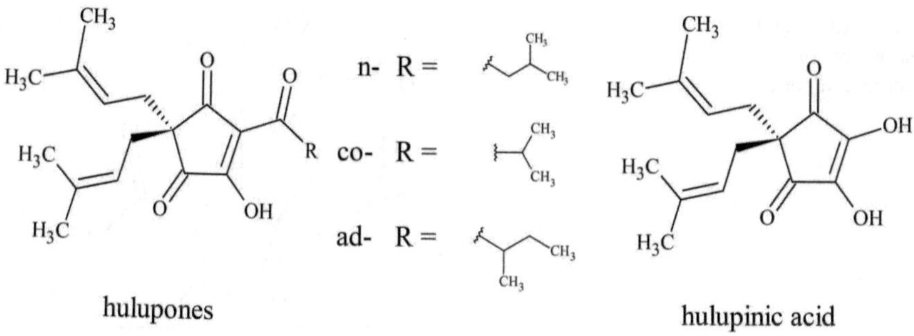

Fig. 9.12 Chemical structures of hulupones and hulupinic acid

The presence and amount of these species into hops was long debated, because researchers would not agree on a uniform method of detection and quantification. This troublesome process resulted in the proposition of very different figures that quantified their influence on beer's bitterness as compared to more known bitter acids (e.g. they have been reported as being half as bitter as iso-α-acids in one instance and twice as bitter in another) (Algazzali & Shellhammer, 2016).

Further research on the topic was conducted in more recent years, establishing that humulinones and hulupones are the most abundant oxidised hop acids in hops aged from 6 to 60 months, and that they can be found in beer at substantial concentrations (Dušek et al., 2014; Krofta et al., 2013).

More oxidation products of β-acids include lupoxes, lupdoxes, lupdeps, lupodols, as well as tricycleoxy-co-lupulones, deriving from three possible kinds of reactions. One is hulupone transformation into hulupinic acids; another is a cyclisation reaction (Haseleu et al., 2009b); the third is the formation of epoxyhulupone from hulupone (Fig. 9.13) (Dušek et al., 2014).

Long-term storage of hops leads to significant formation of all the cyclisation products of β-acids. In particular, degradation of colupulone and lupulones results in the formation of nortricyclo-, dehydrotricyclo-, tricyclo-, hydroperoxytricyclo-, and hydroxytricyclo- derivatives (Hao et al., 2020).

Mimicking of the wort boiling conditions led to the discovery of the aforementioned cyclic oxidation products. All tricyclic compounds induced a long-lasting, lingering and pronounced bitterness feeling on the posterior tongue and throat, as compared to the "clear bitter taste" of their progenitors (Haseleu et al., 2009a, b).

9.2.2.1.5 Uncharacterised Soft Resins

This fraction is non-specific and consists of the soft resins remaining after separation of α- and β-acids by precipitation and crystallisation. It contains non-specified compounds, together with essential oils and hop wax that could not be completely separated. This uncharacterised portion can be further divided into α- and β-soft

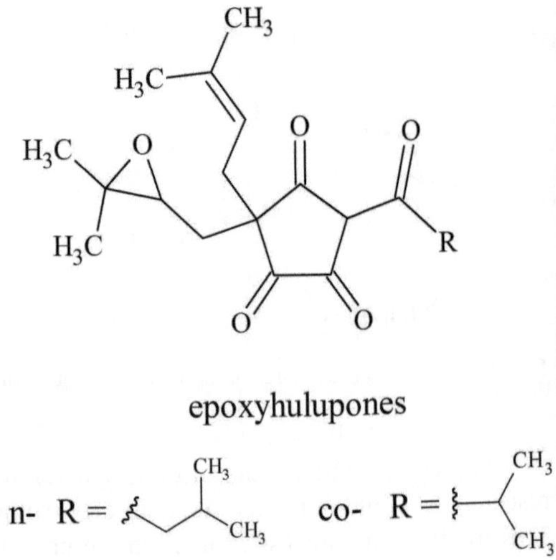

Fig. 9.13 Chemical structures of epoxyhulupones

resins, implying that they contain substances that may later be identified as arising from α- or β-acids, respectively (Almaguer et al., 2014).

Bitter acids are essential for conferring bitterness to beer. Since their loss starts as soon as hops are harvested with different rates depending on hop variety, form, storage conditions, and time of storage, knowing the extent of such losses is pivotal for the brewing industry. In fact, this could affect marketing decisions in times of abundance followed by scarcity, to allow for thoughtful utilisation of long-stored hops (Rutnik et al., 2023).

9.2.2.2 Polyphenols

Polyphenols, or phenolic compounds (PCs), constitute a chemically heterogeneous class of compounds synthesised by plants and broadly distributed in their tissues. According to Taiz and Zeiger (2002), plants produce almost 10,000 different polyphenols playing a wide variety of roles: defense against herbivores and pathogens, mechanical support, attraction of pollinators and fruit dispersers, absorption of harmful ultraviolet radiation, inhibition of the growth of nearby competing plants, etc. (Taiz & Zeiger, 2002).

The type and concentration of these compounds vary depending on the species of the plant and are influenced by several factors, including the growing region, harvesting methods, and degree of ageing. Hop cones are no exception; they contain a complex mixture of polyphenolic compounds, whose composition is highly dependent on variety and abiotic factors (Almaguer et al., 2014; Eßlinger, 2009; Karabín et al., 2016; Stewart & Priest, 2006). Despite the heterogeneity of the polyphenols' structures, they all share a common element, known as phenol: an aromatic ring containing one hydroxy functional group (Fig. 9.14).

Fig. 9.14 Structural element of polyphenols

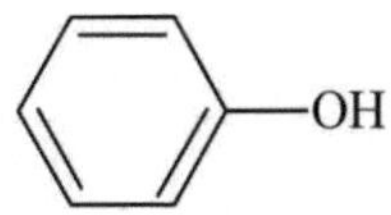

From a biochemical point of view, most polyphenols are synthesised through the shikimic acid pathway, a seven-step metabolic pathway in which simple carbohydrates are converted into the three aromatic amino acids: L-phenylalanine (PHE), L-tyrosine (TYR), and L-tryptophan (TRP). L-phenylalanine is further transformed into various PCs, ranging from simpler phenolic acids to more complex compounds such as flavonoids, up to even more complex structures like lignans (Fig. 9.15) (Taiz & Zeiger, 2002).

In the shikimic acid pathway, the first step entails an enzyme-driven condensation between phosphoenolpyruvate, derived from glycolysis, and erythrose-4-phosphate, derived from the pentose phosphate pathway. The resulting product, 3-deoxy-D-arabinoheptulosonate-7-phosphate, undergoes cyclisation through a series of enzyme-catalysed transformations, namely oxidation, phosphate elimination, reduction, ring opening and internal aldol condensation. Eventually, the cyclic structure is dehydrated and reduced by a bifunctional enzyme (namely, 3-dehydroquinate dehydratase/shikimate dehydrogenase) to form the shikimic acid, from which the whole sequence of reactions is named (Marchiosi et al., 2020). The shikimic acid is then converted to chorismic acid, the biosynthetic precursor of aromatic amino acids.

The subsequent phenylpropanoid pathway belongs to secondary metabolism and begins with the deamination of PHE performed by the phenylalanine ammonia-lyase (PAL). Since this first step occurs at a strategic branching point between primary and secondary metabolism, it represents an important regulatory phase in the polyphenols' biosynthesis. The product, *trans*-cinnamic acid, is successively hydroxylated to form *p*-coumaric acid by the cinnamic acid 4-hydroxylase (C4H), a cytochrome P450 monooxygenase in plants (Liu et al., 2021; Saito et al., 2013). In addition, *p*-coumaric acid can be directly synthesised from tyrosine by the action of tyrosine ammonia-lyase (TAL) (Zagoskina et al., 2023).

Both *trans*-cinnamic acid and *p*-coumaric acid are hydroxycinnamic acids (HCAs) and are classified as phenylpropanoids due to their structure, which consists of a phenol ring and a three-carbon side chain. These compounds serve as key building blocks in the biosynthesis of the different classes of phenolic compounds. Hydroxycinnamic acids, in fact, do not usually accumulate in their free form within plant tissues as they are subjected to further transformations. Such transformations are facilitated by the activation of *p*-coumaric acid through the addition of coenzyme A (CoA), converting it into the corresponding CoA thioester (HCA-CoA). This reaction, catalysed by 4-coumaric acid-CoA ligase (4CL), marks the conclusion of the shikimic acid pathway (El-Seedi et al., 2018). The main transformations of both HCA and HCA-CoA include: β-oxidation, which leads to the formation of hydroxybenzoic acids; reduction, resulting in cinnamic alcohols—some of which (namely coniferyl, coumaryl, and sinapyl alcohols) are involved in lignin

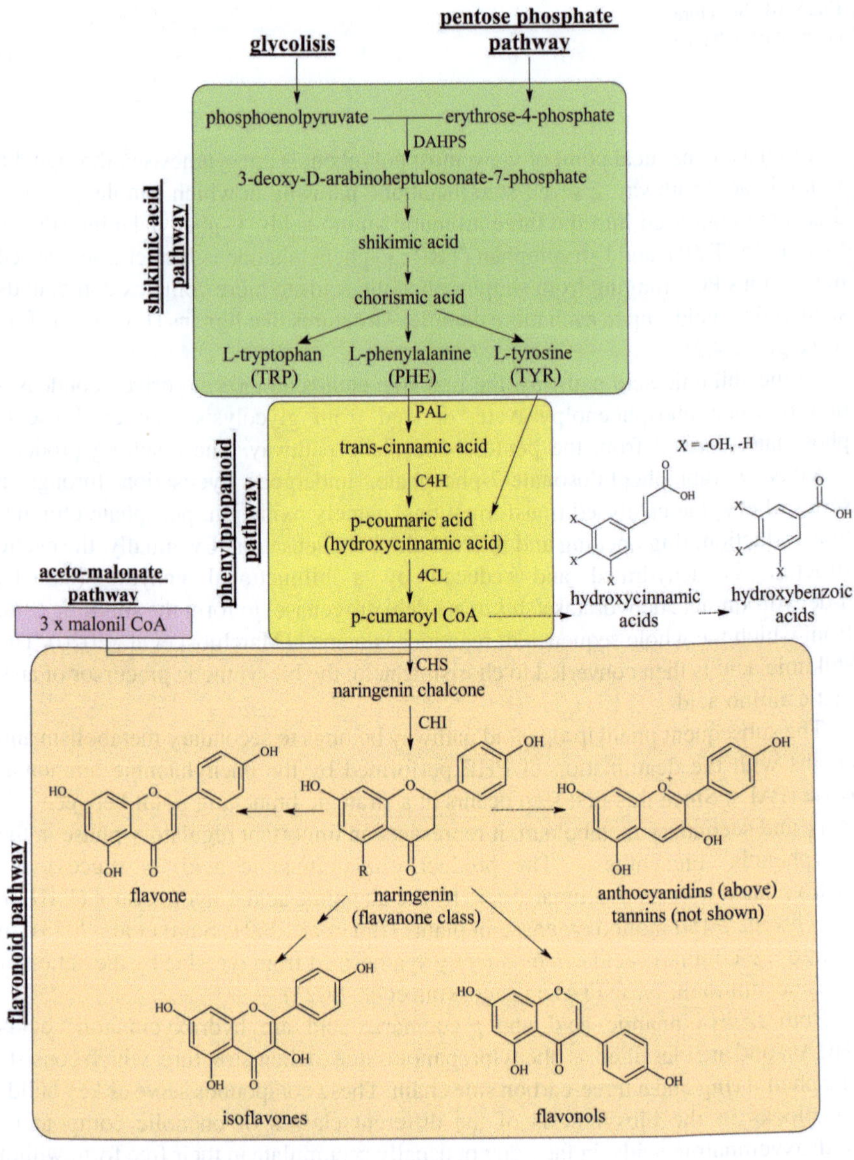

Fig. 9.15 Schematic biosynthetic pathway of PCs

biosynthesis; formation of acyl derivatives and complex esters; synthesis of coumarins (Dewick, 1995; Marchiosi et al., 2020; Zagoskina et al., 2023). Alternatively, HCA-CoA can enter in the flavonoid pathway serving as precursor of flavonoids, one of the largest classes of plant PCs. Their basic scaffold consists of 15 carbon atoms arranged in two aromatic rings, referred to as ring A and ring B, connected by a three-carbon bridge (C6-C3-C6) (Fig. 9.16), except for stilbenes (flavonoids derivatives), which have a C6-C2-C6 structure.

Ring A is synthesised from the condensation of three malonyl-CoA molecules deriving from the aceto-malonate pathway, while ring B, along with the three-carbon bridge, stems from the shikimate pathway from *p*-coumaroyl-CoA (Nabavi et al., 2020; Taiz & Zeiger, 2002). The condensation of ring A and ring B is catalysed by chalcone synthase and leads to the formation of naringenin chalcone. Chalcone synthase, as well as the aforementioned PAL, is a key enzyme for the regulation of plants' secondary metabolism pathways. Naringenin chalcone subsequently undergoes cyclisation to form a third ring (i.e. ring C) becoming naringenin, major representative of flavanones. This latter is further modified by tailoring reactions, including condensation, isomerisation, oxidation and reduction, to define the final class of flavonoids. The basic flavonoid carbon skeleton may have numerous substituents. Hydroxyl groups are usually present at positions 4, 5, and 7, but they may also be found at other positions. Sugars are very common; in fact, most flavonoids exist naturally in the form of glycosides. Whereas both hydroxyl groups and sugars make flavonoids water-soluble, other substituents, such as methyl ethers or modified isopentyl units, enhance their lipophilicity (Taiz & Zeiger, 2002). Depending on the oxidation degree of the heterocyclic ring and the number of hydroxyl or methyl groups on the benzene ring, flavonoids can be categorised into various subclasses.

Tang et al. (2019) found 117 polyphenols compounds in hop cones; they categorised them into six classes, including 30 phenolic acids, 61 flavonoids, 4 lignans, 2 stilbenes, 1 hydroxybenzaldehyde, and 19 other polyphenols (Tang et al., 2019). According to Biendl and Pinzl (2007), the most representative phenolic class in dried hop cones (10% residual moisture) is flavanols, followed by prenylflavonoids, flavonols, and phenolic acids (Biendl & Pinzl, 2007).

In the next paragraphs of this section, we will only focus on those most represented in hops, consisting of four main groups: flavonols and flavonol glycosides, flavan-3-ols and proanthocyanidins, flavanones, and prenylflavonoids (Fig. 9.17) (Liu et al., 2021).

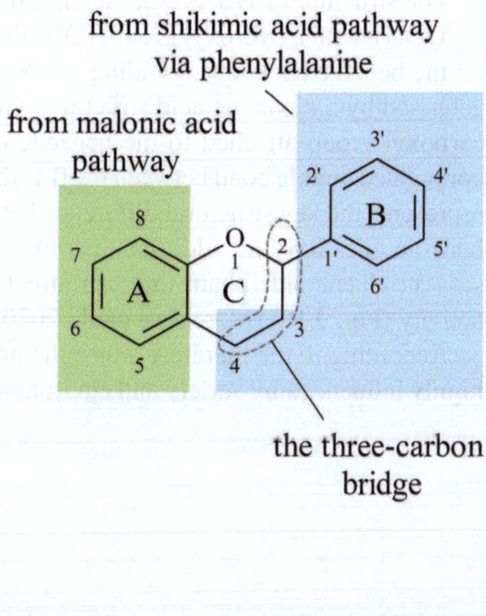

Fig. 9.16 Basic scaffold of flavonoids

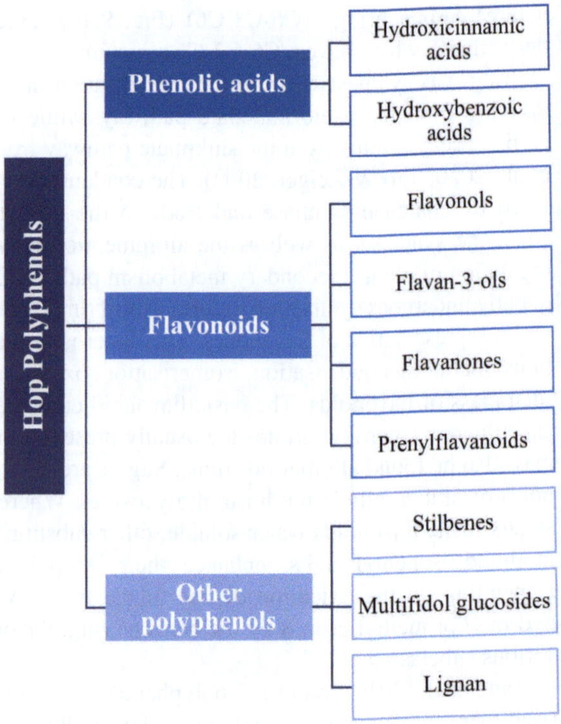

Fig. 9.17 Main classes of hop polyphenols

9.2.2.2.1 Phenolic Acids

Phenolic acids are important representatives of polyphenols found in hop cones, although they are typically present in relatively low concentrations. These acids are usually categorised in two main classes: hydroxybenzoic acids (HBAs) and hydroxycinnamic acids (HCAs) (Fig. 9.18).

The structure of HBAs consists of a benzene ring attached to a carboxyl group (-COOH), with a hydroxyl group (-OH) that can be substituted at different positions on the benzoic backbone, yielding various derivatives. In contrast, HCAs are characterised by a cinnamic acid structure, which has a three-carbon chain bearing the carboxyl group attached to the benzene ring. This side chain, characterised by a conjugated double bond between the β-carbon and the α-carbon (–CH=CH–COOH), represents the key structural difference between the two classes of acids—hydroxybenzoic ones have no additional carbon chain, while hydroxycinnamic acids possess an alkene side chain that can affect their chemical reactivity and biological activity (Fig. 9.18) (Marchiosi et al., 2020).

According to the literature, since the phenolic acids' presence in hop strobiles is highly influenced by variety and environmental factors, their concentration can vary

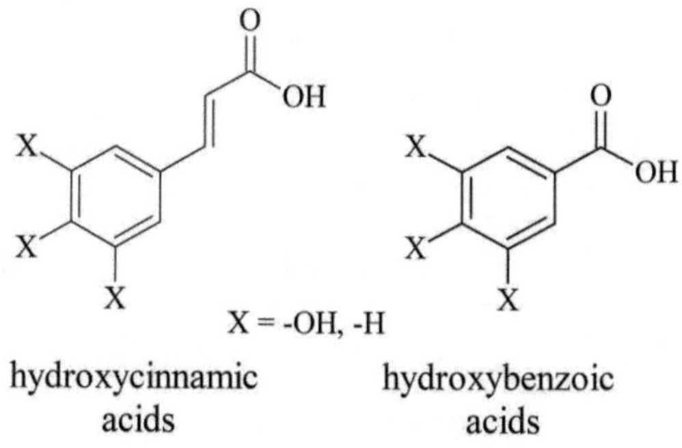

Fig. 9.18 Generic structure of hydroxycinnamic acids (left) and hydroxybenzoic acids (right)

significantly, accounting from 0.05% to 13% of the total phenolic compound content (Carbone & Gervasi, 2022; Inui et al., 2017; Nagasako-Akazome et al., 2007; Telichowska et al., 2020). Tang et al. (2019) conducted a comprehensive characterisation of hop polyphenols, using liquid chromatography coupled with electrospray ionization quadrupole time-of-flight mass spectrometry. Regarding polyphenolic acids, they tentatively identified five hydroxybenzoic acids, fifteen hydroxycinnamic acids, and ten additional hydroxy monocarboxylic acid derivatives. These latter do not belong to the two main families of phenolic acids, as their carboxyl group is positioned at the end of a saturated side chain of one to five carbon atoms. The major phenolic acids they found in hops are chlorogenic acid (16.48 ± 0.03 mg/g of dry sample), gallic acid (3.41 ± 0.02 mg/g of dry sample), caftaric acid (0.72 ± 0.01 mg/g of dry sample), and syringic acid (0.03 ± 0.01 mg/g of dry sample) (Tang et al., 2019). Research by Santarelli et al. (2022) asserted that the most abundant hydroxycinnamic acid in hop cones is chlorogenic acid (varying between 0.612 and 0.877 mg/g of dry sample), followed by lesser amounts of other representatives of this class, such as ferulic, chicoric, caffeic and *p*-coumaric acid, and various derivatives of all of these (Fig. 9.19). As for hydroxybenzoic acids, protocatechuic, syringic, ellagic and *p*-hydroxybenzoic acids were determined in quantities ranging from about 0.262 to 0.654 mg/g of dry sample, similarly followed by their respective derivatives (Fig. 9.19) (Santarelli et al., 2022). On the other hand, Proestos et al. (2009) found ferulic acid to be the most abundant hydroxycinnamic acid in hop cones (0.064 mg/g dry sample), followed by vanillic acid as the major representative of hydroxybenzoic acids (0.029 mg/g dry sample) (Proestos & Komaitis, 2009).

The presence of a wide range of phenolic acids, along with the heterogeneity of their content and profile, as reported in the existing literature, is indicative of the significant impact of hop variety, environmental growing conditions, and diverse extraction methodologies on the phenolic composition of hops.

Notably, phenolic acids in plants can be found both in their free form and as esters bound with alcohols, amino alcohols, carbohydrates and acids (Bandyukova, 1983; Goleniowski et al., 2013). In hops, Tang et al. (2019) found some phenolic

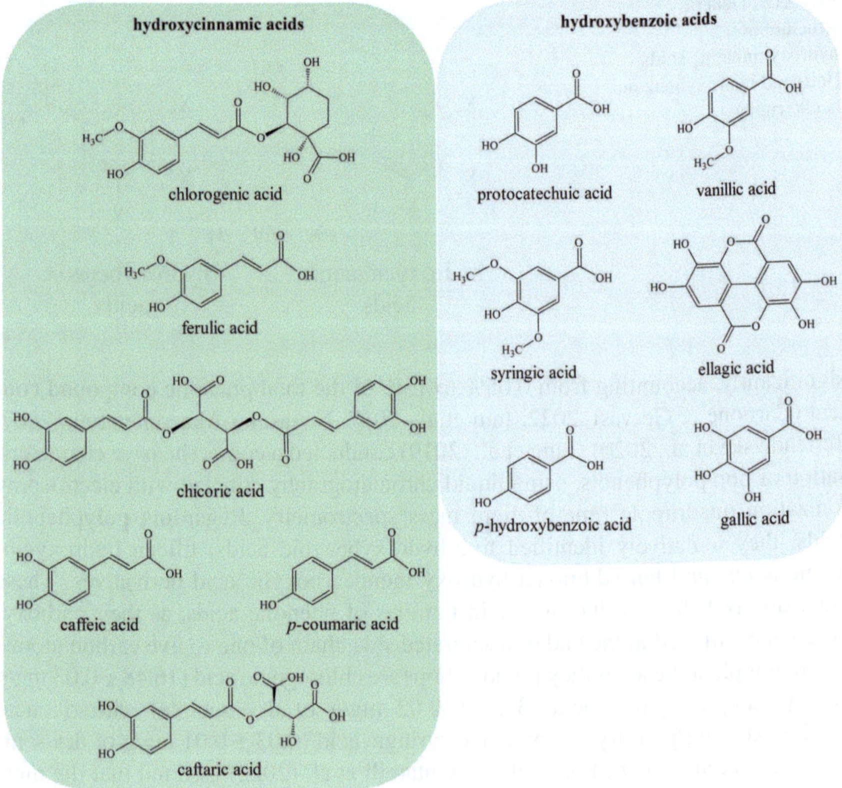

Fig. 9.19 Structures of phenolic acids found in hop cones

acids derivatives conjugated with glucuronide and glucoside moieties at position 3 or 4 (e.g. *p*-coumaric acid 4-O-glucoside, ferulic acid 4-O-glucuronide, ferulic acid 4-O-glucoside, protocatechuic acid 4-O-glucoside, etc.) (Tang et al., 2019).

9.2.2.2.2 Flavonoids

Flavonoids constitute a wide group of secondary metabolites in hops, known both for their significant biological activities and for their role as precursors of biologically active compounds (Fig. 9.17) (Karabin et al., 2015). Flavonoid content in hop cones is very variable and depends on the already mentioned factors affecting all PCs. It is therefore not surprising that their average content hugely varies depending on the literature studies taken in consideration. For conventional, commercially grown hops, the flavonoid content can be up to 5 g/100 g of hop cone dry matter (Jelínek et al., 2010).

Flavonols and Flavonol Glycosides

Flavonols are a prominent subclass of PCs found in hop cones. Their core structure consists of a flavone skeleton (the characteristic three-ring aromatic structure of flavonoids) with a ketone group on C4, a double bond between C2 and C3 and hydroxyl groups (-OH) as substituents (Fig. 9.20) (Karabin et al., 2015).

Flavonols often naturally occur as glycosides, with sugar units attached to the hydroxyl groups—typically at position 3, more rarely at C7; they can bear one (monoglycosylate), two (diglycosylate) or three (triglycosylate) (Dušek et al., 2021).

The sugar component usually consists of D-glucose or L-rutinose. In plant materials, including hop cones, flavonol glycosides are usually found in higher concentrations than their respective aglycones, i.e. the non-glycosylated flavonoids. Glycosylation can alter the solubility and bioavailability of flavonols, influencing their physiological effects (Karabin et al., 2015; Manach et al., 2004). More in detail, aglycones possess higher activity but more unstable structures. Indeed, sugar moieties have been shown to reduce the activity of flavonoids but improve their stability (Xie et al., 2022).

The most significant aglycones of flavonols in hops include quercetin, kaempferol, and myricetin, followed by rhamnetin (Fig. 9.20). Regarding their glycoside compounds, several derivatives have been identified in strobiles, including quercetin 3-O-rutinoside (rutin), quercetin 3-O-(6″-O-malonylglucoside), quercetin 3-O-glucoside, kaempferol 3-O-rutinoside, kaempferol 3-O-(6″-O-malonylglucoside) and kaempferol 3-O-glucoside (Fig. 9.21) (Dušek et al., 2014). Tang et al. (2019) identified two isorhamnetin derivatives, five kaempferol derivatives, four myricetin derivatives, and two quercetin-3-O derivatives (Tang et al., 2019). Less common flavonols, such as morin, which lacks a hydroxyl group at the C5 position, have also been detected in hops (Jelínek et al., 2010). Segawa et al. (2006)

Fig. 9.20 Structures of common flavonols: kaempferol, quercetin, myricetin and rhamnetin

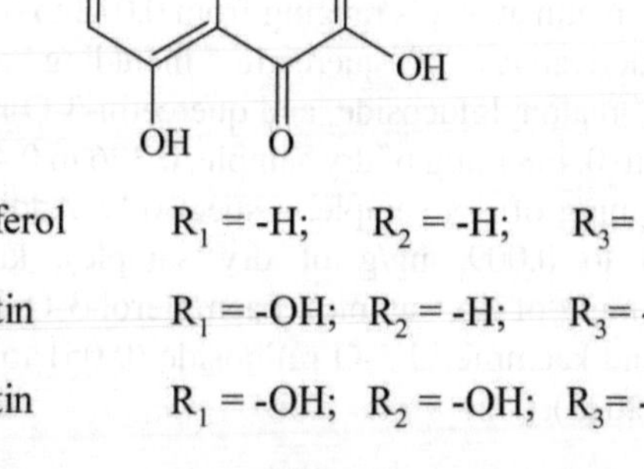

kaempferol R_1 = -H; R_2 = -H; R_3= -OH

quercetin R_1 = -OH; R_2 = -H; R_3= -OH

myricetin R_1 = -OH; R_2 = -OH; R_3= -OH

rhamnetin R_1 = -OH; R_2 = -H; R_3= $-OCH_3$

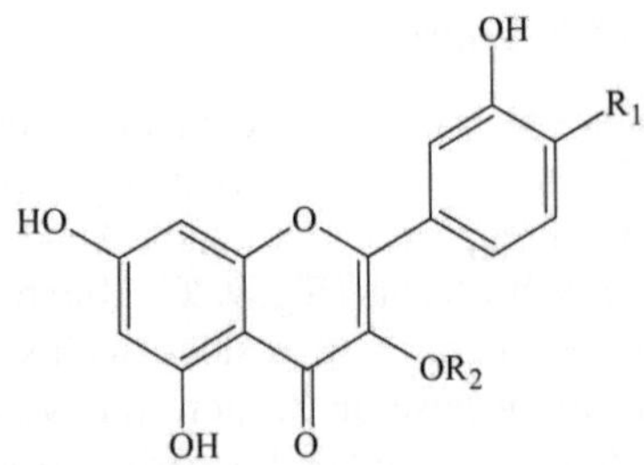

kaempferol-3-O-rutinoside: R_1 = -H, R_2 = -rutinose

kaempferol-3-O-(6"-O-malonylglucoside): R_1 = -H, R_2 = -6-malonyl-β-D-glucopyranose

kaempferol-3-O-glucoside: R_1 = -H, R_2 = -β-D-glucopyranose

quercetin-3-O-rutinoside: R_1 = -OH, R_2 = -rutinose

quercetin-3-O-(6"-O-malonylglucoside): R_1 = -OH, R_2 = -6-malonyl-β-D-glucopyranose

quercetin-3-O-glucosyde: R_1 = -OH, R_2 = -β-D-glucopyranose

Fig. 9.21 Chemical structures of flavonol glycosydes

identified several quercetin glycosides in water extracts of hop cones, including isoquercetin (quercetin-3-O-glucoside), rutin (quercetin-3-O-rutinoside), and isoquercetin malonate, as well as kaempferol glycosides, such as astragaline, kaempferol rutinoside, and astragaline malonate (Segawa et al., 2006).

Kováčová et al. (2011) analysed the content of rutin, quercetin, and kaempferol in hop extracts, finding that quercetin and kaempferol were more soluble in methanol than in water. They quantified rutin at 0.246 mg/g of dry sample, quercetin at 0.155 mg/g of dry sample, and kaempferol at 0.126 mg/g of dry sample (Kováčová et al., 2011). As with phenolic acids, the data available in the literature can significantly vary due to differences in cultivar and environmental factors affecting the plant matrix. Calvert et al. (2025) examined three different hop cultivars, quantifying quercetin at levels ranging from 0.012 to 0.089 mg/g of dry sample. The glycoside derivatives of quercetin, including quercetin-3-O-glucoside, quercetin-3-O-6″-malonylglucoside, and quercetin-3-O-rutinoside, were found to range from 0.283 to 0.438 mg/g of dry sample, 0.326 to 0.481 mg/g of dry sample, and 0.144 to 0.281 mg/g of dry sample, respectively. Additionally, they quantified kaempferol (0.008 to 0.009 mg/g of dry sample), kaempferol-3-O-glucoside (0.075 to 0.273 mg/g of dry sample), kaempferol-3-O-galactoside (0.034 mg/g of dry sample), and kaempferol-3-O-rutinoside (0.051 to 0.473 mg/g of dry sample) (Calvert et al., 2025).

Flavanols

Flavanols, also known as flavan-3-ols or catechins, are a complex subclass of flavonoids, ranging from simple monomer to oligomers, called proanthocyanidins, and polymers, referred to as condensed tannins (Almaguer et al., 2014). The chemical structure of flavonols comprises the characteristic three-ring aromatic structure of flavonoids, with a hydroxyl group (-OH) at the C3 position. The distinctive feature of flavan-3-ols is the lack of both the carbonylic group at C4 and the double bond between C2 and C3. The carbon atoms at positions 2 and 3 are two chiral centres, thus flavan-3-ols can have up to four isomers if the substituents permit such configurations (Crozier et al., 2012). In fact, depending on the angle between the -OH group (attached to the C3 atom) and the plane of the C ring, as well as on the angle of the B ring (bonded at the C2 position) relative to the plane of the C ring, catechin has four diastereoisomers: (+) or (2R,3S)-catechin, (−) or (2R,3R)-epicatechin, (−) or (2S,3R)-catechin, and (+) or (2S,3S)-epicatechin. The brackets contain numbers that indicate the site of the chiral centres, and the letters refer to the direction of the bonds: namely, 'R' indicates dashed wedge bonds (projecting away from the viewer) and 'S' refers to solid wedge bonds (projecting toward the viewer), as illustrated in Fig. 9.22 (Anitha et al., 2020).

The structure of proanthocyanidins and tannins mainly depends on the type of flavan-3-ol unit, the degree of polymerisation, and the spatial configuration (Taylor et al., 2003). The most common proanthocyanidin monomeric units in hop include catechin, epicatechin, gallocatechin and epigallocatechin (Fig. 9.22); the first two can serve either as extension or terminal unit, while the latter are found only within the polymer as they cannot serve as ending units (Li & Deinzer, 2006).

Flavanols are the second most representative group of polyphenolic compounds in hop cones after prenylflavonoids, the most abundant one being (+)-catechin. Other highly concentrated flavanols are (−)-epicatechin and (+)-gallocatechin (Biendl & Pinzl, 2013).

Li and Deinzer (2006) isolated several flavanols and proanthocyanidin oligomers from hops, including catechin, epicatechin, procyanidin B1 (epicatechin-(4β → 8)-catechin), procyanidin B2 (epicatechin-(4β → 8)-epicatechin), procyanidin B3 (catechin-(4β → 8)- catechin), procyanidin B4 (catechin-(4β → 8)-

(+) or (2R,3S) catechin R_1 = -OH; R_2 = -H; R_3= -H

(-) or (2R,3R) epicatechin R_1 = -H; R_2 = -OH; R_3= -H

(+) or (2R,3S) gallocatechin R_1 = -OH; R_2 = -H; R_3= -OH

(-) or (2S,3R) catechin R_1 = -H; R_2 = -OH; R_3= -H

(+) or (2S,3S) epicatechin R_1 = -OH; R_2 = -H; R_3= -H

(-) or (2S,3R) gallocatechin R_1 = -H; R_2 = -OH; R_3= -OH

Fig. 9.22 Structures of common flavon-3-ols

epicatechin), and proanthocyanidin C2 (catechin-(4β → 8)-catechin-(4β → 8)- catechin). The same authors (2009) also found small quantities of prodelphinidin B3 (polymeric tannin composed of gallocatechin, galloepicatechin-(4β → 8)-catechin) and propelargonidin B (condensed tannin formed from epiafzelechin) in hop cones in a successive study (Li & Deinzer, 2006, 2009).

In hop, the amount of total proanthocyanidins ranges from 0.5 to 5% on a dry weight basis, always depending on the variety, geographic origin, freshness, harvesting procedure, and other external factors. According to Briggs (2004), the average quantity of (+)-catechin is 0.817–2.821 mg/g of dry sample, while the amount of proanthocyanidin B3 is 0.428–1.472 mg/g of dry sample, and that of proanthocyanidin C2 is 0.287–0.875 mg/g of dry sample (Briggs, 2004).

Hop proanthocyanidins have gained attention in the brewing industry because they enhance the stabilisation of organoleptic properties as well as contribute to beer astringency and bitterness. However, proanthocyanidins have also been found to be partly responsible for haze formation. All these effects of hop proanthocyanidins on beer quality seem to depend on their affinity for proteins (Li & Deinzer, 2006).

The chemical structure and the hydroxyl group substitution of flavonols are the major factors providing proanthocyanidins with the ability to scavenge free radicals and inhibit lipid peroxidation. These capacities are highly related to important health benefits. Recent studies have shown that proanthocyanidins can inhibit the growth of various human cancer cells and exhibit anti-inflammatory, anti-infectious, and cardioprotective properties (Wang et al., 2011). For example, in human prostate and breast cancer cells, the flavanols catechin and epicatechin have exhibited anti-oxidative and anti-inflammatory effects, as well as inhibition of telomerase (Parekh et al., 2023). Moreover, epicatechin acts as an anti-ageing agent and has been proven able to suppress the growth of various types of cancer cells. It also exerts neuroprotective effects (Corrêa et al., 2018; Kampa et al., 2007). According to Biendl and Pinzl (2013), since the interactions may strengthen the effects of individual components, a combination of hop proanthocyanidins can have stronger antioxidative effect than single flavanols and proanthocyanidins (Biendl & Pinzl, 2013).

Flavanones

The chemical structure of flavanones consists of a flavone skeleton without the double bond between C2 and C3. As a result, they have a chiral centre at C2 position (Karabin et al., 2015). The most important hop flavanone is naringenin (5,7,4′-trihydroxyflavanone) (Fig. 9.23).

Jelínek et al. (2010) identified it in hop with a relative abundance ranging from 0.2 and 0.4% depending on the cultivar (Jelínek et al., 2010). Although flavanones are not the primary focus of research due to their relatively low abundance in hops, they are important precursors of other biologically active compounds, such as prenylnaringenin.

Fig. 9.23 Naringenin chemical structure

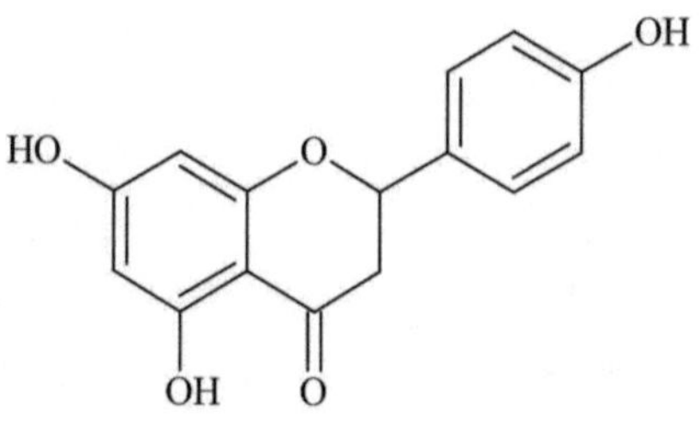

Prenylflavonoids

Undoubtedly, prenylflavonoids represent the most important and studied hop polyphenol class of the last 20 years due to their health-promoting properties. As opposed to the other polyphenols, they are contained in lupulin (Carbone & Gervasi, 2022). Their structure is characterised by at least one lipophilic prenyl substituent in the aromatic system of rings typical of flavonoids. It was established that this side chain significantly influences the biological activity of the corresponding flavonoids, likely due to the increased lipophilicity. In fact, it enhances the binding affinity towards biological membranes, conferring important biological activities to prenylflavonoids with subsequent benefits for human health (Tronina et al., 2020). Depending on the quantity, the position and the modification (e.g. cyclization and hydroxylation) of the prenyl groups, prenylflavonoids are classified as prenylflavanones, prenylchalcones etc. In hop strobiles, the major representatives are two prenylchalcones—xanthohumol (XN) and desmethylxanthohumol (DMX)—and three prenylflavanones—isoxanthohumol (IX), 6-prenylnaringenin (6-PN), and 8-prenylnaringenin (8-PN) (Fig. 9.24).

XN is the main prenylflavonoid in hops, accounting for 82–89% of the total prenylated flavonoids amount (Stevens, 1967) with a relative concentration ranging from 0.18% to 1.00% of hop cones dry matter (Patzak et al., 2015). XN is a source of IX, both in beer and in the human stomach; in fact, the free hydroxyl group can undergo cyclisation resulting in IX, thanks to the high temperature during the boiling phase or to the acidic conditions of the medium within the stomach during digestion (Girisa et al., 2021; Nikolic et al., 2005). Another compound that tends to isomerise is DMX, which, with two free ortho-hydroxyl groups in C10, isomerises into 6- and 8-PN (Carbone & Gervasi, 2022). The latter is known to be the most potent phytoestrogen known to date (Przybyś & Skomra, 2020). The term 'phytoestrogens' refers to a group of plant-derived substances that bear a structural and functional similarity to oestrogens. From a scientific perspective, the term denotes any extragonadal oestrogen of plant origin.

As already observed for the other class of polyphenols, the content of prenylated flavonoids in hop cones can vary profoundly in relation to genetic, environmental, and agronomic factors. Maliar et al. (2017) analysed the XN and DMX contents of eight hop cultivars, recording XN content ranging from 0.4 and 0.9% and DMX

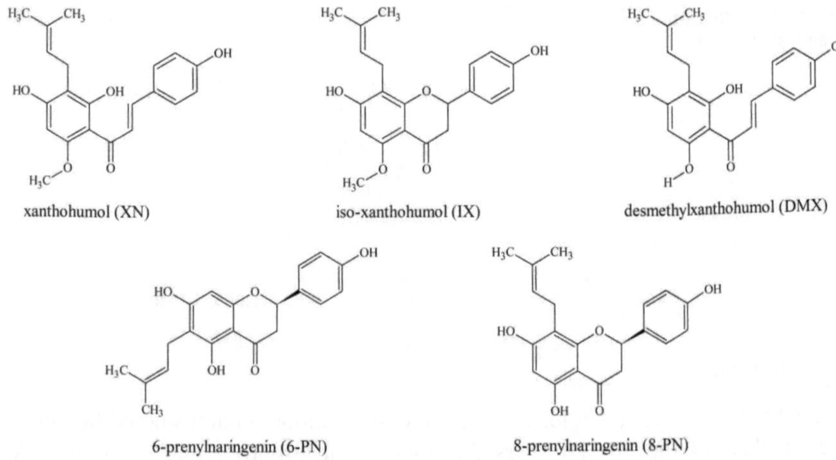

Fig. 9.24 Chemical structures of hop's major prenylflavonoids

contents from 0.1 and 0.3% of the weight of hop cone dry matter depending on the investigated genotype (Maliar et al., 2017).

9.2.2.2.3 Stilbenes

The structure of stilbenes (IUPAC name: 1,2-diarylethenes) consists in two phenyl moieties linked by a two-carbon methylene group (Fig. 9.25).

The meta-positions in ring A are usually substituted by two -OH groups, while various positions in ring B may be substituted by -OH and -OCH_3. As illustrated in Fig. 9.25, stilbenes have the capacity to exist in both the cis- and trans- configurations. Furthermore, they can occur in both their free and glycosylated forms (Singla et al., 2019). In 2005, Callemien and colleagues first identified three stilbenes in hops, namely *trans*-resveratrol, *trans*-piceid, and *cis*-piceid (Fig. 9.26), the latter two being stereoisomers of the glycosylated form of resveratrol (Callemien et al., 2005).

As reviewed by Almaguer et al. (2014) *cis*-resveratrol was not found in fresh hop cones and pellets of any of the tested varieties; nevertheless, later studies demonstrated that *cis*-resveratrol stems from *cis*-piceid during storage (Almaguer et al., 2014). Although stilbene concentration in hops is influenced by several factors (as are other polyphenols), in general hop varieties with lower α-acid content tend to have higher resveratrol amounts, with only some exceptions. Jerkovic et al. (2005) quantified stilbenes in nine different hop cultivars, finding a total amount of stilbenes ranging from 0.007 to 0.015 mg/g, with *trans*-resveratrol in the range 0.0002–0.001 mg/g, *cis*-piceid in the range 0.002–0.006 mg/g, and *trans*-piceid in the range 0.004–0.009 mg/g (Jerkovic et al., 2005).

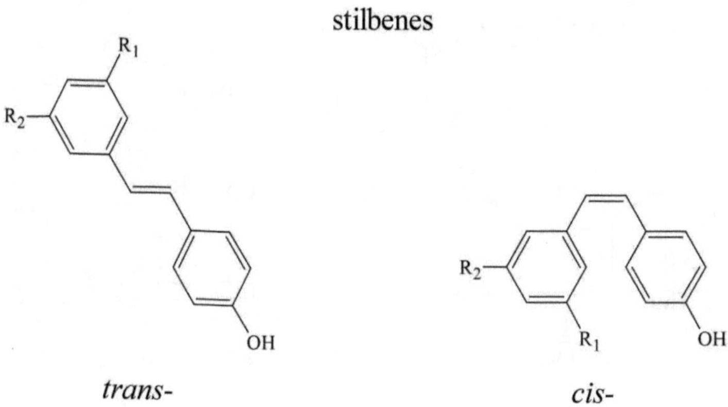

Fig. 9.25 Chemical structure of stilbenes

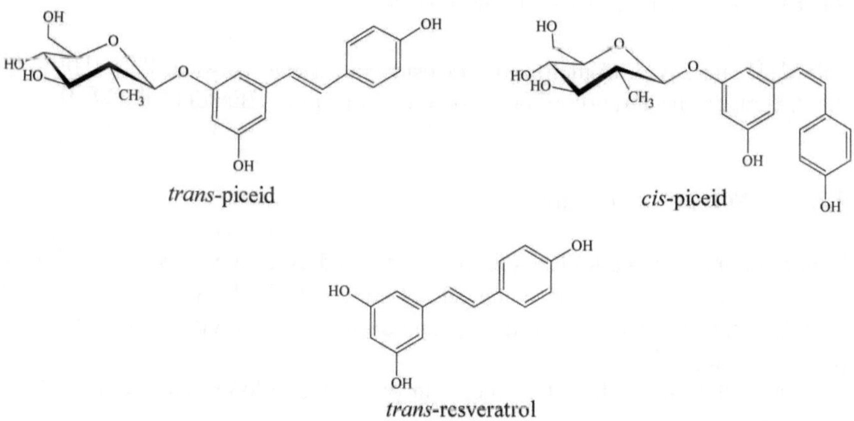

Fig. 9.26 Structure of hop stilbenes

9.2.2.2.4 Multifidol Glucosides

Multifidol glucoside is the glycosylated form of the 2-methylbutyryl-phloroglucinol (Fig. 9.27).

This compound has been first isolated from the latex of *Jatropha multifida L.* (Euphorbiaceae), so it has been called multifidol, referring to the aglycon. However, multifidols do not occur in their free form since they are bounded to a sugar moiety, namely glucopyranoside. These compounds are acylphloroglucinol derivatives as the α-acids and β-acids of hops. For this reason, depending on their structure, multifidol glucosides were named assigning the prefix 'co-', 'n-' or 'ad-', according to the nomenclature of hop bitter acids (Fig. 9.27) (Almaguer et al., 2014). Biendl et al. (2023) investigated the multifidol glucosides content in 34 varieties of hop. They identified co-multifidol glucoside in a range between of

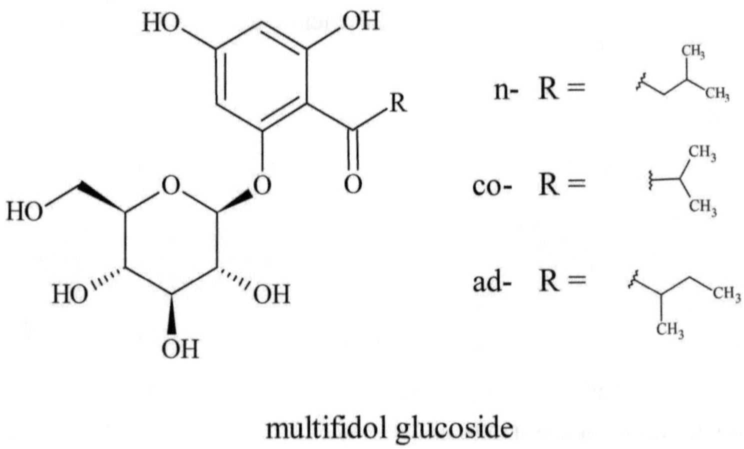

Fig. 9.27 Chemical structure of multifidol glucoside

0.06–1.55 mg/g and ad-multifidol glucoside in a range between 0.03–0.6 mg/g; their average concentrations seem to be varietal features (Biendl et al., 2023).

9.2.2.3 Non-phenolic Pigments

Natural pigments in biological systems are defined as those synthesised and accumulated in, or excreted from, living cells (Hendry, 1996). They are bioactive compounds with relevant antioxidant potential against hydroperoxide generation (Loranty et al., 2010).

Among the most relevant pigments in plants are chlorophyll and carotenoids (Fig. 9.28).

They occur in permanent complexes with proteins into chloroplasts (characteristic organelle of plants responsible for various essential functions, including photosynthesis, lipid metabolism, and starch and amino acid biosynthesis (Armbruster et al., 2011)).

Chlorophylls are classified as porphyrin pigments that present two forms in nature: chlorophyll *a*, the prevalent form, and chlorophyll *b* (Fig. 9.28). The former has a blue-green colour, responsible for the green of plants, while the latter tends towards yellow. They differ in that chlorophyll *a* presents a methyl moiety whereas chlorophyll *b* has an aldehyde group (Kobus-Cisowska et al., 2019).

Carotenoids are a class of hydrocarbons known as carotenes, comprising their oxygenated compounds (xanthophylls). Their carbon skeleton consists of eight isoprenoid units, derived from the lipophilic molecule of lycopene I (Fig. 9.28). They are responsible for the yellow to red hue in plants and are involved in photosynthesis (Spurgeon & Porter, 1980). Some compounds pertaining to this class are correlated to the vision process in animals and humans. They are vitamin A precursors and present antioxidant properties (Telichowska et al., 2020).

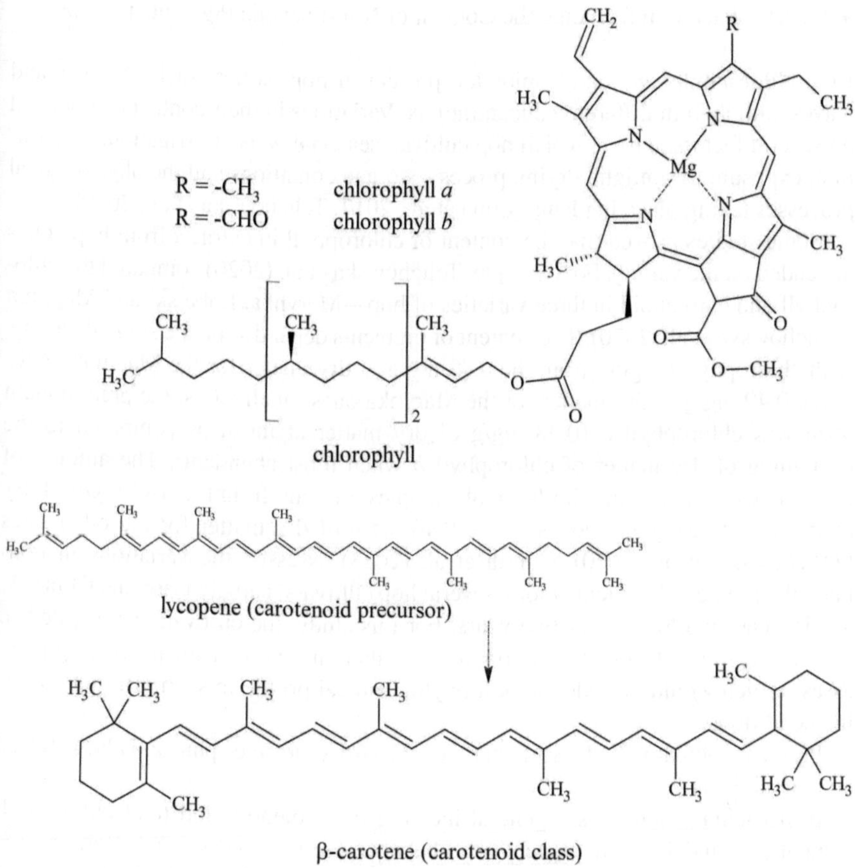

Fig. 9.28 Chemical structure of chlorophyll (top), lycopene (middle) and β-carotene (carotenoid class) (bottom)

These pigments have previously been associated with positive effects in humans: for example, chlorophyll appears to have protective effects against multiple disorders such as cataracts, neurodegenerative diseases, atherosclerosis, osteoporosis, and oxidative stress (Ferruzzi & Blakeslee, 2007). Chlorophyll can manifest both antioxidant and pro-oxidative effects (Maroneze et al., 2019). This is due to its ability to easily exchange magnesium ions with other divalent metals, such as iron, copper or zinc, and results in colour changes (Telichowska et al., 2020).

The radical-scavenging properties of carotenoids make them suitable for preventing various forms of cancer and diseases related to ageing (Manzoor et al., 2021). Regular carotenoid intake can in fact reduce the risk of oxidative stress-related diseases (Loranty et al., 2010).

9.2.2.3.1 Factors Influencing the Content of Non-Phenolic Pigments in Hops

Both chlorophylls and carotenoids are present in hop tissues, such as cones and leaves, although in different concentrations. Variations in their content can depend on several factors, among which hop cultivar, seasonality, heat, irrigation, fertilisation, exposure to sunlight, drying process, storage conditions and the physiological processes taking place in plants (Lima et al., 2017; Telichowska et al., 2020).

Some studies proved that the content of chlorophyll in extracts from hop cones depended on the variety. For example, Telichowska et al. (2020) compared the chlorophyll and carotenoid in three varieties of hop—Marynka, Lubelski and Magnum (Telichowska et al., 2020). The content of pigments depended on the studied variety, with chlorophyll ranging from the 0.42 mg/g of dry matter for the Magnum cones to the 0.49 mg/g of dry matter for the Marynka ones; in all cases, the predominant form was chlorophyll *a* (0.38 mg/g of dry matter at most, as compared to the 0.11 mg/g of dry matter of chlorophyll *b* when most abundant). The amount of carotenoids, instead, was similar in all samples, ranging from the 0.45 mg/g of dry matter for the Magnum cones to the 0.46 mg/g of dry matter for the other ones (Telichowska et al., 2020). Contin et al. (2023) assessed the variations in non-phenolic pigments' content among several hop cultivars, namely Cascade, Chinook, Triple Pearl and Saaz, over two years. For this study, the cultivars were tested in Brazil—in tropical climate—as opposed to their northern hemisphere origin, to assess which would provide the best phytochemical profile in such different growing conditions.

Pigment content varied significantly over a two-year time span, and also differed among cultivars.

In particular, there was a general increase in carotenoid and total chlorophyll content (1.7 and 1.3-fold, respectively) between the first and second year, and the ratio of chlorophyll *a* to chlorophyll *b* grew over the same time span; both instances are true for every variety. Among cultivars, differences were only observed in the second year, where the carotenoid/chlorophyll ratio was higher in the Chinook and Triple Pearl varieties as compared to Cascade (Contin et al., 2023).

Among different abiotic factors, heat has been shown to limit the photosynthetic activity of plants. In fact, excessive temperatures can damage components of the photosynthetic system. Eriksen et al. (2020) assessed how heat affected several photosynthetic and physiological traits of different hop cultivars, as changes to pigment ratios could indicate damage to the light-harvesting complexes (Eriksen et al., 2020). The selected hop varieties were studied in order to assess the most resilient to high temperatures, so that farmers and breeding programs could draw from this data and choose the most convenient cultivars. In fact, most plants have an optimum temperature for photosynthesis, even though they can adapt to a range of temperatures (Berry & Bjorkman, 1980). Damages due to heat to some parts of the photosynthetic system can produce reactive oxygen species in the plant, determining an increase in lipophilic antioxidants such as carotenoids (Das & Roychoudhury, 2014). This effect manifested, in fact, in some of the cultivars studied by Eriksen et al. (2020); they were particulatly resilient to heat stress, but it is not clear whether carotenoids are the main factor involved in the phenomenon (Eriksen et al., 2020).

Rodolfi et al. (2022) exploited chlorophyll content in hop to observe the impact of different fertilisation treatments on leaf functionality (Rodolfi et al., 2022); as a matter of fact, this tool is used as an indirect way to assess plant productivity, vegetative production, plant stress and nitrogen status (Cendrero-Mateo et al., 2016). In particular, the authors subjected Cascade hop plants to four fertilisation plans, all based on organic fertilisers. The highest chlorophyll value in hop leaves was found upon administration of the treatment with the highest nitrogen content. Furthermore, a significant increase in chlorophyll content was observed when a mixture of elements (Mn, Mg, Cu, Zn and Ca) used in chlorophyll biosynthesis was administered to the plants. The carotenoid content, too, was influenced by the type of treatment, and resulted the highest upon application of the same conditions that enhanced the chlorophyll amount, thus proving that both nitrogen and metallic micronutrients significantly affect the plants' pigment content (Rodolfi et al., 2022).

9.2.2.4 Essential Oils

Essential oils (EOs) are complex mixtures of volatile, lipophilic, and odoriferous substances that give aromatic plants their characteristic odour. The relative concentrations and ratios between specific compounds determine the bouquet of the aroma (e.g. citrussy, herbal, floral, etc.) (Rutnik et al., 2022b). However, an odour cannot usually be attributed to a single compound or amount of substance, but rather to a series of synergistic effects that influence the role played by different species, even at very low concentrations (Dietz et al., 2020; Rutnik et al., 2022b). In general, EOs are colourless or pale yellow, liquid at room temperature, and less dense than water (de Sousa et al., 2023). In hops, they include over a thousand compounds, contained in lupulin, and make up 0.5% to 3% (v/w) of the whole cone, constituting its main aroma component (Urban et al., 2013). From a chemical perspective they can be classified into two groups: terpenoids and non-terpenoids. In hops, terpenoids are the major representatives (Eri et al., 2000), accounting for up to 90% of the essential oil fraction (G. Wang et al., 2008). Most of them (about 50–80%) belong to the hydrocarbon class (Carbone & Gervasi, 2022; Kowalska et al., 2022; Rutnik et al., 2022b), the other main classes being oxygenated compounds and sulfur compounds (Fig. 9.29).

Terpenoids (or isoprenoids) are derivatives of terpenes with oxygenated functional groups such as ketone, hydroxy, aldehyde, ether or carboxylic acid moieties. Terpenes are derived from the condensation of two or more units of isoprene, a five-carbon (C5) building block with the branched carbon skeleton of isopentane and the addition of two non-adjacent double bonds (Fig. 9.30).

Terpenes' chemical structure may range from linear to mono- or poly-cyclic and according to the number of C5 units they contain, they are classified in monoterpenes (two C5 units), sesquiterpenes (three C5 units), diterpenes (four C5 units),

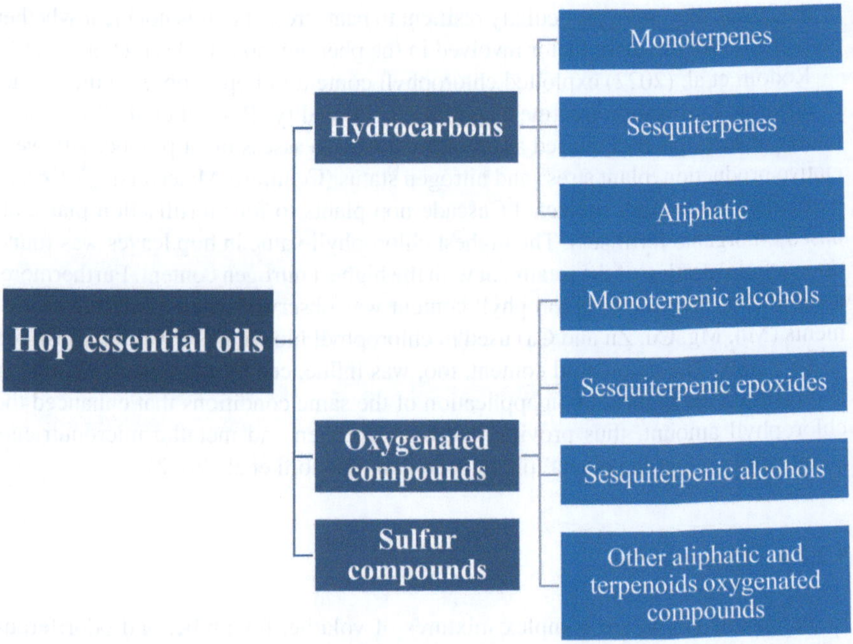

Fig. 9.29 Classification of hop's essential oil compounds

Fig. 9.30 Chemical structure of isoprene, the building block of terpenes

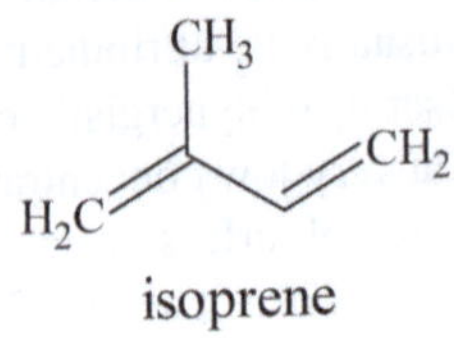

triterpenes (six C5 units), tetraterpenes (eight C5 units), and polyterpenoids ($C5_n$, where $n > 8$) (Taiz & Zeiger, 2002).

Terpenes are biosynthesised from primary metabolites through two different pathways—or a combination of them - each involving several enzymatic reactions: the mevalonic acid (MVA) pathway, active in all eukaryotes and localised in the cytosol, and the methylerythritol phosphate (MEP) pathway, which is restricted to a selection of organisms (higher plants) and localised in the plastid compartment. Both processes produce isopentenyl diphosphate (IPP) and its isomer dimethylallyl diphosphate (DMAPP), which are the universal precursors of essential oils (de Sousa et al., 2023; Rehman et al., 2016).

In the MVA pathway, three molecules of acetyl-CoA are subjected to a two-step enzymatic condensation that leads to the formation of an intermediate. This species is subsequently reduced by an enzyme during the main regulatory step of this route, in order to form the mevalonic acid. This acid is a key compound that is then phosphorylated, decarboxylated, and dehydrated to yield IPP, which can be reversibly isomerised to DMAPP (Bergman et al., 2019; Taiz & Zeiger, 2002).

The MEP pathway involves a sequence of enzymatic reactions, during which the product of the condensation of glyceraldehyde-3-phosphate and pyruvate is reduced, phosphorylated, cyclised and then reduced again by two different enzymes to produce IDP and DMAPP in a ratio of approximately 1:5 (Bergman et al., 2019).

Isopentenyl Diphosphate and its isomer combine to form larger molecules such as geranyl diphosphate (GPP), the 10-carbon precursor of nearly all the monoterpenes. Moreover, GPP can link to another molecule of IPP to give the 15-carbon compound farnesyl diphosphate (FPP), the precursor of nearly all the sesquiterpenes. Addition of yet another molecule of IPP gives the 20-carbon compound geranylgeranyl diphosphate (GGPP), the precursor of the diterpenes. Eventually, FPP and GGPP can dimerise to give the triterpenes (C30) and the tetraterpenes (C40), respectively (Fig. 9.31).

Essential oils tend to fully develop in the later stages of hop ripening, due to their slow biosynthesis (Carbone & Gervasi, 2022). Their composition is impacted by the same factors as other secondary metabolites, but in this case the degree of ripeness plays a more central role. In fact, not only does the amount of essential oils increase over time, but the proportions of their components also vary depending on the stage of ripeness. Namely, oxygenated compounds are the first to be produced, followed by sesquiterpenes and, lastly, monoterpenes, as the flowers ripen. Some chemicals, namely myrcene, can be used as an indicator of hop ripeness. Interestingly, the humulene:caryophyllene ratio remains constant throughout ripeness and is a varietal characteristic (Eyres & Dufour, 2009). Another factor hugely impacting essential oil composition is storage. As use of hops is not always directly subsequent to harvesting, but can also follow relatively long storage periods, this factor will be further discussed in a specific section after classification of the compounds present in hop's essential oils.

9.2.2.4.1 Hydrocarbons

Hydrocarbons present scarce solubility in aqueous media; most are lost by evaporation—especially during some phases of hop processing and when participating to beer brewing. This class of compounds can be further divided into monoterpene and sesquiterpene hydrocarbons. Note that aliphatic hydrocarbons (such as isoprene) are also present (Carbone & Gervasi, 2022).

Monoterpene Hydrocarbons

Monoterpenes are a class of biological molecules deriving from a combination of two molecules of isoprene (5-carbon units, as reported in Fig. 9.30) (Taiz & Zeiger, 2002).

The most representative compound belonging to this class is myrcene (Fig. 9.32), accountable for 30–60% of total oil content. It is responsible for the pungent, vegetal smell of fresh hop, and is the progenitor of other aromatic compounds, namely monoterpene alcohols (Carbone & Gervasi, 2022). Myrcene easily undergoes

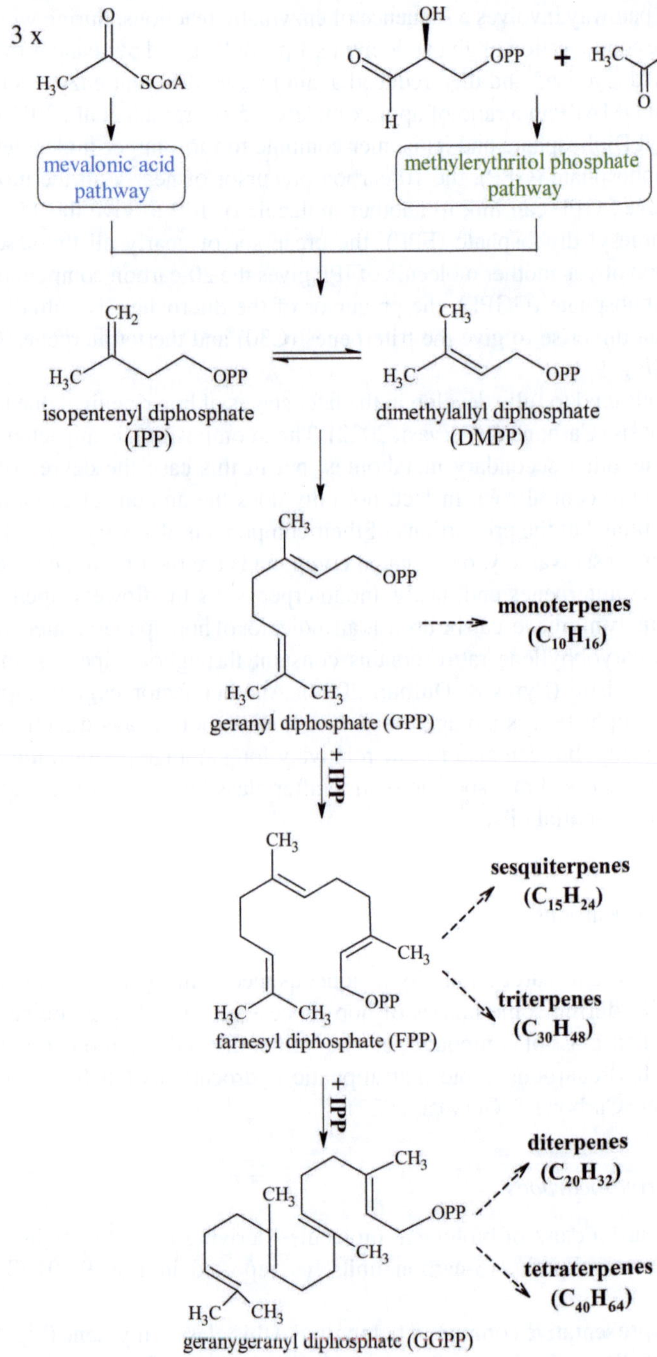

Fig. 9.31 Schematic pathway of the main essential oils

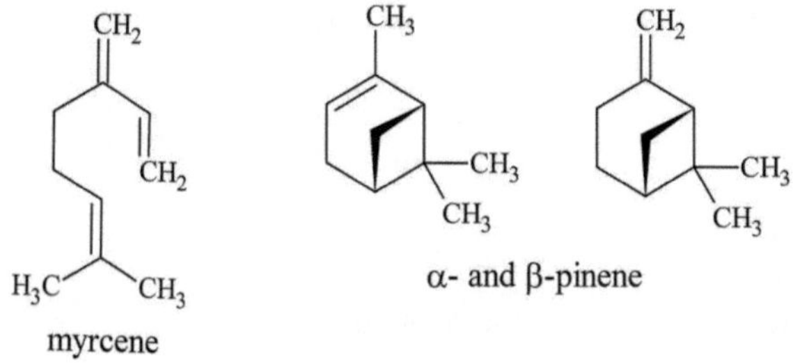

Fig. 9.32 Chemical structure of myrcene and α- and β- pinene

oxidation and polymerisation; this, and its pungent smell, make it undesired in significant percentages (Kowalska et al., 2022). Other typical hop monoterpenes include α- and β-pinene (Fig. 9.32), sabinene, δ-3-carene, camphene, *p*-cymene, (Z)- and (E)-β-ocimene, α- and γ-terpinene, δ-terpinolene and limonene (Eyres & Dufour, 2009).

Sesquiterpene Hydrocarbons

Sesquiterpenes are compounds deriving from the combination of three C5 units of isoprene (Fig. 9.30) (Taiz & Zeiger, 2002). They are less volatile and oxidation-prone as compared to monoterpenes. Humulene has a delicate, refined and elegant scent; it is most abundant sesquiterpene in hops, particularly in aromatic ones (Carbone & Gervasi, 2022) (Fig. 9.33). In fact, it has been described has imparting a soft-woody, reminding of fresh earth flavour to hop (Jirovetz, 2006).

Other acyclic, monocyclic, bicyclic and tricyclic sesquiterpenes include caryophyllene and farnesene (Fig. 9.33), (Z)- and (E)-α-bergomotene, cadinenes, muurolenes, selinenes, ylangenes, copaenes, germacrenes, selinadienes and many others (Carbone & Gervasi, 2022; Moir, 2000). α-humulene, β-caryophyllene and its derivative β-farnesene are foremost players in the determination of hops' scent (Kowalska et al., 2022). α-humulene and β-caryophyllene, in particular, deliver spicy and woody tones (Rutnik et al., 2022a), while β-farnesene has a woody, herbal, and citrus aroma (Rutnik et al., 2022b).

9.2.2.4.2 Oxygenated Compounds

These make up about 30% of hop's essential oils; they include: monoterpene alcohols, sesquiterpene epoxides and sesquiterpene alcohols (Eyres & Dufour, 2009), and aliphatic species such as aliphatic alcohols, ketones, aldehydes, esters and acids. In general, they are more soluble in aqueous media as compared to hydrocarbons, so they tend to have a greater impact on the aroma of beer (Carbone & Gervasi, 2022). In general, oxidation and hydrolysis products of sesquiterpenes have been linked to "noble" hop characters in beer; other oxygenated compounds are implicated as potent odorants in hop essential oils (Eyres & Dufour, 2009).

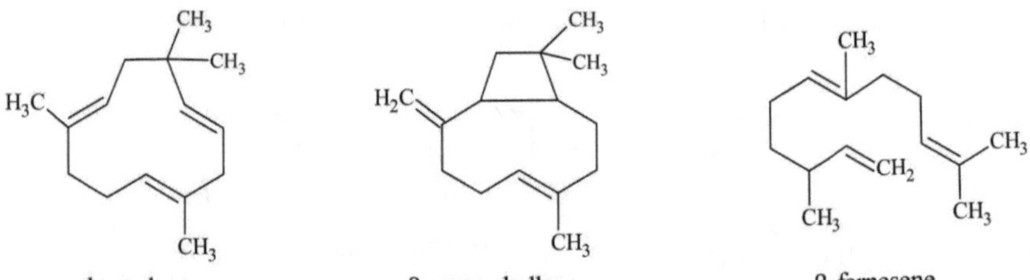

Fig. 9.33 Chemical structures of α-humulene, β-caryophyllene and β-farnesene

Monoterpene Alcohols

This class of compounds usually originates from myrcene. It includes geraniol and linalool, nerol, α-terpineol, borneol, fenchol, myrtenol and limonene-10-ol (Eyres & Dufour, 2009) (Fig. 9.34).

Linalool, in particular, is renowned as key aroma contributor, since it has good solubility and delivers floral, citrus, and fresh notes (Rutnik, Ocvirk, et al., 2022). Geraniol is also a significant aroma contributor, but it is more cultivar-specific than linalool, so its levels in hop oils vary (Takoi et al., 2010).

Sesquiterpene Epoxides and Alcohols

Sesquiterpenes undergo autoxidation and rearrangement, producing a number of reaction products that tend to increase over the storage period. The foremost exponents of the sesquiterpene epoxides are humulene epoxide II (I and III are also present, but in lesser amounts) and caryophyllene oxide (Rutnik et al., 2022b) (Fig. 9.35).

It is hypothesised that this latter is responsible for the warm, soft-woody note of hop flavour, while humulene epoxide II imparts a creamy, fatty, buttery scent (Jirovetz, 2006).

Diepoxides are also found in traces in fresh hops, and accumulate during storage (Lam et al., 1986). Humulene diepoxide A is the most representative compound for this category.

Both epoxides and diepoxides are further hydrolysed and rearranged to ketones and alcohols. These latter, in particular, include humulenol II, humulol, caryophyllenol and caryolan-1-ol, nerolidol, and lastly farnesol, cadinol and eudesmol isomers (Fig. 9.35) (Eyres & Dufour, 2009).

Aliphatic and Terpenoid Ketones, Aldehydes, Esters, Alcohols and Acids

Hop oil includes homologous series of aliphatic compounds (Fig. 9.36).

Among ketones, the most significant representatives are methyl ketones, with the most abundant compound being 2-undecanone (Fig. 9.36) (Eyres & Dufour, 2009).

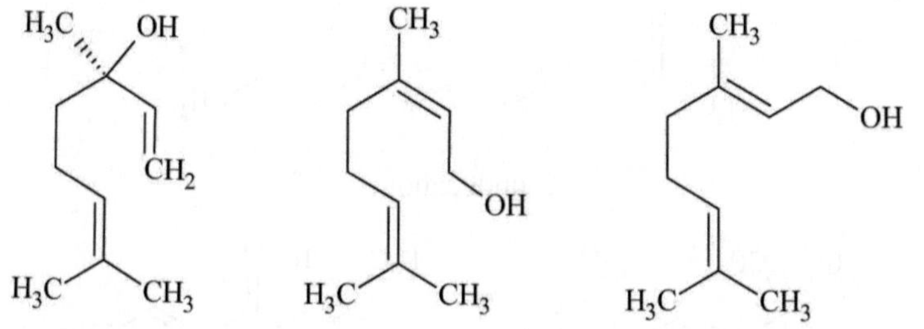

linalool nerol geraniol

Fig. 9.34 Chemical structures of significant representatives of monoterpene alcohols, namely linalool, nerol and geraniol

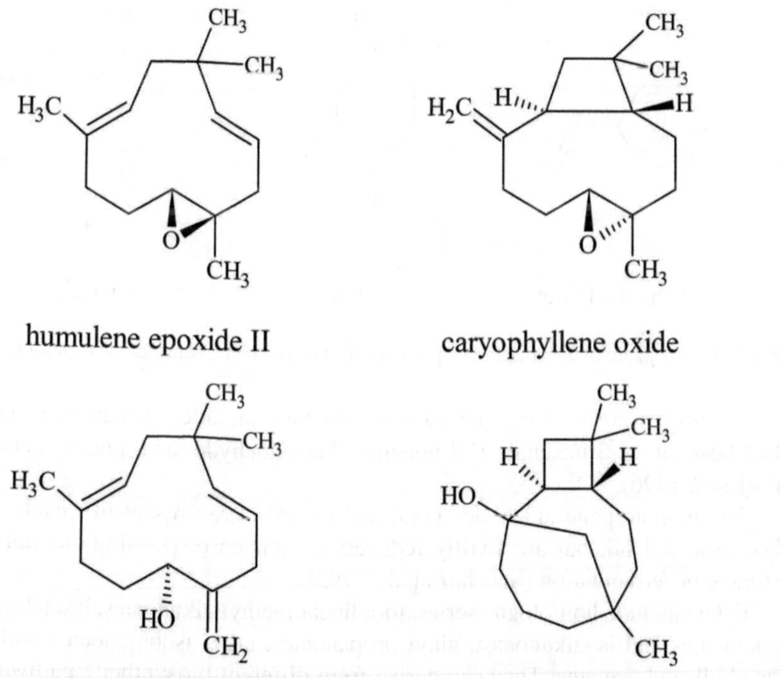

Fig. 9.35 Chemical structures of humulene epoxide II, caryophyllene oxide, humulenol II and caryophyllenol

Ketones can also derive from terpenes. β-ionone and β-damascenone (Fig. 9.36), deriving from β-carotene degradation, are found at trace levels in hop essential oil (Sell, 2003), while humulenone and humuladienone result from the oxidation reactions of humulene.

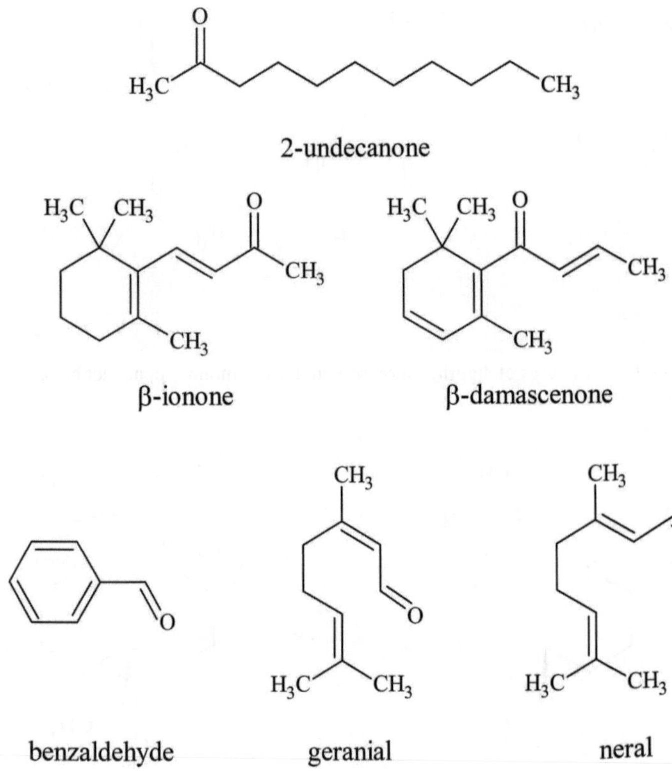

Fig. 9.36 Structure of 2-undecanone, β-ionone, β-damascenone, benzaldehyde, geranial, neral

Aldehydes are scarcely represented, but they include alkanals (e.g. nonanal), E-2-hexenal, Z-3-hexenal, E-2-nonenal, benzaldehyde and phenylacetaldehyde (Nijssen, 1996).

The monoterpene aldehydes neral and geranial are present in freshly distilled hop essential oil, but are swiftly reduced to their corresponding alcohols during storage or fermentation (Sanchez et al., 1992).

Esters include homologue series, too: linear methyl alkanoates, branched methyl alkanoates and isoalkanoates, alkyl propanoates, alkyl isobutanoates and unsaturated alkenyl acetates. They can derive from different biosynthetic pathways (e.g.: several methyl esters are by-products of fatty acid biosynthesis; branched chain compounds likely arise from pathways leading to the formation of the carbon skeletons of amino acids; more compounds derive from transesterification) (Briggs, 2004).

Other significant aliphatic esters not belonging to homologous series are: methyl Z-4-decenoate, 2-methylpropyl 2-methyl-propanoate, 3-methylbutyl 3-methylbutanoate and 2-methylbutyl 3-methyl-butanoate.

Esters of the terpene alcohols are derived from nerol, geraniol and linalool.

Oxygen heterocyclic compounds in hop essential oil include (Z)- and (E)-linalool oxide, rose oxide and cyclic ethers (Eyres & Dufour, 2009).

Alcohols are both straight- and branched-chain. Some examples are 1- and 2-alkanols, and 2-methyl-3-buten-2-ol. The amount of biosynthetic alcohols tends to decrease over time, while the oxidation-derived one rises (Eyres & Dufour, 2009).

Acids in hop essential oil are connected to aged hops as α- and β-acid products. Humulone and lupulone undergo cleavage of the acyl side chains generating 3-methylbutanoic acid; the same occurs to ad-humulone and ad-lupulone, generating 2-methylbutanoic acid, and to co-humulone and co-lupulone, generating 2-methylpropanoic acid. They are all accountable for the cheesy aroma of aged hops (Benitez et al., 1997).

Photo-oxidation of the ring structure of bitter acids also produces 4-methyl-3-pentenoic acid. Other acids include decanoic acid and Z-4-decenoic acid (Eyres & Dufour, 2009).

9.2.2.4.3 Sulfur Compounds

Sulfur compounds considerably concur to hop odour, in spite of their scarce concentrations (ng/g), due to their very low perception thresholds (Lafontaine et al., 2019).

Several volatile sulfur compounds in hops have been identified as belonging to the group of thiols, polysulfides, S-methyl thioesters, sesquiterpene episulfides (homologues of sesquiterpene oxides, with the oxygen atom replaced by sulfur), and polyfunctional thiols (Kammhuber et al., 2017) (Fig. 9.37). Polyfunctional mercaptan precursors are usually non-aromatic cysteine and glutathione complexes, present in hops in much higher concentrations than their free forms, at least in the

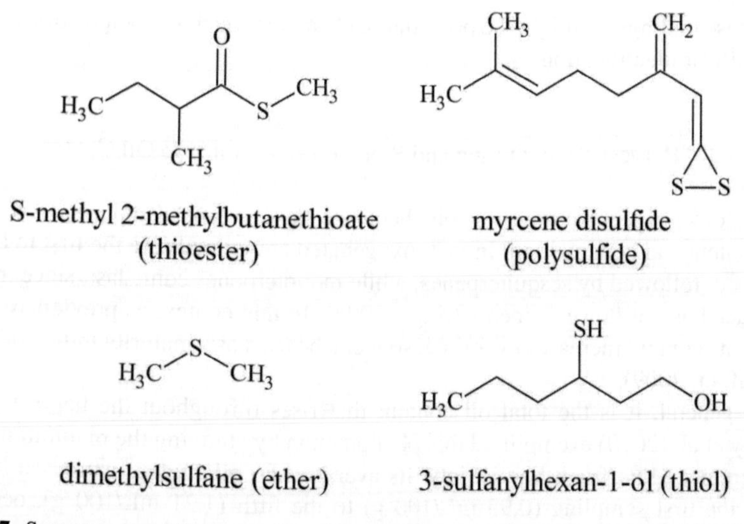

Fig. 9.37 Some representative hop sulfur compounds

early stages of harvest (Rettberg et al., 2018). Indeed, free thiols increase during maturation at the expense of their precursor (Lafontaine et al., 2019).

The most abundant representatives of sulfur-compounds are S-methylthiomethyl thioesters (Fig. 9.37) (Carbone & Gervasi, 2022). They are characterised by an unpleasant aroma of cheese and cooked vegetables, cabbage and onion. Nevertheless, other sulfurous substances from the group of polyfunctional thiols are characterised by pleasant aromas of tropical fruit or blackcurrant. Some examples of such thiols are 3-sulfanyl-4-methylpentan-1-ol, 3-sulfanyl-4-methylpentylacetate, and 4-sulfanyl-4-methylpentan-2-one (Krofta et al., 2021).

Up until the 1980s, the composition of the sulfur substances of hop oils was significantly influenced by agricultural practices applied during cultivation and the subsequent processing. For instance, the utilisation of sulfur-based pesticides has been demonstrated to enhance the levels of certain sulfur derivatives of sesquiterpenes within hops (Sharpe & Peppard, 1977; Suggett et al., 1979). In contrast, contemporary hop oil composition is predominantly influenced by varietal characteristics, as evidenced by Kammhuber et al. (2017) and corroborated by Krofta et al. (2021); the latter, in particular, exploited sulfur-containing compounds to identify 14 Czech hop cultivars growing in conditions with a very limited use of sulfur (Kammhuber et al., 2017; Krofta et al., 2021).

The authors were able to perform GC analysis of the essential oils contained in the various hop cultivars; some compounds were identified, such as S-methyl thiohexanoate—the dominant sulfur component of hop oils, S-methylthio-isovalerate—also identified in all the hops of the study, and dimethyltrisulfan—only pinpointed in small quantities in 'Saaz' and genetically related cultivars. Nonetheless, a great number of other sulfur compounds could not be identified upon comparison with the available chromatographic standards, so the study provides something closer to a varietal fingerprint rather than a detailed analysis of individual substances (Krofta et al., 2021).

These findings highlight the potential of characteristic thiol composition as a tool for cultivar identification.

9.2.2.4.4 Harvest Time, Drying and Storage Effects on Hop Oil Content

As for other aspects of hop cone biochemistry, harvest time also plays a role in hop oil content and composition. In fact, oxygenated compounds are the first to be synthesised, followed by sesquiterpenes, while monoterpenes come last, since they are produced as the flowers ripen (Briggs, 2004). In this contex, as previously stated, myrcene content increases over time, so it can be used as a maturity indicator (Eyres & Dufour, 2009).

In general, it is the total oil content that rises throughout the harvest period. Bailey et al. (2010) exemplified this phenomenon by studying the maturation of the Hallertauer Mittelfrueh hop variety. Its average hop oil content increased by 30% from the first sampling (0.93 mL/100 g) to the fifth (1.21 mL/100 g), occurring about a month later (Bailey et al., 2009).

This increase over the harvest period is not only significant for the brewing itself, but it also represents an economic factor for breweries. In fact, the late-harvested hops rated better in aroma quality, and so did the beers brewed with them (Bailey et al., 2009).

However, after hop harvesting and processing, the levels of essential oil start to drop, and its composition to vary.

Several research groups have investigated the changes in the composition of hop essential oil over extended periods of time. In short, their findings highlighted: an increase in the oxidation products of the initial components; an alteration of the aromatic bouquets; a dependence of such variations on the cultivar; essential oil losses ranging from 4.4 to 90% depending on the storage time and conditions, with the most prominent alterations happening straight after harvesting (Canbaş et al., 2001; Mikyška & Krofta, 2012; Srečec et al., 2009; Tembo, 2020; Veríssimo et al., 2022).

All the studies agreed in that proper storage conditions can slow down the losses and changes, but not prevent them completely.

Rutnik and colleagues (2022b) recently investigated the effect of room temperature *vs* cold storage on both hop cones and pellets of six different varieties of hops over two years, monitoring eight different compounds (Rutnik et al., 2022b). Some hop cultivars lost a higher amount of oil (i.e. Celeia, Bobek and Styrian Wolf), independently of other factors, implying that some varieties are inherently more stable than others, especially in the first phase after harvesting (Rutnik et al., 2022b).

The comparison between cones and pellets yielded very interesting observations. Pelleting is a common practice to increase stability and reduce storage space of hops. The former is known to be due to a thin film forming around pellets, preventing oxidation and evaporation. However, this is only true if no access to oxygen is allowed: in fact, pelletisation entails crushing of the cones with subsequent rupture of lupulin glands, putting the oils and oxygen into direct contact. Consequently, hop cones are more stable in aerobic conditions. Furthermore, it appears that more complex changes occur than a simple change of shape when pelletising, probably due to the process temperature impacting the enzymes responsible for the biotransformation of oils (Rutnik et al., 2022b).

Another finding is that myrcene is a good indicator for length and quality of the storage. This compound steadily declines over time, implying that it is not formed after harvesting, but rather converted into derivatives and oxidation products (Rutnik et al., 2022b).

In general, essential oil compounds are volatile, so low temperatures and proper storage conditions would be expected to be sufficient for optimal preservation of hop; nonetheless, another very significant factor for correct preservation is oxygen access. In all cases, in fact, oil losses are much higher if oxygen is accessible (Rutnik et al., 2022b).

Raut and colleagues (2021) studied the effects of storage on essential oil both prior to and after drying (Raut et al., 2021). In general, drying determines losses of essential oil of up to 40% and alters the colour of the samples. As mentioned, myrcene is one of the most abundant compounds in hop essential oil and undergoes a series of changes that transform it into cyclic and oxidised products. In addition, its starting content has also been found to depend on the harvest period, since it increases over time. Its amount was monitored in the cited study, as was that of some of its derivatives.

The authors compared the content of some compounds of hop's essential oil in samples treated differently to assess the oil's composition upon different kinds of storage. Namely, the hop samples analysed were: fresh; frozen when fresh; stored for 5 hours at room temperature; frozen after storage of 5 hours at room temperature; stored for 24 hours at room temperature. In all cases, the samples were analysed both prior to and after drying for 210 minutes.

Linalool, for example, has been found to increase both in frozen and room temperature stored samples, both prior to and after drying. In particular, its increase during the latter phase is also due to conversion of its glycosides, a form in which linalool can be found in fresh hop. Geraniol was also monitored in this study, but it presented slightly different trends as compared to linalool. Namely, it was typically higher if hops were not dried; plus, it decreased over time in all cases but one. Interestingly, in fact, the sample stored at room temperature for 24 hours after drying presented a spike increase in geraniol content. Geraniol forming as a myrcene oxidation derivative partly explains its raise in stored hops after drying, probably due to an enhancement in enzymatic conversion. Unlike for linalool, though, its glycosidic bonds are heat stable, so cleavage of these precursors is probably not responsible for geraniol's increase. Other processes concurring to this phenomenon are still to be investigated. Caryophyllene and humulene probably undergo similar degradation and conversion processes during pre-drying—thus increasing their total amount, whereas the samples dried straight after harvesting have the conversion process inhibited—thus presenting lower amounts of the components.

The authors established that pre-drying storage periods hugely impact the oil content, as did freezing of the samples. In general, they recommend to maintain the storage period to a minimum (Raut et al., 2021).

9.2.2.4.5 Terpenoids' Content in Beer and Their Interaction with Yeast

It is well-known that beer possesses a multifaceted bouquet of flavours and odours stemming from its raw materials and dependent on complex physical, chemical, and biological changes that happen during the whole production process. Among all these variables, hop plays a pivotal role due to its essential oil content. As seen, the main representatives of hop's aromatic fraction are terpenoids, that have a wide spectrum of odour profiles encompassing resinous, floral, fruity, minty, and more. Hop terpenoids are classified in three groups: hydrocarbons, oxygenated compounds and sulfur compounds. Hydrocarbon terpenes are usually desired in beer as they impart interesting shades of flavour. For instance, α-humulene releases a balsamic-like scent, and β-caryophyllene has been shown to produce a clove-like odour (Almaguer et al., 2014; Eyres & Dufour, 2009). However, these compounds are characterised by high volatility and low solubility, which leads to their easy evaporation during the boiling of the wort. Consequently, they are present in low concentrations in the finished beer (Almaguer et al., 2014). On the other hand, terpene alcohols, which belong to the oxygenated compounds, exhibit enhanced hydrophilic properties, facilitating the retention within wort and, thus, within the final product. Linalool, in particular, proposed by Steinhaus and Schieberle (2000) as one of the most potent odorants in hop cones, has been detected in various beers (Steinhaus & Schieberle, 2000). It is known for its ability to impart floral and citrus notes (Almaguer et al., 2014), thus ranking among the most flavour-active compounds in beer. However, geraniol—along with geraniol acetate—also contributes to this kind of odorous nuances, while 1-hexanal produces green and grassy flavour profiles and 2-methylbutyric acid contributes to the cheesy aroma characteristics of beer (Lewis & Young, 2012). Furthermore, it is important to acknowledge that the interaction between hop oxygenated terpenoids and yeast is another foremost player in shaping the final bouquet of beer. For instance, as demonstrated by King and Dickinson (2000, 2003), yeast possesses the ability to isomerise, reduce or esterify monoterpene alcohols, yielding a variety of other terpenoid products. The authors found that geraniol can be transformed into β-citronellol, and concurrently nerol can be converted into linalool and α-terpineol. Linalool, in turn, can undergo partial cyclisation to α-terpineol. In the authors' fermentation model, the decrease of geraniol corresponds to an increase of β-citronellol, suggesting that the contents of these compounds in finished beer might depend on the initial amount of geraniol in wort (King & Dickinson, 2000, 2003).

Takoi et al. (2010) observed that the mild increase of β-citronellol during the total fermentation period did not correspond to a drop in geraniol content. This finding suggests the presence of an alternative, yet unidentified, mechanism responsible for β-citronellol production, distinct from the traditional conversion of geraniol to β-citronellol (Takoi et al., 2010).

As mentioned before, hop oils contain a small amount of sulphur compounds; however, owing to their very low perception thresholds, they have the potential to become a problem in beer (Almaguer et al., 2014). Indeed, in most cases, sulfur compounds result in undesirable flavour notes in the final product. Nonetheless, in some instances they can impart positive nuances on the sensory profile (Lafontaine et al., 2019). For example, 4-sulfanyl-4-methylpentan-2-one (4MMP) has been identified as an originator of extremely strong fruity aroma of blackcurrants in American and Australian cultivars 'Cascade', 'Simcoe', 'Summit', 'Apollo' and 'Topaz', presenting a sensory threshold of 1.5 ng/L in beer (Kishimoto et al., 2008).

As seen, polyfunctional mercaptans are found in hops both free and as precursors bound to cysteine and glutathione. Yeast has the key enzyme (β-lyase) to cleave the complexes, releasing free thiols into beer (Kankolongo Cibaka et al., 2016). This ability has already been extensively explored in wine industry studies, finding that different strains of wine yeast determine the release of different amounts of volatile mercaptan (Fracassetti & Vigentini, 2018). In beer, *Saccharomyces cerevisiae* lysis of mercaptan precursors occurs during the fermentation of wort by a similar mechanism (Y. Wang et al., 2024). Nonetheless, both the process of release and the transformation of mercaptan in wort carried out by yeast must still be fully elucidated.

Regardless of the subclass of terpenoids, it is crucial to remember that sensory characteristics are not solely dependent on the amount of individual species, but rather on synergistic behaviours of volatile compounds. Furthermore, it is important to remember that, during the fermentation stage, yeast enzymes catalyse the transformation of hop essential oils into different compounds, thereby modulating flavour characteristics and intensity. It has been hypothesised that the extent and nature of these transformations are strain-specific, resulting in beers exhibiting a wide range of concentrations of hop-related volatile organic compounds. Nevertheless, the mechanism of the biotransformations, the differences between the activity of diverse yeast strains and the impact on the sensory perception of beer flavour remains a poorly understood but very interesting topic (Kumar et al., 2023).

9.2.3 Hop Enzymes

Hop plant synthetises different bioactive compounds, like bitter acids, terpenoids and prenylflavonoids, through complex biosynthetic pathways that are mediated by specific enzymes.

In this regard, Hong et al. (2021) provided a comprehensive review of the key enzymes involved in the biosynthesis of important phytochemical compounds in hops (Hong et al., 2021). In their analysis, the authors referred to the primary enzymes involved in the three main pathways of the phytochemical composition of hops, namely those related to the biosynthesis of bitter acids, terpenoids and phenylpropanoids.

The production of bitter acids involves some important enzymes, like valerophenone synthase (VPS) and prenyltransferases (PTs). VPS, unique to hops, is involved in the formation of phlorisovalerophenone, which is a precursor in the bitter acid pathway, while PTs facilitate the prenylation steps that are essential for bitter acid biosynthesis. VPS, whose gene is specifically expressed in the lupulin gland, is a type III polyketide synthase, a homodimeric enzyme with subunits of 45 kDa. It has a pI of 6.1, showing similarity to plant chalcone synthases (CHS) (Hong et al., 2021; Okada et al., 2001). It is interesting to note that VPS shows both VPS and CHS activities, although the latter is much weaker than its VPS activity (Okada et al., 2001).

Terpene biosynthesis in plants is a complex process involving two distinct pathways: the plastidial methylerythritol phosphate (MEP) pathway and the cytosolic mevalonate (MVA) pathway. In hops, the MEP pathway appears to be predominant in terpenoid synthesis occurring in glandular trichomes. It has been determined that several enzymes are critical in the MEP pathway to produce terpene compounds, including:

- 1-deoxyxylulose-5-phosphate synthase (DXS): this is considered a major rate-determining enzyme in the initial stage of the MEP pathway.

- 4-hydroxy-3-methylbut-2-enyl diphosphate synthase (HDS) and 4-hydroxy-3-methylbut-2-enyl diphosphate reductase (HDR): these enzymes are involved in the final stages of the MEP pathway, producing IPP and DMAPP.
- Geranyl diphosphate synthase (GPPS): this enzyme catalyses the condensation of IPP and DMAPP to form GPP, a precursor for monoterpenes. In hops, the small subunit of GPPS acts as a regulator of the activity of the functional heterodimer
- Monoterpene synthases (MTS): these enzymes, including HlMTS1, HlMTS2, and HlMTS3, generate volatile terpenoids from GPP.

In the terpenoid biosynthetic pathway, MTS, as well as sesquiterpene synthases (STS), are very important. MTS enzymes, like monoterpene synthase 2, convert geranyl pyrophosphate to monoterpenes like myrcene, which is a pivotal factor in the genesis of the distinctive aroma of hops. In a similar manner, STS enzymes, such as sesquiterpene synthase 1 (HlSTS1), function by converting farnesyl pyrophosphate to sesquiterpenes, including caryophyllene and humulene. These enzymes play a pivotal role in the flavour profile of hops and their potential medical applications (Wang et al., 2008).

As previously reported, XN is a prenylflavonoid that is found in the lupulin glands of hop cones. It is present in small amounts along with desmethylxanthohumol. These compounds serve as precursors for isomeric flavanones. Key enzymes are found in their biosynthetic pathways, including cinnamate 4-hydroxylase, which catalyses the first oxidative step in the phenylpropanoid pathway, converting *trans*-cinnamate to *p*-coumarate. The next key enzyme is chalcone synthase (CHS_H1), which is hop-specific and essential for chalcone formation (Novak et al., 2006). This enzyme has a molecular weight of 42.5 kD and shows specific activity for 4-coumaroyl-CoA. It has been determined that CHS_H1 is expressed in glandular trichomes during the process of hop cone ripening. Finally, we have chalcone isomerase-like proteins and O-methyltransferase, which are present in different types in hops:

- OMT1: specific for methylating desmethylxanthohumol to xanthohumol.
- OMT2: has broader substrate specificity but does not produce xanthohumol.
- OMT3: primarily involved in flavonoid metabolism.

The essential oil and resins present in hops have been shown to have a direct impact on the aroma profiles of beer, but hop proteins can play a significant role, too. Some of these proteins, in fact, are enzymes, which are typically deactivated during boiling. However, the application of the dry-hopping technique has been demonstrated to facilitate the extraction of these enzymes in their active forms, thereby potentially impacting secondary fermentation and resulting in a phenomenon referred to as hop creep (Klimczak et al., 2023).

Hop creep is defined as the unintended secondary fermentation that occurs in beer following dry hopping, with the result that carbonation, alcohol content and residual sweetness are altered. This phenomenon is due to enzymes, mainly amylases, which hydrolyse residual starches and dextrins into fermentable sugars. These sugars can then be metabolised by yeast, producing increased CO_2 and ethanol levels post-packaging. It is of particular concern in non-pasteurised and unfiltered beers, as evidenced by the craft beer sector, since it increases the risk of over carbonation. This, in turn, has a detrimental effect on the sensory characteristics of the beer and creates a risk of gushing (Klimczak et al., 2023).

The primary enzymes responsible for hop creep are α-amylase and β-amylase, which are naturally present in hops and become active under brewing conditions. In their study, Kirkpatrick and Shellhammer (2018) identified various starch-degrading enzymes in Cascade hop pellets, including α-amylase, β-amylase, amyloglucosidase and limit dextrinase (dextrin 6-α-glucanohydrolase). The latter enzymes exhibited the lowest activities (Kirkpatrick & Shellhammer, 2018).

Currently, studies on this topic are still scarce and leave plenty of room for future research.

9.3 Phytochemical Composition of Other Hop Tissues

In addition to the cones, which are the most valuable part of the plant, several researchers have recently proposed a circular approach to the cultivation. This approach involves enhancing the biochemical makeup of other plant tissues, such as stems, seeds and sprouts.

9.3.1 Hop Stems

It is estimated that approximately 75% of the biomass produced by hop cultivation consists of stems and leaves, with stem production calculated at around 600–900 g per plant (on a dry basis) (Kanai et al., 2021) (Fig. 9.38), which represents a significant concern for their disposal.

Even in earlier times, when the need for the recovery of agro-industrial biomass was not yet strongly felt, there was concern about the virtuous recovery of these wastes. In their interesting study, Limousine et al. (2023) reported on the historical uses of hop stems (HS) in Scandinavia between the fifteenth and nineteenth centuries. These uses primarily involved the use of hop stems in textile production and the production of cushions on Danish farms (Limousin et al., 2023).

Hop stems have a complex chemical composition with potential applications in several areas (Rubira et al., 2024) (Fig. 9.39).

The main constituents of hop stems include lignocellulosic materials such as cellulose, hemicellulose and lignin, which form the structural framework of the plant (Szymańska-Chargot et al., 2022). Cellulose, a polysaccharide composed of β-(1 → 4)-linked glucose units, is the most abundant component and contributes to the stem's rigidity and tensile strength. Hemicellulose, a heterogeneous group of polysaccharides including xylans, mannans and glucans, provides flexibility and

Fig. 9.38 Hop stems and leaves (photo: Federico Vincenti)

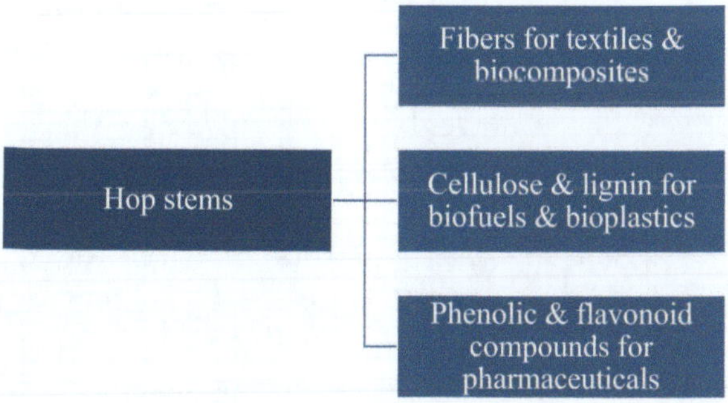

Fig. 9.39 Schematic representation of the fractionation of the different components of hop stems in a circular economy perspective

acts as a matrix surrounding the cellulose fibres. Lignin, a complex polymer, strengthens the cell wall and provides resistance to microbial degradation (Figs. 9.40, 9.41, and 9.42).

Rubira et al. (2024) analysed the chemical composition of hop pruning (HP), underlining the presence of polysaccharides (close to 50% of the weight), Klason lignin (17%) and uronic acids (8%), while ash and protein were found in lower quantities (Rubira et al., 2024). According to the authors, the composition of the product was comparable to that of other agricultural by-products, such as rape and

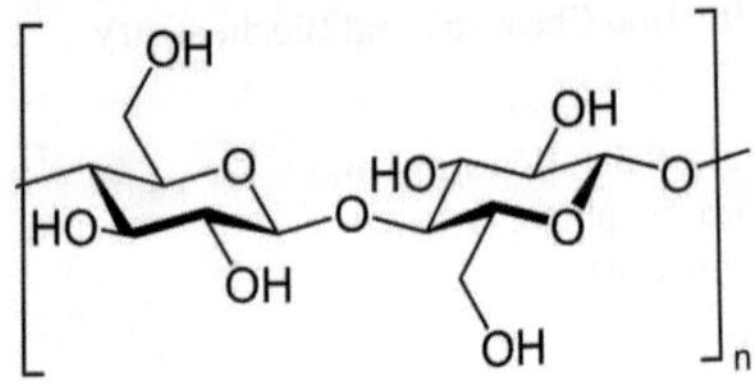

Fig. 9.40 Structure of cellobiose (cellulose structural motif) (https://commons.wikimedia.org/wiki/User:NEUROtiker)

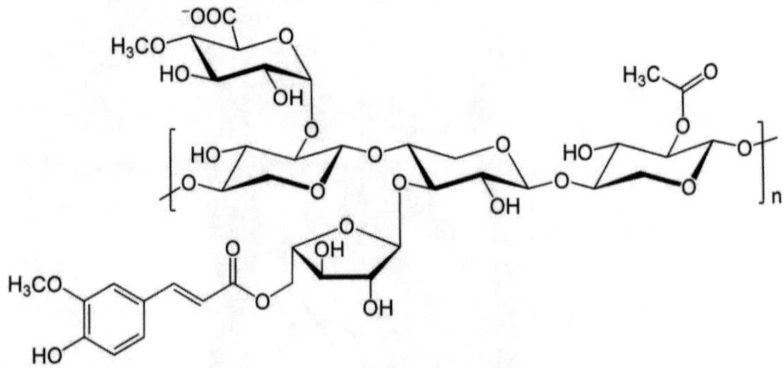

Fig. 9.41 Structure of xylan (https://it.m.wikipedia.org/wiki/File:Xylan.svg)

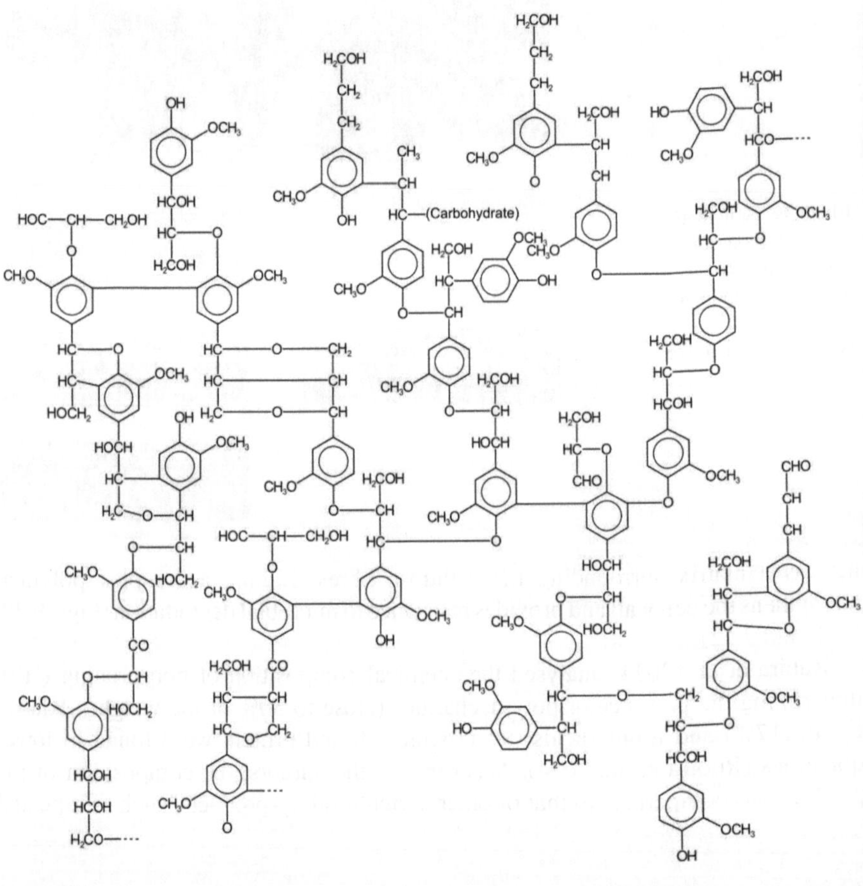

Fig. 9.42 Structure of lignin (https://commons.wikimedia.org/wiki/File:Lignin_structure.svg)

wheat straw. Furthermore, a substantial quantity of polyphenols was reported, amounting to approximately 3.3 kilograms per 100 kilograms of HP, following both conventional and innovative extraction of HP, although significant variations in the overall phytochemical profiles of the extracts were observed. The most representative polyphenols, regardless of the process applied, were catechins and phenolic acids.

In their study, Limousin et al. (2023) provided information on the chemical composition of hop stems, with a particular focus on the fibres. Their findings revealed that hop fibres have a unique composition compared to other bast fibres, such as hemp. The authors reported that hop fibres had a relatively high lignin content of approximately 20%, which is significantly higher than the 3–10% lignin content typically found in hemp. However, hop fibres showed a significantly lower hemicelluloses content compared to hemp. The combined content of xylans and mannanes was found to be approximately 5%, whereas in hemp it is typically 15–20%. Furthermore, the raw hop fibres exhibited relatively high glucan sugar levels (approximately 46%), indicating a substantial cellulose content, whereas the presence of galacturonan, rhamnann, and arabinan sugars (approximately 4.5% and 2.4%, respectively) pointed out a significant pectin content. It is noteworthy that the high lignin content of this by-product, which, according to the authors, is comparable to that of wood, was attributable to the structure of the hop stem, which is more similar to that of woody plants than to that of herbaceous plants (Limousin et al., 2023).

In a separate study, Szymańska-Chargot et al. (2022) applied Fourier transform infrared spectroscopy (FTIR) to evaluate the molecular fingerprinting of hop stems from the Magnum variety (Szymańska-Chargot et al., 2022), thereby highlighting the presence of pectic and hemicellulosic polysaccharides, in addition to lignin. The raw material was then characterised for its proximal composition, evidencing an overall dry matter content of approximately 27%, a cellulose content of 29%, a hemicellulose content of 11%, and a lignin content of approximately 9%.

Kanai et al. (2021) reported a cellulose content of 44%, a hemicellulose content of 13%, and a lignin content of 26% for HS. However, in this case, the authors did not provide information about the hop variety used (Kanai et al., 2021). The authors also noted that the composition of the studied HS was similar to that of softwood stems.

It is important to note that the differences observed in the different studies may be due to factors such as variety, agricultural practices, post-harvest processing and the effects of soil and climate affecting the chemical composition of hop stems.

9.3.2 Hop Leaves

It is estimated that hop leaves account for approximately 25% of the plant, calculated on a dry basis, depending on the variety under consideration (Ceh et al., 2007; Sarraf et al., 2013) (Fig. 9.43), representing an unexploited reservoir of bioactive

Fig. 9.43 Hop leaves (photo: Katya Carbone)

valuable molecules. Recent studies, in fact, have highlighted the rich chemical composition of this hop by-product.

In this regard, Mitton et al. (2025) reported the chemical composition (both primary and secondary metabolites) of Cascade hop leaves used as a natural antioxidant in rough scad (*Trachurus lathami*) fish patties (Mitton et al., 2025). The proximate composition of hop leaves is shown in Fig. 9.44.

Furthermore, the authors analysed the lipid profile (percentage of total fatty acids) of hop leaves, reporting the following composition:

- Saturated fatty acids: 21.5%
- Monounsaturated fatty acids: 9.3%
- Polyunsaturated fatty acids: 52.7% (omega-3 fatty acids: 2.2%, omega-6 fatty acids: 50.5%)

The most abundant fatty acids detected were palmitic acid (16:0; 18.6%) and γ-linolenic acid (18:3 n6; 40.5%), while total essential amino acids (TEAA) accounted for 4.5%, with isoleucine and leucine the most abundant amino acids found in hop leaves (Fig. 9.45).

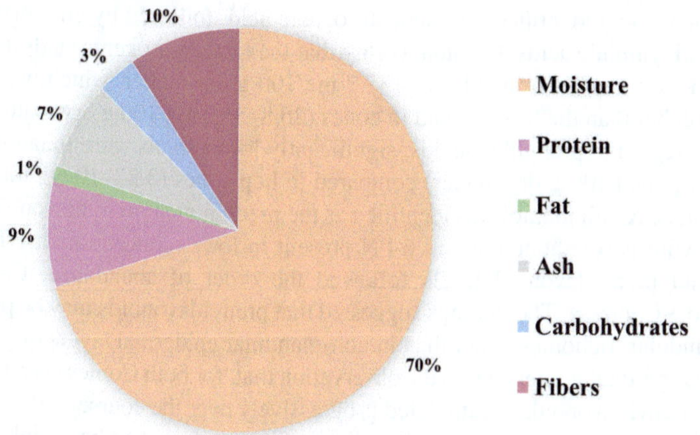

Fig. 9.44 Proximate composition (%) of Cascade hop leaves

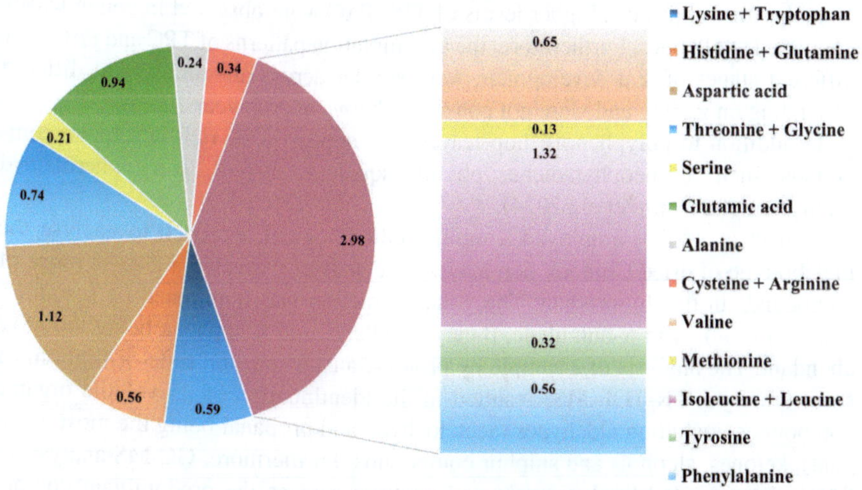

Fig. 9.45 Amino acid composition of Cascade leaves. (Adapted form Mitton et al. (2025))

In a recent study, Calvert et al. (2025) investigated the phenolic composition of hop leaves from three commercially varieties (Calypso, Cascade, and Contessa) cultivated in Washington's Yakima Valley over two harvest years and at different developmental stages post-flowering (Calvert et al., 2025). The study found that hop leaf phenolic profiles exhibit considerable structural diversity, differing significantly from that of the cones. The most abundant subgroups were identified as kaempferol/quercetin 3-O-glycosides and chlorogenic acids, with phenolic acids, procyanidins, prenylflavonoids, and bitter resins also present. In the study, the most abundant chlorogenic acid identified was neochlorogenic acid, followed by chlorogenic and 5-O-feruloylquinic acids. It is noteworthy that the total chlorogenic acids measured in hop leaves ranged from 11.8 to 135.7 mg/100 g dry matter, which was significantly higher than the levels found in cones (26.8–55.2 mg/100 g dry matter).

With regard to prenylflavonoids, significantly lower levels were found in leaves (<300 mg total/100 g dry matter) compared to hop cones (3.37–10.6 g total/100 g dry matter). Xanthohumol was identified as the predominant prenylflavonoid in hop leaves, with isoxanthohumol and 6-PN present in lower concentrations. The total resin and prenylflavonoid levels followed the order of abundance: Cascade > Calypso > Contessa. The findings suggested that prenylflavonoids may be present in leaf glandular trichomes rather than in non-glandular epidermal layers of green leaf tissue. Of particular interest was the observation that, for both Contessa and Cascade leaves, prenylflavonoids accumulated progressively over the course of development for both crop years. Finally, the phenolic profile was found to be mainly variety-dependent, with some crop year and developmental stage effects observed. Total polyphenol and proanthocyanidin content (TPC and PAC, respectively) were found to vary depending on factors such as variety, crop year, and stage of development. Notably, for Calipso cv., higher levels of TPC/PAC were observed in young leaves compared to old ones. Furthermore, the accumulation patterns of TPC and PAC over different stages of leaf development were similar across all samples but differed depending on variety and were not consistent between crop years.

In addition to polyphenols, hop leaves also appear to be rich in terpene compounds such as neophytadiene, phytol, squalene, lupeol, α-tocopherol and β-amyrone (Sabbatini et al., 2024).

Taiti et al. (2023) employed a multi-methodological approach to analyse the metabolome of dried Chinook hop leaves. The analysis revealed a diverse range of compounds in this by-product. The volatile fraction was dominated by sesquiterpenes (85.8%), with α-humulene, β-caryophyllene, and α-copaene being the most abundant. The analysis of a sample by proton-transfer-reaction time-of-flight mass spectrometry (PTR-TOF-MS) resulted in the identification of 25 volatile organic compounds, including aldehydes (acetaldehyde and propanal being the most abundant), ketones, alcohols and sulphur compounds. Furthermore, GC-MS analysis of fatty acids revealed linolenic acid and palmitic acid as the predominant components, whereas the methanolic extract analysis revealed a substantial sugar content, with sucrose being the predominant component. In addition, the analysis of the hydroalcoholic extract via NMR identified various amino acids, organic acids, and sugars, with glutamate and asparagine being the most abundant amino acids (Taiti et al., 2023). However, it should be noted that the study focused on a single hop cultivar. This narrow focus may limit the generalisability of the findings to other hop varieties.

Chinook fresh hop leaves were investigated by Vitalini et al. (2023), who employed different analytical techniques to characterise the volatile and non-volatile compounds present in these plant tissues (Vitalini et al., 2023). Sesquiterpenoids were found to be the dominant compounds in the leaves, comprising 92.4% of the total volatile composition in unpowdered material and 94.6% in powdered one. The most abundant compounds in the leaves were found to be β-caryophyllene and humulene. In unpowdered leaves, β-caryophyllene reached 20.2% and humulene 33.8%. In powdered leaves, the presence of β-caryophyllene increased dramatically to 71.6%, while humulene decreased to 10.0%. The analysis also revealed the presence of other significant compounds in the leaves, including α-copaene, γ-muurolene, germacrene D, β-eudesmene, valencene, α-muurolene, and δ-cadinene. Monoterpenoids were present in much smaller quantities, making up only 3.4% of the composition in unpowdered leaves and 4.0% in powdered ones. Some compounds, such as 1-hepten-3-ol, styrene, and α-terpineol, were found only in the leaves and not in the inflorescences. The study also noted that the chemical composition of the leaves differed from that of the inflorescences, with leaves generally having fewer components but higher levels of certain sesquiterpenes. Additionally, the authors performed a derivatisation process on the methanolic extracts to analyse non-volatile components, revealing the presence of organic acids, sugars (fructofuranose and tagato furanose being the most abundant), and sugar alcohols (e.g. pinitol).

The authors also analysed the volatile chemical composition of Chinook hop leaf aqueous extracts (LAE), identifying five volatile compounds:

- Pentane, 1-methoxy- (17.5%)
- 1-butanol, 3-methyl- (16.6%)
- β-myrcene (8.3%)
- Linalool (49.6%)
- Humulene (8.0%)

The total monoterpenoid content in LAE was 57.9%, while sesquiterpenoids made up 8.0% of the composition. Other compounds not classified as terpenes accounted for 34.1% of the volatile composition in LAE.

Hop leaves have been found to be a significant source of non-phenolic pigments and free soluble thiols. In their analysis, Macchioni et al. (2021) examined the content of these metabolites in dried leaves from five hop varieties cultivated in Italy (Macchioni et al., 2021). The obtained results demonstrated a range of variation in chlorophyll *a* concentration, ranging from 104 to 441 μg/g, chlorophyll *b* concentration, ranging from 285 to 883 μg/g, and carotenoid concentration, ranging from 91 to 266 μg/g. These concentrations were observed to be independent from the variety of hop and the drying process employed on the hop leaves analysed. With regard to free soluble thiols, the author noted a range of variation between 45 and 180 mg/g.

9.3.3 Hop Sprouts

Hop sprouts (Fig. 9.46), defined as the young shoots that emerge from the plant rhizomes, have garnered increased attention due to their substantial content of bioactive compounds and essential nutrients, rendering them a promising candidate for nutraceutical and functional food applications (Rossini et al., 2020).

Chemically, hop sprouts exhibit a high proportion of dietary fibre (e.g. cellulose, hemicellulose, and pectins), contributing to their structural integrity and potential prebiotic effects. They are also rich in phenolic compounds, including flavonoids (e.g. quercetin, kaempferol, rutin) and phenolic acids (e.g. ferulic acid, caffeic acid), which exhibit strong antioxidant properties.

Furthermore, they contain significant levels of terpenes and bitter acids, such as humulones and lupulones, although in lower concentrations compared to mature hop cones. Their mineral composition includes good levels of potassium, calcium, magnesium, and iron. Hop shoots have low fat and sodium levels but are rich in dietary fibre and vitamin B9. In addition, hop shoots contain lower levels of oxalic acid and higher levels of vitamin C compared to other foodstuffs (Rossini et al., 2020). At present, they are primarily utilised for culinary applications in all hop-producing countries, thereby creating avenues for hop growers to diversify their products and income streams.

To the best of our knowledge, Maietti et al. (2017) were the first to conduct a research study on the metabolite profiling and antioxidant activity of young shoots of wild *Humulus lupulus* L. (Maietti et al., 2017).

The objective of the study was to identify and quantify flavonoids in hop shoots, and to compare their composition with antioxidant activity across samples from

Fig. 9.46 Hop sprouts. (Photo: Katya Carbone)

different locations in Northern Italy. The most abundant compounds found were kaempferol derivatives, including kaempferol-3-O-(6′-O-malonyl)-glucoside (or galactoside), kaempferol-3-O-glucoside, and kaempferol-3-O-(6′-O-malonyl)-neohesperidoside. The total flavonol content exhibited significant variation among the samples obtained from diverse geographical locations, ranging from 2698 ± 185 to 517 ± 47.5 μg/g (fresh weight). Regarding the antioxidant potential of the analysed samples, the results showed a strong correlation between total flavonol content and antioxidant activity, with an r^2 value of 0.9577 for the photochemiluminescence assay.

In a separate study, Rossini et al. (2020) investigated the potential of hops as a dual-purpose crop, focusing on shoot production and their nutritional composition under Mediterranean climatic conditions. The study was conducted on three commercial hop cultivars, namely Cascade, Challenger, and Hallertauer Magnum, which were cultivated in Italy (vintage: 2017, 2018) (Rossini et al., 2020), investigating the shoot proximate composition. The findings indicated that the moisture content was notably high (averaging 89.5%), a result that was found to be significantly influenced by the availability of groundwater. Regardless of the genotype or crop year, the authors reported the average ash content to be approximately 9.6% on a dry basis. Ether extract content was reported as approximately 3.4%, crude protein content as approximately 27%, and crude fibre content as approximately 13%. As indicated by the reported results, it is evident that all the traits under investigation were found to be significantly influenced by both genotype and year. Furthermore, it was observed that Cascade exhibited the optimal quality traits in terms of nutritional composition.

In relation to the chemical composition of hop shoots, Vidmar et al. (2019) analysed the moisture, total phenolics and microelements of Slovenian white hop shoots from five commercial cultivars (Aurora, Celeia, Dana, Hallertauer Magnum and Savinjski golding) (Vidmar et al., 2019). The water content of the samples varied across cultivars and years, ranging from 72.1% to 92.1%. According to the authors, white hop shoots were found to exhibit superior antioxidant properties in comparison to hop cones and leaves, with a total polyphenol content ranging from approximately 0.6 to 1.5 mg chlorogenic acid equivalents per g of dry matter, as observed for Hallertauer Magnum and Savinjski golding, respectively. However, the authors emphasised that these values can be influenced by cultivar, year of production, weather conditions, and soil characteristics.

In their study, Rodolfi et al. (2022) used nuclear magnetic resonance (^{1}H-NMR) to profile the metabolites of green sprouts from two different hop varieties, Cascade and Nugget, cultivated in Italy (Rodolfi et al., 2022) (Fig. 9.47).

Of particular interest was the finding that the most prevalent compound in both varieties was trigonelline (Fig. 9.6), a substance also identified in seedless wild hop cones by Carbone et al. (2020).

Trigonelline is a hormone that plays a pivotal role in various physiological processes in plants, including the regulation of the cell cycle, nodulation, and oxidative stress responses. It has been demonstrated to facilitate plant survival and promote growth. Furthermore, this alkaloid has been demonstrated to exert several biological

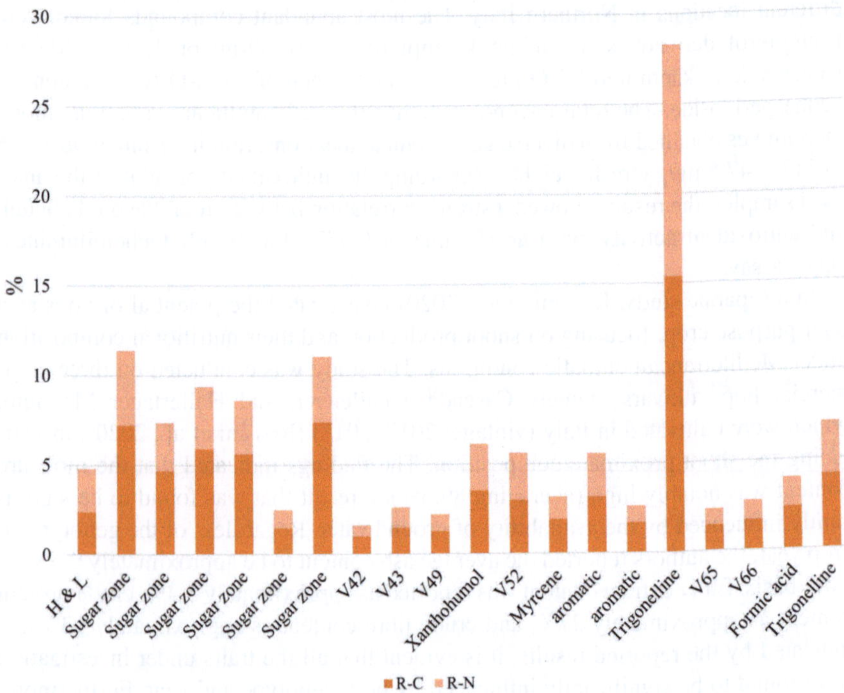

Fig. 9.47 ^{1}H-NMR chemical composition of green hop sprouts. (Adapted from Rodolfi et al. (2022)). H + L: α- and β-acid series; V: unknown. It should be noted that the graph only includes signals greater than 1 ppm

functions. For instance, it has been used as an anti-diabetic agent, improving glucose tolerance in obese diabetics. It has also been proposed as a neuroprotective agent (Liang et al., 2023). In addition, trigonelline has been shown to be involved in various contexts as an antibacterial, antiviral, sedative, anti-migraine, and anticancer agent (Nguyen et al., 2024). These findings therefore provide a new perspective on the classification of hop green sprouts as functional foods.

9.3.4 Hop Seeds

Hop seeds possess a complex chemical composition, making them a valuable but underexplored component of the hop plant.

To the best of our knowledge, only one study has been carried out to date to explore the phytochemical profile of hop seeds (Alonso-Esteban et al., 2019). This research paper investigates the phenolic composition and bioactive properties of hop seeds using LC-DAD-ESI/MS^{n}. The findings revealed that they contain significant amounts of (+)-catechin and (−)-epicatechin (Fig. 9.22), totalling 17.6 ± 0.7 mg/g of extract. The seed extract exhibited antioxidant activity, particularly in terms of inhibiting the formation of thiobarbituric acid reactive substances and demonstrated cytotoxic effects against different human tumour cell lines, without exhibiting toxicity towards normal PLP2 cells. Furthermore, the extract displayed remarkable antimicrobial properties, outperforming positive controls against most tested bacteria and fungi of the genus *Penicillium*.

9.4 Conclusions and Future Perspectives

This chapter elucidates the intricate and diverse phytochemical composition of hops, encompassing both primary and secondary metabolites across various plant tissues. The cone is the most valuable and extensively studied component of the hop plant, containing essential compounds such as bitter acids, essential oils, and polyphenols, which are crucial for brewing and have demonstrated numerous bioactivities. However, recent research has identified the potential of other hop tissues, including stems, leaves, sprouts, and seeds, as sources of valuable bioactive molecules. The chemical profile of hops is influenced by a combination of genetic, environmental, and agronomic factors. Advancements in analytical techniques have enabled more detailed characterisation of hop metabolites, providing deeper insights into their chemistry and biological activities. This enhanced knowledge has implications not only for the brewing industry but also for potential applications in nutraceuticals, pharmaceuticals, and other sectors. However, further investigation is required to optimise the extraction and utilisation of bioactive compounds from hop stems, leaves, sprouts, and seeds. This research could lead to the development of novel value-added products and contribute to a more circular economy approach in hop cultivation.

References

Algazzali, V., & Shellhammer, T. (2016). Bitterness intensity of oxidized hop acids: Humulinones and Hulupones. *Journal of the American Society of Brewing Chemists, 74*(1), 36–43. https://doi.org/10.1094/ASBCJ-2016-1130-01

Almaguer, C., Schönberger, C., Gastl, M., Arendt, E. K., & Becker, T. (2014). Humulus lupulus—A story that begs to be told. A review. *Journal of the Institute of Brewing, 120*(4), 289–314. https://doi.org/10.1002/jib.160

Alonso-Esteban, J. I., Pinela, J., Barros, L., Ćirić, A., Soković, M., Calhelha, R. C., Torija-Isasa, E., De Cortes Sánchez-Mata, M., & Ferreira, I. C. F. R. (2019). Phenolic composition and antioxidant, antimicrobial and cytotoxic properties of hop (Humulus lupulus L.) seeds. *Industrial Crops and Products, 134*, 154–159. https://doi.org/10.1016/j.indcrop.2019.04.001

Anitha, S., Krishnan, S., Senthilkumar, K., & Sasirekha, V. (2020). Theoretical investigation on the structure and antioxidant activity of (+) catechin and (−) epicatechin—A comparative study. *Molecular Physics, 118*(17), e1745917. https://doi.org/10.1080/00268976.2020.1745917

Armbruster, U., Pesaresi, P., Pribil, M., Hertle, A., & Leister, D. (2011). Update on chloroplast research: New tools, new topics, and new trends. *Molecular Plant, 4*(1), 1–16. https://doi.org/10.1093/mp/ssq060

Astray, G., Gullón, P., Gullón, B., Munekata, P. E. S., & Lorenzo, J. M. (2020). Humulus lupulus L. as a natural source of functional biomolecules. *Applied Sciences, 10*(15), 5074. https://doi.org/10.3390/app10155074

Bailey, B., Schönberger, C., Drexler, G., Gahr, A., Newman, R., Pöschl, M., & Geiger, E. (2009). The influence of hop harvest date on hop aroma in dry-hopped beers. *Technical Quarterly of the Master Brewers Association of the Americas, 46*, 1–7.

Bailey, B., Schönberger, C., Drexler, G., Gahr, A., Newman, R., Pöschl, M., & Geiger, E. (2010). The influence of hop harvest date on flavor stability in dry-hopped beers. *Technical Quarterly, 46* (January 2009), 1–7. https://doi.org/10.1094/TQ-47-1-0219-01

Bandyukova, V. A. (1983). Phenolic acids of plants and their esters and glycosides. *Chemistry of Natural Compounds, 19*(3), 251–260. https://doi.org/10.1007/BF00579752

Benitez, J. L., Forster, A., De Keukeleire, D., Moir, R., Sharpe, F. R., Verhagen, L. C., & Westwood, K. T. (1997). *Hops and hop products* (European brewery convention technology and engineering forum, Ed.). Fachverlag Hans Carl.

Bergman, M. E., Davis, B., & Phillips, M. A. (2019). Medically Useful Plant Terpenoids: Biosynthesis, Occurrence, and Mechanism of Action. *Molecules, 24*(21), Article 21. https://doi.org/10.3390/molecules24213961

Berry, J., & Bjorkman, O. (1980). Photosynthetic response and adaptation to temperature in higher plants. *Annual Review of Plant Physiology, 31*(1), 491–543. https://doi.org/10.1146/annurev.pp.31.060180.002423

Biendl, M., & Pinzl, C. (2007). *Arzneipflanze Hopfen: Anwendungen - Wirkungen - Geschichte*. Dt. Hopfenmuseum.

Biendl, M., & Pinzl, C. (2013). *Hops and health: Uses, effects, history*. German Hop Museum.

Biendl, M., Engelhard, B., Forster, A., Gahr, A., Lutz, A., Mitter, W., & Sch, R. (2015). *Hops: Their cultivation, composition and usage* (1. Aufl). Fachverlag Hans Carl.

Biendl, M., Ritter, S., & Schmidt, C. (2023). Monitoring of glycosidically bound polyphenols in hops and hop products using LC-MS/MS technique. *Journal of the American Society of Brewing Chemists, 81*(1), 45–53. https://doi.org/10.1080/03610470.2021.2024112

Bocquet, L., Sahpaz, S., Hilbert, J. L., Rambaud, C., & Rivière, C. (2018). Humulus lupulus L., a very popular beer ingredient and medicinal plant: Overview of its phytochemistry, its bioactivity, and its biotechnology. *Phytochemistry Reviews, 17*(5), 1047–1090. https://doi.org/10.1007/s11101-018-9584-y

Bravi, E., Perretti, G., Buzzini, P., Della Sera, R., & Fantozzi, P. (2009). Technological steps and yeast biomass as factors affecting the lipid content of beer during the brewing process. *Journal of Agricultural and Food Chemistry, 57*(14), 6279–6284. https://doi.org/10.1021/jf9007423

Briggs, D. E. (Ed.). (2004). *Brewing: Science and practice*. CRC Press.

Bruner, J., Marcus, A., & Fox, G. (2023). Changes in Diacetyl and amino acid concentration during the fermentation of dry-hopped beer: A look at twelve *saccharomyces* species and strains. *Journal of the American Society of Brewing Chemists, 81*(2), 242–254. https://doi.org/10.1080/03610470.2022.2078946

Callemien, D., Jerkovic, V., Rozenberg, R., & Collin, S. (2005). Hop as an interesting source of resveratrol for brewers: Optimization of the extraction and quantitative study by liquid chromatography/atmospheric pressure chemical ionization tandem mass spectrometry. *Journal of Agricultural and Food Chemistry, 53*(2), 424–429. https://doi.org/10.1021/jf040179n

Calvert, D., Dew, T., Gadon, A., Gros, J., & Cook, D. (2025). Valorisation of hop leaves for their bioactive compounds: Identification and quantification of phenolics across different varieties, crop years and stages of development. *Food Chemistry, 465*, 142005. https://doi.org/10.1016/j.foodchem.2024.142005

Canbaş, A., Erten, H., & Özşahin, F. (2001). The effects of storage temperature on the chemical composition of hop pellets. *Process Biochemistry, 36*(11), 1053–1058. https://doi.org/10.1016/S0032-9592(01)00134-0

Carbone, K., & Gervasi, F. (2022). An updated review of the genus Humulus: A valuable source of bioactive compounds for health and disease prevention. *Plants, 11*(24), 3434. https://doi.org/10.3390/plants11243434

Carbone, K., Macchioni, V., Petrella, G., & Cicero, D. O. (2020). Exploring the potential of microwaves and ultrasounds in the green extraction of bioactive compounds from Humulus lupulus for the food and pharmaceutical industry. *Industrial Crops and Products, 156*, 112888. https://doi.org/10.1016/j.indcrop.2020.112888

Carson, J. F. (1951). The structure of Humulone and Lupulone as revealed by Ozonization[1a]. *Journal of the American Chemical Society, 73*(10), 4652–4654. https://doi.org/10.1021/ja01154a046

Carson, J. F. (1952). The alkaline isomerization of humulone. *Journal of the American Chemical Society, 74*(18), 4615–4620. https://doi.org/10.1021/ja01138a052

Cattoor, K., Bracke, M., Deforce, D., De Keukeleire, D., & Heyerick, A. (2010). Transport of hop bitter acids across intestinal Caco-2 cell monolayers. *Journal of Agricultural and Food Chemistry, 58*(7), 4132–4140. https://doi.org/10.1021/jf904079h

Ceh, B., Kac, M., Košir, I. J., & Abram, V. (2007). Relationships between xanthohumol and polyphenol content in hop leaves and hop cones with regard to water supply and cultivar. *International Journal of Molecular Sciences, 8*(9), 989–1000. https://doi.org/10.3390/i8090989

Cendrero-Mateo, M. P., Moran, M. S., Papuga, S. A., Thorp, K. R., Alonso, L., Moreno, J., Ponce-Campos, G., Rascher, U., & Wang, G. (2016). Plant chlorophyll fluorescence: Active and passive measurements at canopy and leaf scales with different nitrogen treatments. *Journal of Experimental Botany, 67*(1), 275–286. https://doi.org/10.1093/jxb/erv456

Chadwick, L. R., Pauli, G. F., & Farnsworth, N. R. (2006). The pharmacognosy of Humulus lupulus L. (hops) with an emphasis on estrogenic properties. *Phytomedicine, 13*(1–2), 119–131. https://doi.org/10.1016/j.phymed.2004.07.006

Chinou, I. M. (2008). Primary and secondary metabolites and their biological activity. In *Thin layer chromatography in phytochemistry* (pp. 59–76). CRC Press.

Clark, S. M., Vaitheeswaran, V., Ambrose, S. J., Purves, R. W., & Page, J. E. (2013). Transcriptome analysis of bitter acid biosynthesis and precursor pathways in hop (Humulus lupulus). *BMC Plant Biology, 13*(1), 12. https://doi.org/10.1186/1471-2229-13-12

Contin, D. R., Habermann, E., De Souza, B. C., Dias De Oliveira, E. A., Martinez, C. A., Vieira, P. C., & Da Costa, F. B. (2023). Exploring the tropical acclimation of European and American hop cultivars (Humulus lupulus L.): Focus on physiology, productivity, and chemical composition. *European Journal of Agronomy, 151*, 126990. https://doi.org/10.1016/j.eja.2023.126990

Cook, A. H., & Harris, G. (1950). The chemistry of hop constituents. Part I. Humulinone, a new constituent of hops. *Journal of the Chemical Society (Resumed)*, 1873. https://doi.org/10.1039/jr9500001873

Corrêa, R. C. G., Peralta, R. M., Haminiuk, C. W. I., Maciel, G. M., Bracht, A., & Ferreira, I. C. F. R. (2018). New phytochemicals as potential human anti-aging compounds: Reality, promise, and challenges. *Critical Reviews in Food Science and Nutrition, 58*(6), 942–957. https://doi.org/10.1080/10408398.2016.1233860

Crozier, A., Clifford, M. N., & Rio, D. D. (2012). Bioavailability of dietary monomeric and polymeric flavan-3-ols. In *Flavonoids and related compounds: Bioavailability and function* (pp. 45–78). Scopus.

Das, K., & Roychoudhury, A. (2014). Reactive oxygen species (ROS) and response of antioxidants as ROS-scavengers during environmental stress in plants. *Frontiers in Environmental Science, 2*. https://doi.org/10.3389/fenvs.2014.00053

De Keukeleire, J., Ooms, G., Heyerick, A., Roldan-Ruiz, I., Van Bockstaele, E., & De Keukeleire, D. (2003). Formation and accumulation of α-acids, β-acids, desmethylxanthohumol, and xanthohumol during flowering of hops (*Humulus lupulus* L.). *Journal of Agricultural and Food Chemistry, 51*(15), 4436–4441. https://doi.org/10.1021/jf034263z

de Sousa, D. P., Damasceno, R. O. S., Amorati, R., Elshabrawy, H. A., de Castro, R. D., Bezerra, D. P., Nunes, V. R. V., Gomes, R. C., & Lima, T. C. (2023). Essential oils: Chemistry and pharmacological activities. *Biomolecules, 13*(7), 1144. https://doi.org/10.3390/biom13071144

Dewick, P. M. (1995). The biosynthesis of shikimate metabolites. *Natural Product Reports, 12*(6), 579–607. https://doi.org/10.1039/np9951200579

Dietz, C., Cook, D., Huismann, M., Wilson, C., & Ford, R. (2020). The multisensory perception of hop essential oil: A review. *Journal of the Institute of Brewing, 126*(4), 320–342. https://doi.org/10.1002/jib.622

Dušek, M., Olšovská, J., Krofta, K., Jurková, M., & Mikyška, A. (2014). Qualitative determination of β-acids and their transformation products in beer and hop using HR/AM-LC-MS/MS. *Journal of Agricultural and Food Chemistry, 62*(31), 7690–7697. https://doi.org/10.1021/jf501852r

Dušek, M., Jandovská, V., Mikyška, A., & Olšovská, J. (2021). Novel and efficient approach to identify hop cultivars (Humulus lupulus L.) using cultivar identification diagram strategy based on fingerprint of flavonol di- and tri-O-glycosides. *European Food Research and Technology, 247*(3), 651–662. https://doi.org/10.1007/s00217-020-03653-2

El-Seedi, H. R., Taher, E. A., Sheikh, B. Y., Anjum, S., Saeed, A., AlAjmi, M. F., Moustafa, M. S., Al-Mousawi, S. M., Farag, M. A., Hegazy, M.-E. F., Khalifa, S. A. M., & Göransson, U. (2018). Hydroxycinnamic acids: Natural sources, biosynthesis, possible biological activities, and roles in Islamic medicine. In *Studies in natural products chemistry* (Vol. 55, pp. 269–292). Elsevier. https://doi.org/10.1016/B978-0-444-64068-0.00008-5

Eri, S., Khoo, B. K., Lech, J., & Hartman, T. G. (2000). Direct thermal desorption−gas chromatography and gas chromatography−mass spectrometry profiling of hop (Humulus lupulus L.) essential oils in support of varietal characterization. *Journal of Agricultural and Food Chemistry, 48*(4), 1140–1149. https://doi.org/10.1021/jf9911850

Eriksen, R. L., Rutto, L. K., Dombrowski, J. E., & Henning, J. A. (2020). Photosynthetic activity of six hop (Humulus lupulus L.) cultivars under different temperature treatments. *HortScience, 55*(4), 403–409. https://doi.org/10.21273/HORTSCI14580-19

Eßlinger, H. M. (2009). *Handbook of brewing*. Wiley-VCH.

Eyres, G., & Dufour, J.-P. (2009). Hop essential oil: Analysis, chemical composition and odor characteristics. In *Beer in health and disease prevention* (pp. 239–254). Elsevier. https://doi.org/10.1016/B978-0-12-373891-2.00022-5

Fang, C., Fernie, A. R., & Luo, J. (2019). Exploring the diversity of plant metabolism. *Trends in Plant Science, 24*(1), 83–98. https://doi.org/10.1016/j.tplants.2018.09.006

Ferruzzi, M. G., & Blakeslee, J. (2007). Digestion, absorption, and cancer preventative activity of dietary chlorophyll derivatives. *Nutrition Research, 27*(1), 1–12. https://doi.org/10.1016/j.nutres.2006.12.003

Fracassetti, D., & Vigentini, I. (2018). Occurrence and analysis of sulfur compounds in wine. In A.M. Jordão & F. Cosme (Eds.), *Grapes and wines—Advances in production, processing, analysis and valorization*. InTech. https://doi.org/10.5772/intechopen.72823

García-Villalba, R., Cortacero-Ramírez, S., Segura-Carretero, A., Martín-Lagos Contreras, J. A., & Fernández-Gutiérrez, A. (2006). Analysis of hop acids and their oxidized derivatives and Iso-α-acids in beer by capillary electrophoresis−electrospray ionization mass spectrometry. *Journal of Agricultural and Food Chemistry, 54*(15), 5400–5409. https://doi.org/10.1021/jf060207x

Girisa, S., Saikia, Q., Bordoloi, D., Banik, K., Monisha, J., Daimary, U. D., Verma, E., Ahn, K. S., & Kunnumakkara, A. B. (2021). Xanthohumol from hop: Hope for cancer prevention and treatment. *IUBMB Life, 73*(8), 1016–1044. https://doi.org/10.1002/iub.2522

Goleniowski, M., Bonfill, M., Cusido, R., & Palazón, J. (2013). Phenolic acids. In K. G. Ramawat & J.-M. Mérillon (Eds.), *Natural products* (pp. 1951–1973). Springer. https://doi.org/10.1007/978-3-642-22144-6_64

Hagemann, M. H., Rigling, M., Mannweiler, S., Born, U., Sprich, E., Milyaev, A., & Zhang, Y. (2024). Insight into the aroma quality of 'Callista' cultivar of hop (Humulus lupulus L.): Impact of harvest timing, year, and location. *Food Research International, 175*, 113776. https://doi.org/10.1016/j.foodres.2023.113776

Hao, J., Speers, R. A., Fan, H., Deng, Y., & Dai, Z. (2020). A review of cyclic and oxidative bitter derivatives of alpha, iso-alpha and beta-hop acids. *Journal of the American Society of Brewing Chemists, 78*(2), 89–102. https://doi.org/10.1080/03610470.2020.1712641

Harris, G., & Tatchell, A. R. (1953). Amino acids and peptides of hops and WORT III. The amino acids of fresh hops. *Journal of the Institute of Brewing, 59*(5), 371–377. https://doi.org/10.1002/j.2050-0416.1953.tb02730.x

Haseleu, G., Intelmann, D., & Hofmann, T. (2009a). Identification and RP-HPLC-ESI-MS/MS quantitation of bitter-tasting β-acid transformation products in beer. *Journal of Agricultural and Food Chemistry, 57*(16), 7480–7489. https://doi.org/10.1021/jf901759y

Haseleu, G., Intelmann, D., & Hofmann, T. (2009b). Structure determination and sensory evaluation of novel bitter compounds formed from β-acids of hop (Humulus lupulus L.) upon wort boiling. *Food Chemistry, 116*(1), 71–81. https://doi.org/10.1016/j.foodchem.2009.02.008

Hendry, G. A. F. (1996). Natural pigments in biology. In G. A. F. Hendry & J. D. Houghton (Eds.), *Natural food colorants* (pp. 1–39). Springer US. https://doi.org/10.1007/978-1-4615-2155-6_1

Hong, K., Wang, L., Johnpaul, A., Lv, C., & Ma, C. (2021). Key enzymes involved in the synthesis of hops phytochemical compounds: From structure, functions to applications. *International Journal of Molecular Sciences, 22*(17), 9373. https://doi.org/10.3390/ijms22179373

Intelmann, D., & Hofmann, T. (2010). On the autoxidation of bitter-tasting Iso-α-acids in beer. *Journal of Agricultural and Food Chemistry, 58*(8), 5059–5067. https://doi.org/10.1021/jf100083e

Intelmann, D., Haseleu, G., Dunkel, A., Lagemann, A., Stephan, A., & Hofmann, T. (2011). Comprehensive sensomics analysis of hop-derived bitter compounds during storage of beer. *Journal of Agricultural and Food Chemistry, 59*(5), 1939–1953. https://doi.org/10.1021/jf104392y

Inui, T., Okumura, K., Matsui, H., Hosoya, T., & Kumazawa, S. (2017). Effect of harvest time on some in vitro functional properties of hop polyphenols. *Food Chemistry, 225*, 69–76. https://doi.org/10.1016/j.foodchem.2017.01.002

Jaskula-Goiris, B., Aerts, G., & De Cooman, L. (2010). Hop α-acids isomerisation and utilisation: An experimental review. *Cerevisia, 35*(3), 57–70. https://doi.org/10.1016/j.cervis.2010.09.004

Jelínek, L., Šneberger, M., Karabín, M., & Dostálek, P. (2010). Comparison of Czech hop cultivars based on their contenst of secondary metabolites. *Czech Journal of Food Sciences, 28*(4), 309–316. https://doi.org/10.17221/65/2010-CJFS

Jerkovic, V., Callemien, D., & Collin, S. (2005). Determination of stilbenes in hop pellets from different cultivars. *Journal of Agricultural and Food Chemistry, 53*(10), 4202–4206. https://doi.org/10.1021/jf050222m

Jirovetz, L. (2006). Antimicrobial testings, gas chromatographic analysis and olfactory evaluation of an essential oil of hop cones (Humulus lupulus L.) from Bavaria and some of its main compounds. *Scientia Pharmaceutica, 74*(4), 189–201. https://doi.org/10.3797/scipharm.2006.74.189

Kammhuber, K., Hundhammer, M., & Weihrauch, S. (2017). Influence of the date of harvest on the sulphur compounds of the "special flavour hop" varieties Cascade, Mandarina Bavaria, Hallertau Blanc, Huell Melon and Polaris. *BrewingScience, 70*, 124–130. https://doi.org/10.23763/BrSc17-12Kammhuber

Kampa, M., Nifli, A.-P., Notas, G., & Castanas, E. (2007). Polyphenols and cancer cell growth. In S.G. Amara, E. Bamberg, B. Fleischmann, T. Gudermann, S.C. Hebert, R. Jahn, W.J. Lederer, R. Lill, A. Miyajima, S. Offermanns, & R. Zechner (Eds.), *Reviews of physiology, biochemistry and pharmacology* (pp. 79–113). Springer Berlin Heidelberg. https://doi.org/10.1007/112_2006_0702

Kanai, N., Nishimura, K., Umetani, S., Saito, Y., Saito, H., Oyama, T., & Kawamura, I. (2021). Upcycling of waste hop stems into cellulose nanofibers: Isolation and structural characterization. *ACS Agricultural Science & Technology, 1*(4), 347–354. https://doi.org/10.1021/acsagscitech.1c00041

Kankolongo Cibaka, M.-L., Decourrière, L., Lorenzo-Alonso, C.-J., Bodart, E., Robiette, R., & Collin, S. (2016). 3-Sulfanyl-4-methylpentan-1-ol in dry-hopped beers: First evidence of glutathione S-conjugates in hop (Humulus lupulus L.). *Journal of Agricultural and Food Chemistry, 64*(45), 8572–8582. https://doi.org/10.1021/acs.jafc.6b03788

Karabin, M., Hudcova, T., Jelinek, L., & Dostalek, P. (2015). Biotransformations and biological activities of hop flavonoids. *Biotechnology Advances, 33*(6, Part 2), 1063–1090. https://doi.org/10.1016/j.biotechadv.2015.02.009

Karabín, M., Hudcová, T., Jelínek, L., & Dostálek, P. (2016). Biologically active compounds from hops and prospects for their use. *Comprehensive Reviews in Food Science and Food Safety, 15*(3), 542–567. https://doi.org/10.1111/1541-4337.12201

King, A. J., & Dickinson, J. R. (2000). Biotransformation of monoterpene alcohols by Saccharomyces cerevisiae, Torulaspora delbrueckii and Kluyveromyces lactis. *Yeast (Chichester, England), 16*(6), 499–506. https://doi.org/10.1002/(SICI)1097-0061(200004)16:6<499::AID-YEA548>3.0.CO;2-E

King, A. J., & Dickinson, J. R. (2003). Biotransformation of hop aroma terpenoids by ale and lager yeasts. *FEMS Yeast Research, 3*(1), 53–62. https://doi.org/10.1016/S1567-1356(02)00141-1

Kirkpatrick, K. R., & Shellhammer, T. H. (2018). Evidence of dextrin hydrolyzing enzymes in cascade hops (*Humulus lupulus*). *Journal of Agricultural and Food Chemistry, 66*(34), 9121–9126. https://doi.org/10.1021/acs.jafc.8b03563

Kishimoto, T., Kobayashi, M., Yako, N., Iida, A., & Wanikawa, A. (2008). Comparison of 4-Mercapto-4-methylpentan-2-one contents in hop cultivars from different growing regions. *Journal of Agricultural and Food Chemistry, 56*(3), 1051–1057. https://doi.org/10.1021/jf072173e

Klimczak, K., Cioch-Skoneczny, M., & Duda-Chodak, A. (2023). Effects of dry-hopping on beer chemistry and sensory properties—A review. *Molecules, 28*(18), 6648. https://doi.org/10.3390/molecules28186648

Kobus-Cisowska, J., Szymanowska, D., Szczepaniak, O. M., Gramza-Michałowska, A., Kmiecik, D., Kulczyński, B., Szulc, P., & Górnaś, P. (2019). Composition of polyphenols of asparagus spears (Asparagus officinalis) and their antioxidant potential. *Ciência Rural, 49*(4), e20180863. https://doi.org/10.1590/0103-8478cr20180863

Kováčová, J., Lehotay, J., Űrgeová, E., Mocák, J., & Čižmárik, J. (2011). Determination of selected flavonoids in hop extract by hplc. *Journal of Liquid Chromatography & Related Technologies, 34*(5), 329–340. https://doi.org/10.1080/10826076.2011.551596

Kowalska, G., Bouchentouf, S., Kowalski, R., Wyrostek, J., Pankiewicz, U., Mazurek, A., Sujka, M., & Włodarczyk-Stasiak, M. (2022). The hop cones (Humulus lupulus L.): Chemical composition, antioxidant properties and molecular docking simulations. *Journal of Herbal Medicine, 33*, 100566. https://doi.org/10.1016/j.hermed.2022.100566

Krofta, K., Vrabcová, S., Mikyška, A., & Jurková, M. (2013). The effect of hop beta acids oxidation products on beer bitterness. *Kvasny Prumysl, 59*(10–11), 306–312. https://doi.org/10.18832/kp2013032

Krofta, K., Patzak, J., & Vondráčková, P. (2021). Sulfur substances of hop oils—Another tool for identification of hop cultivars. *Acta Horticulturae, 1328*, 153–161. https://doi.org/10.17660/ActaHortic.2021.1328.21

Kumar, A., Warburton, A., Silcock, P., Bremer, P. J., & Eyres, G. T. (2023). Yeast strain influences the hop-derived sensory properties and volatile composition of beer. *Food, 12*(5), 1064. https://doi.org/10.3390/foods12051064

Lafontaine, S., Varnum, S., Roland, A., Delpech, S., Dagan, L., Vollmer, D., Kishimoto, T., & Shellhammer, T. (2019). Impact of harvest maturity on the aroma characteristics and chemistry of Cascade hops used for dry-hopping. *Food Chemistry, 278*, 228–239. https://doi.org/10.1016/j.foodchem.2018.10.148

Lam, K. C., Foster, R. T., & Deinzer, M. L. (1986). Aging of hops and their contribution to beer flavor. *Journal of Agricultural and Food Chemistry, 34*(4), 763–770. https://doi.org/10.1021/jf00070a043

Lewis, M. J., & Young, T. W. (2012). *Brewing*. Springer Science & Business Media.

Li, H.-J., & Deinzer, M. L. (2006). Structural identification and distribution of proanthocyanidins in 13 different hops. *Journal of Agricultural and Food Chemistry, 54*(11), 4048–4056. https://doi.org/10.1021/jf060395r

Li, H.-J., & Deinzer, M. L. (2009). Proanthocyanidins in hops. In *Beer in health and disease prevention* (pp. 333–348). Elsevier. https://doi.org/10.1016/B978-0-12-373891-2.00032-8

Liang, Y., Dai, X., Cao, Y., Wang, X., Lu, J., Xie, L., Liu, K., & Li, X. (2023). The neuroprotective and antidiabetic effects of trigonelline: A review of signaling pathways and molecular mechanisms. *Biochimie, 206*, 93–104. https://doi.org/10.1016/j.biochi.2022.10.009

Lima, A., Pereira, J. A., Baraldi, I., & Malheiro, R. (2017). Cooking impact in color, pigments and volatile composition of grapevine leaves (Vitis vinifera L. var. Malvasia Fina and Touriga Franca). *Food Chemistry, 221*, 1197–1205. https://doi.org/10.1016/j.foodchem.2016.11.039

Limousin, R., Guendoul, F., Vibert, C., Antoun, K., Ségovia, C., Ziegler-Devin, I., & Brosse, N. (2023). Hop stems, a new fibrous bioresource. *Industrial Crops and Products, 201*, 116966. https://doi.org/10.1016/j.indcrop.2023.116966

Liu, W., Feng, Y., Yu, S., Fan, Z., Li, X., Li, J., & Yin, H. (2021). The flavonoid biosynthesis network in plants. *International Journal of Molecular Sciences, 22*(23), 12824. https://doi.org/10.3390/ijms222312824

Loranty, A., Rembiałkowska, E., Rosa, E. A. S., & Bennett, R. N. (2010). Identification, quantification and availability of carotenoids and chlorophylls in fruit, herb and medicinal teas. *Journal of Food Composition and Analysis, 23*(5), 432–441. https://doi.org/10.1016/j.jfca.2010.01.007

Macchioni, V., Picchi, V., & Carbone, K. (2021). Hop leaves as an alternative source of health-active compounds: Effect of genotype and drying conditions. *Plants, 11*(1), 99. https://doi.org/10.3390/plants11010099

Maietti, A., Brighenti, V., Bonetti, G., Tedeschi, P., Prencipe, F. P., Benvenuti, S., Brandolini, V., & Pellati, F. (2017). Metabolite profiling of flavonols and in vitro antioxidant activity of young shoots of wild Humulus lupulus L. (hop). *Journal of Pharmaceutical and Biomedical Analysis, 142*, 28–34. https://doi.org/10.1016/j.jpba.2017.04.043

Maliar, T., Nemeček, P., Ürgeová, E., Maliarová, M., Nesvadba, V., Krofta, K., Vulganová, K., Krošlák, E., & Kraic, J. (2017). Secondary metabolites, antioxidant and anti-proteinase activities of methanolic extracts from cones of hop (Humulus lupulus L.) cultivars. *Chemical Papers, 71*(1), 41–48. https://doi.org/10.1007/s11696-016-0034-2

Manach, C., Scalbert, A., Morand, C., Rémésy, C., & Jiménez, L. (2004). Polyphenols: Food sources and bioavailability. *The American Journal of Clinical Nutrition, 79*(5), 727–747. https://doi.org/10.1093/ajcn/79.5.727

Manzoor, M., Singh, J., Gani, A., & Noor, N. (2021). Valorization of natural colors as health-promoting bioactive compounds: Phytochemical profile, extraction techniques, and pharmacological perspectives. *Food Chemistry, 362*, 130141. https://doi.org/10.1016/j.foodchem.2021.130141

Marchiosi, R., Dos Santos, W. D., Constantin, R. P., De Lima, R. B., Soares, A. R., Finger-Teixeira, A., Mota, T. R., De Oliveira, D. M., Foletto-Felipe, M. D. P., Abrahão, J., & Ferrarese-Filho, O. (2020). Biosynthesis and metabolic actions of simple phenolic acids in plants. *Phytochemistry Reviews, 19*(4), 865–906. https://doi.org/10.1007/s11101-020-09689-2

Maroneze, M. M., Zepka, L. Q., Lopes, E. J., Pérez-Gálvez, A., & Roca, M. (2019). Chlorophyll oxidative metabolism during the phototrophic and heterotrophic growth of Scenedesmus obliquus. *Antioxidants, 8*(12), 600. https://doi.org/10.3390/antiox8120600

Mikyška, A., & Krofta, K. (2012). Assessment of changes in hop resins and polyphenols during long-term storage: Changes in hop resins and polyphenols during long-term storage. *Journal of the Institute of Brewing, 118*(3), 269–279. https://doi.org/10.1002/jib.40

Mitton, F. M., Turina, Y., Kulisz, N., Vittone, M., & Massa, A. (2025). Hop (Humulus lupulus) leaves as a functional ingredient to mitigate lipid oxidation and preserve PUFAs in rough scad patties. *Food Chemistry Advances, 6*, 100865. https://doi.org/10.1016/j.focha.2024.100865

Moir, M. (2000). Hops—A millennium review. *Journal of the American Society of Brewing Chemists, 58*(4), 131–146. https://doi.org/10.1094/ASBCJ-58-0131

Nabavi, S. M., Šamec, D., Tomczyk, M., Milella, L., Russo, D., Habtemariam, S., Suntar, I., Rastrelli, L., Daglia, M., Xiao, J., Giampieri, F., Battino, M., Sobarzo-Sanchez, E., Nabavi, S. F., Yousefi, B., Jeandet, P., Xu, S., & Shirooie, S. (2020). Flavonoid biosynthetic pathways in plants: Versatile targets for metabolic engineering. *Biotechnology Advances, 38*, 107316. https://doi.org/10.1016/j.biotechadv.2018.11.005

Nagasako-Akazome, Y., Honma, D., Tagashira, M., Kanda, T., Yasue, M., & Ohtake, Y. (2007). Safety evaluation of polyphenols extracted from hop bracts. *Food and Chemical Toxicology, 45*(8), 1383–1392. https://doi.org/10.1016/j.fct.2007.01.019

Neve, R. A. (with ebrary, Inc). (1991). *Hops*. Springer. https://doi.org/10.1007/978-94-011-3106-3

Nezi, P., Cicaloni, V., Tinti, L., Salvini, L., Iannone, M., Vitalini, S., & Garzoli, S. (2022). Metabolomic and proteomic profile of dried hop inflorescences (Humulus lupulus L. cv. Chinook and cv. Cascade) by SPME-GC-MS and UPLC-MS-MS. *Separations, 9*(8), 204. https://doi.org/10.3390/separations9080204

Nguyen, V., Taine, E. G., Meng, D., Cui, T., & Tan, W. (2024). Pharmacological activities, therapeutic effects, and mechanistic actions of Trigonelline. *International Journal of Molecular Sciences, 25*(6), 3385. https://doi.org/10.3390/ijms25063385

Nijssen, L. M. (1996). *Volatile compounds in food: Qualitative and quantitative data* (Centraal Instituut voor Voedingsonderzoek, Ed.; 7. ed).

Nikolic, D., Li, Y., Chadwick, L. R., Pauli, G. F., & Van Breemen, R. B. (2005). Metabolism of xanthohumol and isoxanthohumol, prenylated flavonoids from hops (*Humulus lupulus* L.), by human liver microsomes. *Journal of Mass Spectrometry, 40*(3), 289–299. https://doi.org/10.1002/jms.753

Novak, P., Krofta, K., & Matousek, J. (2006). Chalcone synthase homologues from Humulus lupulus: Some enzymatic properties and expression. *Biologia Plantarum, 50*(1), 48–54. https://doi.org/10.1007/s10535-005-0073-y

Obata, T. (2019). Metabolons in plant primary and secondary metabolism. *Phytochemistry Reviews, 18*(6), 1483–1507. https://doi.org/10.1007/s11101-019-09619-x

Okada, Y., Yamazaki, Y., Suh, D.-Y., Sankawa, U., & Ito, K. (2001). Bifunctional activities of Valerophenone synthase in hop (*Humulus lupulus* L.). *Journal of the American Society of Brewing Chemists, 59*(4), 163–166. https://doi.org/10.1094/ASBCJ-59-0163

Oosterveld, A. (2002). Characterization of hop pectins shows the presence of an arabinogalactan-protein. *Carbohydrate Polymers, 49*(4), 407–413. https://doi.org/10.1016/S0144-8617(01)00350-2

Parekh, N., Garg, A., Choudhary, R., Gupta, M., Kaur, G., Ramniwas, S., Shahwan, M., Tuli, H. S., & Sethi, G. (2023). The role of natural flavonoids as telomerase inhibitors in suppressing cancer growth. *Pharmaceuticals, 16*(4), Article 4. https://doi.org/10.3390/ph16040605

Patzak, J., Krofta, K., Henychová, A., & Nesvadba, V. (2015). Number and size of lupulin glands, glandular trichomes of hop (*Humulus lupulus* L.), play a key role in contents of bitter acids and polyphenols in hop cone. *International Journal of Food Science & Technology, 50*(8), 1864–1872. https://doi.org/10.1111/ijfs.12825

Proestos, C., & Komaitis, M. (2009). Antioxidant capacity of hops. In *Beer in health and disease prevention* (pp. 467–474). Elsevier. https://doi.org/10.1016/B978-0-12-373891-2.00045-6

Przybyś, M., & Skomra, U. (2020). *Hops as a source of biologically active compounds*. https://doi.org/10.26114/PJA.IUNG.438.2020.43.09

Raut, S., Von Gersdorff, G. J., Münsterer, J., Kammhuber, K., Hensel, O., & Sturm, B. (2021). Influence of pre-drying storage time on essential oil components in dried hops (*Humulus lupulus* L.). *Journal of the Science of Food and Agriculture, 101*(6), 2247–2255. https://doi.org/10.1002/jsfa.10844

Rehman, R., Hanif, M. A., Mushtaq, Z., & Al-Sadi, A. M. (2016). Biosynthesis of essential oils in aromatic plants: A review. *Food Reviews International, 32*(2), 117–160. https://doi.org/10.1080/87559129.2015.1057841

Rettberg, N., Thörner, S., & Garbe, L. A. (2013). Starting hop lipidomics—Isolation and characterization of non-polar, neutral and polar hop lipids. *Brewing Science, 66*, 176–184.

Rettberg, N., Biendl, M., & Garbe, L.-A. (2018). Hop aroma and hoppy beer flavor: Chemical backgrounds and analytical tools—A review. *Journal of the American Society of Brewing Chemists, 76*(1), 1–20. https://doi.org/10.1080/03610470.2017.1402574

Riedl, W. (1952). Konstitution und Synthese der Hopfenbitterstoffe *d*, *l* -Humulon und Lupulon sowie einiger Analoga (V. Mitteil). *Chemische Berichte, 85*(7–8), 692–710. https://doi.org/10.1002/cber.19520850706

Rigby, F. L., & Bethune, J. L. (1952). Cohumulone, a new hop constituent. *Journal of the American Chemical Society, 74*(23), 6118–6119. https://doi.org/10.1021/ja01143a517

Rigby, F. L., & Bethune, J. L. (1955). Components of the Lead-precipitable fraction of Humulus lupulus. Adhumulone. *Journal of the American Chemical Society, 77*(10), 2828–2830. https://doi.org/10.1021/ja01615a047

Rodolfi, M., Rinaldi, M., Caligiani, A., Paciulli, M., Lolli, V., Chiancone, B., & Ganino, T. (2022). Hop green sprouts preservation and valorisation as semi-finished and finished products: Impact of different treatments on microstructural, physical and chemical traits. *European Food Research and Technology, 248*(5), 1203–1215. https://doi.org/10.1007/s00217-021-03956-y

Rossini, F., Virga, G., Loreti, P., Provenzano, M. E., Danieli, P. P., & Ruggeri, R. (2020). Beyond beer: Hop shoot production and nutritional composition under Mediterranean climatic conditions. *Agronomy, 10*(10), 1547. https://doi.org/10.3390/agronomy10101547

Rubira, A., Rodríguez-Rebelo, F., Del-Río, P. G., Romaní, A., & Gullón, B. (2024). Valorization of hop (Humulus lupulus L.) pruning to produce valuable compounds using two biorefinery strategies: Conventional processing and microwave-assisted autohydrolysis. *Industrial Crops and Products, 219*, 119174. https://doi.org/10.1016/j.indcrop.2024.119174

Rutnik, K., Knez Hrnčič, M., & Jože Košir, I. (2022a). Hop essential oil: Chemical composition, extraction, analysis, and applications. *Food Reviews International, 38*(sup1), 529–551. https://doi.org/10.1080/87559129.2021.1874413

Rutnik, K., Ocvirk, M., & Košir, I. J. (2022b). Changes in hop (Humulus lupulus L.) oil content and composition during long-term storage under different conditions. *Food, 11*(19), 3089. https://doi.org/10.3390/foods11193089

Rutnik, K., Ocvirk, M., & Košir, I. J. (2023). The stability of hop (Humulus lupulus L.) resins during long-period storage. *Plants, 12*(4), 936. https://doi.org/10.3390/plants12040936

Sabbatini, G., Mari, E., Ortore, M. G., Di Gregorio, A., Fattorini, D., Di Carlo, M., Galeazzi, R., Vignaroli, C., Simoni, S., Giorgini, G., Guarrasi, V., Chiancone, B., Leto, L., Cirlini, M., Del Vecchio, L., Mangione, M. R., Vilasi, S., Minnelli, C., & Mobbili, G. (2024). Hop leaves: From waste to a valuable source of bioactive compounds—A multidisciplinary approach to investigating potential applications. *Heliyon, 10*(18), e37593. https://doi.org/10.1016/j.heliyon.2024.e37593

Saito, K., Yonekura-Sakakibara, K., Nakabayashi, R., Higashi, Y., Yamazaki, M., Tohge, T., & Fernie, A. R. (2013). The flavonoid biosynthetic pathway in Arabidopsis: Structural and genetic diversity. *Plant Physiology and Biochemistry, 72*, 21–34. https://doi.org/10.1016/j.plaphy.2013.02.001

Sanchez, N. B., Lederer, C. L., Nickerson, G. B., Libbey, L. M., & McDaniel, M. R. (1992). Sensory and analytical evaluation of beers brewed with three varieties of hops and an unhopped beer. In *Developments in food science* (Vol. 29, pp. 403–426). Elsevier. https://doi.org/10.1016/B978-0-444-88834-1.50037-5

Santarelli, V., Neri, L., Carbone, K., Macchioni, V., & Pittia, P. (2022). Use of conventional and innovative technologies for the production of food grade hop extracts: Focus on bioactive compounds and antioxidant activity. *Plants, 11*(1), Article 1. https://doi.org/10.3390/plants11010041

Sarraf, C., Desjardins, Y., Leonhart, S., Gosselin, A., & Gosselin, G. (2013). Agronomic and nutraceutical potential of hops (Humulus lupulus L.) grown in Québec, Canada. *Acta Horticulturae, 1010*, 155–161. https://doi.org/10.17660/ActaHortic.2013.1010.18

Schulz, C., Fritz, N., Sommer, T., Krofta, K., Friedland, K., & Pischetsrieder, M. (2018). Activation of membrane-located Ca2+ channels by hop beta acids and their tricyclic transformation products. *Food Chemistry, 252*, 215–227. https://doi.org/10.1016/j.foodchem.2018.01.073

Segawa, S., Yasui, K., Takata, Y., Kurihara, T., Kaneda, H., & Watari, J. (2006). Flavonoid glycosides extracted from hop (*Humulus lupulus* L.) as inhibitors of chemical mediator release from human basophilic KU812 cells. *Bioscience, Biotechnology, and Biochemistry, 70*(12), 2990–2997. https://doi.org/10.1271/bbb.60384

Sell, C. S. (2003). *A fragrant introduction to Terpenoid chemistry*. The Royal Society of Chemistry. https://doi.org/10.1039/9781847550019

Sharp, D. C., Townsend, M. S., Qian, Y., & Shellhammer, T. H. (2014). Effect of harvest maturity on the chemical composition of cascade and willamette hops. *Journal of the American Society of Brewing Chemists, 72*(4), 231–238. https://doi.org/10.1094/ASBCJ-2014-1002-01

Sharpe, F. R., & Peppard, T. L. (1977). Formation of sesquiterpene episulphides under mild conditions and their occurrence in the essential oil of hops.

Singla, R. K., Dubey, A. K., Garg, A., Sharma, R. K., Fiorino, M., Ameen, S. M., Haddad, M. A., & Al-Hiary, M. (2019). Natural polyphenols: Chemical classification, definition of classes, subcategories, and structures. *Journal of AOAC International, 102*(5), 1397–1400. https://doi.org/10.5740/jaoacint.19-0133

Spurgeon, S. L., & Porter, J. W. (1980). Carotenoids. In *Lipids: Structure and function* (pp. 419–483). Elsevier. https://doi.org/10.1016/B978-0-12-675404-9.50020-5

Srečec, S., Rezić, T., Šantek, B., & Marić, V. (2009). Hop pellets type 90: Influence of manufacture and storage on losses of α-acids. *Acta Alimentaria, 38*(1), 141–147. https://doi.org/10.1556/AAlim.2008.0014

Srivastava, S. (2024). A review on biosynthetic pathways in plants. *International Journal of Research Publication and Reviews, 5*(2), 668–673. https://doi.org/10.55248/gengpi.5.0224.0422

Steinhaus, M., & Schieberle, P. (2000). Comparison of the most odor-active compounds in fresh and dried hop cones (Humulus lupulus L. variety spalter select) based on GC-olfactometry and odor dilution techniques. *Journal of Agricultural and Food Chemistry, 48*(5), 1776–1783. https://doi.org/10.1021/jf9905141

Stevens, R. (1967). The chemistry of hop constituents. *Chemical Reviews, 67*(1), 19–71. https://doi.org/10.1021/cr60245a002

Stewart, G. G., & Priest, F. G. (Eds.). (2006). *Handbook of brewing*. CRC Press. https://doi.org/10.1201/9781420015171

Suggett, A., Moir, M., and Seaton, J.C. (1979). *The role of sulphur compounds in hop flavour*. Paper presented at: 17th EBC Congress (West Berlin).

Sun, S., Wang, X., Yuan, A., Liu, J., Li, Z., Xie, D., Zhang, H., Luo, W., Xu, H., Liu, J., Nie, C., & Zhang, H. (2022). Chemical constituents and bioactivities of hops (*Humulus lupulus L.*) and their effects on beer-related microorganisms. *Food and Energy Security, 11*(2), e367. https://doi.org/10.1002/fes3.367

Szymańska-Chargot, M., Cieśla, J., Pękala, P., Pieczywek, P. M., Oleszek, W., Żyła, M., Szkopek, Z., & Zdunek, A. (2022). The influence of high-intensity ultrasonication on properties of cellulose produced from the hop stems, the byproduct of the hop cones production. *Molecules, 27*(9), 2624. https://doi.org/10.3390/molecules27092624

Taiti, C., Di Matteo, G., Spano, M., Vinciguerra, V., Masi, E., Mannina, L., & Garzoli, S. (2023). Metabolomic approach based on analytical techniques for the detection of secondary metabolites from Humulus lupulus L. dried leaves. *International Journal of Molecular Sciences, 24*(18), 13732. https://doi.org/10.3390/ijms241813732

Taiz, L., & Zeiger, E. (2002). *Plant physiology*. Sinauer Associates.

Takoi, K., Itoga, Y., Koie, K., Kosugi, T., Shimase, M., Katayama, Y., Nakayama, Y., & Watar[illegible] J. (2010). The contribution of geraniol metabolism to the citrus flavour of beer: Synergy [illegible] geraniol and β-citronellol under coexistence with excess linalool. *Journal of the Institute [illegible] Brewing, 116*(3), 251–260. https://doi.org/10.1002/j.2050-0416.2010.tb00428.x

Tang, J., Dunshea, F. R., & Suleria, H. A. R. (2019). LC-ESI-QTOF/MS characterization of pheno[illegible]lic compounds from medicinal plants (hops and Juniper berries) and their antioxidant activit[illegible] *Food, 9*(1), 7. https://doi.org/10.3390/foods9010007

Taniguchi, Y., Matsukura, Y., Ozaki, H., Nishimura, K., & Shindo, K. (2013). Identification a[illegible] quantification of the oxidation products derived from α-acids and β-acids during storage [illegible]

hops (Humulus lupulus L.). *Journal of Agricultural and Food Chemistry, 61*(12), 3121–3130. https://doi.org/10.1021/jf3047187

Taniguchi, Y., Taniguchi, H., Matsukura, Y., Kawachi, Y., & Shindo, K. (2014a). Structural elucidation of Humulone autoxidation products and analysis of their occurrence in stored hops. *Journal of Natural Products, 77*(6), 1252–1261. https://doi.org/10.1021/np4008427

Taniguchi, Y., Taniguchi, H., Yamada, M., Matsukura, Y., Koizumi, H., Furihata, K., & Shindo, K. (2014b). Analysis of the components of hard resin in hops (*Humulus lupulus* L.) and structural elucidation of their transformation products formed during the brewing process. *Journal of Agricultural and Food Chemistry, 62*(47), 11602–11612. https://doi.org/10.1021/jf504394h

Taniguchi, Y., Yamada, M., Taniguchi, H., Matsukura, Y., & Shindo, K. (2015). Chemical characterization of beer aging products derived from hard resin components in hops (*Humulus lupulus* L.). *Journal of Agricultural and Food Chemistry, 63*(46), 10181–10191. https://doi.org/10.1021/acs.jafc.5b04138

Taylor, A. W., Barofsky, E., Kennedy, J. A., & Deinzer, M. L. (2003). Hop (Humulus lupulus L.) proanthocyanidins characterized by mass spectrometry, acid catalysis, and gel permeation chromatography. *Journal of Agricultural and Food Chemistry, 51*(14), 4101–4110. https://doi.org/10.1021/jf0340409

Telichowska, A., Kobus-Cisowska, J., Stachowiak, B., Byczkiewicz, S., Zawadzka, A., & Szulc, P. (2020). Comparative quantitative study on phytochemical contents of hops (Humulus lupulus L.). *Journal of Research and Applications in Agricultural Engineering, 65*(2), 18 23.

Tembo, G. M. (2020). Effects of storage conditions on alpha acid degradation of Indiana Grown Hops (Humulus lupulus). *Purdue University*, *December*. /articles/thesis/Effects_of_storage_conditions_on_alpha_acid_degradation_of_Indiana_Grown_Hops_Humulus_lupulus_/13360154/1

Tronina, T., Popłoński, J., & Bartmańska, A. (2020). Flavonoids as phytoestrogenic components of hops and beer. *Molecules, 25*(18), Article 18. https://doi.org/10.3390/molecules25184201

Urban, J., Dahlberg, C. J., Carroll, B. J., & Kaminsky, W. (2013). Absolute configuration of beer's bitter compounds. *Angewandte Chemie International Edition, 52*(5), 1553–1555. https://doi.org/10.1002/anie.201208450

Van Cleemput, M., Cattoor, K., De Bosscher, K., Haegeman, G., De Keukeleire, D., & Heyerick, A. (2009). Hop (*Humulus lupulus*)-derived bitter acids as multipotent bioactive compounds. *Journal of Natural Products, 72*(6), 1220–1230. https://doi.org/10.1021/np800740m

Velde, N. V. D., & Verzele, M. (1986). High performance liquid chromatography of hop and beer extracts with photodiode array detection. *Journal of the Institute of Brewing, 92*(6), 584–587. https://doi.org/10.1002/j.2050-0416.1986.tb04457.x

Veríssimo, L. S., Ferreira, A., Pinheiro, P. F., & Ribeiro, J. S. (2022). Chemometric studies of hops degradation at different storage forms using UV-vis, NIRS and UPLC analyses. *Brazilian Journal of Food Technology, 25*, e2021093. https://doi.org/10.1590/1981-6723.09321

Verzele, M. (1958). Posthumulone, A new "alpha acid". *Bulletin Des Sociétés Chimiques Belges, 67*(5–6), 278–279. https://doi.org/10.1002/bscb.19580670511

Verzele, M., & Govaert, F. (1949). On the constitution of the β hops acid. *Bulletin Des Sociétés Chimiques Belges, 58*(10–12), 432–449. https://doi.org/10.1002/bscb.19490581003

Verzele, M., & De Keukeleire, D. (1991). *Chemistry and analysis of hop and beer bitter acids*. Elsevier.

Vidmar, M., Abram, V., Čeh, B., Demšar, L., & Poklar Ulrih, N. (2019). White hop shoots production in Slovenia: Total phenolics, microelements and pesticides content from five commercial cultivars. *Food Technology and Biotechnology, 57*(4). https://doi.org/10.17113/ftb.57.04.6269

Vitalini, S., Di Martile, M., Cicaloni, V., Iannone, M., Salvini, L., Del Bufalo, D., Iriti, M., & Garzoli, S. (2023). Volatile and non-volatile content determination and biological activity evaluation of fresh Humulus lupulus L. (cv. Chinook) leaves and inflorescences. *Separations, 10*(2), 91. https://doi.org/10.3390/separations10020091

Wang, G., Tian, L., Aziz, N., Broun, P., Dai, X., He, J., King, A., Zhao, P. X., & Dixon, R. A. (2008). Terpene biosynthesis in glandular trichomes of hop. *Plant Physiology, 148*(3), 1254–1266. https://doi.org/10.1104/pp.108.125187

Wang, Y., Chung, S.-J., Song, W. O., & Chun, O. K. (2011). Estimation of daily proanthocyanidin intake and major food sources in the U.S. diet. *The Journal of Nutrition, 141*(3), 447–452. https://doi.org/10.3945/jn.110.133900

Wang, Y., Song, J., & Nie, C. (2024). A review of biotransformation of sulfur compounds in beer brewing. *BIO Web of Conferences, 111*, 01018. https://doi.org/10.1051/bioconf/202411101018

Wieland, H. (1925a). Über die chemische Natur der Hopfenharz-Säuren (I.). (In Gemeinschaft mit Walter Schneider und Erwin Martz). *Berichte Der Deutschen Chemischen Gesellschaft (A and B Series), 58*(1), 102–112. https://doi.org/10.1002/cber.19250580124

Wieland, H. (1925b). Über die chemische Natur der Hopfenharz-Säuren (II.). (In Gemeinschaft mit Erwin Martz und Heinz Hoek.). *Berichte Der Deutschen Chemischen Gesellschaft (A and B Series), 58*(9), 2012–2017. https://doi.org/10.1002/cber.19250580912

Wöllmer, W. (1925). Über die Bitterstoffe des Hopfens. *Berichte Der Deutschen Chemischen Gesellschaft (A and B Series), 58*(4), 672–678. https://doi.org/10.1002/cber.19250580407

Xie, L., Deng, Z., Zhang, J., Dong, H., Wang, W., Xing, B., & Liu, X. (2022). Comparison of flavonoid O-glycoside, C-glycoside and their Aglycones on antioxidant capacity and metabolism during in vitro digestion and in vivo. *Food, 11*(6), 882. https://doi.org/10.3390/foods11060882

Zagoskina, N. V., Zubova, M. Y., Nechaeva, T. L., Kazantseva, V. V., Goncharuk, E. A., Katanskaya, V. M., Baranova, E. N., & Aksenova, M. A. (2023). Polyphenols in plants: Structure, biosynthesis, abiotic stress regulation, and practical applications (review). *International Journal of Molecular Sciences, 24*(18), 13874. https://doi.org/10.3390/ijms241813874

Zanoli, P., & Zavatti, M. (2008). Pharmacognostic and pharmacological profile of Humulus lupulus L. *Journal of Ethnopharmacology, 116*(3), 383–396. https://doi.org/10.1016/j.jep.2008.01.011

Zhang, G., Zhang, N., Yang, A., Huang, J., Ren, X., Xian, M., & Zou, H. (2021). Hop bitter acids: Resources, biosynthesis, and applications. *Applied Microbiology and Biotechnology, 105*(11), 4343–4356. https://doi.org/10.1007/s00253-021-11329-4

Zhou, P., Kang, J., Cheng, Q., Chen, M., Xie, Y., & Zhou, H. (2022). Therapeutic potential of traditional Chinese medicine against atherosclerosis: Targeting trimethylamine N-oxide. *Phytomedicine, 104*, 154305. https://doi.org/10.1016/j.phymed.2022.154305

Chapter 10
Sustainable Hop Extracts for Food and Non-Food Applications

Veronica Santarelli and Katya Carbone

10.1 Introduction

Plants represent a valuable source of fibre and bioactive compounds for a range of sectors, including pharmaceuticals, cosmetics, and agriculture. In the botanical kingdom, hops constitute a copious reservoir of bioactive compounds, a number of which are exclusive to this species. A notable example is 8-prenylnaringenin (8PN), which is recognised as the most effective phytoestrogen identified to date (Carbone & Gervasi, 2022). The extraction process represents the initial stage in the use, quantification, and identification of these compounds, and it is widely regarded as the most significant phase in these procedures (Tušek et al., 2018). In recent years, there has been a notable development of methodologies for extracting bioactive compounds from plants. However, there is a growing interest in the development of sustainable methods. The objective is twofold: first, to minimise or eliminate the use of organic solvents, which are detrimental to the environment; and second, to enhance the safety and quality of extracts. In addition, there is a concerted effort to employ extraction techniques that can reduce waste production and energy consumption while maintaining cost-effectiveness and efficiency (Da Silva et al., 2022). This shift has led to the emergence of "green extractions" (GEs), a term that encompasses methodologies and technologies aimed at minimizing the environmental impact of extraction processes while maximizing efficiency. The adoption of GE methods, in fact, not only aligns with global sustainability goals but also offers

V. Santarelli
Department of Bioscience and Technology for Food, Agriculture and Environment, University of Teramo, Teramo, Italy

K. Carbone (✉)
CREA – Research Centre for Olive, Fruit and Citrus Crops, Rome, Italy
e-mail: katya.carbone@crea.gov.it

K. Carbone (ed.), *All About Hops: The Crop, its Cultivation, and its Uses in Brewing and Beyond*, https://doi.org/10.1007/978-3-031-96606-4_10

economic benefits by reducing operational costs and enhancing the market appeal of products (Chemat et al., 2019a, b; Herrero, 2024).

GEs focus on employing renewable energy sources, reducing waste generation, and implementing eco-friendly practices throughout the extraction lifecycle. This approach encompasses a range of techniques such as using environmentally friendly solvents, optimizing extraction conditions to minimize energy consumption, and implementing closed-loop systems to recycle resources and minimize waste (More et al., 2022).

10.2 Conventional and Innovative Extraction Techniques

Extraction techniques can be classified into conventional and innovative methods. Conventional extraction techniques include distillation, and solid-liquid extractions (SLE), such as percolation, maceration and mechanical stirring extraction (Da Silva et al., 2022).

However, these techniques have several drawbacks, making them economically inefficient and environmentally unsustainable due to their significant time and energy consumption as well as the use of polluting solvents (Santarelli et al., 2021).

The exploitation of polyphenols as bioactive compounds at various commercial levels has motivated scientists to explore eco-friendly, efficient, and cost-effective extraction techniques based on a green extraction approach. GEs are promising for the extraction of thermolabile bioactive compounds because of their advantages over conventional, time-consuming, and laborious extraction techniques, such as reduced solvent use time and energy consumption and higher recovery rates with lower operational costs (Ameer et al., 2017).

Modern GE techniques include Supercritical Fluid Extraction (SFE), Pressurized Liquid Extraction (PLE), High Hydrostatic Pressure Extraction (HHPE), Ultrasound-Assisted Extraction (UAE), Microwave-Assisted Extraction (MAE), Pulsed Electric Field Extraction (PEFE) (Ameer et al., 2017; Mena-García et al., 2019; Rodríguez García & Raghavan, 2022).

Moreover, recent developments in integrating the principles of green chemistry into analytical methods have prompted the creation and adoption of novel solvents to replace traditional organic solvents, known for their high volatility, flammability, and toxicity. Over the past two decades, supercritical fluids, ionic liquids, and deep eutectic solvents have become the most actively investigated as potential green solvents. Utilising the solvation properties of these innovative solvents and their synthetic versatility, modern literature has highlighted a wide range of applications in the field of green analytical chemistry (Choi & Verpoorte, 2019; Pacheco-Fernández & Pino, 2019).

Table 10.1 presents a comprehensive overview of the various green extraction methods, focusing on their respective advantages, limitations, and areas of application.

Table 10.1 Comparison of different green extraction techniques: main advantages, disadvantages, and applications

Extraction method	Advantages	Disadvantages	Applications	Extracted compounds	References
SFE	Reduced extraction time Highly suitable for the extraction of solid/liquid compounds with lower volatility Cost-effective handling Ease of recovering the used solvent The use of co-solvents with co-solutes results in enhanced efficiency	Not suitable for polar molecules CO_2 cannot always be used as an extraction solvent (due to its nonpolar nature for polar solutes) Complex equipment	Extraction of natural products and thermosensitive compounds from a wide variety of plant matrices Recovery of byproducts from fruits and vegetables	Phenolic compounds Prenylflavonoids α-acids and β-acids Essential oils Carotenoids Sugars	Ameer et al. (2017), Knez Hrnčič et al. (2019), and Rodríguez García and Raghavan (2022)
FLE	Reduced extraction time and solvent consumption Improved solvent penetration into the solid matrix, increased mass transfer rate and diffusion High extraction yield Employing solvents with different polarities enables the extraction of polar and non-polar compounds Allows substituting hazardous solvents with environmentally friendly ones	High cost of equipment The application of mild temperatures may lead to the degradation of thermolabile compounds and a decrease in selectivity	Extraction of bioactive compounds from a wide range of vegetables, spices, and fruits	Phenolic compounds Anthocyanins Flavonoids Terpenoids Essential oils Amino acids Melanoidins Carbohydrates	Donn et al. (2024), Esquivel-Hernández et al. (2017), Lefebvre et al. (2021), Mena-García et al. (2019), and Picot-Allain et al. (2021)

(continued)

Table 10.1 (continued)

Extraction method	Advantages	Disadvantages	Applications	Extracted compounds	References
HHPE	Enhances mass transfer rates Low energy consumption High extraction yields High extraction selectivity Shortened processing time Can be used for polar and non-polar compounds	Great inceptive investment Difficulty of maintenance	Extraction of bioactive compounds from vegetables, roots, fruits, and waste peel	Polyphenols Flavonoids Anthocyanins Carotenoids Essential oils Sugars	Oreopoulou et al. (2019), Rifna et al. (2023), and Rodríguez García and Raghavan (2022)
UAE	High extraction yield Short extraction time Easy to use Low solvent usage Low energy and electricity consumption Appropriated for thermo-sensitive compounds Low initial investment and low operating cost	Limited extraction efficiency Degradation of bioactives due to oxidative pyrolysis caused by hydroxyl (OH^-) radicals during cavitation	Extraction of bioactive compounds from a wide range of vegetables (leaves and inflorescence), spices, officinalis herbs, roots, fruits, and food by-products	Phenolic compounds Anthocyanins Carotenoids Essential oils α-acids and β-acids Volatile organic compounds	Ameer et al. (2017), Bitwell et al. (2023), Herrero (2024), Rifna et al. (2023), Rodríguez García and Raghavan (2022), and Shen et al. (2023)

MAE	High extract quality High selectivity High extraction yield Short extraction time	Expensive equipment Unsuitable for heat-sensitive compounds	Extraction of bioactive compounds from a wide range of vegetables, spices, officinalis herbs, roots, and fruits	Phenolic compounds Anthocyanins Essential oils α-acids and β-acids	Ameer et al. (2017), Bitwell et al. (2023), Herrero (2024), Rifna et al. (2023), and Rodríguez García and Raghavan (2022)
PEFE	Short extraction time High extraction efficiency High selectivity Low energy consumption High extract quality	High maintenance Accurate control parameters	Extraction of bioactive compounds from a wide range of vegetables (leaves and inflorescence), spices, officinalis herbs, roots, fruits, and food by-products	Phenolic compounds Phytosterols α-acids and β-acids Terpenoids Anthocyanins Carbohydrates	Poojary et al. (2020), and Rodríguez García and Raghavan (2022)

(continued)

Table 10.1 (continued)

Extraction method	Advantages	Disadvantages	Applications	Extracted compounds	References
ILs DESs NaDESs	Biodegradability Recyclability Low costs Low toxicity DESs or NADESs are easily synthesized It is possible to fine-tune physicochemical characteristics for a specific purpose	For ILs synthesis, organic solvents are required to dilute the components High density and viscosities of DESs/NADESs: lower extraction efficiencies, the energy required for stirring and pumping; low diffusion rate and mass transfer, decreasing its solvation ability NADESs have almost zero vapour and recovery of the target compound could be difficult	Extraction of bioactive compounds from different vegetable matrices: herbs, flowers, fruits, and food waste	Phenolic compounds Carotenoids α-acids and β-acids Carbohydrates Anthocyanins Essential oils	Benvenutti et al. (2019), El Achkar et al. (2021), and Herrero (2024)

SFE Supercritical Fluid Extraction, *PLE* Pressurized Liquid Extraction, *HHPE* High Hydrostatic Pressure Extraction, *UAE* Ultrasound-Assisted Extraction, *MAE* Microwave-Assisted Extraction, *PEFE* Pulsed Electric Field Extraction, *ILs* Ionic liquids, *DESs* Deep Eutectic Solvents, *NaDESs* Natural Deep Eutectic Solvents

The following paragraphs provide a brief overview of the principles behind the different GE technologies, and their applications in the production of green hop extracts are examined.

10.2.1 Supercritical Fluid Extraction (SFE)

Supercritical fluid extraction (SFE) is an innovative method for extracting compounds with food-enhancing and health-promoting properties from a wide range of plants and biomaterials. This extraction technique offers efficiency, cost-effectiveness, and promising avenues for resource recovery while maintaining the quality of extracted products (Essien et al., 2020).

A supercritical (sc) fluid can be defined as a fluid existing in a phase that possesses the characteristics of both liquids (solubilization capacity) and gases (diffusion, surface tension, viscosity, etc.). This condition can be reached when a substance is exposed to temperatures and pressures exceeding its critical point (Fig. 10.1). More in detail, the critical pressure (Pc) is the minimum pressure required to liquefy a gas at its critical temperature, and the critical temperature (Tc) of a gas is the temperature at which the gas remains in a gaseous state until additional pressure is applied to induce liquefaction (Ameer et al., 2017; Bitwell et al., 2023; More et al., 2022).

The most widely used supercritical solvent is CO_2 (sc-CO_2) because of its advantageous properties, namely its low cost and the fact that it has been designated as Generally Recognised as Safe (GRAS) by both the European Food Safety Authority

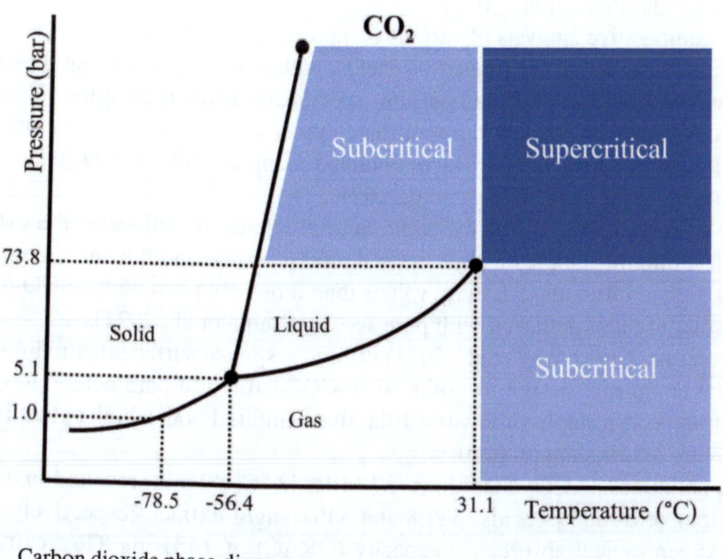

Fig. 10.1 Carbon dioxide phase diagram

(EFSA) and the United States Food and Drug Administration (FDA) (Ameer et al., 2017). Furthermore, CO_2 is easily volatilized once system pressure is released, thus providing a solvent-free high-purity extract. However, due to the low polarity of CO_2, its extraction capability is optimal for non-polar compounds. To address this limitation and enhance polarity, a co-solvent is employed (Chemat et al., 2020). Ethanol is the most widely used co-solvent for SFE (More et al., 2022).

Briefly, during the SFE process, plant material was fed into an extraction vessel. The desired extraction conditions were maintained by the operation of a pressure release valve and temperature controllers attached to the extraction vessel (Ameer et al., 2017). The target compounds dissolved when the supercritical state of CO_2 was achieved. Supercritical CO_2 and dissolved compounds were, then, passed to the separators, and the extract was collected.

At the end of the process, CO_2 is restored and cycled or released into the surroundings (Chemat et al., 2020; More et al., 2022).

The factors that influence SFE efficiency include the correct selection of the supercritical fluid, sample solubility, polarity, moisture content, and particle size (Ameer et al., 2017).

SFE is one of the most widely employed GE techniques for the extraction of bioactive compounds from hops (Table 10.2). Notably, sc-CO_2 exhibits high selectivity for lupulones (Rodrigues Arruda, 2022).

Pinto et al. (2023) recently studied the impact of temperature and pressure variations on the extraction process of hop cones using sc-CO_2. Researchers successfully obtained rich bitter acid extracts using sc-CO_2 extraction process at a temperature of 40 °C and a pressure of 20 MPa. The extraction time was reduced to 40 min. However, the study also revealed that the interaction between pressure and temperature significantly influenced the extraction yield, despite not affecting the purity of α- and β-acids (Pinto et al., 2023).

In a comparative analysis of various compressed fluid extraction methods, Veiga et al. (2021) evaluated the impact on yields, antioxidant activities, total phenolics, and flavonoid content of Brazilian hop, specifically investigating the influence of ethyl acetate as a co-solvent in a semi-batch process on extraction time and extract characteristics compared to extracts obtained using sc-CO_2, sc-CO_2 with ethanol, and compressed propane (Veiga et al., 2021).

The findings revealed that the addition of ethyl acetate enhanced the extraction yield, resulting in samples with increased total phenolic and flavonoid contents, as well as higher antioxidant activity values than those obtained with sc-CO_2/ethanol, and similar to those achieved with pure sc-CO_2 (Veiga et al., 2021).

In contrast, Nagybákay et al. (2021) utilised response surface methodology with a central composite design to optimise sc-CO_2 extraction parameters, to obtain a hop extract from a single variety (cv. Ella) that exhibited both a high extraction yield and in vitro antioxidant properties.

The optimized sc-CO_2 conditions (37 MPa, 43 °C, 80 min) resulted in high concentrations of α- and β- acids (522.8 and 345.0 mg/g extract, respectively) with in vitro oxygen radical absorbance capacity (ORAC) of 1481 mg TE/g extract. It is noteworthy that this method granted a threefold increase in extraction yield,

Table 10.2 Key Studies on Supercritical Fluid Extraction (SFE) of Bioactive Compounds from Hops: Extraction Parameters and Major Findings

Extraction method	Plant matrix	Evaluated parameters	Tested condition	Optimum condition	Main goals/conclusion	References
sc-CO_2	Hop cones	BAs	40 or 60 °C 15–20–25–30 MPa	40 °C 20 MPa 40 min	A hop extract rich in BAs was obtained by using the optimized condition Higher yields of extracts and BAs were achieved during the constant extraction phase (15 min) and the falling extraction phase (40 min)	Pinto et al. (2023)
sc-CO_2 + co-solvent	Hop flowers	EY TPC TFC AA	sc-CO_2 sc-CO_2 + ethanol sc-CO_2 + ethyl acetate 40–60–80 °C 15–20–25 MPa	sc-CO_2 + ethyl acetate 80 °C 25 MPa	The use of ethyl acetate enhances the EY and produces samples with higher TPC, TFC, and AA values compared to sc-CO_2 + ethanol Ethanol (used as a co-solvent) increased the overall extraction yield but reduced the selectivity for phenolic compounds	Veiga et al. (2021)
sc-CO_2	Hop pellets	ORAC BAs TCl and TCa Volatile Composition	10 MPa; 40 °C; 300 min 12.5 MPa; 40 °C; 300 min 15 MPa; 40 °C; 300 min 37 MPa; 43 °C; 80 min	37 MPa 43 °C 80 min	Under the optimized conditions, the EY and antioxidant recovery were up to 3 times higher than the traditional commercial one-stage SFE-CO_2 process (10–15 MPa, 40 °C) Reduction in the extraction time of BAs from hop pellet	Nagybákay et al. (2021)

(continued)

Table 10.2 (continued)

Extraction method	Plant matrix	Evaluated parameters	Tested condition	Optimum condition	Main goals/conclusion	References
sc-CO_2	Hop pellets	EY BAs	CO_2 Propane Sulfur hexafluoride (SF_6) DME 50–100–150 bar 20–40–60–80 °C	DME α-acids 40 °C; 100 bar β-acids 60 °C; 50 bar < EY 60 °C; 150 bar	The highest EY was obtained using DME, followed by propane under the same conditions The EY of bitter acids is different depending on the solvent, the temperature, and pressure conditions DME showed a high temperature and pressure variability with a remarkable effect on the EY and concentration of BAs SF_6 was demonstrated to be an ineffective solvent for extracting hop resins	Bizaj et al. (2021)
sc-CO_2	Hop pellets	EY XN EOs	CO_2 Propane DME 50–100–150 bar 20–40–60–80 °C	DME was the best extraction solvent for XN < EY 40 °C; 50 bar	The different extraction solvents and process conditions affect the volatile composition of hop extracts CO_2 proved ineffective for extracting XN in the experimental range of temperature (20 °C to 80 °C) and pressure (100 to 150 bar) adopted	Bizaj et al. (2022)

Two-Step sc-CO_2	Hop cones	XN Volatile compounds	Optimization of 2-step extraction: 1st step: CO_2; 50 °C 250 bar 2nd step: CO_2:Ethanol (99:1; v/w); 85 °C; 800 bar	The extract from the 1st step comprises BAs and terpene derivatives The final extract was enriched with XN	This method allows to obtain extract with enhanced composition of the bioactive compounds	Klimek et al. (2021)
sc-CO_2	Hop cone residue	AA Cytotoxic activity EY Antimicrobial activity TBARS OxHLIA Terpene and BAs composition	Static extraction period 20 min Solvent flow rate of 2.0 ± 0.1 mL/min 40–50–60 °C 10–15–20 Mbar 5–7.5–10 % Ethanol (co-solvent)	40 °C 10 Mbar 5–10% Ethanol	The optimized conditions allow for higher concentrations of specific compounds, such as α-copaene, γ-elemene, nerolidyl acetate, linalyl isobutyrate, α-acorenol, and geranyl isovalerate, along with enhanced antioxidant potential, lipophilic composition and biological activity linked to the extraction method and hop variety The percentage of ethanol used as a co-solvent could influence the extract composition and biological activity	Inumaro et al. (2023)

sc supercritical, *BAs* Bitter acids, *EY* Extraction Yield, *TPC* Total Phenolic Content, *TFC* Total Flavonoid Content, *AA* Antioxidant Activity, *ORAC* Oxygen Radical Absorbance Capacity, *TCl* total chlorophylls, *TCa* total carotenoids, *DME* dimethyl ether, *XN* Xanthohumol, *EOs* Essential Oils, *TBARS* Analysis of the thiobarbituric acid reactive substances, *OxHLIA* Oxidative haemolysis inhibition assay antioxidant recovery (389.8 mg TE/g pellets), and extraction of bitter acids (228.4 mg/g pellets), while requiring a substantially shorter extraction time compared to the commercially employed one-stage sc-CO_2 process (10–15 MPa and 40 °C) (Nagybákay et al., 2021).

Ella hop sc-CO_2 extracts also contained a high relative content of the major identified volatiles (including β-pinene, β-myrcene, β-humulene, α-humulene, α-selinene, and methyl-4-decenoate).

In their study, Bizaj et al. (2021, 2022) examined the efficiency of sc-CO_2 in extracting bitter acids from hops using different solvents, including carbon dioxide (CO_2), propane, sulfur hexafluoride (SF_6), and dimethyl ether (DME), at both low and high density (Bizaj et al., 2021, 2022). The study investigated the impact of process parameters on the total extraction yield and the content of bitter acids in the hop extracts. The findings revealed that the extraction yield of bitter acids was significantly influenced by the choice of solvent, as well as the temperature and pressure applied during extraction. Specifically, CO_2 and propane exhibited solvating capabilities comparable to DME, while SF_6 demonstrated lower extraction performance under the same conditions. The optimal conditions for extraction, which granted the highest yield and concentration of bitter acids, were determined to be DME at 40 °C and 100 bar. The optimal operating conditions for DME extraction were determined to be 40 °C and 100 bar for α-acids (maximum concentration 9.6%), 60 °C and 50 bar for β-acids (4.5%), and 60 °C and 150 bar for maximum extraction yield (25.6%).

In their second study, the authors also assessed the effect of SFE on the extraction of essential oils (EOs) and Xanthohumol (XN) in extracts obtained from Slovenian hops, utilizing fluids with varying polarity. The extraction process was carried out semi-continuously, employing sub- and supercritical fluids with differing polarities: CO_2 and propane as non-polar agents, and DME as the polar solvent, as well as different temperatures (from 20 to 80 °C) and pressure conditions (from 100 to 150 bar). The results highlighted that the highest concentration of XN was achieved using DME. Furthermore, hop oil derived from CO_2 extracts (under specific conditions) exhibited the highest relative concentrations of linalool, β-caryophyllene, and α-humulene. On the other hand, oil obtained from propane extracts showed the highest content of the other five selected components: myrcene, geraniol, farnesene, α-selinene, and δ-cadinene. The relative content of the investigated EO components in DME extracts was similar to that found in propane extracts (Bizaj et al., 2022).

Klimek et al. (2021) proposed an efficient extraction method for the isolation of bioactive compounds from the Polish hop variety "Marynka". Results showed that the first extract, obtained under conditions of CO_2 at 50 °C and 300 bar, comprised α-acids, β-acids, terpene derivatives, and pure XN, exhibiting a higher yield compared to alternative methods (Klimek et al., 2021). Subsequently, the residues remaining after the first extraction were subjected to a second step using a combination of CO_2 and ethanol at 85 °C and 800 bar, thus obtaining an extract enriched in XN.

Supercritical fluid-assisted extraction was used to extract bioactive compounds not only from hop cones/pellets but also from hop cone residue from craft breweries (spent hops), with promising results. In fact, in their research, Inumaro et al. (2023) aimed to employ supercritical fluid-assisted extraction on spent hops, with the aim to determine the optimal extraction parameters (Inumaro et al., 2023). The assessment of the yield, the extraction of compounds, and the measurement of antioxidant activities were conducted via in vitro chemical and cell-based assays. In addition, the investigation encompassed the evaluation of antimicrobial and cytotoxic potentials. The authors investigated several parameters including temperature (ranging from 40 to 60 °C), pressure (from 10 to 20 MPa), and ethanol percentage as a co-solvent (ranging from 5% to 10%).

It was found that a temperature of 40 °C, a pressure of 10 MPa, and an ethanol concentration of 5% yielded higher concentrations of specific aromatic compounds, including α-copaene, γ-elemene, γ-muurolene, α-muurolene, and caryophyllene oxide, along with enhanced antioxidant activity, although the overall yields were lower. Conversely, maintaining the same temperature and pressure but increasing the ethanol concentration to 10% resulted in elevated levels of other compounds, such as nerolidyl acetate, linalyl isobutyrate, α-acorenol, and geranyl isovalerate, as well as an increase in antioxidant activity within the chemical systems. The study's findings demonstrate the effectiveness and high yield of modifier-enhanced sc-CO_2 extraction in retrieving compounds from hop cone residue used in craft beer production. Additionally, the study shows that the chemical composition and mass percentage of extracts obtained from hop cone residue are comparable to those of virgin hop cones, emphasising their potential as significant co-products.

10.2.2 Pressurized Liquid Extraction (PLE)

PLE is another GE technique mostly used for the extraction of natural products from food and plant sample matrices. This technique is also known as pressurized fluid extraction (PFE), pressurized solvent extraction (PSE), accelerated solvent extraction (ASE), or enhanced solvent extraction (ESE). When PLE is carried out with water as extraction solvent, it is also called subcritical water extraction (SWE), superheated water extraction (SHWE) or pressurized hot-water extraction (PHWE) (Bitwell et al., 2023; Ummat et al., 2021).

In PLE, a rapid extraction rate is achieved due to the combination of two factors: the interaction between the liquid solvent and matrix molecules, and the increase in temperature and pressure to facilitate efficient extraction of the target compounds (Ameer et al., 2017).

The extraction process involves the use of organic solvents, which are heated to temperatures above their boiling points (ranging from 20 to 200 °C) and maintained under elevated pressures (34–204 bar) to ensure their liquid state is sustained. Increasing the temperatures of solvents above their atmospheric boiling points allows for the increase of solubility and mass transfer rates between the plant matrix and the selected solvent.

Consequently, the enhanced diffusivity of the solvent gives rise to more pronounced extraction kinetics (Mena-García et al., 2019).

PLE is regarded as a green extraction method because it uses smaller amounts of organic solvents (~10–30% w/v). This reduces the amount of waste and the need to produce more organic solvents (Esquivel-Hernández et al., 2017).

The factors that influence the efficiency of PLE include matrix characteristics (e.g. particle size), operating conditions (e.g. pressure and temperature), the solvent-to-matrix ratio, the extraction time, mixing and flow rate.

In a recent study, Pichon et al. (2025) investigated the advancement of green solvent extraction techniques to optimize industrial-scale extraction processes. The study evaluated different extraction methods, including SM, TSE, ultrasound-assisted extraction (UAE), pressurized liquid extraction (PLE), and MAE. Additionally, a mixed approach combining UAE and SM was analysed. The extraction techniques were compared based on extraction yield, total flavonoid, polyphenol, and chlorophyll content, as well as metabolite profiling, LC-MS/MS quantification, and molecular networking analysis. Among the tested methods, PLE exhibited superior extraction efficiency, achieving higher yields with reduced processing time compared to maceration. The extraction yields demonstrated a positive correlation with increased concentrations of α/β hop acids, polyphenols, and flavonoids. Notably, PLE resulted in enhanced recovery of flavonoids, polyphenols, and xanthohumol compared to simple maceration.

10.2.2.1 Superheated Water Extraction (SHWE)

Superheated Water Extraction (SHWE), also known as pressurized hot water extraction (PHWE) or subcritical water extraction (SWE), is a water-based extraction method of particular interest due to the low cost of water, its non-toxic and non-flammable properties, and its status as an environmentally friendly solvent. PHWE is based on the dramatic change in the dielectric constant of water at high temperatures (Lefebvre et al., 2021).

The use of PHWE is contingent on the transformation of water above its atmospheric boiling point of 100 °C at 0.1 MPa, yet below its critical threshold of 374 °C at 22.1 MPa, into its non-polar form. This process facilitates the dissolution of non-polar phytochemicals (Bitwell et al., 2023).

Furthermore, elevated temperatures promote the mobility of water molecules, thereby reducing intermolecular hydrogen bonding and inducing a lower polarity (Essien et al., 2020; Lefebvre et al., 2021). However, it should be noted that the selectivity behaviour of water is subject to variation due to a number of factors, including temperature, extraction time/extraction cycle, the chemical structure of the compounds of interest, and the solvent to matrix ratio (Lefebvre et al., 2021).

The SWE process involves the transfer of solutes from various active sites within the sample matrix to the extraction medium. This transfer occurs via diffusion, partitioning equilibrium and convection mechanisms (Essien et al., 2020). The efficiency of PHWE can be significantly enhanced through its integration with alternative extraction techniques (e.g. pulsed electric field) (Bitwell et al., 2023).

10.2.2.2 Subcritical CO_2 Extraction

Subcritical CO_2 (subCO_2) extraction is an advanced green technique for selectively recovering phytocompounds, operating at moderate temperatures and pressures (10–80 bar and 25–45 °C) to maintain CO_2 in a liquid-like state. Milder extraction conditions increase selectivity by modulating the density and polarity of CO_2 through pressure changes, thus enabling the separation of target compounds. In addition to the advantageous characteristics (non-toxic, non-flammable, recyclable, cost-effective, easy extract separation, absence of residues) associated with subcritical CO_2 extraction, the principal benefit of reducing the extraction temperature lies in the preservation of volatile compounds susceptible to thermal degradation or volatilization. This, in turn, leads to increased yield due to the heightened solvent density (Fischer et al., 2023).

Most extraction studies in the literature employ CO_2 in the supercritical phase, while the application of subcritical conditions to extract the compounds of interest from hops remains largely unexplored.

In this context, Fischer et al. (2023) attempted to obtain and characterise essential oils and extracts from hop pellets of different hop varieties (El Dorado, Polaris, Hallertau Blanc and Callista) obtained using hydrodistillation and subCO_2 extraction methods (Fischer et al., 2023). The extraction process utilising subCO_2 exhibited superior yield and efficiency in the extraction of phenolic compounds and flavonoids from hops when compared with the hydrodistillation method. Furthermore, the extracts exhibited superior antioxidant activity in comparison to essential oils, attributed to the presence of phenolic compounds and flavonoids. The essential oils were found to contain β-myrcene, α–humulene, and β-caryophyllene as the primary constituents. However, extracts obtained with subCO_2 showed a chemical composition that was partially distinct from that of essential oils, with larger molecules such as oxygenated and non-oxygenated sesquiterpenes contributing to a distinct chemical profile.

10.2.3 High Hydrostatic Pressure Extraction (HHPE)

High hydrostatic pressure extraction (HHPE) is a technique used for the extraction of bioactive compounds from plant matrices which involves mass transport phenomena (Oreopoulou et al., 2019) (Fig. 10.2).

HHP treatment takes place at pressure ranges from 100 to 1000 MPa, and at temperatures ranging from 0 to 120 °C, but the most investigated treatment parameters are 100–800 MPa and 20–60 °C. During the process, the product was placed in high-pressure-resistant packaging, which was then placed inside a chamber to

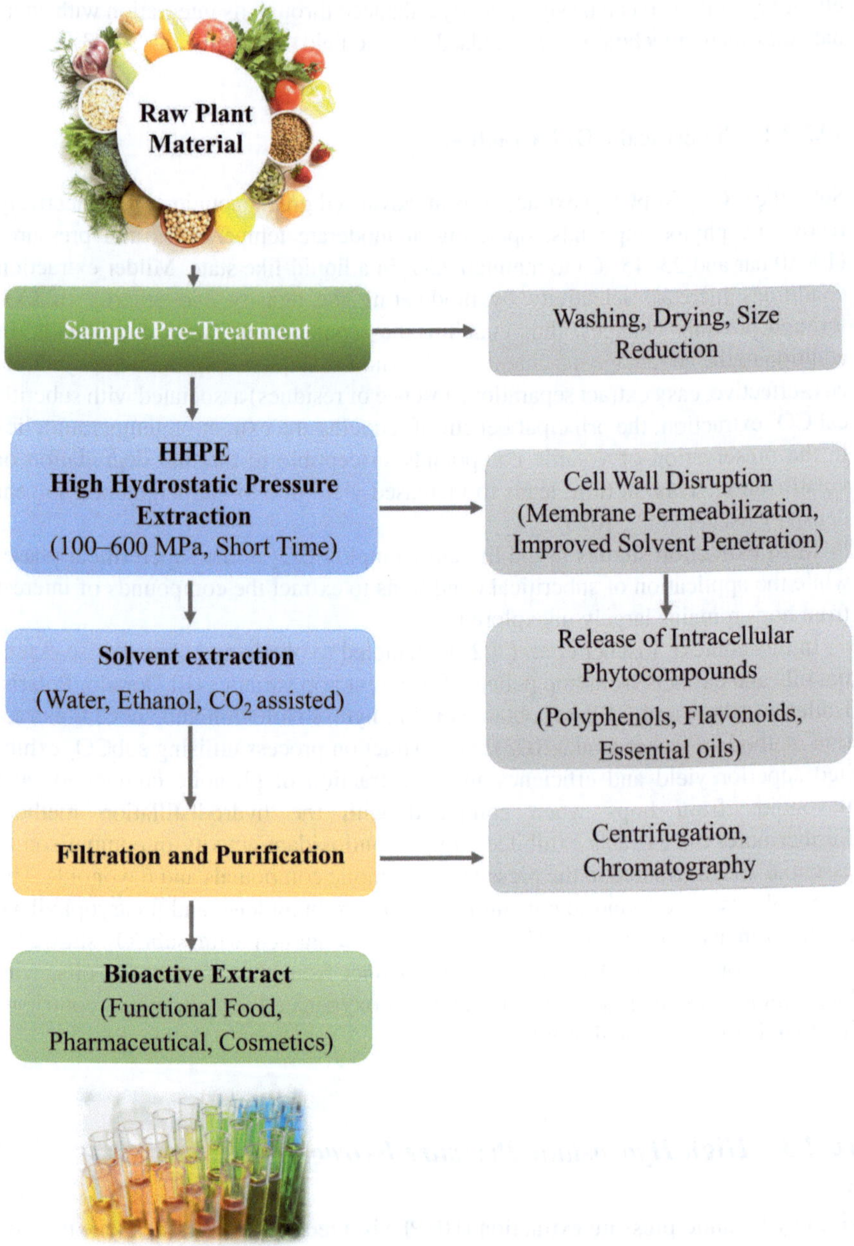

Fig. 10.2 High hydrostatic pressure scheme for extraction of phytocompounds from plant matrix

which the selected pressure was applied for several minutes. The product is then kept under hydrostatic pressure until final decompression (Aganovic et al., 2021).

The process of enhanced mass transfer is observed in response to the damage caused to the cell membrane, thereby increasing its permeability. This, in turn, facilitates the diffusion of secondary metabolites through the extraction solvent. Furthermore, the disruption of weak interactions between phenolics and cell walls has been shown to favour their release (Rifna et al., 2023). In particular, the increase in free phenolic content can be attributed to cell decompartmentalization, which results in the release of phenolics from plant tissue, thereby enhancing extraction yield. Furthermore, the increase in bound phenolic compounds, in addition to decompartmentalization, is likely due to the heightened activity of enzymes involved in the hydrolysis of proanthocyanidins, phenolic acids, and hydrolysable tannins. These compounds are esterified-bound and glycosylated-bound, and it is these that contribute to the overall increase in phenolic content (Navarro-Baez et al., 2022).

The impact of high hydrostatic pressure on plant tissue, organelles, cell walls and membranes is contingent on the magnitude of the applied pressure. For instance, the implementation of HHP at 100 MPa enhances the release of phenolic compounds that are bound to the cell wall through its modification. However, at higher pressure levels ranging from 350 to 600 MPa, the cell wall undergoes collapse, leading to an enhancement in the extractability of phenols due to a loss of cell wall integrity.

Concerning hops, Krofta and co-authors (2018) investigated the effects of high HPPE on the composition of secondary metabolites in hop homogenates, focusing on several Czech hop varieties, including Agnus, Vital, and Saaz (Krofta et al., 2018). The study compared the content and composition of hop oils, α- and β-acids, and prenylflavonoids in dried hops, frozen homogenates, and HPP-treated homogenates. The study pointed out that hop homogenates contained more hop oils (27% on average) compared to dried hops. HPPE led to a significant decrease in fatty acid methyl and thioesters fractions (80% and 100% respectively) in the essential oils composition. However, other bioactive compounds from the group of resins and prenylflavonoids were retained to a greater extent in HPP homogenates than in dried hops. The study also examined the impact of storage conditions on these compounds, finding that low temperatures and an oxygen-free atmosphere were effective for preserving raw hops and hop products. Interestingly, the authors observed demethylation of methyl esters in hop oils following HPP treatment across all crops and varieties, suggesting a potential enzymatic reaction triggered by high pressure. A comparison was drawn between this finding and similar changes observed in hops attacked by powdery mildew, indicating a potential stress response mechanism in the plant. This opens up new avenues for understanding hop biochemistry and hop processing techniques.

In their recent study, Santarelli et al. (2021) reported that the hop extract obtained through HHPE, despite the shorter extraction time (approximately 8 min in total), exhibited hydroxybenzoic acid concentrations similar to those found in the extract obtained by stirring at 25 °C for 30 min, and comparable levels of hydroxycinnamic acids, rutin, and xanthohumol to those found in the extract obtained via UAE for 60 min (Santarelli et al., 2021).

To the best of our knowledge these are the only studies on hop HHPE that have been published thus far, which creates an opportunity for further research in this field.

10.2.4 *Ultrasound-Assisted Extraction (UAE)*

The utilisation of ultrasounds in extraction procedures is characterised by a multi-faceted approach, marked by its versatility, simplicity, and safety. This method is notable for its rapidity and its eco-friendliness. It involves a significant reduction in the consumption of time, energy and costly organic solvents. In contrast to conventional extraction techniques, ultrasound-assisted extraction (UAE) is also notable for its cost-effectiveness (Ameer et al. 2017).

In terms of application, ultrasonic technology can be divided into two distinct categories: low-intensity ultrasonics (5–10 MHz, along with a relatively low power level of less than 1 W cm^{-2}) and high-intensity ultrasonics (20–100 kHz, and substantial power levels ranging from 10 to 1000 W/cm^2). Unlike low-intensity ultrasonics, high-intensity ultrasonics are characterized by their destructive capabilities. The physical mechanisms that occur during UAE include fragmentation, erosion, the sonocapillary effect, sonoporation, local shear stress, and detexturation (Chemat et al., 2017; Shen et al., 2023). The cavitation effect (Fig. 10.3) is widely regarded as the most significant of the various ultrasonic effects. The process involves the generation, growth and collapse of bubbles. It releases energy and generates localised high temperatures and pressures (4700 °C and 1000 atm) (Rodríguez García & Raghavan, 2022). This phenomenon facilitates the breakdown of cell walls, enabling the extraction of compounds of interest (Shen et al., 2023).

In order to carry out UAE, two types of ultrasonic equipment are typically used: bath-type ultrasonic equipment (indirect extraction equipment) and probe-type ultrasonic equipment (direct extraction equipment).

The UAE is influenced by a number of key factors. These include physical factors, such as power and frequency, the ultrasonic working cycle, duration, transducer efficiency, intensity, and the shape and size of ultrasonic reactors. In addition,

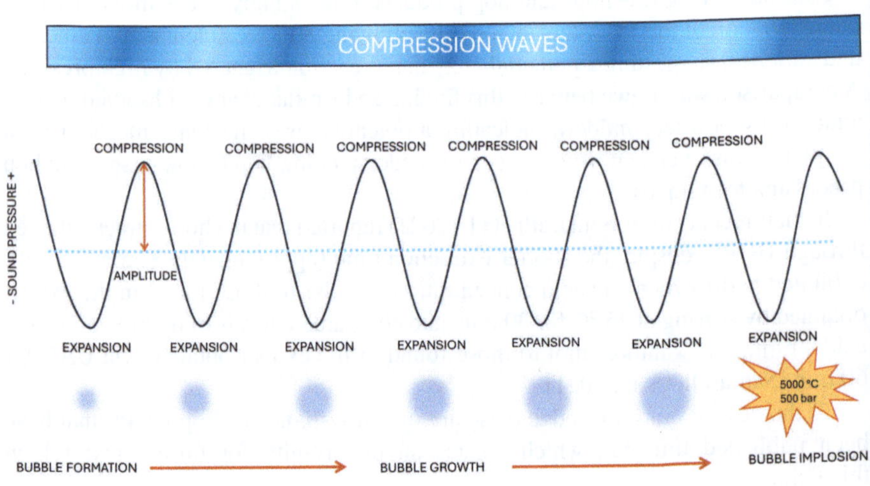

Fig. 10.3 Cavitation phenomenon

medium-related factors such as solvent, temperature, the presence of dissolved gases, external pressure, and matrix characteristics are also taken into account (Chemat et al., 2017; Shen et al., 2023).

There are a number of studies in the literature on the use of ultrasound for the extraction of bioactive compounds from hop cones, and how this compares with other commonly used methods (conventional and non-conventional).

In this regard, Almeida et al. (2020) analysed the bioactive compounds and antioxidant activities of Brazilian hop (*Humulus lupulus* L.) extracts, with a particular emphasis on the comparison of Cascade phenolic profile from both Brazil and USA. In order to obtain the bioactive compounds, the researchers employed UAE, using a Central Composite Rotatable Design (CCRD) to optimise the extraction conditions (Almeida et al., 2020).

The optimal extraction conditions were determined to be a solution of 49% ethanol at 52 °C and a solid/liquid ratio of 1 g *per* 34 mL. Under these conditions, the Brazilian hop exhibited a higher content of total phenolic compounds (TPC) and total flavonoids (TFC) compared to the USA hops, and also demonstrated higher antioxidant activity as measured by $ABTS^{+\bullet}$ and $DPPH^{\bullet}$ assays. The main phenolic compounds identified in both hop extracts were the flavonoids isoquercitrin and quercetin.

In order to facilitate better interpretation of the results, it should be noted that the Brazilian hops were freeze-dried after harvest, whereas the US hops were pelleted. This discrepancy in processing may potentially influence the obtained results.

Santarelli and co-authors (2021, 2023) investigated and compared conventional and non-conventional extraction methods to produce a hop extract rich in bioactive compounds in two different studies. Conventional extractions were conducted using dynamic maceration at two different temperatures (25 and 60 °C), while non-conventional methods involved two ultrasound systems: a laboratory bath (US; 100 W) and a high-power ultrasound bath (HPUS; 800 W); and high-hydrostatic pressure extraction (HHPE; 600 MPa for 5 min).

In their first study, the investigation concentrated specifically on the effect of different extractions on phenolic compounds and antioxidant capacity, while their subsequent study investigated the impact of extraction methods on bitter acids content and volatile organic compounds (VOCs). The initial findings of the first study indicated that the extraction method, temperature, and time have a significant impact on the phenolic profile, with notable effects on the total phenolic and flavonoid content, antiradical capacity, chlorophylls, and total carotenoid content. Overall, US demonstrated higher extraction efficiency for the majority of the compounds studied in comparison to the other extraction methods (Santarelli et al., 2021). The findings from their second study emphasise the significant influence that different extraction methods and operating conditions can have on the chemical composition of hop extracts. This results in hop products with different chemical compositions (Santarelli et al., 2023). Specifically, the combination of a 50% v/v aqueous ethanolic solution with UAE has been shown to facilitate the production of hop extracts rich in bitter acids and a wide range of volatile organic compounds (VOCs). UAE has been found to enhance the extraction of esters and monoterpenes, including β-myrcene, β-pinene, and myrtenol. In contrast, HHP has demonstrated limited effectiveness in extracting both bitter acids and VOCs under the employed conditions. The most appropriate extraction method and duration depend on the specific target compounds desired for extraction.

In their study, Carbone et al. (2020) investigated the use of microwaves (MW) and ultrasounds (US) to improve the extraction efficiency of bioactive compounds from hop cones (Carbone et al., 2020). Water, ethanol, and their mixtures were utilized as food-grade solvents. While UAE exhibited lower overall extraction efficiency compared to microwave-assisted extraction (MAE) ($p < 0.01$), it significantly enhanced the extraction of bitter acids. In particular, the extraction efficiency of US showed a significant increase with higher concentrations of organic solvent. Notably, the combination of US with ethanol resulted in an α-acid yield approximately 62% greater than that obtained through conventional solid-liquid extraction using toluene. The lower extraction efficiency of UAE compared to MAE can be attributed to the ability of microwave application to reach higher temperatures in a shorter time frame, promoting a faster release of target compounds into the solvent. These results highlight the potential of UAE in optimizing the recovery of specific bioactive compounds, particularly bitter acids, with solvent composition playing a crucial role in extraction efficiency.

The study carried out by Kamelan et al. focused on the extraction of bioactive compounds from hops, specifically the aerial parts of the stem and leaves (Kamelan et al., 2023). Extraction was performed using 80% ethanol, with a weight-to-solvent volume ratio of 1:10; the mixture was then subjected to sonication (5, 10, and 15 min) at different temperatures (25, 35, and 45 °C), and ultrasound intensities (20%, 60%, and 100%). The UAE was carried out with a 400-watt power using an H-type probe. The study employed the response surface methodology (RSM) based on the Box-Behnken design. The optimal extraction conditions were determined through numerical and graphical optimization, revealing that the ideal parameters were 12.73 minutes of extraction time, 25°C, and 100% ultrasound intensity. The results confirmed that this statistical approach was effective for modeling the extraction of bioactive compounds from hops. A study conducted by Ramazzina et al. (2022) investigated the relationship between green hop cone extracts (HGEs) and their cytoprotective or cytotoxic effects on human colonocytes, employing multivariate statistical techniques (Ramazzina et al., 2022). The HGEs were obtained using UAE and MAE with food-grade solvents (ethanol and a 50:50 ethanol-water mixture). In relation to the UAE of hop cones, the principal findings indicated that UAE employing an ethanol-water mixture (UAE_mix) did not demonstrate any cytotoxic effects on Caco2 cells across the concentration range that was examined, whereas UAE utilising pure ethanol (UAE_EtOH) exhibited strong cytotoxic properties, with a substantial reduction in cell viability commencing from very low concentrations. Furthermore, both UAE_mix and UAE_EtOH extracts increased reactive oxygen species (ROS) production in Caco2 cells, with UAE_EtOH having a more pronounced effect. Moreover, the latter extract also showed the lowest overall phenol concentration compared to other tested extraction methods. Finally, neither

UAE_mix nor UAE_EtOH extracts caused significant changes in the mRNA levels of the phase II enzymes UGT1A1 and GSTP1.

Table 10.3 provides a thorough investigation of the evaluated parameters, the tested and optimal extraction conditions, and the key findings reported in the literature regarding the use of UAE for the extraction of the compound of interest from hop.

10.2.5 Microwave-Assisted Extraction (MAE)

Microwave-assisted extraction (MAE) has emerged as a rapid and efficient technique for the isolation of phytocompounds from plant materials (Mungwari et al., 2025). The method utilises microwave radiation to generate heat within the sample, thereby enhancing the diffusion of solvents into plant cells and accelerating the release of bioactive compounds. The process is based on the interaction between electromagnetic waves with polar solvent molecules through dipole rotation and ionic conduction, as represented in Fig. 10.4 (Mena-García et al., 2019).

In comparison with conventional extraction techniques, MAE has been shown to reduce extraction time, solvent consumption, and energy usage, while maintaining or even enhancing the yield and quality of target compounds such as phenolics, flavonoids, alkaloids, and terpenoids. Furthermore, MAE offers enhanced control over extraction parameters, such as temperature, pressure, and solvent composition, ensuring reproducibility and scalability. These advantages position MAE as a promising alternative to traditional extraction methods in the field of phytochemistry, particularly for the large-scale production of natural products with pharmaceutical and nutraceutical applications. However, it is imperative to optimise operational parameters to ensure selectivity and efficiency, as excessive microwave exposure may lead to the degradation of sensitive compounds.

The microwave frequency ranges from 300 MHz to 300 GHz (Bagade & Patil, 2021; Bitwell et al., 2023); however, only a few frequencies are used for scientific and industrial purposes, i.e. 0.915 and 2.45 GHz (Vernès, 2019). MAE can be performed with or without the use of solvents (Rodríguez García & Raghavan, 2022). During the extraction process, microwaves are employed to heat the extraction solvent and plant tissue, thereby enhancing the extraction kinetics. This is achieved by directly impacting polar molecules. The MAE process involves the diffusion of solvents into the sample, followed by the separation of solute from the functional site, and finally the release of solutes into the solvents. Heating is directly proportional to the dielectric constant of the solvents, and the solvent's viscosity significantly affects the extraction process, with lower viscosity facilitating the dispersion of ions and hence the solvation (Bitwell et al., 2023).

In contrast, solvent-free or solid-state MAE (SF-MAE) operates without the use of any liquid solvent, relying solely on the interaction of microwave energy with the plant matrix. This method induces localised heating within the plant material itself, causing thermal degradation of cell walls and the release of phytochemicals directly

Table 10.3 Key studies on the use of Ultrasound Assisted Extraction (UAE) to extract compounds of interest from hops

Plant matrix	Evaluated parameters	Tested condition	Optimum condition	Main goals/conclusion	References
Brazilian freeze-dried hop cones (BH) USA cone pellet (UP)	TPC TFC Individual phenolic compounds AA	37 kHz 30 min CCRD was employed to optimize the extraction conditions for phenolic compounds: Ethanol 43%–77%; temperature 33–67 °C; solid:liquid ratio 17–33 g/ml	Ethanol 49% 52 °C Solid:liquid ratio of 1:34	The TPC, TFC, and AA revealed that BH was different from UP. Notably, the BH showed a high concentration of bioactive compounds and AA The main phenolic compounds identified in the extracts were flavonoids (isoquercitrin and quercetin)	Almeida et al. (2020)
Hop cones	TPC BAs Individual α- and β-acids Individual phenols XN ^{1}H-NMR analysis AA	25 °C 30 min Solid:liquid ratio of 1:15 (w/v) Ethanol 100%; Water 100%; Ethanol:water (50:50 v/v)	Solid:liquid ratio of 1:15 (w/v) Higher yields of TPC were detected in ethanol extract regardless of the used extraction method	UAE extraction efficiency increased with increasing percentages of organic solvent UAE with ethanol 100% has led to an α-acid yield of about 62% higher than conventional solid-liquid extraction with toluene	Carbone et al. (2020)

Dried hop cones	TPC TFC AA Chlorophyll α, β and TCa Phenolic compounds XN	100 W 800 W 15–30–60–120 min Ethanol 100% Ethanol:water (50:50 v/v) Water 100% Matrix:solvent ratios (1:10; 1:20; 1:50 w/v)	100 W 30 min Ethanol 50% Matrix:solvent ratio 1:50 (w/v)	UAE (100 W) with ethanol 50% for 30 min was the most effective extraction for almost all the studied compounds Phenolic profile, TPC, TFC, chlorophylls, and TCa were significantly and differently affected by extraction method, solvent, and time	Santarelli et al. (2021)
Dried hop cones	VOCs BAs Individual α- and β-acids	100 W 800 W 15–30–60 min Ethanol 100% Ethanol:water (50:50 v/v) Water 100% Matrix:solvent ratios (1:10; 1:20; 1:50 w/v)	100 W 30 min Ethanol 50% Matrix:solvent ratio 1:50 (w/v)	UAE enabled the extraction of a high amount of esters and monoterpenes Different extraction conditions affect the composition of BAs and VOCs of hop extracts	Santarelli et al. (2023)

(continued)

Table 10.3 (continued)

Plant matrix	Evaluated parameters	Tested condition	Optimum condition	Main goals/conclusion	References
Hops Including the aerial parts of the stem and leaves Without cone flowers	TPC AA BAs	Matrix:solvent ratios 1:10 w/v 5, 10, and 15 min 25, 35, and 45 °C Different intensities (20, 60, and 100%)	12.73 min 25 °C US intensity of 100%	The extraction conditions were optimized with the response level statistical method based on the Baxbenken scheme The obtained results confirmed the appropriateness of the selected statistical method for modelling the extraction conditions of bioactive compounds from hops	Kamelan et al. (2023)
Hop cones	AA Total Flavans TPC Phenolic compounds XN BAs Biological activity of hop extracts in cell culture ATPlite assay Reactive oxygen species determination PCR	UAE: 40 kHz Ethanol:water (50:50 v/v) Ethanol 100% 30 min 25 °C MAE: 2.45 GHz 400W Ethanol:water (50:50 v/v) 75 °C 5 min	UAE: 40 kHz Ethanol:water (50:50 v/v) 30 min 25 °C	Only the UAE with Ethanol 50% samples did not cause cytotoxicity The different cytotoxicity exerted by the hop extract could be attributed to the different qualitative and quantitative bioactive profiles of each extract The use of multivariate statistical techniques allowed the authors to study the possible influence of the individual analysed compounds on the effects recorded *in vitro*, which were suitably explained considering 2D models based on inhibitor – promoting factors	Ramazzina et al. (2022)

TPC Total Phenolic Content, *TFC* Total Flavonoid Content, *AA* Antioxidant Activity, *CCRD* Central Composite Rotational Design, *BAs* Bitter acids, *XN*

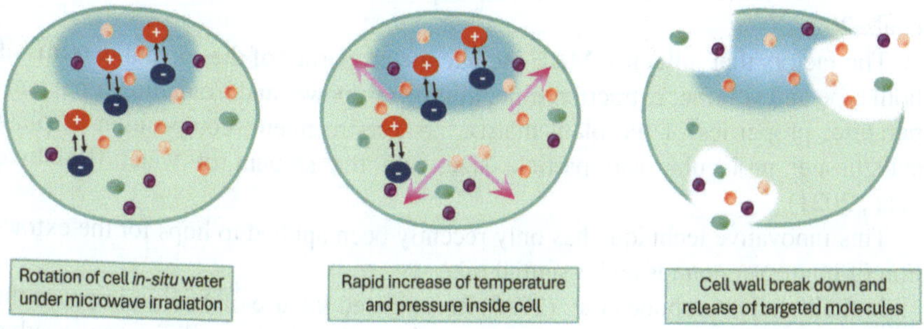

Fig. 10.4 The dynamics of Microwave-Assisted Extraction

into the vapour phase, which can then be condensed and collected (Bagade & Patil, 2021; Bitwell et al., 2023).

The utilisation of SF-MAE is particularly suitable for the extraction of heat-sensitive and volatile compounds (e.g. essential oils), as the controlled microwave heating process ensures minimal degradation and loss during the procedure (Boukhatem et al., 2022). De facto, while this technique offers an eco-friendly alternative, reducing the need for organic solvents and minimising environmental impact, solvent-assisted MAE remains more widely used due to its ability to extract a broader range of compounds in a more controlled manner. Both methods have been shown to reduce extraction time and solvent consumption compared to conventional techniques, with solvent-free MAE offering significant advantages in terms of sustainability and cost-effectiveness. However, the choice between these two approaches depends on the specific targeted phytocompounds and the need for selective extraction, as well as the optimization of operational parameters such as microwave power, exposure time, and sample-to-solvent ratio to maximize yield and minimize compound degradation.

The equipment required for MAE comprises four principal components:

1. A microwave generator, which is a magnetron that produces microwaves;
2. A waveguide that disseminates microwaves to the microwave cavity;
3. A circulator that facilitates the transmission of microwaves;
4. An applicator that determines the test parameters (Bagade & Patil, 2021).

Depending on the used vessel, MAE can be classified into closed (multimode) and open systems (monomode). In the first system, extraction is carried out under controlled conditions of temperature and pressure. Microwaves (MW) are diffused from a cavity magnetron and radiated in all directions to interact with plant samples positioned in extraction vessels within a closed-vessel chamber. In the second system, the extraction vessel is exposed to microwave radiation (focused radiation). A circular metallic waveguide directs the focused microwaves towards the extraction vessel inside the microwave cavity. This interaction promotes the initiation of mass transfer between the solute and extractant upon solvation (Ameer et al., 2017; Rifna et al., 2023).

The factors that influence MAE include the properties of the solvent in terms of both type and volume; extraction time; microwave power and irradiation time; temperature; properties of the plant matrix, i.e. pre-treatment; composition of plant cells/tissue; molecular and particle size; and the amount of water (Lefebvre et al., 2021).

This innovative technique has only recently been applied to hops for the extraction of phytocompounds and essential oils.

In their study, Carbone et al. (2020) investigated the use of MW to improve the extraction efficiency of bioactive compounds from hop cones, utilizing water, ethanol, and their mixtures as food-grade solvents (Carbone et al., 2020). The results revealed that MAE using ethanol was the most effective method, providing the shortest extraction time and the highest antiradical capacity of the extracts. Moreover, MAE with ethanol produced the highest yield of xanthohumol, whereas MW showed better extraction efficiency for all types of tested bitter acids regardless of the used solvent. ^{1}H-NMR analysis indicated that the composition of the final extract was primarily determined by solvent polarity rather than the extraction technique itself. The fingerprinting of aqueous hop extracts via ^{1}H-NMR revealed sugars, amino acids, and other metabolites, including trigonelline – a biologically significant alkaloid – as the main components. Interestingly, the study found that MAE extracts had higher antiradical capacity compared to UAE samples, regardless of the used solvent and the in vitro test. The authors also highlighted that MW extraction efficiency is influenced by the polarity of the solvent. Ethanol, despite having a lower dielectric constant than water, has a higher dissipation factor, leading to a higher heating rate and enhanced MW extraction efficiency.

Lamberti et al. (2021) presented a study on the microwave-assisted hydrodistillation (MAHD) of hop terpenes, focusing on the scaling up of the process from laboratory to pilot scale. The research investigated the extraction of volatile fractions from three types of hops: fresh (FH), dried (DH), and pelletized (PH), with the objective of optimising the extraction protocol using microwaves to reduce energy and time consumption. The pilot-scale ETHOS XL reactor exhibited substantial enhancements in extraction yields, particularly for dried and pelletized hops. Pellet productivity was quadrupled, while dry cone yields nearly doubled in comparison to small-scale projections. The pilot reactor also demonstrated notable energy savings, with the absorbed kJ mL^{-1} of volatile fraction being reduced by almost half. Furthermore, the study revealed that the extraction process is non-linear, with the relative terpene ratios changing over time. Towards the end of the extraction process, the collected product becomes progressively enriched with the sesquiterpene fraction. This finding suggests the possibility of selecting different product fractions according to desired flavour profiles, which could have significant implications for the brewing industry and other applications of hop extracts. However, the study would be improved by the inclusion of a comprehensive economic analysis comparing the proposed MAHD method to traditional extraction methods, which would be crucial for industry adoption (Lamberti et al., 2021).

In a separate study, Tyśkiewicz et al. (2018) presented a comparative analysis of two extraction methods for obtaining essential oils from hops supercritical CO_2 extract: microwave-assisted hydrodistillation (MAHD) and conventional hydrodistillation (HD). The study focused on the Polish hops variety 'Marynka' and aimed to determine the optimal conditions for maximum essential oil production (Tyśkiewicz et al., 2018). The study found that MAHD at 335 W microwave power for 30 min with a water-to-raw material ratio of 8:3 yielded the best results. The essential oils obtained were found to be rich in β-myrcene, ranging from 74.13% to 89.32% by weight. The authors also compared the efficiency of MAHD and HD methods, revealing that MAHD achieved similar extraction yields in significantly shorter times and with lower energy consumption. For instance, MAHD required 30 min, whereas HD required 276 min, and used 0.17 kWh/g of essential oil, as opposed to HD's 1.89 kWh/g.

In a circular economy approach, Rubira et al. (2024) investigated the valorisation of hop pruning to produce valuable compounds using two biorefinery strategies. The first strategy involves conventional processing with mild water extraction followed by dilute acid pretreatment, while the second strategy utilises microwave-assisted autohydrolysis as an innovative approach (Rubira et al., 2024). With regard to the use of microwaves, the obtained results demonstrated that the microwave-assisted autohydrolysis strategy (200 °C for 5 min) enabled the simultaneous extraction of antioxidant phenolics (31.74 mg GAE/g raw material and 61.06 mg rutin equivalents (RE)/g raw material) and xylooligosaccharides and xylose (5.50 g/L). It was observed that both strategies yielded similar amounts of phenolic compounds (approximately 3.3 kg per 100 kg of hop pruning). However, a divergence was noted in the phytochemical profiles, with the conventional method yielding more flavonoids and the microwave method producing more phenolic acids and aldehydes. In addition, microwave-assisted autohydrolysis was less severe than the conventional method, resulting in lower solubilization but also less formation of pseudo-lignin. However, it should be noted that the study compares an optimised conventional method with a non-optimised microwave-assisted method, which may not provide a fair comparison of their true potential.

In a recent study, Pichon et al. (2025) evaluated different green solvent extraction methods for optimizing the industrial-scale extraction of bioactive compounds from hop. Among the tested techniques, MAE demonstrated high efficiency in extracting key metabolites (Pichon et al., 2025). In particular, MAE resulted in significantly increased yields of flavonoids and polyphenols (two- and threefold) and xanthohumol (sixfold) compared to standard maceration. The method provided enhanced extraction efficiency while maintaining a desirable metabolite profile. LC-MS/MS analysis confirmed the presence of targeted compounds, aiding in technique selection. Overall, MAE proved to be a highly effective extraction method, offering improved yields and compound concentrations.

As illustrated in Table 10.4, a range of conditions are applied in order to extract the main bioactive compounds from hops using microwaves.

Table 10.4 Key studies on the use of MAE (microwave-assisted extraction) to extract compounds of interest from hops

Extraction method	Plant matrix	Evaluated parameters	Tested condition	Optimum condition	Main goals/conclusion	References
MAE	Hop cones	TPC BAs Individual α- and β-acids Individual phenols XN ^{1}H-NMR analysis AA	400 W 75 °C 5 min Solid:liquid ratio of 1:15 (w/v) Ethanol 100% Water 100%; Ethanol:water (50:50 v/v)	400 W 75 °C 5 min Solid:liquid ratio of 1:15 (w/v) Ethanol 100%	MAE showed the strongest extraction power for almost all the investigated bioactive compounds ^{1}H-NMR analyses indicated that solvent polarity had a more significant impact on the overall composition of the final extract than the extraction technique	Carbone et al. (2020)
MAHD	Fresh Hop (FH) Dried Hop (DH) Hop Pellet (HP)	Volatile profile	Optimization of the two used tools and the scale-up transposition Two instruments were tested: ETHOS X and ETHOS XL (Milestone srl, Bergamo, Italy), and MADH under vacuum	Myrcene concentration was significantly affected by hop type, with the highest levels in FH The concentration of sesquiterpenes (caryophyllene, humulene, and farnesene) increased relative to monoterpenes as extraction time was extended	The extraction procedure can be optimized based on the nature of the matrix	Lamberti et al. (2021)
MAHD	Hop cones	EY Oil composition	275–295–335–395 W	335 W 30 min Water:raw material ratio of 8:3 The essential oils obtained with MAHD were enriched in β-myrcene	As the optimum temperature is achieved, extraction efficiency decreases Higher microwave power led to a reduction in extraction time	Tyśkiewicz et al. (2018)

MAE	HP	Polysaccharides Oligomers/monomers TPC TFC Individual phenolic compounds AA	Sequential fractionation of HP with MWE and DAP MWE: 110 °C 30–60 min Liquid:solid ratio of 15 (g/g) Optimization of DAP: by RSM 110–130 °C 15–60 min HCl concentration 0.25–1.75% w/w MAE autohydrolysis: 300 W 170 °C (5–20 min) 180 °C (5–20 min) 190 °C (5–12 min) 200 °C (2–10 min)	MWE: 110°C 30 min LSR 15 g/g DAP: HCl 1.6932% 123.5 °C 59.6 °C LSR 8 g/g MAE: 200°C 5 min LSR 15 mL/g	A MWE followed by DAP was an alternative to MAE autohydrolysis Conventional aqueous extract presented higher content of flavonoids and higher recovery of hemicellulose-derived saccharide recovery MAE had higher content of phenolic acids and aldehydes	Rubira et al. (2024)

(continued)

Table 10.4 (continued)

Extraction method	Plant matrix	Evaluated parameters	Tested condition	Optimum condition	Main goals/conclusion	References
MAE	Dried hop cones	EY TPC TFC TCl and TCa Individual BAs XN	Solid:liquid ratio of 1:10 (w/v) SM: stirring at 150 rpm 2 h TSE: Hexane:ethyl acetate:acetonitrile:water (2:3:3:2) stirred 2 h 25 °C MAE: Ethanol 100% 150 W 75 °C 5 min 75 °C 15 min	The findings were assessed and compared using desirability optimization as a statistical approach to identify the most promising extraction methods	MAE resulted in a 1.9-fold and 1.59-fold enhancement in TPC at 5 minutes and 15 minutes, respectively, compared to TSE MAE resulted in a slight increase of 1.58-fold at 5 minutes followed by 15 minutes, and 1.23-fold compared to TSE Both MAE 5 min and MAE 15 min were the most effective for extracting lupulone and co-lupulone MAE resulted in significantly increased EY of flavonoids and polyphenols (2- and 3-fold) and XN (6-fold) compared to standard maceration	Pichon et al. (2025)

MAE Microwave-Assisted Extraction, *HHPE* High Hydrostatic Pressure Extraction, *RSM* Response Surface Methodology, *MADH* Microwave-Assisted Hydrodistillation (HHPE+MAE), *TPC* Total Phenolic Content, *BAs* Bitter acids, *XN* Xanthohumol, *AA* Antioxidant Activity, *EY* Extraction Yield, *HP* hop pruning, *TFC* Total Flavonoid Content, *MWE* Mild Water Extraction, *DAP* Dilute Acid Pretreatment, *TCl* Total Chlorophyll, *TCa* Total Carotenoids, *SM* Simple Maceration, *TSE* Triphasic Solvent Extraction

10.2.6 *Pulsed Electric Field Extraction (PEFE)*

The application of a pulsed electric field (PEF) to the plant matrix enhances solvent extraction efficiency by facilitating diffusion and mass transfer through a phenomenon known as "electroporation" or "electropermeabilisation" of the membrane. The prevailing theory behind electroporation induced by PEF is the "transmembrane potential ($\Delta\Phi$) breakdown model". According to this model, the application of external PEF to a biological cell triggers a transmembrane potential ($\Delta\Phi$) across its semipermeable cell membrane. If the applied field strength (E) is higher than the physiological transmembrane potential (critical transmembrane voltage), the membrane loses its semipermeable nature, leading to cell lysis (Poojary et al., 2020; Tylewicz, 2020) (Fig. 10.5).

PEF membrane electroporation is achieved by applying pulses of moderate-to-high electric field strength to the samples positioned between two electrodes within a conducting medium.

This process can be performed in both reversible and irreversible modes. It is important to note that no electroporation will occur if the external electric field is lower than the critical electric field ($E_e < E_c$). Conversely, poration will be reversible if the external electric field is equal to or slightly higher than the critical electric field ($E_e > E_c$). In reversible electroporation, temporary electroporation occurs in the presence of an electric field, and upon removal of the electric field, the cell membrane returns to its original structure. Irreversible or permanent electroporation, instead, occurs only in circumstances where the magnitude of the external electric field significantly exceeds the critical electric field value ($E_e \gg E_c$) (Naliyadhara et al., 2022; Tylewicz, 2020).

The factors that affect PEF can be categorised as follows (Poojary et al., 2020):

- Pulse parameters, including electric field strength, pulse duration, pulse polarity, and pulse frequency, which have the capacity to affect the degree of electroporation;

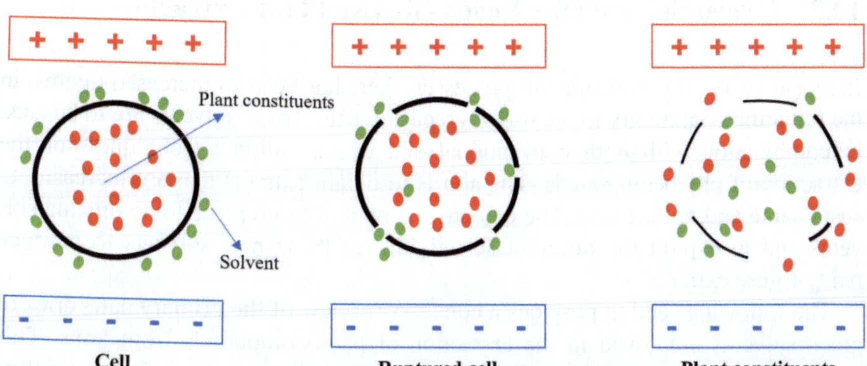

Fig. 10.5 Schematic electroporation mechanism for extraction

- Tissue parameters, encompassing the physicochemical properties of the cell or tissue, such as its size, membrane structure and thickness, intracellular electrolytes, ionic mobility, and water content;
- Media parameters, including ionic strength and conductivity of the medium.

PEF is primarily employed as a non-thermal treatment of liquid foods, with the objective of inactivating microorganisms (Ntourtoglou et al., 2020). Additionally, it is being adopted by food industries to augment the efficiency of conventional drying methods (e.g. freezing, drying, freeze-drying, and osmotic dehydration) by accelerating mass transfer processes (Neri et al., 2021). Furthermore, it has been demonstrated to enhance the extraction yield of bioactive compounds and essential oils (Ntourtoglou et al., 2020).

In extraction applications, the electric field strength may vary between 0.1 and 20 kV cm^{-1}, while the pulse duration falls in the range of nanoseconds to milliseconds (Poojary et al., 2020). The application of electrical pulses to plant tissues can lead, in fact, to electrical breakdown and structural damage of cell membranes, which cause the diffusion of cellular metabolites into the extracellular medium and therefore an increase in extraction yield.

To the best of our knowledge, literature only comprises a single study that addresses the application of PEF technology in the context of hops. In their study, Ntourtoglou et al. (2020) aimed to extract α-acids and β-acids from two different hop varieties using PEF (Ntourtoglou et al., 2020). According to the findings, samples treated with PEF exhibited higher concentrations of humulones and lupulones. Under PEF conditions (1.5 $kVcm^{-1}$; 15 μs and 1800 pulses), the extraction of total bitter acids and sesquiterpenes increased by approximately 1.3 times, whereas the treatment notably enhanced the extraction of α-acids by 21–100% and β-acids by 9–120%. The quantity of extracted acids depended on the used solvent and the duration of treatment. Furthermore, hops demonstrated retention of their antiradical activity, which, in certain instances, was also subject to enhancement.

10.3 Green Solvents for Ready-To-Use Plant Extracts

In recent years, alongside the GE processes, there has been an increased interest in the scientific community in the use of green solvents. These solvents are to be used in combination with both conventional and unconventional techniques for the extraction of phytocompounds. The aim is to obtain extracts that are increasingly sustainable and tailor-made. The objective is twofold: to replace toxic organic solvents and to exploit the different selectivities of these new solvents to produce ready-to-use extracts.

The following section provides a concise overview of the primary categories of green solvents employed in the extraction of phytocompounds from hops. This overview is accompanied by a brief exposition of the characteristics that define these solvents.

10.3.1 *Ionic Liquids (Ils) and Deep Eutectic Solvents (DESs)*

Ionic liquids (ILs) are defined as compounds that are composed entirely of ions and possess a melting point below 100 °C. They have specific physico-chemical properties such as extremely low vapour pressure, high thermal stability, high conductivity, and a wide range of electrochemical and polarity properties (Choi & Verpoorte, 2019; Lei et al., 2017).

In recent times, a new class of green solvents has emerged, which has gained significant popularity: Deep Eutectic Solvents (DESs). These are a class of ionic liquid analogues that share many characteristics with ILs but are distinct in their composition and properties. The formation of DESs occurs through eutectic mixtures of Lewis or Brønsted acids and bases, comprising a variety of anionic and/or cationic species (Smith et al., 2014). As a consequence, DESs are defined as binary or ternary mixtures of compounds such as organic salts (quaternary ammonium or phosphonium salt) or hydrogen bond donors (HBDs) that are capable of associating with each other through intramolecular hydrogen bonds (Cvjetko Bubalo et al., 2018; El Achkar et al., 2021). These solvents are frequently categorised within the IL family and are designated as ILs analogues (Kaur et al., 2020). However, it is important to emphasise that, despite the physical properties of DESs and ILs being similar in a broader sense, molecular-level interactions and structural organisation in these solvents are observed to be quite different. It has been demonstrated that the combination of ionic and molecular species in DESs yields a more structurally complex liquid, with contributions from hydrogen bonding as well as electrostatic forces (Kaur et al., 2020).

An emerging class of DESs is that of the so-called Natural Deep Eutectic Solvents (NADESs), which are derived from the combination of naturally occurring primary metabolites such as organic acids, sugars, alcohols and amino acids (Herrero, 2024; Mena-García et al., 2019). These solvents are formed through hydrogen bonding interactions between natural HBDs and Hydrogen Bonding Acceptors (HBAs), resulting in a eutectic mixture with a melting point that is significantly lower than that of its individual components. Figure 10.6 shows the most common HBDs and HBAs used in the synthesis of NADES.

In recent years, there has been a considerable increase in the focus on NADESs due to their distinctive physicochemical properties, which include elevated thermal stability, diminished volatility, adaptable viscosity, and the capacity to dissolve a broad spectrum of polar and non-polar compounds. The biocompatibility, minimal toxicity and biodegradability of these solvents position them as a sustainable alternative to conventional organic solvents in a wide range of applications, including the extraction of bioactive compounds. It has been demonstrated that these solvents exhibit a high degree of polarity, thus facilitating the extraction of both hydrophilic and hydrophobic compounds that are present within the complex matrix of plants (Hikmawanti et al., 2021). However, it should be noted that the efficiency of extraction can be influenced by a number of factors, including the chemical composition of HBA and HBD, the ratio between them, and their physico-chemical characteristics such as pH, polarity, and viscosity. Furthermore, the water content is found to play a crucial role (Macchioni et al., 2021).

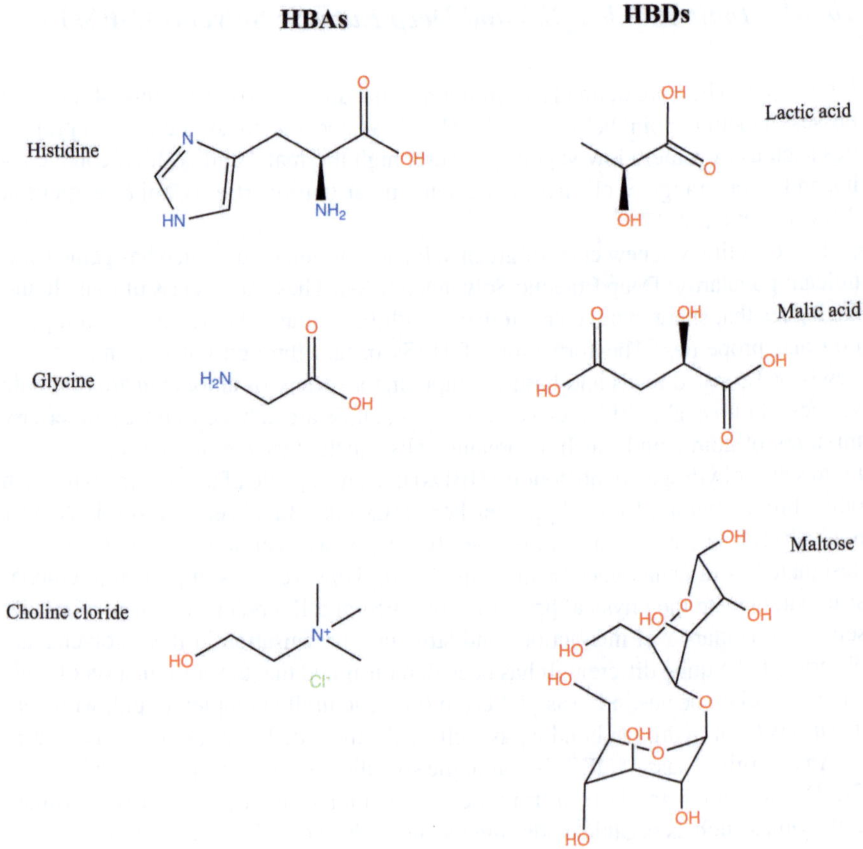

Fig. 10.6 Visual representation of the chemical structures of common natural HBDs and HBAs

In relation to hops, DESs derived from choline chloride as the HBA to extract XN and bitter acids from Slovenian hops of the Aurora variety were the subject of a study by Metaj et al. (2023). In this study, researchers compared various extraction methods using both classical organic solvents and DESs based on choline chloride. A range of extraction methods were examined, including Soxhlet extraction and ultrasounds. Ten solvents were prepared and assessed, with choline chloride-phenol (ChCl-Phe) proving the most efficacious for the extraction of hop bitter acids and XN. It was found to be comparable to, and in some cases superior to, diethyl ether, a classic hop polyphenol extraction solvent.

In their study, Macchioni et al. (2021) explored the use of lactic acid-based deep natural eutectic solvents (LaDES) for the extraction of bioactive compounds from hop cones (Macchioni et al., 2021). The study compared different LaDES combinations - lactic acid:sucrose (L1), lactic acid:urea (L2), and lactic acid:glycine (L3) - with a conventional hydroalcoholic solvent (80% acidified methanol) as a control. The researchers investigated the extraction efficiency of these solvents for various hop bioactive compounds, including polyphenols, bitter acids, and pigments, as well as the antimicrobial and antioxidant properties of the obtained green extracts. The results showed that different LaDES combinations were more effective at extracting specific compounds. L3 demonstrated the highest extraction efficiency for polyphenols and exhibited the strongest antioxidant activity. L1 was most effective for extracting bitter acids and XN, while L2 showed the best results for non-phenolic pigments like chlorophylls and carotenoids. All LaDESs extracts displayed significant antimicrobial activity against both Gram-positive and Gram-negative bacteria, outperforming the conventional solvent in many cases. The findings demonstrate that by carefully selecting the hydrogen bond donor/acceptor pair in LaDES, it is possible to tailor the extraction process to target specific classes of bioactive compounds. This approach not only allows for more efficient extraction but also enables simultaneous formulation of phytoextracts that could be directly incorporated into food, dermatological, or cosmetic products. Furthermore, the study suggests a potential synergistic antimicrobial effect between LaDES and hop constituents, particularly against Gram-negative bacteria, which are often resistant to hop compounds.

In a separate study, Lakka et al. (2019) investigated the extraction of polyphenols from hops using a NADES composed of glycerol and L-alanine, optimising the extraction conditions (Lakka et al., 2019). The variables considered were the proportion of DES/water, the liquid-to-solid ratio, and the stirring speed. Under the optimized conditions, the achieved yield of total polyphenols was 118.97 ± 8.27 mg gallic acid equivalents *per* g of dry mass. The researchers also examined the impact of ultrasonication as a pretreatment step on the extraction kinetics at different temperatures (40–80 °C).

The newly synthesised NADES demonstrated superior extraction efficiency in comparison to conventional solvents such as methanol and ethanol. Interestingly, ultrasound-assisted pretreatment was found to be temperature-dependent, suppressing polyphenol extraction up to 70 °C but enhancing it at 80 °C. Furthermore, the study highlighted an unexpected trend in extraction rates. Specifically, the authors observed that the extraction kinetics did not follow the typical Arrhenius behaviour (i.e. an increase in reaction speed with increasing temperature), which would be unusual and contradictory to previous studies on polyphenol extraction. The authors suggested that this anti-Arrhenius behaviour could be due to the partial disintegration of NADES at higher temperatures, which would lead to a lower solubilisation capacity of polyphenols. Further investigation is necessary to fully comprehend the mechanisms underlying the observed behaviour.

Spent hops are an important by-product of the brewing industry and are rich in valuable ingredients such as proteins and XN. Grudniewska and Popłoński (2020) studied four choline chloride-based DESs that were assessed for their ability to extract XN from spent hops (Grudniewska & Popłoński, 2020). These DESs incorporated glycerol, ethylene glycol, propylene glycol, or lactic acid as a HBD. XN-rich precipitates were obtained by introducing water (acting as an antisolvent) into the DES extracts. The most effective extraction was achieved using a DES comprising choline chloride and propylene glycol (1:2 mol/mol), with 5% (w/w) water, a 1:50 ratio of spent hops to DES (w/w), and a 3:1 ratio of antisolvent to DES (v/w), at 60 °C for 1 h.

Furthermore, the authors pointed out that this DES could be recovered and reused for at least three cycles without a significant reduction in the XN extraction yield.

In a separate study, Grudniewska and Pastyrczyk (2023) developed an eco-friendly method for simultaneously extracting proteins and XN from spent hops (SH) using choline chloride-based DESs and water as an antisolvent to obtain enriched precipitates, which could be considered as 'dual-function' products with potential use in nutraceuticals and functional food ingredients (Grudniewska & Pastyrczyk, 2023).

10.3.2 Other Green Solvents

In addition to super and subcritical fluids, green solvents such as hot water can also be used effectively to recover hop bioactive compounds.

In their study, Gil-Ramírez et al. (2012) obtained an isoxanthohumol (IX)-enriched hop extract using Pressurized Hot Water Extraction (PHWE) (Gil-Ramírez et al., 2012). The study also identified the presence of other compounds in the extracts, including phenolic compounds and bitter acids. The obtained extract also demonstrated anti-inflammatory activity in vitro, showing effects similar to those of indomethacin on certain cytokines (cytokines IL-10 and TNF-α). However, it should be noted that while high temperatures are known to promote the formation of IX, they can also potentially lead to the degradation of other beneficial compounds. This aspect was not thoroughly addressed in the study and requires further investigation for the applicability of the obtained results.

Rapinel et al. (2020) reported that 2-methyloxolane (also known as 2-methyltetrahydrofuran or 2-MeTHF) is a promising green solvent with the potential to replace hexane in the production of green aromas from hops. This could offer the industry the opportunity to create cleaner and more innovative products (Rapinel et al., 2020). The study's findings, based on COSMO-RS simulations, showed that the primary aromatic molecules in hops, particularly the oxygenated compounds, exhibited enhanced solubility in 2-methyloxolane compared to n-hexane. Extractions with 2-methyloxane produced higher yields (20.2% compared to 16.6% with hexane) and showed faster kinetics. The study also found that both solvents produced similar profiles, consisting mainly of α- and β-acids. However, the 2-methyloxane extract exhibited a higher percentage of lupulone (60% compared to 41% in hexane) and a reduced presence of degradation products. In addition, sensory evaluation found that both extracts had similar olfactory profiles. The study offers encouraging results for the utilisation of 2-methyloxolane in hop aroma extraction. Nevertheless, further research and validation may be required, as the study also demonstrated that the solubility predictions of COSMO-RS are not always in agreement with the chemical compositions determined by GC-MS. This discrepancy calls into question the reliability of the theoretical model in predicting practical outcomes.

Table 10.5 summarises the evaluated parameters, extraction conditions, and principal findings from the literature concerning the use of ILs and deep eutectic solvents in hop extraction.

Table 10.5 The use of Deep Eutectic Solvents (DESs) and Ionic Liquids (ILs): extraction conditions and main results

Extraction method	Plant matrix	Evaluated parameters	Tested condition	Optimum condition	Main goals/conclusion	References
DESs	Hop pellet	Individual BAs XN	ChCl-Glu 2:1:1 ChCl-Fruc 2:1:1 ChCl-Lac 1:1 ChCl-TA 2:1 ChCl-GL 1:1 ChCl-EG 1:2 ChCl-U 1:1 ChCl-DMU 1:2 ChCl-Phe 11.3 ChCl-BSA 1:1 80 °C 500–600 rpm OSE with organic solvent or DES-OSE 40 min 125 rpm UAE with organic solvent or DES-UAE 30 min 25 °C UHE with organic solvent or DES-UHE Amplitude 50% Cycle setting 0.6	The type of extraction method can influence the efficiency of extraction for both BAs and XN ChCl-Phe was found to be the most effective solvent for extracting both hop BAs and XN	With the exception of ChCl-Phe, ChCl-TA demonstrated extraction efficiencies similar to those of pure organic solvents, even then, only when UHE was employed The most efficient DES solvent is ChCl-Phe, with extraction efficiency for OSE and UAE being comparable to, or even surpassing, that of diethyl ether. The extraction efficiency of ChCl-Phe for XN also varies depending on the used technique, decreasing in the order of OSE > UAE > UHE The second most effective DES is ChCl-DMU, with extraction efficiencies of 82% of ChCl-Phe in the OSE technique, 62% in UAE, and 67% in UHE	Metaj et al. (2023)

(continued)

Table 10.5 (continued)

Extraction method	Plant matrix	Evaluated parameters	Tested condition	Optimum condition	Main goals/conclusion	References
DESs	Hop cones	TPC PAC TCl and TCa Individual phenols and XN Individual BAs AOC EE	Lac:Sucr molar ratio 4:1 water 20% Lac:U molar ratio 3:1 water 20% Lac:Gly molar ratio 3:1 water 20% UAE: 40 kHz 25 °C 30 min	Lac:Sucr showed the highest EE against hop BAs Lac:U extracted chlorophylls and carotenoids Lac:Gly extracted more polyphenols with high AOC and TPC	Appropriate selection of the HBD/HBA pair enabled the enrichment of hop extracts with diverse phytochemical classes	Macchioni et al. (2021)
DESs	Hop pellet	TPC TFC TF Reducing Power AOC	DES: glycerol (HBD), L-alanine (HBA) CDES in aqueous mixtures 55–70–85% w/w LSR 20–40–60 mL/g Speed of stirring: 200–500–800 rpm HBD:HBA varying from 5:1 to 13:1 *UAE:* 50 Hz 550 W	Among all DESs tested, GL-Ala gave a significantly higher TPC yield CDES profoundly affect TPC yield, and the optimal level was 80% (w/w) to be the most effective DES/water proportion Optimal LSR was 59 mL/g 0–min ultrasonication gave statistically higher TPC	DES/water proportion, the liquid-to-solid ratio and the stirring speed were demonstrated to be statistically significant UAE incorporated into the process as a pretreatment step was shown to significantly change the kinetic pattern of polyphenol extraction.	Lakka et al. (2019)

DESs	Spent hops	XN XN EY	Water content: 5–10–20–30–40% (w/w) ChCl-Eg with 10% water (w/w) ChCl-GL with 5% water (w/w) ChCl-Pg with 5% water (w/w) ChCl-Lac with 5% water (w/w) Ratios of SH to DES: 1:10–1:20–1:40–1:50 (w/w) Ratios of antisolvent to DES: 3:1–4:1–5:1 (v/w) 40–50–60–70–80 °C for 1 h 15–30–60–120 min at 60 °C	DES composed of ChCl-Pg (1:2 mol/mol) with 5% (w/w) water, 1:50 SH/DES (w/w), 3:1 antisolvent/DES (v/w) at 60 °C for 1 h allowed to extract a high amount of XN	The DES-based method appears to be an effective alternative to conventional XN extraction methods	Grudniewska and Popłoński (2020)
DESs	Spent Hops (SH)	Protein XN IR ^{13}C CP-MAS NMR Elemental analysis Electrophoretic analysis	DES1: ChCl-EG water content 10% (w/w) DES2: ChCl-GL water content 5% (w/w) DES3: ChCl-Pg water content 5% (w/w) DES4: ChCl-Lac water content 5% (w/w) Molar ratio (HBA:HBD) 1:2	DES-4 allowed to obtain higher XN EY Protein and XN-rich precipitates were obtained by adding water to the DES extracts	It is possible to obtain the simultaneous extraction of XN and proteins from SH using choline chloride-based DESs NMR and IR data indicated that most of the polysaccharides were not extracted by DESs	Grudniewska and Pastyrczyk (2023)

(continued)

Table 10.5 (continued)

Extraction method	Plant matrix	Evaluated parameters	Tested condition	Optimum condition	Main goals/conclusion	References
PHWE		TPC Anti-inflammatory activity IXN XN BAs	Two extraction procedures 1) individual extraction with water or ethanol and 2) sequential extraction experiments with ASE Water: 1500 psi 50–100–150–200 °C 30 min Ethanol: 1500 psi 150 °C 20 min Sequential extraction: hexane, ethanol water 1500 psi 150 °C 20 min	Subcritical water at 150 °C was used to selectively isolate IXN	The water-based extraction method proved to be more selective than the subcritical ethanol or sequential extraction methods using different solvents The extract obtained with PHWE exhibited *in vitro* anti-inflammatory activity	Gil-Ramirez et al. (2012)

2–MeTHF	Hop pellet	Aroma profile EY	SLE: 17 g of hop pellets 175 mL of 2-MeTHF 2h Soxhlet Extraction: 30 g of hop pellets 200 ml 2-MeTHF/hexane 6h	SLE2-MeTHF	2-MeTHF resulted in higher extraction yields Both solvents exhibited a similar chemical profile 2-MeTHF extract contained a higher proportion of lupulone compared to the hexane extract	Rapinel et al. (2020)

DESs Deep eutectic solvents, *BAs* Bitter acids, *XN* Xanthohumol, *OSE* Orbital Shaker Extraction, *UAE* Ultrasound-Assisted Extraction, UHE Ultrasonic Homogenizer Extraction, *ChCl* choline chloride, *Eg* ethylene glycol, *Gly* Glycine, *Pg* propylene glycol, *Lac* lactic acid, *Fruct* fructose, *TA* Tartaric Acid, *U* Urea, *DMU* 1,3-dimethylurea, *Phe* Phenol, *BSA* Benzenesulfonic acid, *2-MeTHF* 2-methyloxolane or 2methyltetrahydrofuran, *TPC* Total Phenolic Content, *PAC* Total Proanthocyanidin Content, *TCl* total chlorophylls, *TCa* total carotenoids, *AOC* Antioxidant Capacity, *EE* Extraction Efficiency, *HBA* hydrogen bond acceptor, *HBD* hydrogen bond donor, *TFC* Total Flavonoid Content, *TF* Total Flavanol, *PR* Determination of the Reducing Power, *CDES* Concentration of DES, *LSR* liquid to solid ratic, *EY* Extraction Yield, *IR* Infra-Red spectroscopy, ^{13}C *CP-MAS NMR* Solid-State analysis, *IXN* isoxanthohumol, *ASE* Accelerated solvent extraction, *PHWE* Pressurised Hot Water Extraction, *SLE* Solid Liquid Extraction, *Glu* Glucose, *GL* Glycerol, *Sucr* Sucrose

10.4 Combined Green Extraction Techniques

In recent research, a novel approach was adopted to enhance the extraction efficiency of bioactive compounds from plant matrices. This involved the integration of complementary extraction technologies to leverage the synergy between different extraction methods (Ummat et al., 2021).

The combination of UAE with MAE through simultaneous irradiation (UMAE) stands out as one of the most promising hybrid techniques known for its rapid and efficient extraction capabilities (Chemat et al., 2017).

In relation to hops, Sanz et al. (2022) proposed a sequential extraction method that utilised UAE and MAE to recover phenolic compounds with antioxidant properties from two varieties of hop dry cones (Hallertau Perle and Nugget) (Sanz et al., 2022). The results demonstrated that the combined process enhanced the recovery of antioxidant compounds by approximately five times, irrespective of the hop variety. In particular, for Hallertau Perle dry cones, UAE (60% ethanol, 55 °C, 30 min) produced an extract containing 160.43 mg GAE/g with an antiradical capacity of about 390 mg Trolox/g extract. Subsequent with MAE (96% ethanol, 190 °C, 5 min) an additional extract Credo sia sbagliata la costruzione ma occorre ricontrollare l'articolo with 77 mg GAE/g and 227 mg Trolox/g was produced. Similarly, Nuggets samples extracted with UAE under similar conditions produced an extract containing 98.80 mg GAE/g and 370 mg Trolox/g. The subsequent MAE increased the phenolic content in the extract to 99 mg GAE/g, equivalent to 369 mg Trolox/g extract.

In another study, Ferreira-Anta et al. (2023) focused on the green extraction of antioxidant fractions from two hop varieties, Nugget and Perle, and the subsequent production of microparticles via the process of spray-drying (Ferreira-Anta et al., 2023). The extraction process was performed sequentially, utilising both UAE and PHW techniques, with the objective of optimising the recovery of bioactive compounds from hops. As reported by the authors, the sequential extraction process (UAE followed by PHW) influenced the yield and composition of the extracts in several ways compared to UAE alone. The total carbohydrate content increased from about 30% after UAE to about 45% after sequential UAE-PHW extraction. Furthermore, after UAE, the glucose fraction was about 13.5% for both varieties. However, after the sequential extraction process, the maximum observed was for xylose, which was 13.5% for both varieties, indicating the solubilisation of the hemicellulose fraction. For the Perle variety, the TEAC value increased from 5.2 after UAE to 6.1 after sequential extraction, while for the Nugget variety, it increased from 2.6 to 4.2.

These results suggest that the sequential extraction process, in particular the PHW step, caused further breakdown of cell wall components, leading to increased extraction of certain compounds and changes in the molecular weight profiles of the extracts.

Kljakić et al. (2024) investigated the composition and potential applications of four hop varieties through various extraction techniques (SFE, UAE and MAE). The authors employed UAE and MAE in isolation and after SFE. This approach is demonstrative of a sequential extraction process, whereby SFE is initially employed to extract lipophilic components, and subsequently, UAE/MAE is utilised to extract the remaining hydrophilic components from the same plant material (Kljakić et al., 2024).

In a recent study, the development of green solvent extraction techniques aimed at adapting industrial-scale extraction processes to diverse plant species was investigated (Pichon et al., 2025). Different extraction methods were assessed: SM, TSE, UAE, PLE, and MAE. Additionally, a combination of UAE and SM was evaluated to enhance results. The extraction methods were compared in terms of yield, total flavonoid, polyphenol, and chlorophyll content, metabolite profiling, LC-MS/MS quantification, and molecular networking. UAE demonstrated high yields with reduced extraction times in comparison to maceration, and this effect was further enhanced by PLE. The extraction yields were positively correlated with elevated concentrations of α/β acids, polyphenols, and flavonoids. MAE and PLE resulted in higher levels of flavonoids, polyphenols, and xanthohumol compared to SE. Among the tested treatments, PLE and UAE combined with SM exhibited the highest overall desirability, followed by UAE+SM. These methods emerged as the most effective approaches, providing an optimal balance between mass yield, metabolite content, and concentration of compounds of interest.

Table 10.6 outlines the principal combined extraction techniques employed to recover compounds of interest from hops.

10.5 Applications of Hop Extracts in the Food and Non-Food Sectors

The subsequent section will examine the various applications of hop extracts in the food and non-food sectors. Phytotherapeutic and pharmacological applications will not be considered as these are addressed in a dedicated chapter of the book.

As previously mentioned in this chapter, hop extracts possess a plethora of interesting properties that could be further exploited in these sectors, such as antimicrobial and antioxidant properties. Green hop extracts are also notable for their ready-to-use properties, which render them suitable for application in a wide range of sectors.

Rosa and Lannes (2024) conducted a review of the potential applications of hop extracts in the food, veterinary and pharmaceutical industries. The review noted that the majority of reported applications are related to the antimicrobial properties of hops, with other pharmacological applications following closely behind (Rosa & Lannes, 2024). The literature study reviewed by the authors included publications from 2012 to 2022.

More recently, Bekiroglu et al. (2025) investigated the effects of incorporating hops extract into durum wheat pasta at two different levels (0.25 and 0.5 g/100 g

Table 10.6 Combined green extraction technique for the extraction of bioactive compounds from hop

Extraction method	Plant matrix	Evaluated parameters	Tested condition	Optimum condition	Main goals/conclusion	References
UMAE	Hallertau Perle hop pellet (PHP) Nugget hop pellet (NHP)	EY TPC α-acids Prenylflavonoids Saccharides and derivatives AA	UAE: 80 kHz Ethanol content 40–60–80% Time for PHP samples 15–30–45 min Time for NHP samples 60–120–180 min MAE: ethanol 96% Time 5 min Temperature 130–190 °C	UAE: 55 °C, 60% ethanol, 30 min, high TCP and AOC for PHP 55 °C, 60% ethanol, 120 min, high TPC for NHP 55 °C, 40% ethanol 180 min, high AOC for NHP MAE: 190 °C for PHP and NHP	A sequential approach combining UAE and MAE enables the production of a hop extract rich in bioactive compounds	Sanz et al. (2022)
UAE+PHWE	PHP NHP	Solid Residue Fraction Characterization (minerals, oligosaccharide fraction, protein) Liquid Fraction Characterization (TPC, AOC)	UAE: solid:water ratio 1:15, w/w 80 kHz amplitude 100% 55 °C 30 min for the PHP sample and 120 min for the NHP sample PHWE: solid:liquid ratio 1:15 190 °C	The UAE process extracted the majority of ashes, carbohydrates, and minerals After the two extractions, a slight increase in TPC and AOC for both varieties was observed	The final results were influenced by the sequential extraction (UAE-PHW)	Ferreira-Anta et al. (2023)

SFE+MAE SFE+UAE	Dried hop cones of four different Slovenian varieties	Lipophilic composition AA Antimicrobial activity Cytotoxic activity	SFE: 55 °C 200 bar CO_2 flow rate of 0.2 kg/h 4 h UAE: 100 Hz 52 °C 30 min MAE: 600 W 30 min	These findings highlight the potential of hops' essential oils and extracts for various applications, with differences in lipophilic composition and biological activity linked to the extraction method and hop variety	All extracts and essential oils exhibited notable antimicrobial activity, with MAE and UAE extracts demonstrating a greater ability to inhibit microbial growth compared to SFE Higher content of trans-caryophyllene in essential oils than in SFE extracts Cytotoxic activity showed that all extracts and EOs showed a high degree of activity but also a high degree of selectivity towards normal cells	Kljakić et al. (2024)

(continued)

Table 10.6 (continued)

Extraction method	Plant matrix	Evaluated parameters	Tested condition	Optimum condition	Main goals/conclusion	References
UAE+SM	Dried hop cones	EY TPC TFC TCl and TCa Individual BAs XN	Solid:liquid ratio of 1:10 (w/v) SM: stirring at 150 rpm 2 h TSE: hexane:ethyl acetate:acetonitrile:water (2:3:3:2) stirred 2 h 25 °C PLE: ASE three static cycles 5 min rinsing volume 75% purge time 60 s 100 bar PLE 25–75–125 °C PLE 75 °C (70% Ethanol) PLE 75 °C (30% Ethanol) MAE: Ethanol 100 % 600 rpm 5–15 min temperature 75 °C 150W	The findings were assessed and compared using desirability optimization as a statistical approach to identify the most promising extraction methods	UAE-P + SM resulted high levels of humulone, co-humulone, and XN The combination with USs enhances the concentration of extracted polyphenols by a factor of 1.5 and 2.06 for UAE-P + SM and UAE-B + SM, respectively The highest extracted carotenoid content was achieved with UAE-P, followed by PLE at 75 °C and UAE-B, showing an increase of 4.4-fold, 4.6-fold, and 4.8-fold, respectively, compared to SE	Pichon et al. (2025)

			UAE: -UAE-B 35 kHz -UAE-P 20 kHz SE (UAE+SM), I step: SM II step: 15 min at 25 °C UAE-B or UAE-P 12 W 100 % amplitude, 0 seconds On/Off. III step: shaking for 2 h at 150 rpm		65.5% < PLE < 77.4%, followed by UAE+SM with 64.5%	

EY Extraction Yield, *TPC* Total Phenolic Content, *AA* Antioxidant Activity, *UAE* Ultrasound-Assisted Extraction, *MAE* Microwave-Assisted Extraction, *AOC* Antioxidant Capacity, *PHWE* Pressurised Hot Water Extraction, *PHW* Pressurised Hot Water, *SFE* Supercritical Fluid Extraction, *EOs* Essential Oils, *TFC* Total Flavonoid Content, *TCl* total chlorophylls, *TCa* tvotal carotenoids, *BAs* Bitter acids, *XN* Xanthohumol, *SM* Simple Maceration, *TSE* Triphasic Solvent Extraction, *PLE* Pressurized Liquid Extraction, *ASE* Accelerated solvent extraction, *UAE-B* Ultrasound Bath, *UAE-P* Ultrasound Probe semolina) for enhancing the nutritional profile and functional properties of pasta while maintaining its quality characteristics (Bekiroglu et al., 2025). The enriched pasta samples were evaluated for various parameters, including cooking properties, colour, texture, predicted glycemic index, and sensory attributes. The addition of hop extract reduced the expected glycaemic index (GI) values by approximately 5% at both tested concentrations compared to the control pasta. However, it also increased the total phenolic content, total flavonoid content, and antioxidant capacity of the samples compared to the control ones.

However, the pasta with hop extract received significantly lower sensory scores than the control pasta and the pasta fortified with grape leaf extract, particularly in terms of taste. This was attributed to the presence of phenolic compounds in the hop extract, which have been shown to contribute to bitterness or unpleasant flavours in certain products. Additionally, the pasta samples fortified with 0.5% hop extract exhibited a higher cooking loss compared to the control ones.

In another recent study, Kyrykbaeva et al. (2025) evaluated the use of natural hop extract as an antimicrobial agent in cheese production, specifically focusing on soft brine cheese (Kyrykbaeva et al., 2025). They analysed the extract's efficacy in inhibiting pathogenic bacteria growth, particularly *Staphylococcus aureus*, and its impact on the physicochemical properties of cheese over a 90-day ripening period. The experimental process involved the incorporation of hop extract at a concentration of 20 μg/mL into the production of cheese. This concentration was determined on the basis of in vitro studies that established the minimum inhibitory concentration (MIC) of the extract against *S. aureus*. The results showed that cheese treated with hop extract exhibited significant reductions in microorganism growth, with bacterial counts decreasing by over one log unit during the storage period.

Moreover, the hop extract also enhanced the cheese's antioxidant properties, potentially classifying it as a functional food.

Another intriguing new application of green hop extracts pertains to the preparation of a pâté formulation. In their study, Bilska et al. (2024) investigated the impact of incorporating hop extracts (HEs; 0.1% and 0.2% on a weight basis), from the Magnum and Lubelski varieties, into pâté-type offal sausages, wherein 20% of the animal fat was substituted with rapeseed oil (Bilska et al., 2024). The results demonstrated that the addition of these extracts, particularly from the Lubelski variety, helped maintain lower pH levels, reduced changes in water activity, and significantly slowed down lipid oxidation compared to the control samples. The research demonstrated that these HEs not only improved the oxidative stability of the modified pâtés but also enhanced their microbiological quality by inhibiting the growth of aerobic microorganisms. This suggests that hop extracts could be a promising natural alternative to synthetic antioxidants in the meat industry, potentially extending product shelf life while meeting consumer demands for cleaner labels and healthier meat products.

Notwithstanding the fact that the study in question examined a variety of chemical and microbiological parameters, it did not include a sensory evaluation of the experimental sausages. This omission gives rise to questions regarding the palatability and consumer acceptance of the modified products.

In their study, Basilio-Cortes et al. (2024) conducted a phytochemical and antifungal activity analysis of hydrothermal bioactive metabolites extracted from hops. Extracts were obtained at two different temperatures, 25 and 65 °C, from two hop varieties, Citra™ and Ekuanot™ (Basilio-Cortes et al., 2024). The researchers found that the extraction temperature significantly influenced the concentration and composition of the bioactive compounds in both hop varieties. The antifungal activity of the extracts was subsequently evaluated both in vitro and in vivo against two common phytopathogenic fungi affecting peppers: *Fusarium oxysporum* and *Alternaria solani*. In vivo tests on bell peppers demonstrated that hop extracts could effectively inhibit fungal growth for up to 5 days under normal environmental conditions without refrigeration, with the Ekuanot™ extract at 65 °C demonstrating the highest antifungal activity against the tested microorganisms.

The study of Arruda et al. (2023), evaluated the functional properties of hop extract rich in β-acids, in the form of potassium salt structures (KBA), for application in sustainable active food packaging (Arruda et al., 2023). Polylactic acid (PLA)-based sheets were incorporated with varying KBA concentrations (0.1–5% w/w) and analysed for structural, mechanical, thermal, and bioactive characteristics. KBA incorporation reduced crystallinity and modified mechanical and thermal properties, with pronounced effects at higher concentrations. Surface hydrophobicity decreased with increasing KBA content, whereas water vapor permeability remained unchanged. Release studies demonstrated Fickian diffusion behaviour, favouring application in fatty food matrices. The addition of KBA enhanced UV-blocking and antioxidant properties. Antibacterial assays revealed significant inhibition of *Staphylococcus aureus* and *Listeria monocytogenes* at 5% KBA concentration. Atomic force microscopy (AFM) confirmed severe bacterial membrane disruption and cell death. These findings highlight the potential of hop β-acids, such as KBA, in developing sustainable active packaging for improved food shelf-life and safety.

A novel coating approach was developed in the study of Carbone et al. (2021) to extend the shelf life of fresh-cut kiwifruit, and its efficacy was assessed over 10 days under refrigerated storage conditions (4 °C) (Carbone et al., 2021). Alginate (2% w/v) was functionalized with bioactive extracts derived from hop cones (HE; 0.5% and 1% v/v) and employed as an edible coating to determine the most effective formulation for maintaining product stability. At the utilised concentrations, the hop extract did not impart any detectable bitterness to the final product. The phytochemical composition of the hop extract was characterized through its primary chemical constituents and analysed via ^{1}H nuclear magnetic resonance (NMR) spectroscopy, identifying the presence of bioactive compounds with antioxidant and antimicrobial properties, such as alpha and beta hop acids and xanthohumol. The physicochemical parameters (e.g., weight loss, soluble solid content, titratable acidity, pH, and colour attributes) and nutraceutical properties (e.g., total polyphenol content, ascorbic acid content, total carotenoids, and chlorophylls) of the coated samples were monitored throughout the storage period.

The findings demonstrated that the integration of hop extracts into the edible coating formulations effectively preserved the quality and nutraceutical properties of fresh-cut kiwifruit during refrigeration, thereby extending its shelf life and enhancing its commercial viability.

10.6 Conclusions and Future Perspectives

The development of eco-friendly extraction techniques has led to significant advancements in the field of bioactive compound extraction from hops, resulting in a reduction in environmental impact, enhanced efficiency, and improved product quality. Various extraction methods, including supercritical fluid extraction, pressurized liquid extraction, high hydrostatic pressure extraction, ultrasound-assisted extraction, microwave-assisted extraction, and pulsed electric field extraction, have demonstrated promising results. The employment of novel green solvents, such as deep eutectic solvents and ionic liquids, has expanded the scope for sustainable hop extraction. The combination of diverse techniques has been shown to yield synergistic effects, enhancing yields and the comprehensiveness of extraction processes. While these methods demonstrate considerable promise, further research is necessary to optimise process parameters, scale up operations, and evaluate economic feasibility for industrial applications. Additionally, further studies are required to understand the impact on bioactivity and stability of the extracted compounds. Green extraction techniques offer a sustainable and efficient approach to obtaining valuable bioactive compounds from hops, likely playing an increasingly important role in the brewing and nutraceutical industries.

References

Aganovic, K., Hertel, C., Vogel, R. F., Johne, R., Schlüter, O., Schwarzenbolz, U., Jäger, H., Holzhauser, T., Bergmair, J., Roth, A., Sevenich, R., Bandick, N., Kulling, S. E., Knorr, D., Engel, K., & Heinz, V. (2021). Aspects of high hydrostatic pressure food processing: Perspectives on technology and food safety. *Comprehensive Reviews in Food Science and Food Safety, 20*(4), 3225–3266. https://doi.org/10.1111/1541-4337.12763

Almeida, A. D. R., Maciel, M. V. D. O. B., Machado, M. H., Bazzo, G. C., De Armas, R. D., Vitorino, V. B., Vitali, L., Block, J. M., & Barreto, P. L. M. (2020). Bioactive compounds and antioxidant activities of Brazilian hop (*Humulus lupulus* L.) extracts. *International Journal of Food Science & Technology, 55*(1), 340–347. https://doi.org/10.1111/ijfs.14311

Ameer, K., Shahbaz, H. M., & Kwon, J. (2017). Green extraction methods for polyphenols from plant matrices and their byproducts: A review. *Comprehensive Reviews in Food Science and Food Safety, 16*(2), 295–315. https://doi.org/10.1111/1541-4337.12253

Arruda, T. R., Bernardes, P. C., Moraes, A. R. F. E., Marques, C. S., Pinheiro, P. F., De Oliveira, T. V., Ferreira, S. O., Naves, E. A. A., & Soares, N. D. F. F. (2023). Beyond brewing: β-acid rich hop extract in the development of a multifunctional polylactic acid-based food packaging. *International Journal of Biological Macromolecules, 228*, 23–39. https://doi.org/10.1016/j.ijbiomac.2022.12.191

Bagade, S. B., & Patil, M. (2021). Recent advances in microwave assisted extraction of bioactive compounds from complex herbal samples: A review. *Critical Reviews in Analytical Chemistry, 51*(2), 138–149. https://doi.org/10.1080/10408347.2019.1686966

Basilio-Cortes, U. A., Ramírez-Rodrigues, M. M., Ramírez-Rodrigues, M. A., González-Mendoza, D., Tzintzun-Camacho, O., Durán-Hernández, D., & González-Salitre, L. (2024). Phytochemical, spectroscopic analysis and antifungal activity on bell peppers of hydrothermal bioactive metabolites of *Humulus lupulus* L. extracts. *Natural Product Research*, 1–12. https://doi.org/10.1080/14786419.2024.2405010

Bekiroglu, H., Ozulku, G., Erol, K. F., Karadag, A., & Sagdic, O. (2025). Pasta with grape leaves and hops extract: Effect on quality properties, predicted glycemic index and in vitro bio-accessibility of phenolics. *Food Bioscience, 64*, 105909. https://doi.org/10.1016/j.fbio.2025.105909

Benvenutti, L., Zielinski, A. A. F., & Ferreira, S. R. S. (2019). Which is the best food emerging solvent: IL, DES or NADES? *Trends in Food Science & Technology, 90*, 133–146. https://doi.org/10.1016/j.tifs.2019.06.003

Bilska, A., Kobus-Cisowska, J., Wojtczak, J., Kowalski, R., & Kaczmarek, E. (2024). Antioxidant activity of Humulus lupulus phenolic hop extracts in creating a new pâté: An element affecting fat stability and microbiological quality during storage. *Molecules, 29*(7), 1561. https://doi.org/10.3390/molecules29071561

Bitwell, C., Indra, S. S., Luke, C., & Kakoma, M. K. (2023). A review of modern and conventional extraction techniques and their applications for extracting phytochemicals from plants. *Scientific African, 19*, e01585. https://doi.org/10.1016/j.sciaf.2023.e01585

Bizaj, K., Škerget, M., Košir, I. J., & Knez, Ž. (2021). Sub- and supercritical extraction of Slovenian Hops (Humulus lupulus L.) aurora variety using different solvents. *Plants, 10*(6), 1137. https://doi.org/10.3390/plants10061137

Bizaj, K., Škerget, M., Košir, I. J., & Knez, Ž. (2022). Hop (Humulus lupulus L.) essential oils and xanthohumol derived from extraction process using solvents of different polarity. *Horticulturae, 8*(5), 368. https://doi.org/10.3390/horticulturae8050368

Boukhatem, M. N., Ferhat, M. A., Rajabi, M., & Mousa, S. A. (2022). Solvent-free microwave extraction: An eco-friendly and rapid process for green isolation of essential oil from lemongrass. *Natural Product Research, 36*(2), 664–667. https://doi.org/10.1080/14786419.2020.1795852

Carbone, K., & Gervasi, F. (2022). An updated review of the genus Humulus: A valuable source of bioactive compounds for health and disease prevention. *Plants, 11*(24), 3434. https://doi.org/10.3390/plants11243434

Carbone, K., Macchioni, V., Petrella, G., & Cicero, D. O. (2020). Exploring the potential of microwaves and ultrasounds in the green extraction of bioactive compounds from Humulus lupulus for the food and pharmaceutical industry. *Industrial Crops and Products, 156*, 112888. https://doi.org/10.1016/j.indcrop.2020.112888

Carbone, K., Macchioni, V., Petrella, G., Cicero, D. O., & Micheli, L. (2021). Humulus lupulus cone extract efficacy in alginate-based edible coatings on the quality and nutraceutical traits of fresh-cut kiwifruit. *Antioxidants, 10*(9), 1395. https://doi.org/10.3390/antiox10091395

Chemat, F., Rombaut, N., Sicaire, A.-G., Meullemiestre, A., Fabiano-Tixier, A.-S., & Abert-Vian, M. (2017). Ultrasound assisted extraction of food and natural products. Mechanisms, techniques, combinations, protocols and applications. A review. *Ultrasonics Sonochemistry, 34*, 540–560. https://doi.org/10.1016/j.ultsonch.2016.06.035

Chemat, F., Abert Vian, M., Ravi, H. K., Khadhraoui, B., Hilali, S., Perino, S., & Fabiano Tixier, A.-S. (2019a). Review of alternative solvents for green extraction of food and natural products: Panorama, principles, applications and prospects. *Molecules, 24*(16), 3007. https://doi.org/10.3390/molecules24163007

Chemat, F., Abert-Vian, M., Fabiano-Tixier, A. S., Strube, J., Uhlenbrock, L., Gunjevic, V., & Cravotto, G. (2019b). Green extraction of natural products. Origins, current status, and future challenges. *TrAC Trends in Analytical Chemistry, 118*, 248–263. https://doi.org/10.1016/j.trac.2019.05.037

Chemat, F., Abert Vian, M., Fabiano-Tixier, A.-S., Nutrizio, M., Režek Jambrak, A., Munekata, P. E. S., Lorenzo, J. M., Barba, F. J., Binello, A., & Cravotto, G. (2020). A review of sustainable and intensified techniques for extraction of food and natural products. *Green Chemistry, 22*(8), 2325–2353. https://doi.org/10.1039/C9GC03878G

Choi, Y. H., & Verpoorte, R. (2019). Green solvents for the extraction of bioactive compounds from natural products using ionic liquids and deep eutectic solvents. *Current Opinion in Food Science, 26*, 87–93. https://doi.org/10.1016/j.cofs.2019.04.003

Cvjetko Bubalo, M., Vidović, S., Radojčić Redovniković, I., & Jokić, S. (2018). New perspective in extraction of plant biologically active compounds by green solvents. *Food and Bioproducts Processing, 109*, 52–73. https://doi.org/10.1016/j.fbp.2018.03.001

Da Silva, R. F., Carneiro, C. N. D. C., De Sousa, C. B., Gomez, F. J. V., Espino, M., Boiteux, J., De Los, Á., Fernández, M., Silva, M. F., & Dias, F. D. S. (2022). Sustainable extraction bioactive compounds procedures in medicinal plants based on the principles of green analytical chemistry: A review. *Microchemical Journal, 175*, 107184. https://doi.org/10.1016/j.microc.2022.107184

Donn, P., Seyyedi-Mansour, S., Chamorro, F., Garcia-Oliveira, P., Echave, J., Perez-Vazquez, A., Barciela, P., Cassani, L., & Prieto, M. A. (2024). Recent advances in understanding the keys factors influencing pressurized liquid extraction of secondary metabolites: A comprehensive review. *IECBM, 2024*, 1. https://doi.org/10.3390/blsf2024035001

El Achkar, T., Greige-Gerges, H., & Fourmentin, S. (2021). Basics and properties of deep eutectic solvents: A review. *Environmental Chemistry Letters, 19*(4), 3397–3408. https://doi.org/10.1007/s10311-021-01225-8

Esquivel-Hernández, D. A., Ibarra-Garza, I. P., Rodríguez-Rodríguez, J., Cuéllar-Bermúdez, S. P., Rostro-Alanis, M. D. J., Alemán-Nava, G. S., García-Pérez, J. S., & Parra-Saldívar, R. (2017). Green extraction technologies for high-value metabolites from algae: A review. *Biofuels, Bioproducts and Biorefining, 11*(1), 215–231. https://doi.org/10.1002/bbb.1735

Essien, S. O., Young, B., & Baroutian, S. (2020). Recent advances in subcritical water and supercritical carbon dioxide extraction of bioactive compounds from plant materials. *Trends in Food Science & Technology, 97*, 156–169. https://doi.org/10.1016/j.tifs.2020.01.014

Ferreira-Anta, T., Torres, M. D., Vilarino, J. M., Dominguez, H., & Flórez-Fernández, N. (2023). Green extraction of antioxidant fractions from Humulus lupulus varieties and microparticle production via spray-drying. *Foods, 12*(20), 3881. https://doi.org/10.3390/foods12203881

Fischer, B., Gevinski, E. V., Da Silva, D. M., Júnior, P. A. L., Bandiera, V. J., Lohmann, A. M., Rigo, D., Duarte, P. F., Franceschi, E., Zandoná, G. P., Rombaldi, C. V., Cansian, R. L., Paroul, N., & Junges, A. (2023). Extraction of hops pelletized (Humulus lupulus) with subcritical CO2 and hydrodistillation: Chemical composition identification, kinetic model, and evaluation of antioxidant and antimicrobial activity. *Food Research International, 167*, 112712. https://doi.org/10.1016/j.foodres.2023.112712

Gil-Ramírez, A., Mendiola, J. A., Arranz, E., Ruíz-Rodríguez, A., Reglero, G., Ibáñez, E., & Marín, F. R. (2012). Highly isoxanthohumol enriched hop extract obtained by pressurized hot water extraction (PHWE). Chemical and functional characterization. *Innovative Food Science & Emerging Technologies, 16*, 54–60. https://doi.org/10.1016/j.ifset.2012.04.006

Grudniewska, A., & Pastyrczyk, N. (2023). New insight for spent hops utilization: Simultaneous extraction of protein and xanthohumol using deep eutectic solvents. *Biomass Conversion and Biorefinery, 13*(16), 14975–14986. https://doi.org/10.1007/s13399-022-03462-5

Grudniewska, A., & Popłoński, J. (2020). Simple and green method for the extraction of xanthohumol from spent hops using deep eutectic solvents. *Separation and Purification Technology, 250*, 117196. https://doi.org/10.1016/j.seppur.2020.117196

Herrero, M. (2024). Towards green extraction of bioactive natural compounds. *Analytical and Bioanalytical Chemistry, 416*(9), 2039–2047. https://doi.org/10.1007/s00216-023-04969-0

Hikmawanti, N. P. E., Ramadon, D., Jantan, I., & Mun'im, A. (2021). Natural Deep Eutectic Solvents (NADES): Phytochemical extraction performance enhancer for pharmaceutical and nutraceutical product development. *Plants, 10*(10), 2091. https://doi.org/10.3390/plants10102091

Inumaro, R. S., Casalvara, R. F. A., Corrêa, R. C. G., Barros, L., Pereira, C., Calhelha, R. M., Heleno, S. A., Pimentel, T. C., Gonçalves, J. E., & Cardozo-Filho, L. (2023). Supercritical fluid-assisted extraction of hop cone residue from craft breweries. *Journal of CO2 Utilization, 72*, 102498. https://doi.org/10.1016/j.jcou.2023.102498

Kamelan, M., Nia, P., Sharayei, P., Elhamirad, A. H., & Saeidi Asl, M. (2023). Optimization of ultrasonic-assisted extraction of the effective compounds from hops (Humulus lupulus). *Journal of Food Science and Technology(Iran), 20*(141). https://doi.org/10.22034/FSCT.20.141.75

Kaur, S., Kumari, M., & Kashyap, H. K. (2020). Microstructure of deep eutectic solvents: Current understanding and challenges. *The Journal of Physical Chemistry B, 124*(47), 10601–10616. https://doi.org/10.1021/acs.jpcb.0c07934

Klimek, K., Tyśkiewicz, K., Miazga-Karska, M., Dębczak, A., Rój, E., & Ginalska, G. (2021). Bioactive compounds obtained from polish "Marynka" hop variety using efficient two-step supercritical fluid extraction and comparison of their antibacterial, cytotoxic, and anti-proliferative activities in vitro. *Molecules, 26*(8), 2366. https://doi.org/10.3390/molecules26082366

Kljakić, A., Ocvirk, M., Rutnik, K., Košir, I. J., Pavlić, B., Mašković, P., Mašković, J., Teslić, N., Stupar, A., Uba, A. I., & Zengin, G. (2024). Exploring the composition and potential uses of four hops varieties through different extraction techniques. *Food Chemistry, 447*, 138910. https://doi.org/10.1016/j.foodchem.2024.138910

Knez Hrnčič, M., Španinger, E., Košir, I. J., Knez, Ž., & Bren, U. (2019). Hop compounds: Extraction techniques, chemical analyses, antioxidative, antimicrobial, and anticarcinogenic effects. *Nutrients, 11*(2), 257. https://doi.org/10.3390/nu11020257

Krofta, K., Houška, M., Mikyška, A., & Čermák, P. (2018). Changes in the composition of hop secondary metabolites induced by high hydrostatic pressure: Changes in the composition of hop secondary metabolites induced by high hydrostatic pressure. *Journal of the Institute of Brewing, 124*(2), 158–172. https://doi.org/10.1002/jib.481

Kyrykbaeva, S., Kalibekkyzy, Z., Kapshakbayeva, Z., Baytukenova, S., Assirzhanova, Z., Baytukenova, S., Mustafayeva, A., Ospanova, B., & Utegenova, A. (2025). Evaluation of antimicrobial efficacy and shelf life of natural hop extract in cheese production. *CyTA - Journal of Food, 23*(1), 2446821. https://doi.org/10.1080/19476337.2024.2446821

Lakka, A., Karageorgou, I., Kaltsa, O., Batra, G., Bozinou, E., Lalas, S., & Makris, D. (2019). Polyphenol extraction from Humulus lupulus (Hop) using a neoteric glycerol/L-alanine deep eutectic solvent: Optimisation, kinetics and the effect of ultrasound-assisted pretreatment. *AgriEngineering, 1*(3), 403–417. https://doi.org/10.3390/agriengineering1030030

Lamberti, L., Grillo, G., Gallina, L., Carnaroglio, D., Chemat, F., & Cravotto, G. (2021). Microwave-assisted hydrodistillation of hop (Humulus lupulus L.) terpenes: A pilot-scale study. *Foods, 10*(11), 2726. https://doi.org/10.3390/foods10112726

Lefebvre, T., Destandau, E., & Lesellier, E. (2021). Selective extraction of bioactive compounds from plants using recent extraction techniques: A review. *Journal of Chromatography A, 1635*, 461770. https://doi.org/10.1016/j.chroma.2020.461770

Lei, Z., Chen, B., Koo, Y.-M., & MacFarlane, D. R. (2017). Introduction: Ionic liquids. *Chemical Reviews, 117*(10), 6633–6635. https://doi.org/10.1021/acs.chemrev.7b00246

Macchioni, V., Carbone, K., Cataldo, A., Fraschini, R., & Bellucci, S. (2021). Lactic acid-based deep natural eutectic solvents for the extraction of bioactive metabolites of Humulus lupulus L.: Supramolecular organization, phytochemical profiling and biological activity. *Separation and Purification Technology, 264*, 118039. https://doi.org/10.1016/j.seppur.2020.118039

Mena-García, A., Ruiz-Matute, A. I., Soria, A. C., & Sanz, M. L. (2019). Green techniques for extraction of bioactive carbohydrates. *TrAC Trends in Analytical Chemistry, 119*, 115612. https://doi.org/10.1016/j.trac.2019.07.023

Metaj, I., Hajdini, D., Gliha, K., Košir, I. J., Ocvirk, M., Kolar, M., & Cerar, J. (2023). Extraction of polyphenols from slovenian hop (Humulus lupulus L.) aurora variety using deep eutectic solvents: Choice of the extraction method vs. structure of the solvent. *Plants, 12*(16), 2890. https://doi.org/10.3390/plants12162890

More, P. R., Jambrak, A. R., & Arya, S. S. (2022). Green, environment-friendly and sustainable techniques for extraction of food bioactive compounds and waste valorization. *Trends in Food Science & Technology, 128*, 296–315. https://doi.org/10.1016/j.tifs.2022.08.016

Mungwari, C. P., King'ondu, C. K., Sigauke, P., & Obadele, B. A. (2025). Conventional and modern techniques for bioactive compounds recovery from plants: Review. *Scientific African, 27*, e02509. https://doi.org/10.1016/j.sciaf.2024.e02509

Nagybákay, N. E., Syrpas, M., Vilimaitė, V., Tamkutė, L., Pukalskas, A., Venskutonis, P. R., & Kitrytė, V. (2021). Optimized supercritical CO2 extraction enhances the recovery of valuable lipophilic antioxidants and other constituents from dual-purpose hop (Humulus lupulus L.) variety Ella. *Antioxidants, 10*(6), 918. https://doi.org/10.3390/antiox10060918

Naliyadhara, N., Kumar, A., Girisa, S., Daimary, U. D., Hegde, M., & Kunnumakkara, A. B. (2022). Pulsed electric field (PEF): Avant-garde extraction escalation technology in food industry. *Trends in Food Science & Technology, 122*, 238–255. https://doi.org/10.1016/j.tifs.2022.02.019

Navarro-Baez, J. E., Martínez, L. M., Welti-Chanes, J., Buitimea-Cantúa, G. V., & Escobedo-Avellaneda, Z. (2022). High hydrostatic pressure to increase the biosynthesis and extraction of phenolic compounds in food: A review. *Molecules, 27*(5), 1502. https://doi.org/10.3390/molecules27051502

Neri, L., Giancaterino, M., Rocchi, R., Tylewicz, U., Valbonetti, L., Faieta, M., & Pittia, P. (2021). Pulsed electric fields (PEF) as hot air drying pre-treatment: Effect on quality and functional properties of saffron (Crocus sativus L.). *Innovative Food Science & Emerging Technologies, 67*, 102592. https://doi.org/10.1016/j.ifset.2020.102592

Ntourtoglou, G., Tsapou, E. A., Drosou, F., Bozinou, E., Lalas, S., Tataridis, P., & Dourtoglou, V. (2020). Pulsed electric field extraction of α and β-acids from pellets of Humulus lupulus (Hop). *Frontiers in Bioengineering and Biotechnology, 8*, 297. https://doi.org/10.3389/fbioe.2020.00297

Oreopoulou, A., Tsimogiannis, D., & Oreopoulou, V. (2019). Extraction of polyphenols from aromatic and medicinal plants: An overview of the methods and the effect of extraction parameters. In *Polyphenols in plants* (pp. 243–259). Elsevier. https://doi.org/10.1016/B978-0-12-813768-0.00025-6

Pacheco-Fernández, I., & Pino, V. (2019). Green solvents in analytical chemistry. *Current Opinion in Green and Sustainable Chemistry, 18*, 42–50. https://doi.org/10.1016/j.cogsc.2018.12.010

Pichon, V., Millot, M., Pinault, E., Hamion, G., Ducret, G., Viana, M., & Mambu, L. (2025). Development and optimization of different extraction and analytical methods on Humulus lupulus L. with molecular networks, quantifications and desirability approaches. *Industrial Crops and Products, 223*, 120166. https://doi.org/10.1016/j.indcrop.2024.120166

Picot-Allain, C., Mahomoodally, M. F., Ak, G., & Zengin, G. (2021). Conventional versus green extraction techniques—A comparative perspective. *Current Opinion in Food Science, 40*, 144–156. https://doi.org/10.1016/j.cofs.2021.02.009

Pinto, M. B. C., Vardanega, R., Náthia-Neves, G., De França, P. R. L., Kurozawa, L. E., Meireles, M. A. A., & Schmidt, F. L. (2023). Novel Brazilian hop (Humulus lupulus L.) extracts through supercritical CO2 extraction: Enhancing hop processing for greater sustainability. *Food Research International, 172*, 113169. https://doi.org/10.1016/j.foodres.2023.113169

Poojary, M. M., Lund, M. N., & Barba, F. J. (2020). Pulsed electric field (PEF) as an efficient technology for food additives and nutraceuticals development. In *Pulsed electric fields to obtain healthier and sustainable food for tomorrow* (pp. 65–99). Elsevier. https://doi.org/10.1016/B978-0-12-816402-0.00004-5

Ramazzina, I., Macchioni, V., & Carbone, K. (2022). Antioxidant and pro-oxidant phytochemicals in ultrasound and microwave assisted extracts from hop cones: A statistical modelling approach. *Food & Function, 13*(18), 9589–9601. https://doi.org/10.1039/D2FO02020C

Rapinel, V., Chemat, A., Santerre, C., Belay, J., Hanaei, F., Vallet, N., Jacques, L., & Fabiano-Tixier, A.-S. (2020). 2-Methyloxolane as a bio-based solvent for green extraction of aromas from hops (Humulus lupulus L.). *Molecules, 25*(7), 1727. https://doi.org/10.3390/molecules25071727

Rifna, E. J., Misra, N. N., & Dwivedi, M. (2023). Recent advances in extraction technologies for recovery of bioactive compounds derived from fruit and vegetable waste peels: A review.

Critical Reviews in Food Science and Nutrition, 63(6), 719–752. https://doi.org/10.1080/10408398.2021.1952923

Rodrigues Arruda, T. (2022). Exclusive raw material for beer production? Addressing greener extraction techniques, the relevance, and prospects of hops (Humulus lupulus L.) for the food industry. *Food and Bioprocess Technology, 15*, 275–305. https://doi.org/10.1007/s11947-021-02716-w

Rodríguez García, S. L., & Raghavan, V. (2022). Green extraction techniques from fruit and vegetable waste to obtain bioactive compounds—A review. *Critical Reviews in Food Science and Nutrition, 62*(23), 6446–6466. https://doi.org/10.1080/10408398.2021.1901651

Rosa, R. S., & Lannes, S. C. D. S. (2024). Hop extracts and their utilizations: Perspectives based on the last 10 years of research. *Brazilian Journal of Pharmaceutical Sciences, 60*, e23905. https://doi.org/10.1590/s2175-97902024e23905

Rubira, A., Rodríguez-Rebelo, F., Del-Río, P. G., Romaní, A., & Gullón, B. (2024). Valorization of hop (Humulus lupulus L.) pruning to produce valuable compounds using two biorefinery strategies: Conventional processing and microwave-assisted autohydrolysis. *Industrial Crops and Products, 219*, 119174. https://doi.org/10.1016/j.indcrop.2024.119174

Santarelli, V., Neri, L., Carbone, K., Macchioni, V., & Pittia, P. (2021). Use of conventional and innovative technologies for the production of food grade hop extracts: Focus on bioactive compounds and antioxidant activity. *Plants, 11*(1), 41. https://doi.org/10.3390/plants11010041

Santarelli, V., Neri, L., Carbone, K., Macchioni, V., Faieta, M., & Pittia, P. (2023). Conventional and innovative extraction technologies to produce food-grade hop extracts: Influence on bitter acids content and volatile organic compounds profile. *Journal of Food Science, 88*(4), 1308–1324. https://doi.org/10.1111/1750-3841.16487

Sanz, V., Torres, M. D., Lopez Vilarino, J. M., & Dominguez, H. (2022). Green extraction of phenolic compounds from Perle Hallertau and Nuggets hop pellets. *Food Bioscience, 50*, 102044. https://doi.org/10.1016/j.fbio.2022.102044

Shen, L., Pang, S., Zhong, M., Sun, Y., Qayum, A., Liu, Y., Rashid, A., Xu, B., Liang, Q., Ma, H., & Ren, X. (2023). A comprehensive review of ultrasonic assisted extraction (UAE) for bioactive components: Principles, advantages, equipment, and combined technologies. *Ultrasonics Sonochemistry, 101*, 106646. https://doi.org/10.1016/j.ultsonch.2023.106646

Smith, E. L., Abbott, A. P., & Ryder, K. S. (2014). Deep Eutectic Solvents (DESs) and their applications. *Chemical Reviews, 114*(21), 11060–11082. https://doi.org/10.1021/cr300162p

Tušek, J., Benković, M., Valinger, D., Jurina, T., Belščak-Cvitanović, A., & Gajdoš Kljusurić, J. (2018). Optimizing bioactive compounds extraction from different medicinal plants and prediction through nonlinear and linear models. *Industrial Crops and Products, 126*, 449–458. https://doi.org/10.1016/j.indcrop.2018.10.040

Tylewicz, U. (2020). How does pulsed electric field work? In *Pulsed electric fields to obtain healthier and sustainable food for tomorrow* (pp. 3–21). Elsevier. https://doi.org/10.1016/B978-0-12-816402-0.00001-X

Tyśkiewicz, K., Gieysztor, R., Konkol, M., Szałas, J., & Rój, E. (2018). Essential oils from Humulus lupulus scCO2 extract by hydrodistillation and microwave-assisted hydrodistillation. *Molecules, 23*(11), 2866. https://doi.org/10.3390/molecules23112866

Ummat, V., Sivagnanam, S. P., Rajauria, G., O'Donnell, C., & Tiwari, B. K. (2021). Advances in pre-treatment techniques and green extraction technologies for bioactives from seaweeds. *Trends in Food Science & Technology, 110*, 90–106. https://doi.org/10.1016/j.tifs.2021.01.018

Veiga, B. A., Hamerski, F., Clausen, M. P., Errico, M., De Paula Scheer, A., & Corazza, M. L. (2021). Compressed fluids extraction methods, yields, antioxidant activities, total phenolics and flavonoids content for Brazilian Mantiqueira hops. *The Journal of Supercritical Fluids, 170*, 105155. https://doi.org/10.1016/j.supflu.2020.105155

Vernès, L. (2019). Development of an innovative process of eco-extraction assisted by ultrasound of food ingredients from spirulina and transposition to industrial scale (Doctoral dissertation, Avignon Université).

Chapter 11
Humulus lupulus L.: Pharmacological Properties and Medicinal Use

Antonella Di Sotto, Ester Percaccio, Annabella Vitalone, Elisabetta Bazzaro, and Silvia Di Giacomo

11.1 Introduction

Humulus lupulus L. became famous worldwide as a preservative and flavouring additive in brewing; indeed, although initially used as an alternative additive to *Myrica gale* L., it was widely adopted in the eighteenth century owing to its better preserving properties (Zanoli & Zavatti, 2008).

Along with the interest in brewing, *H. lupulus* was also used for medicinal purposes in various cultures and centuries, as food and as a source of materials for making fabric and paper, packing fragile cargo (Korpelainen & Pietiläinen, 2021; Wilson, 1975). Numerous folk customs and beliefs were also associated with this plant: hops were thought to keep demons away at night and hop gardens were considered as sites to dispose of spiritually impure items, such as water to wash the deceased. In Northern Europe, hop gardens were believed to be inhabited by elves; therefore, these areas were avoided after sundown (Korpelainen & Pietiläinen, 2021; Wilson, 1975).

Female strobili or hops or cones are mainly used in medicine and for brewing; however, other parts of the hop plant have also been utilised historically (Zanoli & Zavatti, 2008).

A. Di Sotto (✉) · E. Percaccio · A. Vitalone · E. Bazzaro
Department of Physiology and Pharmacology "V. Erspamer", Sapienza University of Rome, Rome, Italy
e-mail: antonella.disotto@uniroma1.it; ester.percaccio@uniroma1.it; annabella.vitalone@uniroma1.it; elisabettabazzaro.guest@uniroma1.it

S. Di Giacomo
Italian National Institute of Health – Unit of Human Nutrition and Health, Department of Food Safety, Nutrition and Veterinary Public Health, Rome, Italy
e-mail: silvia.digiacomo@iss.it

K. Carbone (ed.), *All About Hops: The Crop, its Cultivation, and its Uses in Brewing and Beyond*, https://doi.org/10.1007/978-3-031-96606-4_11

The earliest recorded medicinal use dates back to ancient European herbal traditions, where hop extracts were employed for anxiety, insomnia, nervousness, inflammation, and digestive ailments (Zanoli & Zavatti, 2008). Similarly, hops were used to relieve insomnia, nervous tension, and stress-related conditions in Ayurvedic medicine, Traditional Chinese Medicine, and aromatherapy. Native American tribes also used hop as an antirheumatic, an analgesic, and a treatment for urinary issues (Zanoli & Zavatti, 2008).

This chapter comprehensively reviews the pharmacological evidence and safety issues related to the medicinal use of hop-based products, as supported by preclinical and clinical studies. Moreover, novel findings suggesting future pharmaceutical and nutraceutical applications are also presented.

11.2 Pharmacological Properties

Biological activities of *Humulus lupulus* L., especially those of the cone-derived extracts, have been investigated in different experimental models, revealing a plethora of promising properties (Fig. 11.1), such as sedative, estrogenic, antioxidant, antimicrobial, antiviral, anti-inflammatory, and chemopreventive (Zanoli & Zavatti, 2008; Carbone & Gervasi, 2022). More recently, some studies also explored the benefits of hop cone extracts in metabolic disorders (Alves et al., 2024; Keskin

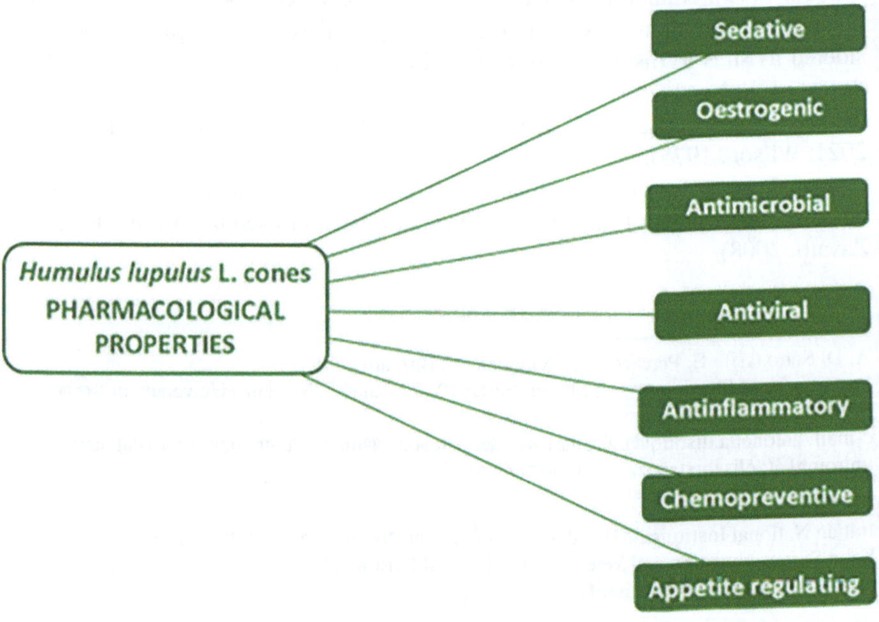

Fig. 11.1 Summary of the most investigated pharmacological properties of the extracts from *Humulus lupulus* L. cones

et al., 2018), highlighting their potential for future nutraceutical applications. Moreover, diverse hop cultivars have been screened to evaluate potential enhancements in their bioactivity profile owing to specific genetic traits and environmental growing conditions (Salihu et al., 2024; Min et al., 2023). Other studies focused on hop-derived pure compounds, highlighting their potential relevance in cancer chemoprevention, neuroprotection, metabolic disorders, and obesity (Ponticelli et al., 2021; Takahashi & Osada, 2017; Liu et al., 2014; Oberbauer et al., 2013; Dostálek et al., 2017; Harish et al., 2021; Jiang et al., 2018).

Details about the most investigated bioactivities, the possible involved compounds and the mechanisms of action are described in the following paragraphs.

11.2.1 Sedative Activity

The sedative effects of *H. lupulus* L. have been recognised since its early cultivation, as it was observed that individuals harvesting and working with the plants often became extremely drowsy during the day (Zanoli et al., 2007). Therefore, hop-based remedies, such as drinks or soft pillows filled with hops, became popular as sedative and sleeping aid. Accordingly, herbal tea and extracts from the hop strobiles have been approved by the German Commission E and by the European Medicine Agency to relieve restlessness, anxiety, and sleep disorders (Zanoli & Zavatti, 2008; Committee on Herbal Medicinal Products (HMPC) of the European Medicines Agency (EMA), 2024a). However, only a few studies investigating sedative and relaxing effects of *H. lupulus* extracts in preclinical models have been performed.

Sahin et al. (2016) studied the ability of an aqueous extract from hop cones to modulate the γ-aminobutyric acid receptor (GABA) neurotransmission in *Xenopus laevis* oocytes. GABA is a crucial amino acid in brain, regulating behaviour and providing major inhibitory control, by the activation of specific $GABA_A$ receptors ($GABA_AR$). Enhancing $GABA_AR$ activity through a pharmacological modulation has been found a suitable strategy to improve sleep disturbances, anxiety disorders, and epilepsy. The Authors highlighted that the aqueous extract from hop cones was a potent modulator of $GABA_AR$, significantly potentiatiating the GABA-induced response mediated by these receptors, as analysed by two-electrode voltage clamp. The extract was also found to contain basal levels of GABA (0.44 μg/mg of extract), although the Authors did not identify the specific compounds responsible for its sedative effects.

Other in vitro studies mainly focused on the spasmolytic effects of hop cones extracts. Particularly, an alcoholic extract (70% v/v ethanol; 1:10 drug–solvent ratio, DSR) of hop strobiles was found to induce spasmolytic effects on isolated smooth muscle from guinea pig intestine, markedly inhibiting contractions induced by acetylcholine and barium chloride; similarly, the extract blocked rat uterus contractions (Caujolle et al., 1969). Moreover, a methanolic extract from hop strobiles was reported to strongly inhibit calcium fluxes in clonal rat pituitary cells

(Blumenthal et al., 2000). Despite this evidence, the specific bioactive compounds were not identified.

Sedative-like properties of hop-based extracts were also reported in animal models; however, old studies carried out in frogs, pigeons, goldfish, and golden carp were considered unreliable owing to methodological shortcomings (Committee on Herbal Medicinal Products (HMPC) of the European Medicines Agency (EMA), 2024a). Hänsel and Wagener (1967) reported a lack of sedative effects and muscle relaxation for diverse hop extracts (extraction solvents: ethanol and methylisobutylketone) as well as for lupulone administered by gavage in mice or rats. By contrast, a reduction in locomotor activity was reported after intraperitoneal administration of aqueous, ethereal, and alcoholic extracts of hops in mice, with the ethereal extract being the most active (Bravo et al., 1974). This evidence suggests that the extraction solvent may significantly affect the sedative effects of the extract, likely due to a different ability to recover the bioactive compounds.

Other studies confirmed the ability of intraperitoneal administration of hop cone ethanolic extracts to suppress spontaneous locomotor activity and to potentiate the pentobarbital-induced sleep. Particularly, the Authors highlighted that the extract exhibited sedative and hypnotic properties at the lower doses, while anticonvulsant and hypothermic effects at the highest doses (Lee et al., 1993a, b). However, these studies did not explore the possible involvement of specific bioactive compounds.

In this respect, Zanoli et al. (2005) showed that a hop CO_2 extract and an α-acids-containing fraction, administered orally in rats, were able to increase the pentobarbital-induced sleeping time, without affecting locomotor activity and inducing anxiolytic effects; moreover, they reduced the period of immobility in the forced swimming test, thus suggesting that α-acids may contribute to the sedation by hop extracts and also act as antidepressant agents.

Similar findings were reported by Schiller et al. (2006), studying the ethanolic and CO_2 extracts from diverse hop varieties, α-acid and β-acid fractions, and pure oil administered orally in mice, although highlighting an impairment of locomotor activity and decreases in body temperature induced by all hop extracts along with the sedating activity. Discrepancy in the results of the diverse Authors may be ascribed to the differences in raw materials, storage, and extraction, which can result in a variable amount of α- and β-acids. Schiller et al. (2006) also revealed for the first time the possible involvement of β-acids in the sedative effects of hop extracts, thus suggesting that they may arise because of subtle interactions among diverse bioactive compounds, particularly phloroglucinol derivatives and volatile compounds. More recently, sedative effects were highlighted by Franco et al. (2012) in common quail (*Coturnix coturnix*) which are similar to humans in the sleep-wake rhythm; indeed, the hop extract decreased the nocturnal activity in the circadian rhythm. However, the possible bioactive compounds were not investigated, as the extracts were unspecified and uncharacterised (Table 11.1).

Regarding the hop varieties, Min et al. (2021) investigated the sleep-promoting effects of alcoholic dry extracts from Saaz and Saphir hop varieties, each containing a different content in α-acids, xanthohumol and humulone, and of their combination (Table 11.1). The samples were tested in a mouse model of

Table 11.1 In vivo studies evaluating the sedative activity of extracts from hop cones in the last fifteen years

Hop sample	Hop varieties	Chemical composition	Animals/model	Treatment/Dose/Duration/ Administration route	Outcome/*Comments*	References
Unidentified dry extract from hop cones	nr	nr	Common quail (Coturnix coturnix)	1, 2 and 11 mg daily/1 week/p.o.	Dose of 2 mg: ↓ nocturnal activity/ *Preservation of the circadian activity/rest rhythm*	Franco et al. (2012)
Alcoholic dry extracts	Saaz Saphir	mg compounds/g Saaz hop extract: α-acids 70.60, xanthohumol 2.51, humulone 47.23 mg compounds/g Saphir hop extract: α-acids 83.94, xanthohumol 4.23, humulone 53.13	Male ICR mice/ none	Saaz and Saphir extracts alone and in combination (ratio of 75:25)/150 mg/ kg/1 weeks/p.o.	↑ GABA in brain ↑ expression of $GABA_A$ and 5-HT1A receptors/ *$GABA_A$ – mediated sleep promoting activity; potency of mixture > extracts alone*	Min et al. (2021)
			Male ICR mice/ Pentobarbital-induced sleep	Saaz and Saphir extracts in combination (ratio of 75:25)/120–150 mg/kg/p.o. Humulone/20 mg/kg/nr Lupulone/20 mg/kg/nr	↓ the sleep latency time ↑ sleep duration/*sleep enhancing activity attributable to $GABA_A$ receptor modulation*	
			SD male rats/ none	Saaz and Saphir extracts alone and in combination (ratio of 75:25)/120–150 mg/ kg/n.r./p.o.	↑ sleep time ↓ awake time ↑ NREM sleep time ↑δ waves/*Increased sleep time attributable to* δ *waves increase/potency of mixture > extracts alone*	
			SD male rats/ Caffeine-induced insomnia	Saaz and Saphir extracts alone and in combination (ratio of 75:25)/120–150 mg/ kg/n.r./p.o.	↓ awake time ↓ REM sleep time by the extracts alone ↑ sleep time/*not significant respect to the control group* ↑ NREM sleep time of the extracts alone and in mixture ↑δ and **θ** waves increase/*Increased sleep time attributable to* δ and θ waves increase/p*otency of mixture > extracts alone*	

(continued)

Table 11.1 (continued)

Hop sample	Hop varieties	Chemical composition	Animals/model	Treatment/Dose/Duration/ Administration route	Outcome/*Comments*	References
Hot water or 70% ethanolic extracts	Amarillo Cascade Centennial Chinook Citra Herkules Mosaic Saaz Saphir Simcoe	Amount of α-acids, β-acids and xanthohumol: ethanolic extracts > hot water extracts Amount of β-acids: Saphir extracts[a] > other varieties. Amount of xanthohumol: highest content in Saphir alcoholic extract (4.23 μg/mg)	Drosophila melanogaster	All hop extracts/2% in agar medium/4 days	↑ total nighttime sleep by Simcoe and Mosaic water extracts and by Saphir and Simcoe alcoholic extracts	Min et al. (2023)
			Male ICR mice/ Pentobarbital-induced sleep	Water extracts from Centennial, Mosaic, and Simcoe varieties and alcoholic extracts from Saphir and Simcoe varieties/200 mg/kg/nr/p.o.	↓ sleep latency ↑ total sleep time by all the extracts except for Centennial	
			Male ICR mice/ Caffeine-induced insomnia	Water extracts from Centennial, Mosaic, and Simcoe varieties and alcoholic extracts from Saphir and Simcoe/200 mg/kg/nr/p.o.	↓ sleep latency time by Saphir ethanolic extract, by Mosaic water extract and by Simcoe water and ethanolic extracts ↑ total sleep time in the Saphir- and Simcoe-treated groups Recover of sleep time by Saphir ethanol extract similar to the benzodiazepine group	
			Male ICR mice/ none	Simcoe water extract and Saphir ethanol extract/200 mg/kg/21 days/p.o.	↑ GABA content in brain by Saphir alcohol extract Improvement of sleep ↑ mRNA expression of $GABA_A$ and $GABA_B$ receptors and 5-HT1A gene expression in brain	

nr not reported, *$GABA_A$* γ-amino butyric acid receptor A, *ICR* Institute of Cancer Research, *GABA* γ-amino butyric acid, *5-HT1A* serotonin 1A receptor, *$GABA_B$* γ-amino butyric acid receptor B

[a]beta-acids 1.81 and 66.37 μg/mg Saphir hop water and ethanol extract, respectively

pentobarbital-induced sleep and in a rat model of caffeine-induced insomnia. The results showed that the treatments, especially the Saaz/Saphir mixture, increased the GABA level in the brain and the expression of $GABA_A$ and 5-HT1A receptors; moreover, they reduced the sleep latency and prolonged sleep duration in pentobarbital-stimulated mice. Similar sleep-promoting effects were highlighted in caffeine-treated rats, in which the mixture showed to prolong the sleep time and non-rapid eye movement (NREM) sleep time. The Authors suggest that humulone and xanthohumol may be involved in the observed effects, although more complex interactions among multiple compounds in the extract are expected.

Min et al. (2023) also assessed the sleep-promoting activity of hot water and 70% ethanol extracts from different varieties of *H. lupulus*, including Amarillo, Cascade, Centennial, Chinook, Citra, Herkules, Mosaic, Saaz, Saphir, and Simcoe, in both invertebrate and vertebrate models (Table 11.1). The ethanolic extracts showed higher contents of α-acid, β-acid, and xanthohumol than the water ones, while the water extracts achieved the highest levels in flavonoids, such as quercetin-3-glucuronide and kaempferol-3-glucoside (Min et al., 2023). When assessed in animal models, Simcoe water extract and Saphir ethanol extract were found to induce the best sleep-improving effects, and to effectively restore sleep behaviour. As for their previous study, Min et al. (2023) associated these effects with an increase in the GABA content and in the mRNA expression of $GABA_A$ and $GABA_B$ receptors in mouse brains.

Regarding the possible bioactive compounds, Meissner and Häberlein (2006) highlighted that xanthohumol, a prenylated flavonoid, possessed sedative properties by modulating the $GABA_AR$ without affecting the benzodiazepine site. More recently, the substance has been reported to be able to modulate $GABA_ARs$, acting at the δ subunit, which represents a key target for other drugs, such as anaesthetics, sleep-promoting drugs, neurosteroids, and alcohol, and at the α6 subunit; this latter subunit could be responsible for behavioural implications in motor coordination (Benkherouf et al., 2020a).

GABA-potentiating effects were also reported for other prenylflavonoids such as isoxanthohumol, 8- prenylnaringenin and 6-prenylnaringenin, the latter being the most potent (Benkherouf et al., 2020b). It seems to act as a positive allosteric modulator, also displaying affinity for the benzodiazepines site; however, the Authors concluded that further studies are needed to confirm the behavioural effects of 6-prenylnaringenin in anxiolysis and sleeping activity (Benkherouf et al., 2020b). Regarding 8- prenylnaringenin, it was shown to possess sedative and panicolytic effects, likely acting as a positive allosteric modulator (Benkherouf et al., 2020b) and inhibiting (especially the R enantiomer) transporters of serotonin, norepinephrine, and dopamine (Bagatin et al., 2014).

Sedative effects were also displayed by lupulone, colupulone, and adlupulone, which also exhibited antidepressant activities (Schulz et al., 2018). Moreover, 2-methyl-3-buten-2-ol, a volatile compound arising from the auto-oxidation of humulone and lupulone during the storage process and identified in the essential oil from hop cones, has been reported to possess hypnotic activity (Wohlfart et al., 1983; Hänsel et al., 1980). Owing to the low levels of 2-methyl-3-buten-2-ol in hops, it is conceivable that other volatile compounds, may act synergistically. For instance, the monoterpene linalool, a typical component in hop essential oil (Rutnik et al., 2021), has demonstrated marked sedative and anticonvulsant effects in animal models (Linck et al., 2009).

Taken together, this evidence supports the hypothesis that multiple compounds, based on their amount and interactions with the other components, may contribute to the sedative effects of the hop extracts. In this respect, Benkherouf et al. (2020a) highlighted that hops prenylflavonoids and humulone positively interacted to increase the $GABA_A$Rs ion modulation, thus suggesting that more than one compound may modulate different binding sites of $GABA_A$R, leading to an enhanced potentiation of its function. Moreover, Min et al. (2023) concluded that flavonoids, such as quercetin-3-glucuronide and kaempferol-3-glucoside, which were the highest identified compounds in the sleep-promoting Simcoe water extract, are expected to be endowed with sedative effects, and to contribute or synergise the other compounds in the phytocomplex. In support, other flavonol glycosides, such as quercetin-3-glucuronide and kaempferol-3-glucoside, were found to be sleeping-promoting agents (Kim et al., 2021a; Li et al., 2017).

Potentiating sedative effects were also reported when hop was combined with other herbal drugs, such as valerian. In this regard, a mixture of an ethanolic extract from Cascade hop, characterised by its xanthohumol amount (827.49 ± 73.79 μg/g extract), and an ethanolic extract from valerian root (valerenic acid amount: 1260.53 ± 50.47 μg/g extract), exhibited a higher $GABA_A$R binding ability than valerenic acid and/or xanthohumol; moreover, the mixture improved sleep-related behaviours (Choi et al., 2017). The Authors also showed that the extracts from Hallertau and Saaz varieties, containing lower levels of xanthohumol, especially Saaz hop, were lacking in sedative effects (Choi et al., 2017). The obtained finding also revealed that, although valerenic acid and xanthohumol were expected to be the main bioactive compounds for the sleep promoting effect of valerian/hop mixture, their ability to bind the benzodiazepine site in $GABA_A$R was significantly lower than that of the extract: this indicates the presence of other sleep promoters in the mixture (Choi et al., 2017). At last, a modulation by valerian/hop mixture in the mRNA levels of dieldrin (Rdl), which is a Drosophila $GABA_A$R, $GABA_B$ receptor 1 and receptor 2, and 5-hydroxytryptamine receptor 1A (5-HT1A) was found (Choi et al., 2017). Based on these results, sedation by hop extracts seems to involve not only the modulation of GABA signalling but also other neurotransmissions, such as serotonin. Butterweck et al. (2007) supposed the involvement of the melatoninergic system in the sedative effect of hops, based on the ability of the melatonin antagonist luzindole to similarly block the hypothermic effects of a hop methanolic extract and melatonin in BL6/C57J mice. Conversely, other Authors strengthened the involvement of a GABA transmission modulation in the sedative effects of hop, while excluding an adenosine A1 receptor (Müller et al., 2002). Further studies are needed to explore and clarify the true mechanisms underlying the sedative effects of hop and its sleep-inducing power.

11.2.2 Oestrogenic Activity

Research into the bioactivities of *H. lupulus* L. extracts revealed that hop cones are a natural source of oestrogen receptor agonists (Koch & Heim, 1953).

In this respect, Liu et al. (2001) highlighted the ability of a methanolic hop extract to bind both oestrogen receptors (Erα and Erβ) to induce alkaline phosphatase activity in Ishikawa cells (human endometrial adenocarcinoma cell line), and to up-regulate the mRNA expression of progesterone receptor and presenilin-2, an oestrogen-inducible gene in S30 cells (breast cancer cell line transfected with Erα). The hop estrogenic activity was confirmed by Overk et al. (2005), using a chloroform partition of a methanol extract from a CO_2-extract from the Nugget hop cultivar. The extract showed binding affinity to both Erα and Erβ receptors, involving 8-prenylnaringenin and isoxanthohumol, but not 6-prenylnaringenin and xanthohumol. Moreover, it activated oestrogen response elements, up-regulated the mRNA expression of the progesterone receptor, and induced alkaline phosphatase activity, usually stimulated by oestrogens in human endometrial cancer cells, primarily involving 8-prenylnaringenin.

Similarly, Krause et al. (2014) showed that a spent hop extract, containing 0.42% 8-prenylnaringenin, 2.18% 6-prenylnaringenin, 35.78% xanthohumol, and 1.35% isoxanthohumol, stimulated alkaline phosphatase activity in Ishikawa cells, a human endometrial adenocarcinoma cell line, achieving about 52% of the level induced by 17β-oestradiol. Under the same conditions, 8-prenylnaringenin exhibited 86% of the activity of 17β-oestradiol, isoxanthohumol was only partially effective, while 6-prenylnaringenin and xanthohumol were not effective. The study concluded that the hop extract acts as a partial agonist at the oestrogen receptor, with 8-prenylnaringenin acting as a full agonist.

Dietz et al. (2017) also studied the potential estrogenic activity of eight hop enriched-extracts obtained through adsorbent-free counter-current separation in Ishikawa cells and murine hepatoma Hepa1c1c7 cells. The extracts enriched in 8-prenylnaringenin induced alkaline phosphatase activity, and isoxanthohumol was found able to synergise the activity of 8-prenylnaringenin, while acting as a weaker estrogenic agent (Dietz et al., 2017). The xanthohumol-enriched extracts exhibited marked cytotoxic effects in Ishikawa cells compared to Hepa1c1c7 cells, in which they increased the level of the chemopreventive marker NQO1 (Dietz et al., 2017). Based on this evidence, the Authors (Dietz et al., 2017) suggested the possibility to design specific hop extracts varying the contents of the oestrogen 8-prenylnaringenin and the chemopreventive compound xanthohumol, achieving “menopausal extract” with balanced estrogenic and chemopreventive activities, and “pre-menopausal extracts” with mainly chemopreventive properties, leading to hop-based supplements with greater efficacy and safety.

The estrogenic activity of 8-prenylnaringenin was also confirmed in animal models, especially in ovariectomised (OVX) rats. Indeed, subcutaneous administration of 8-prenylnaringenin resulted in estrogenic responses, such as increased uterine vascular permeability and vaginal epithelial mitosis (Milligan et al., 2002); high doses of 8-prenylnaringenin also increased uterine weight and serum prolactin, vaginal epithelial hyperplasia and mammary gland secretion in OVX rats (Rimoldi et al., 2006). A reduction in hot flushes, similar to oestradiol was reported too (Bowe et al., 2006). Hümpel et al. (2005) demonstrated that 8-prenylnaringenin fully prevented trabecular bone loss in OVX rats while inducing only minimal, dose-independent uterotrophic effects at bone-effective doses, in contrast to the potent uterine stimulation observed with estradiol, thus supporting its potential relevance in hormone replacement therapy for menopause. Despite these effects, the potency of 8-prenylnaringenin is significantly lower than oestradiol, likely as a result of its rapid metabolism in vivo (Pohjanvirta & Nasri, 2022). Indeed, pharmacokinetic studies in vitro of 8-prenylnaringenin reveal a good absorption likely via passive diffusion, and significant metabolism into glucuronides and sulfates (van Breemen et al., 2014).

11.2.3 Antibacterial, Antifungal, and Antiviral Activity

Extensive research investigated the antibacterial power of hop extracts and essential oils, especially against Gram-positive bacteria, highlighting humulone, lupulone, xanthohumol, and 6-prenylnaringenin as the possible bioactive compounds (Teuber & Schmalreck, 1973; Mizobuchi & Sato, 1984). Membrane leakage, due to the interaction of the hydrophobic parts of the molecules with the bacterial cell wall, is considered the major mechanism of these antibacterial effects (Teuber & Schmalreck, 1973).

In this respect, Langezaal et al. (1992) showed that the chloroform extracts and essential oils obtained by hydrodistillation from 13 hop cultivars were effective against the Gram-positive bacteria *Bacillus subtilis* and *Staphylococcus aureus* but not against the Gram-negative *Escherichia coli*. Similarly, Flieger et al. (2024) highlighted that water, DMSO, 80% ethanol, and acetone extracts from spent hops, obtained as waste product from supercritical carbon dioxide hop extraction, possessed Gram-positive antimicrobial activity, without significant effects on Gram-negative bacteria and yeasts, with the DMSO extract being the most potent.

A hop CO_2 extract, containing 50% humulone and lupulone, was also effective against *Propionibacterium acnes* and *Staphylococcus aureus*, suggesting a possible application in the treatment of acne (Weber et al., 2019). Di Lodovico et al. (2020) showed a marked antimicrobial activity of a hop hydroalcoholic extract, containing 4.7% α-acids and 6.1% β-acids, against *Staphylococcus epidermidis*, *S. aureus* and *Cutibacterium acnes*. The antibacterial activity against *Staphylococcus aureus* was also confirmed by Kolenc et al. (2022) and by Khaliullina et al. (2024) for acetone-extracts. A marked antimicrobial effect, comparable to reference antibiotics and showing synergy with amikacin, ciprofloxacin, and ceftriaxone was highlighted too (Khaliullina et al., 2024). Similarly, Rozalski et al. (2013) highlighted a promising antibacterial activity of a hop extract containing 51% xanthohumol against *Staphylococcus aureus* (MSSA and MRSA strains); the extract also inhibited biofilm formation and reduced biofilm viability, suggesting its potential in treating resistant bacterial infections.

Another study assessed the antimicrobial activity of essential oils and methanolic extracts of six hop varieties against eight human cocci pathogenic strains: *Staphylococcus aureus* MRSA (ATCC 43300), *S. aureus* MRSA (29213), S. *aureus* MSSA (ATCC 29213), *S. epidermidis* (ATCC 12228), *Enterococcus faecalis* (ATCC 29212), *E. faecalis* VRE (ATCC 51299), *E. faecium* (ATCC 19434), and *Micrococcus luteus* (ATCC 10240), highlighting a marked activity of the extracts, with *M. luteus* being the most susceptible. This effect was correlated with xanthohumol concentration, underscoring its role as a bioactive compound (Piasecki et al., 2023).

Two hop essential oil fractions, containing one myrcene and linalool and another-containing β-caryophyllene, β-copaene, β-farnesene, and α-humulene as major compounds, were also reported to be effective against the Gram-negative *Escherichia coli* (Jeliazkova et al., 2018).

Essential oils and chloroform extracts from different hop cultivars exhibited antifungal activity against *Trichophyton mentagrophytes* var. interdigitale but not against the yeast *Candida albicans* (Langezaal et al., 1992). Similarly, Moureu et al. (2023) highlighted the antifungal activity of crude hydro-ethanolic extracts and dichloromethane sub-extracts from various hop parts against *Venturia inaequalis*, suggesting their potential for agricultural application. The hop bitter acids were also found to possess antifungal activity, with humulones more potent than lupulones, while prenylchalcones resulted ineffective (Mizobuchi & Sato, 1985).

Hop extracts were also assessed against viral infections although in a few studies. Buckwold et al. (2004) found lacking antiviral activity of hop extracts and its chalcones against different RNA and DNA viruses, such as influenza A and B viruses, HIV, rhinovirus, RSV, and yellow fever virus. By contrast, a xanthohumol-enriched extract showed weak to moderate antiviral activity, suggesting that xanthohumol could be involved in the observed effect.

More recently, Di Sotto et al. (2018) reported a promising antiviral activity of a commercial hop hydroalcoholic dry extract, standardised to contain 0.4% flavonoids, against various influenza A virus strains in MDCK and A549 cells. The extract inhibited PR8 viral replication and reduced viral proteins in MDCK cells and reduced viral titre by approximately 80% for pH1N1 and around 60% for other strains in A549 cells. Additionally, it significantly restored GSH levels in infected cells, suggesting that the effects of hop extract may arise from an interference with the redox imbalance caused by influenza virus. Accordingly, a further study confirmed the antiviral effect of this extract against PR8/H1N1 influenza strain in A549 lung and BEAS-2B bronchial cells (Percaccio et al., 2023) and showed that the treatment restored the cellular redox state and enhanced antioxidant defence by increasing Nrf2 expression, thus confirming the previous hypothesis about the ability of hop to stimulate the host cell defences and affect viral replication. The antiviral effect of the hop extract has been associated with phenolic compounds; indeed, antiviral and anti-influenza properties have been reported for rutin, quercetin, syringic acid, and gallic acid (Percaccio et al., 2023); however, further studies are needed to clarify the role of other compounds, such as prenylflavonoids and bitter acids, and to assess their possible interaction in the phytocomplex.

11.2.4 Antioxidant Activity

Numerous studies explored the antioxidant properties of *H. lupulus* L. extracts, focusing on various hop parts, extraction methods, and chemical compositions. The radical scavenging properties towards the synthetic chromogenic DPPH and ABTS radicals were reported for a hydroalcoholic hop extract, characterised as containing rutin and syringic acid as the major phenolic compounds. Based on the diverse neutralization mechanisms of the tested radicals, the higher potency against ABTS suggests the involvement of an electron transfer in the scavenger activity of the extract (Di Sotto et al., 2018).

DPPH radical scavenging abilities were also reported for an hydroalcoholic extract from *H. lupulus* cones, characterised by high levels of gallic acid, quercetin, and kaempferol derivatives. However, its activity was lower than that of hexane and methanol extracts, suggesting that more effective antioxidant compounds were extracted by methanol and hexane (Önder et al., 2013). Similar results were also reported for methanolic, ethanolic, and hot water extracts of hop strobiles from different cultivars, while the extracts from leaves showed a lower radical scavenger power (Abram et al., 2015; Lyu et al., 2022; Liu et al., 2007; Almeida et al., 2020). Interestingly, Mudura et al. (2010) showed that the antiradical activity was higher in hop cones from aromatic cultivars than from bitter ones. Similarly, an increased antiradical power was highlighted in hop cone extracts obtained by ultrasound-assisted extractions and using a eutectic solvent mixture of lactic acid: glycine (Santarelli et al., 2022; Macchioni et al., 2021).

Radical scavenging properties were also reported for the methanolic extracts of hop young shoots of diverse ecotypes (Castelmassa, Cologna, Santa Maria in Punta, and Vicchio), containing quercetin and kaempferol glycosides as the main compounds (Maietti et al., 2017) and of hop seed, characterised by the content of catechin and epicatechin (Alonso-Esteban et al., 2019).

Other antioxidant effects, including ferric reducing power and lipoperoxidation inhibition were also reported for hop cone extracts (Di Sotto et al., 2018; Abram et al., 2015; Keskin et al., 2019). The antioxidant properties of some hop cone extracts were also found to be involved in their cytoprotection against both pro-oxidant stressors and viral infections; indeed, the extracts were able to lower the intracellular levels of reactive oxygen species and to increase the levels of reduced glutathione, likely as a consequence of a stimulation of the Nrf2 (nuclear factor erythroid 2-related factor 2)-mediated response (Di Sotto et al., 2018; Percaccio et al., 2023).

Among the possible bioactive compounds, prenylated chalcones (i.e., xanthohumol) exhibited strong antioxidant activities, likely due to their prenyl group; moreover, they possess chelating ability, which may inhibit ROS generation (Bartmańska et al., 2013). Similarly, polyphenols, such as quercetin, chlorogenic acid, syringic acid, benzoic acid, catechin, and epigallocatechin, as well as bitter acids, are known antioxidant agents (Liu et al., 2007; Mladěnka et al., 2011; Inui et al., 2017). This evidence strengthens the hypothesis that the bioactivities of hop cone extracts arise from the contribution of all components of the phytocomplex.

11.2.5 Anti-inflammatory Activity

Preclinical studies demonstrated the anti-inflammatory activities of various hop extracts, particularly those rich in polyphenols and flavonoids, although with differences due to hop varieties, extraction method, and chemical composition.

Sangiovanni et al. (2019) studied the anti-inflammatory activity of a hydroalcoholic extract from hop cones (cv. Cascade), containing 1.86% xanthohumol A and 0.28% xanthohumol D in AGS gastric cancer cells, highlighting the ability of the extract to reduce the IL-8 secretion and NF-κB nuclear transcription. This activity was retained after digestion; in contrast, the water extract of hop pellets was ineffective. Anti-inflammatory effects were also reported for an extract from dried spent hops, rich in polyphenols and flavonols, in LPS-activated RAW 264.7 murine macrophages (Caban et al., 2020). Indeed, the extract was found to reduce the mRNA expression of pro-inflammatory factors (i.e., IL-6, TNF-α, COX-2, and NF-κB) and to decrease the protein expression of iNOS and NO production; it also inhibited the AGS cell viability, suggesting dual anti-inflammatory and antiproliferative effects (Caban et al., 2020). Anti-inflammatory effects were also highlighted for ethanolic extracts from hop pellet in LPS-activated RAW 264.7 murine macrophages (Wu et al., 2020) and for extracts from dried hop cones of Hallertau Mittelfrüh, Columbus, Nugget, Chinook, and Cascade varieties, containing prenylated phloroglucinols, phenolics, chalcones, and flavones, in pooled peripheral venous blood from healthy volunteers (Nicácio et al., 2022). Finally, Velot et al. showed the anti-inflammatory power of the methanolic extract of the fresh leaves of Cascade hop, (Velot et al., 2022), which contained oxidised β-acids, hulupone, adhulupone, and feruloyl quinic acid, in chondrocytes stimulated with interleukin-1β. Indeed, at nontoxic concentration, it reduced mPGES-1, PGE2, and NO release and the effect increased when the extract was encapsulated in rapeseed lecithin nanoliposomes.

Other studies also demonstrated that hop bitter acids prevented inflammation and cognitive decline in a mouse model of Alzheimer disease and aging (Ano et al., 2017). Moreover, both hop extract and bitter acids have been found able to reduce paw swelling induced by zymosan in female C57BL/6 mice (Van Cleemput et al., 2009; Hougee et al., 2006). The swelling reduction by hop bitter compounds was similar to that induced by celecoxib, although of lower potency; however, it was also accompanied by reduced joint degradation, likely acting upstream of the NF-κB pathway (Tripp et al., 2010). The available evidence strengthens the ability of both iso-α-acids and reduced iso-α-acids to modulate inflammation by acting on several targets, with a well-tolerated profile (Ponticelli et al., 2021). Further studies are needed to clarify the contribution of other compounds in the anti-inflammatory properties of hop extracts and to harness them for nutraceutical purposes.

11.2.6 Chemopreventive and Antiproliferative Activity

Some studies highlighted the suppressing effects of various hop extracts in diverse cancer cell lines. Hudcová et al. (2014) reported the antiproliferative activity of residual material obtained from the extraction of the hop Vital variety with supercritical carbon dioxide, containing significant amounts of 8-prenylnaringenin, xanthohumol, and isoxanthohumol, in human HT-29 and SW620, colon cancer cell lines, while being less toxic to noncancerous IEC-6 cells. Accordingly, the dried spent hop extract, rich in total polyphenolic content, flavonols, and proanthocyanidins, reduced invasion, migration, and VEGFA mRNA expression in SW-480 cells; moreover, at high concentrations, it lowered the mRNA expression of MMP-2, MMP-9, and HIF1a. Similar effects were found in HT-29 cells, suggesting that the extract was able to inhibit cancer cell invasion and migration by downregulating pro-metastatic markers (Caban et al., 2022). Hemachandra et al. (2012) also suggested that an ethanol extract from the Nugget hop cultivar, containing xanthohumol and 8-prenylnaringenin, could inhibit proliferation in oestrogen receptor-negative breast epithelial cells, based on its ability to affect oestrogen metabolism and colony formation in MCF-10A cell line.

More studies focused on the antiproliferative and chemopreventive power of prenylflavonoids, especially xanthohumol (Jiang et al., 2018; Venturelli et al., 2016). These compounds exhibited promising antioxidant and anti-inflammatory effects, and were shown to modulate the carcinogen metabolism, through the inhibition of phase 1 metabolic enzymes and the activation of phase 2 detoxifying ones. Their chemopreventive power seems also to arise from the inhibition of tumour initiation and a direct inhibition of cancer cell growth (Fig. 11.2), with low toxicity in healthy tissues (Jiang et al., 2018; Venturelli et al., 2016; Girisa et al., 2021). These findings warrant further research to fully understand the mechanisms underlying the chemopreventive power of hop extracts and their phytochemicals, and to harness these products clinically.

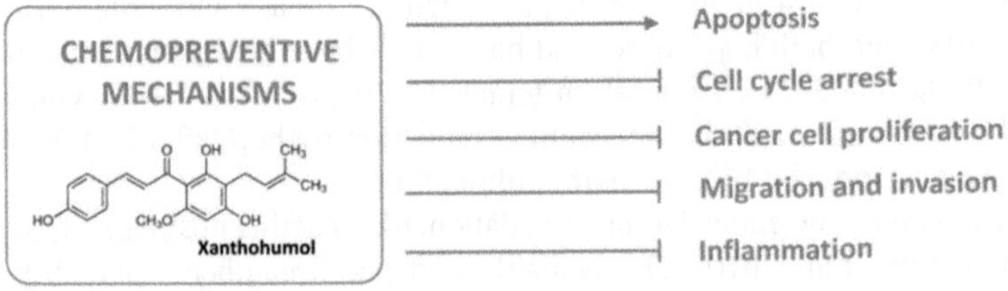

Fig. 11.2 Scheme of the major mechanisms underlying the chemopreventive properties of prenylflavonoids from *Humulus lupulus* L. cones

11.2.7 Other Pharmacological Activities

The extracts from hop cones were also investigated for digestive and metabolic-regulating properties. An in vivo study demonstrated that a 95% ethanol extract of dried hop flowers was able to protect hepatocyte from the CCl_4-induced hepatotoxicity in mice, likely affecting oxidative stress and inflammation, highlighting their therapeutic promise for liver diseases (Zhou et al., 2018).

Despite its traditional use as a digestive remedy, specific studies are scanty. Kurasawa et al. (2005) investigated the digestive power of hop using a pylorus-ligated model in rats. They found that orally administered hop-dried extract increased gastric juice volume without altering acidity, despite a null effect of intragastric administration. This suggests that bitterness of hop triggers gastric secretion via the cephalic phase, likely through the modulation of the cholinergic nervous system. Indeed, hop bitter acids activate bitter taste-sensing type 2 receptors (TAS2R) in the oral cavity, thus stimulating salivation and secretion of gastric juice, as well as secretion of mucopolysaccharide-rich mucus, thereby facilitating digestion. Moreover, Yamazaki et al. showed that matured hop bitter acids, which are oxidized derivatives of bitter acid, can activate specific human TAS2R1, 8, 10, which mediate calcium response and CCK production in enteroendocrine cells (Yamazaki et al., 2020). This mechanism suggests a possible modulation of extra-oral TAS2R by hop bitter, which can be involved in metabolic regulating effects (Tuzim & Korolczuk, 2021). In this respect, hop bitter compounds, especially α-acids and iso-α-acids, have been shown to modulate weight gain, lipid metabolism, glucose homeostasis, insulin sensitivities, and inflammation by acting on different targets (Ponticelli et al., 2021). For instance, it has been discovered that hop iso-α-acids are able to influence the activity of both PPAR-α, involved in lipid metabolism, and PPAR-γ, highly expressed in adipose tissue where it regulates the genes responsible for adipocyte growth, differentiation, and lipid storage (Ponticelli et al., 2021). Body weight reduction was reported in both mice treated with isomerised hop compounds and in ovariectomised mice after hop extract administration (Yajima et al., 2005; Hamm et al., 2019). They have also been shown to improve blood lipid status by activating PPAR-α, thus leading to the up-regulation of genes involved in fatty acid uptake, microsomal ω-oxidation, peroxisomal, and mitochondrial β-oxidation (Shimura et al., 2005). An improvement in the insulin resistance and hyperglycaemia in mice has been observed as well (Yajima et al., 2005). These findings suggest a possible interest in hop extracts and their bitter acids for the management of obesity and metabolic syndrome; however, further research is needed to optimise their possible use in clinical settings.

11.3 Clinical Trials

Most available clinical trials investigated the efficacy of hop-based formulations in the management of menopausal discomforts, including bone mineral density (Table 11.2). A multicentre open non-controlled study with post-menopausal

Table 11.2 Clinical trials investigating the efficacy of hop-based formulation in various physiological and pathological contexts

Type of study/*Aim of the study*	Subjects (age)/n.	Treatment/Duration	Type of hop extract/Chemical characterization	Outcome/*Comments*	References
Hop extract – based formulation					
Randomised, clinical trial/*Estrogenic power of hop-based formulation*	Post-menopausal women (>40 y)/n. 63 (n. 33 Hop group; n. 30 oestradiol)	Hop extract-based gel (Barij Essence, Tehran, Iran)/vaginal application/ daily for 7 days and then twice weekly for 2 months 0.625 mg Oestradiol-based gel/vaginal application/ over two 28-day cycles (21 days of therapy and seven days rest)	Nr	No statistically significant differences in FSFI scores between the hop and oestradiol groups/*improvement of the sexual dysfunction by hop extract-based gel with no adverse events*	Vahedpoorfard et al. (2023)
Double-blind, placebo-controlled, randomised trial/*Anti-osteoporotic power of hop-based formulation in menopausal discomforts*	Postmenopausalosteopenic women (50–85 y)/n. 100 (n. 50 Hop group)	Hop extract-based capsule (Lifenol)/undefined dose/ one capsule daily/48 weeks/per os Calcium and vitamin D3 capsules/500 mg and 400 IU/2 capsules daily/48 weeks/per os	Supercritical CO_2/ ethanolic dried hop cones extract/ Substances per capsule: 8-PN 100 μg 6-PN 110 μg X 1.25 mg IX 2.94 mg	↑ total body BMD and quality of life Lacking effects on gut microbiome α-diversity and SCFA levels ↑ *Turicibacter* and *Shigella* in gut microbiome/*Gut microbiome modifications were associated with total body BMD*	Lecomte et al. (2023)
Randomised, controlled trial/*Role of hop-based formulation in menopausal discomforts*	Postmenopausal (40–60 y) and premenopausal women)/n. 120 (n. 60 Hop group)	Powdered hop cones-based tablets/650 mg of which 500 mg of hop daily/per os/90 days	Nr	↓ early menopausal symptoms ↓ number of hot flashes	Aghamiri et al. (2016)

Prospective, randomised, double-blind, placebo-controlled study/*Role of hop-based formulation in menopausal discomforts*	Menopausal women (45–60 y)/n. 67 (undefined n. per group)	Low dose hop extract capsules/444 mg containing 120 mg of hop extract, corresponding to 100 µg of 8-PN/daily/per os/12 weeks High dose hop extract capsules/444 mg containing 300 mg of hop extract, corresponding to 250 µg of 8-PN/daily/per os/12 weeks	Supercritical CO_2 extract of spent hop cones previously exposed to aqueous ethanol extraction/ standardised on 8-PN	↓ Kupperman index after 6 weeks with low dose hop extract capsules ↓ hot flush score after 12 weeks/*Favourable effects on vasomotor symptoms and other menopausal discomforts*	Heyerick et al. (2006)
Multicentre, open, non-controlled design study/*Role of hop-based formulation in menopausal discomforts*	Post-menopausal women with urogenital atrophy	Gel containing hop extract-based phytoestrogens, hyaluronic acid, liposomes, and Vitamin E/2.5 g/vaginal application/daily for 1 week and then twice a week for 11 weeks	Nr	↓ vaginal dryness and atrophic vaginitis symptoms (itching, burning, dyspareunia, vaginal inflammation/oedema, rash, vaginal abrasions and disepithelialisation) since the first week Good safety profile/*Effective and safe towards genital atrophy in post-menopausal women*	Morali et al. (2006)

(continued)

Table 11.2 (continued)

Type of study/*Aim of the study*	Subjects (age)/n.	Treatment/Duration	Type of hop extract/Chemical characterization	Outcome/*Comments*	References
Randomised, crossover clinical trial/*Role of bitter hop-based formulation for modulating energy intake, appetite, and hormonal responses*	Healthy-weight men (18–55 y)/n. 19	Amarasate targeted for gastric release/500 mg daily/per os/nr Amarasate targeted for duodenal release/500 mg daily/per os/nr Canola oil (placebo)/500 mg daily/per os/nr	Supercritical CO_2 bitter extract of hop cones, Pacific Gem variety/51.5% α-acids (21.1% cohumulone; 22.3% humulone; 8.2% adhumulone); 28.3% β-acids (19.7% colupulone; 6.0% lupulone; 3.1% adlupulone)	↓ ad libitum energy intake ↑ pre-lunch ghrelin secretion ↑ post-prandial cholecystokinin, glucagon-like peptide 1, and peptide YY ↓ insulin, glucose-dependent insulinotropic peptide, and pancreatic polypeptide Gastrointestinal discomfort assessed by subjective measures No effects on glycemia and on subjective measures of appetite or meal palatability/*Reduced food intake, suppressed appetite, and release of appetite-suppressing gut hormones*	Walker et al. (2022)
Randomised, double-blind, cross-over study/*Role of bitter hop-based formulation in modulating appetite and food cravings*	Adult women (18–40 y)/n. 30	Amarasate/250 mg twice during a 24 h fast/per os/3 occasions Amarasate/125 mg twice during a 24 h fast/per os/3 occasions Canola oil (placebo)/undefined dose during a 24 h fast/per os/3	Supercritical CO_2 bitter extract of hop cones, Pacific Gem variety/51.5% α-acids (21.1% cohumulone; 22.3% humulone; 8.2% adhumulone); 28.3% β-acids (19.7% colupulone; 6.0% lupulone;	↓ appetite and cravings for food Loose stools (3 cases) Heartburn (1 case)	Walker et al. (2024)

Randomised, double-blind study/*Role of bitter hop-based formulation for modulating upper gut motility, gut hormone secretion and energy intake*	Healthy-weight men (22–28 y)/n. 21	Bitter hop extract/100, 250 mg/intraduodenal administration/single dose Bitter hop extract/250 mg/intraduodenal or intragastric administration/single dose	Supercritical CO_2 bitter extract of hop cones, Pacific Gem variety/Nr	Weak effects at high dose on pyloric pressures stimulation and peptide YY, no effect on antral or duodenal pressures, cholecystokinin, appetite, gastrointestinal symptoms or energy intake/*No effects on appetite or energy intake except for weak effects by intraduodenal administration of hop extract 250 mg*	Bitarafan et al. (2023)
Randomised, double-blind, placebo-controlled parallel group study/*Role of matured hop-based formulation in reducing fat*	Healthy overweight women and men (20–65 y)/n. 200 (n. 100 Hop group; n. 100 placebo group)	Matured hop extract-based beverage/350 mL daily containing 35 mg of mature hop bitter acids/per os/12-week	Aqueous matured hop extract obtained by oxidised hop/18.3% mature hop bitter acids. α-, β- and iso-α-acids undetectable	↓ visceral fat area after 8 and 12 weeks ↓ total fat area after 12 weeks ↓ body fat ratio No adverse events related to the circulatory, blood and urine parameters/*Safe reduce of body fat*	Morimoto-Kobayashi et al. (2016)

(continued)

Table 11.2 (continued)

Type of study/*Aim of the study*	Subjects (age)/n.	Treatment/Duration	Type of hop extract/Chemical characterization	Outcome/*Comments*	References
Randomised, placebo-controlled, double-blind, crossover design/*Role of hop dry extract ability to modulate self-reported depression, anxiety and stress symptoms*	Healthy young adults reporting mild depression, anxiety and stress (almost 25 y)/n. 36 (2 groups of n. 18)	Capsules containing 0.2 g of Melcalin hop extract or placebo capsules/2 capsules daily/per os/2 cycles of 4 weeks separated by a 2-week wash-out	Nr	No effects on body weight and morning circulating cortisol ↓ DASS-21 anxiety, depression and stress scores/*Beneficial effects of hops on anxiety and mood disorders, according to the German Commission E*	Kyrou et al. (2017)
Hop extracts in multicomponent dietary supplements					
Multicentre, randomised, double-Blind, placebo-controlled clinical trial/*Role of hop-based formulation in menopausal discomforts*	Women with moderate or severe menopausal symptoms/n. 78 (n. 38 Hop group)	Combined soy and hop extract/190 mg/12 weeks	Nr	↓ fatigue, paraesthesia, arthralgia, and myalgia, palpitation and vaginal dryness, ↓ urine N-telopeptide (≥50 y) No changes in endometrial thickness and hormonal profiles No serious adverse events/ *Soy - hop mixture is safe and effective against menopausal symptoms*	Kim et al. (2021b)

Double-Blind, randomised placebo-controlled trial/*Role of hop-based multicomponent dietary supplement to modulate sleep quality*	Healthy adults with mild insomnia (18–65 y)/n. 171 (n. 78 LZComplex3)	LZComplex3 tablets containing hop cones extract 500 mg, Lactium 75 mg, Sour date seed extract 4.5 g, magnesium oxide 81.7, Vitamin B6 10 mg/2 tablets daily/per os/7 day	Nr	No difference between hop and placebo groups in Pittsburgh Sleep Quality Index (PSQI) score, daytime functioning, physical fatigue, anxiety, cognitive performance, and stress symptoms	Scholey et al. (2017)
Parallel group, double- blind, randomised controlled trial/*Sedative-hypnotic power of hops-based multicomponent dietary supplement*	Adults with primary insomnia (20–80 y)/n. 78 (n. 39 NSF-3 group; n. 39 Zolpidem group)	NSF-3 tablet containing hops extract 30 mg (0.35% rutin), valerian extract 300 mg (standardised to 0.8% total valerinic acid), passion flower extract 80 mg (4% isovitexin)/1 tablet daily/per os/2 weeks Zolpidem tablet/10 mg daily/per os/2 weeks	Undefined hop extract/0.35% rutin	↑ total sleep time, sleep latency in both groups ↓ nightly awakenings and insomnia severity index scores in both groups No statistically significant difference between the groups Mild and no serious adverse events/*NSF-3 is a safe and effective short-term alternative to zolpidem for primary insomnia*	Maroo et al. (2013)

(continued)

Table 11.2 (continued)

Type of study/*Aim of the study*	Subjects (age)/n.	Treatment/Duration	Type of hop extract/Chemical characterization	Outcome/*Comments*	References
Randomised placebo controlled trial/*Role of hop-based multicomponent dietary supplement to modulate sleep quality*	Adults with a chronic primary insomnia (25–65 y)/n. 101 (n. 51 Cyclamax group)	Cyclamax capsule containing Houblon (Humulus lupulus) 50 mg, Soy oil 260 mg, Cade oil 173 mg. and Soy lecithin 6 mg/2 capsules daily/1 month	Nr	↑ quality of sleep, sleep efficiency in both groups No effects on melatonin metabolism	Cornu et al. (2010)
Multicentre, randomised, placebo-controlled trial/	184 adults (110 women, 74 men; mean age of 44.3 years) with mild insomnia	Tablets containing standardised methanolic extracts of valerian (187 mg) and hops (41.9 mg)/two nightly/28 days	Nr	↑ quality of sleep/*Valeriana-hop combination as an adjuvant treatment for mild insomnia*	Morin et al. (2005)
Randomised, placebo-controlled trial/*Role of hops-based multicomponent dietary supplement in counteracting caffeine-induced insomnia*	Healthy male volunteers (21/32 y)/n. 47 (n. 16 Ze91019 group; n. 16 caffeine group)	Ze91019 tablet containing 60 mg hop and 250 mg valerian standardised extracts/2 tablets/once daily/per os Ze91019 tablet containing 60 mg hop and 250 mg valerian standardised extracts/2 tablets/once daily/per os + caffeine tablets/200 mg/once daily/per os	Methanolic hop extract/0.48% flavonoids; bitter acids not detected; presence of hydrophilic aminoacids and lignans	↓ or inhibition of caffein-induced arousal after 2 or 6 tablets of Ze91019/*Ze91019 was hypothesised to compete with caffeine to act via central adenosine mechanism*	Schellenberg et al. (2004)

n number, *y* years old, *nr* not reported, *8-PN* 8-prenylnaringenin, *IU* international unit, *6-PN* 6-prenylnaringenin, *X* xanthohumol, *IX* isoxanthohumol, *BMD* bone mineral density, *SCFA* short-chain fatty acid, *DASS-21* Depression Anxiety Stress Scale-21 women suffering from urogenital atrophy found that a gel containing hop extract-based phytoestrogens, hyaluronic acid liposomes, and Vitamin E reduced vaginal dryness and symptoms of atrophic vaginitis, also resulting well tolerated by patients over 12 weeks (Morali et al., 2006). In the same year, a prospective randomised double-blind placebo-controlled study, involving 67 menopausal women, assessed the efficacy of low and high doses of hop extract capsules, standardised for the amount of 8-prenylnaringenin, in improving the menopausal symptoms. The authors concluded that a daily intake over 12 weeks of this formulation improved hot flushes, vasomotor symptoms, and other menopausal symptoms (Heyerick et al., 2006).

Other trials were carried out to assess the possible benefit of hop formulations in menopausal discomforts; among them, a randomised controlled trial, with 120 post-menopausal and premenopausal women, reported that powdered hop cones-based tablets reduced early menopausal symptoms and the number of hot flashes over 90 days (Aghamiri et al., 2016). Vahedpoorfard et al. (2023) also reported that a hop extract-based gel improved sexual dysfunction in post-menopausal women like an oestradiol-based gel, albeit without adverse events, suggesting a potential interest as a safe alternative to hormonal treatments. Lastly, Lecomte et al. (2023) demonstrated that a hop extract-based capsule (Lifenol®) significantly increased the bone mineral density and quality of life over 48 weeks in postmenopausal osteopenic women, without affecting gut microbiome α-diversity or SCFA levels. They also revealed an increase in *Turicibacter* and *Shigella* species in the gut microbiome, which could be associated with the bone mineral density improvements.

A combination of soy and hop extract (Table 11.2) was also assessed for menopausal discomforts in a multicentre randomised double-blind placebo-controlled clinical trial, involving 78 women with moderate or severe menopausal symptoms, highlighting a reduction in fatigue, paraesthesia, arthralgia, myalgia, palpitation, and vaginal dryness without changing endometrial thickness or hormonal profiles (Kim et al., 2021b).

Some studies also evaluated the role of bitter hop-based formulation in modulating energy intake, appetite, and hormonal responses, reducing body fat, regulating appetite and food intake, enhancing mood, and improving anxiety and stress symptoms (Table 11.2). Particularly, a randomised crossover clinical trial, involving 19 healthy-weight men tested a bitter hop extract (Amarasate®), targeted for gastric or duodenal release. The study highlighted a reduction in *ad libitum* energy intake and insulin levels, with an increase in pre-lunch ghrelin secretion, post-prandial cholecystokinin, glucagon-like peptide 1, and peptide YY, without effects on glycemia or subjective measures of appetite (Walker et al., 2022). Amarasate® was also found to reduce appetite and food cravings, albeit with some gastrointestinal discomfort (loose stools and heartburn) in a recent trial involving 30 adult women (Walker et al., 2024). Similarly, a matured hop extract-based beverage was found to be able to reduce visceral fat area, total fat area, and body fat ratio after 8 and 12 weeks in healthy overweight individuals, without adverse events related to blood and urine parameters (Morimoto-Kobayashi et al., 2016). By contrast, another trial found that a bitter hop extract had weak effects on pyloric pressures and peptide YY stimulation, without affecting antral or duodenal pressures, cholecystokinin, appetite, gastrointestinal symptoms, and energy intake (Bitarafan et al., 2023). Based on this conflicting evidence, further studies involving more patients are needed to clarify the possible role of hop formulations in the control of appetite and food intake in eating disorders.

Finally, Kyrou et al. (2017) assessed the capsules containing Melcalin hop extract in 36 healthy young adults with mild depression, anxiety, and stress (Table 11.2), highlighting beneficial effects on mood disorders, as confirmed by Depression Anxiety Stress Scale-21, without affecting body weight or morning cortisol levels.

Combinations of hop extracts and other herbal drugs were assessed for sleep disorders and insomnia (Table 11.2). Morin et al. (2005) demonstrated that a valerian-hop combination and diphenhydramine might be useful as an adjuvant treatment for mild insomnia. Similarly, Schellenberg et al. (2004) highlighted that Ze91019 tablets containing hop and valerian standardised extracts inhibited caffeine-induced arousal in healthy male volunteers, suggesting a possible interference with the central adenosine mechanisms of caffeine.

Cyclamax capsules, containing a hop extract in combination with soy oil, Cade oil and soy lecithin, improved sleep quality and efficiency in adults with chronic primary insomnia, without affecting melatonin metabolism (Cornu et al., 2010). NSF-3 tablet containing hops extract, valerian extract, and passion flower extract increased total sleep time and sleep latency, reduced nightly awakenings and insomnia severity index scores in adults with primary insomnia (Maroo et al., 2013). By contrast, LZComplex3 tablets, containing a hop cones extract, lactium, sour date seed extract, magnesium oxide and vitamin B6, did not modify sleep quality, daytime functioning, physical fatigue, anxiety, cognitive performance, or stress symptoms with respect to placebo (Scholey et al., 2017).

Overall, these findings confirm the tolerability of the hop extracts in humans, both as single treatments and in combination with other agents. However, despite some promising evidence, the true clinical efficacy remains to be clarified due to several limitations, including the small number of participants in the studies, variations in the formulations of hop extracts tested, and challenges in defining the efficacy of hop extracts when used in combination with other agents. Further large-scale, standardised studies are needed to better understand the nutraceutical potential of hop extracts.

11.4 *Humulus lupulus* L. in Traditional Medicines

Although hops have been used in brewing beer for their bitter taste and preservative qualities for over a millennium, their medicinal and tonic properties have also been recognised and utilised from early times. In traditional medicine, hop extracts were often used to address anxiety, insomnia, digestive issues, and to relieve toothache, earache and neuralgia (Wilson, 1975). Their sedative effects were harnessed in various cultures, such as Native American tribes, Traditional Chinese Medicine and

Ayurveda, through teas and infusions, to induce relaxation and aid sleep (Blumenthal et al., 2000; Committee on Herbal Medicinal Products (HMPC) of the European Medicines Agency (EMA), 2024b). Native American tribes also used a poultice of heated hops to treat pneumonia and a decoction of hops was recommended for intestinal pain and fevers. Additionally, hops have been used as diuretic, antirheumatic, analgesic, antispasmodic and anaphrodisiac remedies (Wilson, 1975). Alcoholic extracts of hops have been also harnessed in China for the treatment of leprosy, pulmonary tuberculosis, acute bacterial dysentery, silicosis and asbestosis with positive outcomes. Additionally, it has been used topically to treat crural ulcers and skin injuries, to alleviate skin irritations and inflammations and to relieve muscle spasms and nerve pain (Wilson, 1975). A folk legend suggests that women who travelled to hop gardens to pick hops would begin menstruating two days after arrival: this led to the use of brewing sludge baths containing about 30% hops for treating gynaecological disorders (Fenselau & Talalay, 1973).

The medicinal parts are glandular hairs separated from the infructescence, whole dried female flowers, fresh cones (preferably with few seeds) collected before seed ripening, and fresh or dried female inflorescences; the dried inflorescences are often used in a comminuted form for tea mixtures or prepared as infusions (Committee on Herbal Medicinal Products (HMPC) of the European Medicines Agency (EMA), 2024b). Lipophilic extracts from hops are utilised in the preparation of bath oils, while hydroalcoholic liquid extracts, sometimes prepared using sweet wine, are used internally as sedatives, typically in combination with other sedative plant extracts like valerian, passion flower, and lemon balm (Committee on Herbal Medicinal Products (HMPC) of the European Medicines Agency (EMA), 2024b).

The use of hops in treating mood disturbances, such as restlessness and anxiety, as well as for sleep disturbances is approved by the German Commission E and European Scientific Cooperative on Phytotherapy (ESCOP) (Blumenthal et al., 2000).

According to the Committee on Herbal Medicinal Products (HMPC) of the European Medicines Agency (EMA), *Humulus lupulus* L. flos are accepted by the Member States as "Traditional herbal medicinal product for relief of mild symptoms of mental stress and to aid sleep" based on long standing use, as documented in literature, including standard herbal reference books (Committee on Herbal Medicinal Products (HMPC) of the European Medicines Agency (EMA), 2024a).

Hop strobile medicines should be used only in adults, elderly, and adolescents over the age of 12 years; the recommended dosages and administration methods for hop strobiles vary depending on the preparation and intended use (Committee on Herbal Medicinal Products (HMPC) of the European Medicines Agency (EMA), 2024a). For mental stress, herbal tea can be prepared by infusing 500 mg of comminuted herbal substance in 150–200 mL of boiling water and consumed up to four times daily; alternatively, 400 mg of powdered herbal substance can be taken twice daily for adults, while 200 mg for adolescents. For liquid extracts and tinctures, about 0.5–2.0 mL three times daily, while 125 mg of dry extracts (4–5:1) up to three times daily are recommended (Committee on Herbal Medicinal Products (HMPC) of the European Medicines Agency (EMA), 2024a). To aid sleep, 500–1000 mg of comminuted herbal substance should be infused in 150–200 mL of boiling water to prepare herbal tea, while 800–2000 mg of powdered herbal substance or 125–250 mg of dry extracts (4–5:1) are recommended; these medicinal products should be consumed 30–60 min before bedtime (Committee on Herbal Medicinal Products (HMPC) of the European Medicines Agency (EMA), 2024a). The use of hop medicinal products should be stopped if symptoms persist for more than two weeks or worsen, while consulting a doctor or a qualified healthcare professional (Committee on Herbal Medicinal Products (HMPC) of the European Medicines Agency (EMA), 2024a).

11.5 Safety and Warnings

Hop-based preparations, both alone and in combination with other herbal drugs, have been shown to be well tolerated, without serious side effects, both traditionally and in clinical trials (Committee on Herbal Medicinal Products (HMPC) of the European Medicines Agency (EMA), 2024b). Some allergic reactions, such as bronchial irritation, dry cough, dyspnoea, and contact dermatitis, particularly among hop processing workers, but not during medicinal use, have been reported (Wilson, 1975).

Owing to the presence of 8-prenylnaringenin in hop preparations, which inhibits ABC transporters, an influence on the pharmacokinetics of co-administered drugs could be expected (Tan et al., 2014; Wesołowska et al., 2010); similarly, the endocrine-disrupting potential of 8-prenylnaringenin suggests potential adverse effects (Pohjanvirta & Nasri, 2022). However, clinical evidence is still lacking and further toxicological investigations are needed in support. Benkherouf et al. (2020a) also highlighted that the hop compound humulone and ethanol may synergistically modulate sleep duration, indicating a potential risk of enhanced alcohol intoxication; this finding warrants further investigation to fully understand the implications.

According to the HMPC of the European Medicines Agency (EMA) (Committee on Herbal Medicinal Products (HMPC) of the European Medicines Agency (EMA), 2024a), the use of hop preparations in children under the age of 12 years, and during pregnancy and lactation is not recommended owing to a lack of sufficient data. Moreover, hop preparations may impair the ability to drive and use machines, therefore, they should be avoided in these conditions.

11.6 Conclusions

Over the years, numerous studies have investigated the biological properties of extracts from *Humulus lupulus* L. cones, mainly aiming to provide scientific basis for the ethnopharmacological and traditional uses of the plant, and to explore novel pharmacological and nutraceutical applications.

Despite the numerous available studies, certain aspects of their clinical efficacy remain unresolved. The challenge arises from the complex composition of hop cones and the variety of the hop cultivars and extracts tested. These factors are not always considered or declared in studies, leading to phytochemical variations and inconsistent biological activity. Furthermore, older studies often did not specify the type of extract, chemical composition, or hop cultivar, complicating comparisons with more recent research.

Researchers also tried to identify the bioactive compounds, to understand their subtle interactions in the phytocomplex and to clarify their mechanisms of action. Hops contain a rich array of phytochemicals, including prenylflavonoids and bitter acids, which have been the primary focus of many studies. However, other phytochemicals, such as volatile compounds, including monoterpenes and sesquiterpenes, should be considered for their possible contribution to the bioactivities of hop extracts.

While studies on pure compounds help elucidate specific mechanisms of action, they often fail to capture the synergistic effects observed in whole extracts, which require further clarification. This growing body of evidence supports the traditional use of hop and opens new avenues for developing innovative treatments and health-promoting products.

Acknowledgment The authors gratefully acknowledge the Enrico and Enrica Sovena Foundation for supporting Elisabetta Bazzaro with a research fellowship.

References

Abram, V., Čeh, B., Vidmar, M., Hercezi, M., Lazić, N., Bucik, V., Možina, S. S., Košir, I. J., Kač, M., Demšar, L., & Poklar, U. N. (2015). A comparison of antioxidant and antimicrobial activity between hop leaves and hop cones. *Industrial Crops and Products, 64*, 124–134.

Aghamiri, V., Mirghafourvand, M., Mohammad-Alizadeh-Charandabi, S., & Nazemiyeh, H. (2016). The effect of Hop (*Humulus lupulus* L.) on early menopausal symptoms and hot flashes: A randomized placebo-controlled trial. *Complementary Therapies in Clinical Practice, 23*, 130–135.

Almeida, A. D. R., Maciel, M. V. D. O. B., Machado, M. H., Bazzo, G. C., de Armas, R. D., Vitorino, V. B., & Barreto, P. L. M. (2020). Bioactive compounds and antioxidant activities of Brazilian hop (*Humulus lupulus* L.) extracts. *International Journal of Food Science & Technology, 55*(1), 340–347.

Alonso-Esteban, J. I., Pinela, J., Barros, L., Ćirić, A., Soković, M., Calhelha, R. C., & Ferreira, I. C. (2019). Phenolic composition and antioxidant, antimicrobial and cytotoxic properties of hop (*Humulus lupulus* L.) Seeds. *Industrial Crops and Products, 134*, 154–159.

Alves, M. D. R., Nascimento, R. P., da Fonseca Machado, A. P., Dos Santos, P., Aledo, E., Morandi Vuolo, M., Cavalheiro, C. O., Giaculi, V. O., Berilli, P., Dos Santos, N. M., & Marostica Junior, M. R. (2024). Hop (*Humulus lupulus* L.) extract reverts glycaemic imbalance and cognitive impairment in an animal model of obesity. *Food & Function, 15*(14), 7669–7680.

Ano, Y., Dohata, A., Taniguchi, Y., Hoshi, A., Uchida, K., Takashima, A., & Nakayama, H. (2017). Iso-acids, bitter components of beer, prevent inflammation and cognitive decline induced in a mouse model of Alzheimer's disease. *The Journal of Biological Chemistry, 292*, 3720–3728.

Bagatin, M. C., Tozatti, C. S., Abiko, L. A., Yamazaki, D. A., Silva, P. R., Perego, L. M., Audi, E. A., Seixas, F. A., Basso, E. A., & Gauze, G.deF. (2014). Molecular docking and panicolytic effect of 8-prenylnaringenin in the elevated T-maze. *Chemical & Pharmaceutical Bulletin, 62*(12), 1231–1237.

Bartmańska, A., Tronina, T., Popłoński, J., & Huszcza, E. (2013). Biotransformations of prenylated hop flavonoids for drug discovery and production. *Current Drug Metabolism, 14*(10), 1083–1097.

Benkherouf, A. Y., Eerola, K., Soini, S. L., & Uusi-Oukari, M. (2020a). Humulone modulation of $GABA_A$ receptors and its role in hops sleep-promoting activity. *Frontiers in Neuroscience, 14*, 594708.

Benkherouf, A. Y., Logrén, N., Somborac, T., Kortesniemi, M., Soini, S. L., Yang, B., Salo-Ahen, O. M. H., Laaksonen, O., & Uusi-Oukari, M. (2020b). Hops compounds modulatory effects and 6-prenylnaringenin dual mode of action on $GABA_A$ receptors. *European Journal of Pharmacology, 873*, 172962.

Bitarafan, V., Fitzgerald, P. C. E., Poppitt, S. D., Ingram, J. R., & Feinle-Bisset, C. (2023). Effects of intraduodenal or intragastric administration of a bitter hop extract (*Humulus lupulus* L.) on upper gut motility, gut hormone secretion, and energy intake in healthy-weight men. *Appetite, 184*, 106490.

Blumenthal, M., Goldberg, A., & Brinckmann, J. (2000). *Herbal medicine: Expanded commission E monographs* (pp. 297–303). Integrative Medicine Communications.

Bowe, J., Feng Li, X., Kinsey-Jones, J., Heyerick, A., Brain, S., Milligan, S., & O'Byrne, K. (2006). The hop phytoestrogen, 8-prenylnaringenin, reverses the ovariectomy-induced rise in skin temperature in an animal model of menopausal hot flushes. *Journal of Endocrinology, 191*, 399–405.

Bravo, L., Cabo, J., Fraile, A., Jimenez, J., & Villar, A. (1974). Pharmacodynamic study of the lupulus' (*Humulus lupulus* L.) tranquilizing action. *Bollettino Chimico Farmaceutico, 113*(5), 310–315.

Buckwold, V. E., Wilson, R. J., Nalca, A., Beer, B. B., Voss, T. G., Turpin, J. A., Buckheit, R. W., 3rd, Wei, J., Wenzel-Mathers, M., Walton, E. M., Smith, R. J., Pallansch, M., Ward, P., Wells, J., Chuvala, L., Sloane, S., Paulman, R., Russell, J., Hartman, T., & Ptak, R. (2004). Antiviral activity of hop constituents against a series of DNA and RNA viruses. *Antiviral Research, 61*(1), 57–62.

Butterweck, V., Brattstroem, A., Grundmann, O., & Koetter, U. (2007). Hypothermic effects of hops are antagonized with the competitive malatonin receptor antagonist luzindole in mice. *Journal of Pharmacy and Pharmacology, 59*, 549–552.

Caban, M., Chojnacka, K., Owczarek, K., Laskowska, J., Fichna, J., Podsędek, A., & Lewandowska, U. (2020). Spent hops (*Humulus lupulus* L.) extract as modulator of the inflammatory response in lipopolysaccharide stimulated raw 264.7 macrophages. *Journal of Physiology and Pharmacology, 70*(1).

Caban, M., Owczarek, K., Chojnacka, K., Podsędek, A., Sosnowska, D., & Lewandowska, U. (2022). Chemopreventive properties of spent hops (Humulus Lupulus L.) extract against angiogenesis, invasion and migration of colorectal cancer cells. *Journal of Physiology and Pharmacology, 73*(3).

Carbone, K., & Gervasi, F. (2022). An updated review of the genus Humulus: A valuable source of bioactive compounds for health and disease prevention. *Plants (Basel), 11*(24), 3434.

Caujolle, F., Pham-Huu-Chanh, Duch-Kan, P., & Bravo-Diaz, L. (1969). Etude de l'action spasmolytique de houblon (*Humulus lupulus*, Cannabinacées) [Spasmolytic action of hop (*Humulus lupulus*, Cannabinacees)]. *Agressologie, 10*(5), 405–410.

Choi, H. S., Ko, B. S., Kim, H. D., Hong, K. B., & Suh, H. J. (2017). Effect of Valerian/Hop mixture on sleep-related behaviors in *Drosophila melanogaster*. *Biological & Pharmaceutical Bulletin, 40*(7), 1101–1110.

Committee on Herbal Medicinal Products (HMPC) of the European Medicines Agency (EMA) Community herbal monograph on Humulus lupulus L., flos. EMA/HMPC/682384/2013

Available online: https://www.ema.europa.eu/en/documents/herbal-monograph/final-community-herbal-monograph-humulus-lupulus-l-flos-revision-1_en.pdf, accessed on 30 July 2024a).

Committee on Herbal Medicinal Products (HMPC) of the European Medicines Agency (EMA). Assessment report on *Humulus lupulus* L., flos EMA/HMPC/418902/2005. Available online: https://www.ema.europa.eu/en/documents/herbal-report/final-assessment-report-humulus-lupulus-l-flos-revision-1_en.pdf, accessed on 30 July 2024b).

Cornu, C., Remontet, L., Noel-Baron, F., Nicolas, A., Feugier-Favier, N., Roy, P., Claustrat, B., Saadatian-Elahi, M., & Kassaï, B. (2010). A dietary supplement to improve the quality of sleep: a randomized placebo controlled trial. *BMC Complementary and Alternative Medicine, 10*, 29.

Di Lodovico, S., Menghini, L., Ferrante, C., Recchia, E., Castro-Amorim, J., Gameiro, P., Cellini, L., & Bessa, L. J. (2020). Hop extract: An efficacious antimicrobial and anti-biofilm agent against multidrug-resistant Staphylococci strains and *Cutibacterium acnes*. *Frontiers in Microbiology, 11*, 1852.

Di Sotto, A., Checconi, P., Celestino, I., Locatelli, M., Carissimi, S., De Angelis, M., Rossi, V., Limongi, D., Toniolo, C., Martinoli, L., Di Giacomo, S., Palamara, A. T., & Nencioni, L. (2018). Antiviral and antioxidant activity of a hydroalcoholic extract from *Humulus lupulus* L. *Oxidative Medicine and Cellular Longevity, 2018*, 5919237.

Dietz, B. M., Chen, S. N., Alvarenga, R. F. R., Dong, H., Nikolić, D., Biendl, M., van Breemen, R. B., Bolton, J. L., & Pauli, G. F. (2017). DESIGNER extracts as tools to balance estrogenic and chemopreventive activities of botanicals for women's health. *Journal of Natural Products, 80*(8), 2284–2294.

Dostálek, P., Karabín, M., & Jelínek, L. (2017). Hop phytochemicals and their potential role in metabolic syndrome prevention and therapy. *Molecules, 22*(10), 1761.

Fenselau, C., & Talalay, P. (1973). Is oestrogenic activity present in hops? *Food and Cosmetics Toxicology, 11*, 597–603.

Flieger, J., Pasieczna-Patkowska, S., Żuk, N., Panek, R., Korona-Głowniak, I., Suśniak, K., Pizoń, M., & Franus, W. (2024). Characteristics and antimicrobial activities of iron oxide nanoparticles obtained via mixed-mode chemical/biogenic synthesis using spent hop (*Humulus lupulus* L.) extracts. *Antibiotics (Basel, Switzerland), 13*(2), 111.

Franco, L., Sanchez, C., Bravo, R., Rodriguez, A., Barriga, C., & Juanez, J. C. (2012). The sedative effects of hops (*Humulus lupulus*), a component of beer, on the activity/rest rhythm. *Acta Physiologica Hungarica, 99*(2), 133–139.

Girisa, S., Saikia, Q., Bordoloi, D., Banik, K., Monisha, J., Daimary, U. D., Verma, E., Ahn, K. S., & Kunnumakkara, A. B. (2021). Xanthohumol from Hop: Hope for cancer prevention and treatment. *IUBMB Life, 73*(8), 1016–1044.

Hamm, A. K., Manter, D. K., Kirkwood, J. S., Wolfe, L. M., Cox-York, K., & Weir, T. L. (2019). The effect of hops (*Humulus lupulus* L.) extract supplementation on weight gain, adiposity, and intestinal function in ovariectomized mice. *Nutrients, 11*, 3004.

Hänsel, R., & Wagener, H. H. (1967). Attempts to identify sedative-hypnotic active substances in hops. *Arzeneimittel-Forshung/Drug Research, 17*, 79–81.

Hänsel, R., Wohlfart, R., & Coper, H. (1980). Versuche, sedativ-hypnotische Wirkstoffe im Hopfen nachzuweisen, II [Sedative-hypnotic compounds in the exhalation of hops]. *Zeitschrift fur Naturforschung Section C, Biosciences, 35*(11–12), 1096–1097.

Harish, V., Haque, E., Śmiech, M., Taniguchi, H., Jamieson, S., Tewari, D., & Bishayee, A. (2021). Xanthohumol for human malignancies: Chemistry, pharmacokinetics and molecular targets. *International Journal of Molecular Sciences, 22*(9), 4478.

Hemachandra, L. P., Madhubhani, P., Chandrasena, R., Esala, P., Chen, S. N., Main, M., et al. (2012). Hops (Humulus lupulus) inhibits oxidative estrogen metabolism and estrogen-induced malignant transformation in human mammary epithelial cells (MCF-10A). *Cancer Prevention Research, 5*(1), 73–81.

Heyerick, A., Vervarcke, S., Depypere, H., Bracke, M., & De Keukeleire, D. (2006). A first prospective, randomized, double-blind, placebo-controlled study on the use of a standardized hop extract to alleviate menopausal discomforts. *Maturitas, 54*(2), 164–175.

Hougee, S., Faber, J., Sanders, A., van den Berg, W. B., Garssen, J., Smit, H. F., & Hoijer, M. A. (2006). Selective inhibition of COX-2 by a standardized CO2 extract of *Humulus lupulus* in vitro and its activity in a mouse model of zymosan-induced arthritis. *Planta Medica, 72*, 228–233.

Hudcová, T., Bryndová, J., Fialová, K., Fiala, J., Karabín, M., Jelínek, L., & Dostálek, P. (2014). Antiproliferative effects of prenylflavonoids from hops on human colon cancer cell lines. *Journal of the Institute of Brewing, 120*(3), 225–230.

Hümpel, M., Isaksson, P., Schaefer, O., Kaufmann, U., Ciana, P., Maggi, A., & Schleunong, W. D. (2005). Tissue specificity of 8-prenylnaringenin: protection from ovariectomy induced bone loss with minimal trophic effects on the uterus. *The Journal of Steroid Biochemistry and Molecular Biology, 97*, 299–305.

Inui, T., Okumura, K., Matsui, H., Hosoya, T., & Kumazawa, S. (2017). Effect of harvest time on some in vitro functional properties of hop polyphenols. *Food Chemistry, 225*, 69–76.

Jeliazkova, E., Zheljazkov, V. D., Kačániova, M., Astatkie, T., & Tekwani, B. L. (2018). Sequential elution of essential oil constituents during steam distillation of hops (*Humulus lupulus* L.) and influence on oil yield and antimicrobial activity. *Journal of Oleo Science, 67*(7), 871–883.

Jiang, C. H., Sun, T. L., Xiang, D. X., Wei, S. S., & Li, W. Q. (2018). Anticancer activity and mechanism of xanthohumol: A prenylated flavonoid from hops (*Humulus lupulus* L.). *Frontiers in Pharmacology, 9*, 530.

Keskin, Ş., Şirin, Y., Çakir, H. E., & Keskin, M. (2018). An investigation of *Humulus lupulus* L.: Phenolic composition, antioxidant capacity and inhibition properties of clinically important enzymes. *South African Journal of Botany, 120*, 170–174.

Keskin, Ş., Şirin, Y., Çakir, H. E., & Keskin, M. (2019). An investigation of *Humulus lupulus* L.: Phenolic composition, antioxidant capacity and inhibition properties of clinically important enzymes. *South African Journal of Botany, 120*, 170–174.

Khaliullina, A., Kolesnikova, A., Khairullina, L., Morgatskaya, O., Shakirova, D., Patov, S., Nekrasova, P., Bogachev, M., Kurkin, V., Trizna, E., & Kayumov, A. (2024). The antimicrobial potential of the hop (*Humulus lupulus* L.) extract against *Staphylococcus aureus* and Oral *Streptococci*. *Pharmaceuticals (Basel), 17*(2), 162.

Kim, S., Hong, K.-B., Jo, K., & Suh, H. J. (2021a). Quercetin-3-Oglucuronide in the ethanol extract of lotus leaf (*Nelumbo nucifera*) enhances sleep quantity and quality in a rodent model via a GABAergic mechanism. *Molecules, 26*(10), 3023.

Kim, H. I., Kim, M. K., Lee, I., Yun, J., Kim, E. H., & Seo, S. K. (2021b). Efficacy and safety of a standardized soy and hop extract on menopausal symptoms: A 12-week, multicenter, randomized, double-blind, placebo-controlled clinical trial. *Journal of Alternative and Complementary Medicine (New York, N.Y.), 27*(11), 959–967.

Koch, W., & Heim, G. (1953). Estrogens in hops and beer; preliminary report. *Munchener Medizinische Wochenschrift, 95*, 845.

Kolenc, Z., Langerholc, T., Hostnik, G., Ocvirk, M., Štumpf, S., Pintarič, M., Košir, I. J., Čerenak, A., Garmut, A., & Bren, U. (2022). Antimicrobial properties of different hop (*Humulus lupulus*) genotypes. *Plants (Basel, Switzerland), 12*(1), 120.

Korpelainen, H., & Pietiläinen, M. (2021). Hop (*Humulus lupulus* L.): Traditional and present use, and future potential. *Economic Botany, 75*(3), 302–322.

Krause, E., Yuan, Y., Hajirahimkhan, A., Dong, H., Dietz, B. M., Nikolic, D., Pauli, G. F., Bolton, J. L., & van Breemen, R. B. (2014). Biological and chemical standardization of a hop (*Humulus lupulus*) botanical dietary supplement. *Biomedical Chromatography: BMC, 28*(6), 729–734.

Kurasawa, T., Chikaraishi, Y., Naito, A., Toyoda, Y., & Notsu, Y. (2005). Effect of *Humulus lupulus* on gastric secretion in a rat pylorus-ligated model. *Biological and Pharmaceutical Bulletin, 28*, 353–357.

Kyrou, I., Christou, A., Panagiotakos, D., Stefanaki, C., Skenderi, K., Katsana, K., & Tsigos, C. (2017). Effects of a hops (*Humulus lupulus* L.) dry extract supplement on self-reported depression, anxiety, and stress levels in apparently healthy young adults: a randomized, placebo-controlled, double-blind, crossover pilot study. *Hormones (Athens, Greece), 16*(2), 171–180.

Langezaal, C. R., Chandra, A., & Scheffer, J. J. (1992). Antimicrobial screening of essential oils and extracts of some *Humulus lupulus* L. cultivars. Pharmaceutisch weekblad. *Scientific edition, 14*(6), 353–356.

Lecomte, M., Tomassi, D., Rizzoli, R., Tenon, M., Berton, T., Harney, S., & Fança-Berthon, P. (2023). Effect of a hop extract standardized in 8-prenylnaringenin on bone health and gut microbiome in postmenopausal women with osteopenia: A one-year randomized, double-blind, placebo-controlled trial. *Nutrients, 15*(12), 2688.

Lee, K. M., Jung, J. S., Song, D. K., Kräuter, M., & Kim, Y. H. (1993a). Effects of *Humulus lupulus* extract on the central nervous system in mice. *Planta Medica, 59*, A691.

Lee, K. M., Jung, J. S., Song, D. K., Krauter, M., & Kim, Y. H. (1993b). Neuropharmacological activity of *Humulus lupulus* extracts. *Korean Journal of Pharmacognosy, 24*, 231–234.

Li, X., Tang, Z., Fei, D., Liu, Y., Zhang, M., & Liu, S. (2017). Evaluation of the sedative and hypnotic effects of astragalin isolated from *Eucommia ulmoides* leaves in mice. *Natural Product Research, 31*(17), 2072–2076.

Linck, V. M., da Silva, A. L., Figueiró, M., Piato, A. L., Herrmann, A. P., Dupont Birck, F., Caramão, E. B., Nunes, D. S., Moreno, P. R., & Elisabetsky, E. (2009). Inhaled linalool-induced sedation in mice. *Phytomedicine, 16*(4), 303–307.

Liu, J., Burdette, J. E., Xu, H., Gu, C., van Breemen, R. B., Bhat, K. P., Booth, N., Constantinou, A. I., Pezzuto, J. M., Fong, H. H., Farnsworth, N. R., & Bolton, J. L. (2001). Evaluation of estrogenic activity of plant extracts for the potential treatment of menopausal symptoms. *Journal of Agricultural and Food Chemistry, 49*, 2472–2479.

Liu, Y., Gu, X. H., Tang, J., & Liu, K. (2007). Antioxidant activities of hops (*Humulus lupulus*) and their products. *Journal of the American Society of Brewing Chemists, 65*(2), 116–121.

Liu, M., Yin, H., Liu, G., Dong, J., Qian, Z., & Miao, J. (2014). Xanthohumol, a prenylated chalcone from beer hops, acts as an α-glucosidase inhibitor in vitro. *Journal of Agricultural and Food Chemistry, 62*(24), 5548–5554.

Lyu, J. I., Ryu, J., Seo, K. S., Kang, K. Y., Park, S. H., Ha, T. H., & Kang, S. Y. (2022). Comparative study on phenolic compounds and antioxidant activities of Hop (*Humulus lupulus* L.) strobile extracts. *Plants, 11*(1), 135.

Macchioni, V., Carbone, K., Cataldo, A., Fraschini, R., & Bellucci, S. (2021). Lactic acid-based deep natural eutectic solvents for the extraction of bioactive metabolites of *Humulus lupulus* L.: Supramolecular organization, phytochemical profiling and biological activity. *Separation and Purification Technology, 264*, 118039.

Maietti, A., Brighenti, V., Bonetti, G., Tedeschi, P., Prencipe, F. P., Benvenuti, S., & Pellati, F. (2017). Metabolite profiling of flavonols and in vitro antioxidant activity of young shoots of wild *Humulus lupulus* L.(hop). *Journal of Pharmaceutical and Biomedical Analysis, 142*, 28–34.

Maroo, N., Hazra, A., & Das, T. (2013). Efficacy and safety of a polyherbal sedative-hypnotic formulation NSF-3 in primary insomnia in comparison to zolpidem: A randomized controlled trial. *Indian Journal Of Pharmacology, 45*(1), 34–39.

Meissner, O., & Häberlein, H. (2006). Influence of xanthohumol on the binding behavior of $GABA_A$ receptors and their lateral mobility at hippocampal neurons. *Planta Medica, 72*(7), 656–658.

Milligan, S., Kalita, J., Pocock, V., Heyerick, A., De Cooman, L., Rong, H., & De Keukeleire, D. (2002). Oestrogenic activity of the hop phyto-oestrogen 8-prenylnaringenin. *Reproduction, 123*, 235–242.

Min, B., Ahn, Y., Cho, H. J., Kwak, W. K., Suh, H. J., & Jo, K. (2021). $GABA_A$ receptor-mediated sleep-promoting effect of Saaz-Saphir hops mixture containing xanthohumol and humulone. *Molecules (Basel, Switzerland), 26*(23), 7108.

Min, B., Ahn, Y., Cho, H. J., Kwak, W. K., Jo, K., & Suh, H. J. (2023). Chemical compositions and sleep-promoting activities of hop (*Humulus lupulus* L.) varieties. *Journal of Food Science, 88*(5), 2217–2228.

Mizobuchi, S., & Sato, Y. (1984). A new flavanone with antifungal activity isolated from hops. *Agricultural and Biological Chemistry, 48*, 2771–2775.

Mizobuchi, S., & Sato, Y. (1985). Antifungal activities of hop bitter resins and related compounds. *Agricultural and Biological Chemistry, 49*, 399–403.

Mladěnka, P., Macáková, K., Filipský, T., Zatloukalová, L., Jahodář, L., Bovicelli, P., Silvestri, I. P., Hrdina, R., & Saso, L. (2011). In vitro analysis of iron chelating activity of flavonoids. *Journal of Inorganic Biochemistry, 105*(5), 693–701.

Morali, G., Polatti, F., Metelitsa, E. N., Mascarucci, P., Magnani, P., & Marrè, G. B. (2006). Open, non-controlled clinical studies to assess the efficacy and safety of a medical device in form of gel topically and intravaginally used in postmenopausal women with genital atrophy. *Arzneimittel-Forschung, 56*(3), 230–238.

Morimoto-Kobayashi, Y., Ohara, K., Ashigai, H., Kanaya, T., Koizumi, K., Manabe, F., Kaneko, Y., Taniguchi, Y., Katayama, M., Kowatari, Y., & Kondo, S. (2016). Matured hop extract reduces body fat in healthy overweight humans: a randomized, double-blind, placebo-controlled parallel group study. *Nutrition Journal, 15*, 25.

Morin, C. M., Koetter, U., Bastien, C., Ware, J. C., & Wooten, V. (2005). Valerian-hops combination and diphenhydramine for treating insomnia: A randomized placebo-controlled clinical trial. *Sleep, 28*, 1465–1471.

Moureu, S., Jacquin, J., Samaillie, J., Deweer, C., Rivière, C., & Muchembled, J. (2023). Antifungal activity of hop leaf extracts and xanthohumol on two strains of *Venturia inaequalis* with different sensitivities to triazoles. *Microorganisms, 11*(6), 1605.

Mudura, E., Tofană, M., Păucean, A., & Socaci, S. (2010). The evaluation of antioxidant capacity of Romanian hops. *Journal of Agroalimentary Processes and Technologies Timişoara, 16*(2), 262–264.

Müller, C. E., Schumacher, B., Brattström, A., Abourashed, E. A., & Koetter, U. (2002). Interactions of valerian extracts and a fixed valerian–hop extract combination with adenosine receptors. *Life Sciences, 71*, 1939–1949.

Nicácio, K. J., Ferreira, M. S., Katchborian-Neto, A., Costa, M. L., Murgu, M., Dias, D. F., Soares, M. G., & Chagas-Paula, D. A. (2022). Anti-inflammatory markers of hops cultivars (*Humulus lupulus* L.) evaluated by untargeted metabolomics strategy. *Chemistry & Biodiversity, 19*(4), e202100966.

Oberbauer, E., Urmann, C., Steffenhagen, C., Bieler, L., Brunner, D., Furtner, T., Humpel, C., Bäumer, B., Bandtlow, C., Couillard-Despres, S., Rivera, F. J., Riepl, H., & Aigner, L. (2013). Chroman-like cyclic prenylflavonoids promote neuronal differentiation and neurite outgrowth and are neuroprotective. *The Journal of Nutritional Biochemistry, 24*(11), 1953–1962.

Önder, F. C., Ay, M., & Sarker, S. D. (2013). Comparative study of antioxidant properties and total phenolic content of the extracts of *Humulus lupulus* L. and quantification of bioactive components by LC–MS/MS and GC–MS. *Journal of Agricultural and Food Chemistry, 61*(44), 10498–10506.

Overk, C. R., Yao, P., Chadwick, L. R., Nikolic, D., Sun, Y., Cuendet, M. A., Deng, Y., Hedayat, A. S., Pauli, G. F., Farnsworth, N. R., van Breemen, R. B., & Bolton, J. L. (2005). Comparison of the in vitro estrogenic activities of compounds from hops (*Humulus lupulus*) and red clover (*Trifolium pratense*). *Journal of Agricultural and Food Chemistry, 53*(16), 6246–6253.

Percaccio, E., De Angelis, M., Acquaviva, A., Nicotra, G., Ferrante, C., Mazzanti, G., Di Giacomo, S., Nencioni, L., & Di Sotto, A. (2023). ECHOPvir: A mixture of echinacea and hop extracts endowed with cytoprotective, immunomodulatory and antiviral properties. *Nutrients, 15*(20), 4380.

Piasecki, B., Biernasiuk, A., & Ludwiczuk, A. (2023). Anti-coccal activity and composition of the essential oils and methanolic extracts obtained from brewing quality *Humulus lupulus* L. hop pellets. *Pharmaceuticals (Basel, Switzerland), 16*(8), 1098.

Pohjanvirta, R., & Nasri, A. (2022). The potent phytoestrogen 8-prenylnaringenin: A friend or a foe? *International Journal of Molecular Sciences, 23*(6), 3168.

Ponticelli, M., Russo, D., Faraone, I., Sinisgalli, C., Labanca, F., Lela, L., & Milella, L. (2021). The promising ability of *Humulus lupulus* L. Iso-α-acids vs. diabetes, inflammation, and metabolic syndrome: A systematic review. *Molecules, 26*(4), 954.

Rimoldi, G., Christoffel, J., & Wuttke, W. (2006). Morphologic changes induced by oral long term treatment with 8-prenylnaringenin in the uterus, vagina, and mammary gland of castrated rats. *Menopause, 13*, 669–677.

Rozalski, M., Micota, B., Sadowska, B., Stochmal, A., Jedrejek, D., Wieckowska-Szakiel, M., & Rozalska, B. (2013). Antiadherent and antibiofilm activity of *Humulus lupulus* L. derived products: new pharmacological properties. *BioMed Research International*, 101089.

Rutnik, K., Knez Hrnčič, M., & Jože Košir, I. (2021). Hop essential oil: Chemical composition, extraction, analysis, and applications. *Food Reviews International*, 1–23.

Sahin, S., Eulenburg, V., Kreis, W., Villmann, C., & Pischetsrieder, M. (2016). Three-step test system for the identification of novel $GABA_A$ receptor modulating food plants. *Plant Foods for Human Nutrition (Dordrecht, Netherlands), 71*(4), 355–360.

Salihu, B., Samarakoon, T., Pulaj, B., Quave, C. L., Mustafa, B., & Hajdari, A. (2024). Analysis of chemical and genetic variability in wild hop (*Humulus lupulus* L.) populations of Kosovo. *Plant Biology (Stuttgart, Germany)*.

Sangiovanni, E., Fumagalli, M., Santagostini, L., Forino, M., Piazza, S., Colombo, E., & Dell'Agli, M. (2019). A bio-guided assessment of the anti-inflammatory activity of hop extracts (*Humulus lupulus* L. cv. Cascade) in human gastric epithelial cells. *Journal of Functional Foods, 57*, 95–102.

Santarelli, V., Neri, L., Carbone, K., Macchioni, V., & Pittia, P. (2022). Use of conventional and innovative technologies for the production of food grade hop extracts: Focus on bioactive compounds and antioxidant activity. *Plants, 11*, 41.

Schellenberg, R., Sauer, S., Abourashed, E. A., Koetter, U., & Brattström, A. (2004). The fixed combination of valerian and hops (Ze91019) acts via a central adenosine mechanism. *Planta Medica, 70*(7), 594–597.

Schiller, H., Forster, A., Vonhoff, C., Hegger, M., Biller, A., & Winterhoff, H. (2006). Sedating effects of *Humulus lupulus* L. extracts. *Phytomedicine, 13*, 535–541.

Scholey, A., Benson, S., Gibbs, A., Perry, N., Sarris, J., & Murray, G. (2017). Exploring the effect of Lactium™ and zizyphus complex on sleep quality: A double-blind, randomized placebo-controlled trial. *Nutrients, 9*(2), 154.

Schulz, C., Fritz, N., Sommer, T., Krofta, K., Friedland, K., & Pischetsrieder, M. (2018). Activation of membrane-located Ca^{2+} channels by hop beta acids and their tricyclic transformation products. *Food Chemistry, 252*, 215–227.

Shimura, M., Hasumi, A., Minato, T., Hosono, M., Miura, Y., Mizutani, S., Kondo, K., Oikawa, S., & Yoshida, A. (2005). Isohumulones modulate blood lipid status through the activation of PPARα. *ochimica et Biophysica Acta (BBA)-Molecular and Cell Biology of Lipids, 1736*, 51–60.

Takahashi, K., & Osada, K. (2017). Effect of dietary purified xanthohumol from hop (*Humulus lupulus* L.) pomace on adipose tissue mass, fasting blood glucose level, and lipid metabolism in KK-Ay mice. *Journal of Oleo Science, 66*(5), 531–541.

Tan, K. W., Cooney, J., Jensen, D., Li, Y., Paxton, J. W., Birch, N. P., & Scheepens, A. (2014). Hop-derived prenylflavonoids are substrates and inhibitors of the efflux transporter breast cancer resistance protein (BCRP/ABCG2). *Molecular Nutrition & Food Research, 58*, 2099–2110.

Teuber, M., & Schmalreck, A. F. (1973). Membrane leakage in *Bacillus subtilis* 168 induced by the hop constituents lupulone, humulone, isohumulone and humulinic acid. *Archiv fur Mikrobiologie, 94*, 159–171.

Tripp, M., Konda, V., Darland, G., Desai, A., Chang, J., Carroll, B., & Bland, J. (2010). Rho-iso-alpha acids and tetrahydro-iso-alpha acids are selective protein kinase inhibitors which potently reduce inflammation in macrophages in vitro and in the collagen-induced rheumatoid arthritis model in vivo. *Acta Horticulturae, 848*, 221.

Tuzim, K., & Korolczuk, A. (2021). An update on extra-oral bitter taste receptors. *Journal of Translational Medicine, 19*(1), 440.

Vahedpoorfard, Z., Ferdosi, S., Rahimi, H., & Motedayyen, H. (2023). Effect of *Humulus lupulus* L. (Hop) on postmenopausal sexual dysfunction: A randomized clinical trial. *International Journal of Clinical Practice, 2023*, 9528335.

van Breemen, R. B., Yuan, Y., Banuvar, S., Shulman, L. P., Qiu, X., Alvarenga, R. F. R., Chen, S. N., Dietz, B. M., Bolton, J. L., Pauli, G. F., et al. (2014). Pharmacokinetics of prenylated hop phenols in women following oral administration of a standardized extract of hops. *Molecular Nutrition & Food Research, 58*(10), 1962–1969.

Van Cleemput, M., Heyerick, A., Libert, C., Swerts, K., Philippé, J., De Keukeleire, D., Haegeman, G., & De Bosscher, K. (2009). Hop bitter acids efficiently block inflammation independent of GRα, PPARα, or PPARγ. *Molecular Nutrition & Food Research, 53*(9), 1143–1155.

Velot, É., Ducrocq, F., Girardeau, L., Hehn, A., Piutti, S., Kahn, C., & Arab-Tehrany, E. (2022). Hop extract anti-inflammatory effect on human chondrocytes is potentiated when encapsulated in rapeseed lecithin nanoliposomes. *International Journal of Molecular Sciences, 23*(20), 12423.

Venturelli, S., Burkard, M., Biendl, M., Lauer, U. M., Frank, J., & Busch, C. (2016). Prenylated chalcones and flavonoids for the prevention and treatment of cancer. *Nutrition, 32*(11–12), 1171–1178.

Walker, E. G., Lo, K. R., Pahl, M. C., Shin, H. S., Lang, C., Wohlers, M. W., Poppitt, S. D., Sutton, K. H., & Ingram, J. R. (2022). An extract of hops (*Humulus lupulus* L.) modulates gut peptide hormone secretion and reduces energy intake in healthy-weight men: a randomized, crossover clinical trial. *The American Journal of Clinical Nutrition, 115*(3), 925–940.

Walker, E., Lo, K., & Gopal, P. (2024). Gastrointestinal delivery of bitter hop extract reduces appetite and food cravings in healthy adult women undergoing acute fasting. *Obesity Pillars, 11*, 100117.

Weber, N., Biehler, K., Schwabe, K., Haarhaus, B., Quirin, K. W., Frank, U., Schempp, C. M., & Wölfle, U. (2019). Hop extract acts as an antioxidant with antimicrobial effects against *Propionibacterium Acnes* and *Staphylococcus Aureus*. *Molecules (Basel, Switzerland), 24*(2), 223.

Wesołowska, O., Wiśniewski, J., Środa, K., Krawczenko, A., Bielawska-Pohl, A., Paprocka, M., Duś, D., & Michalak, K. (2010). 8-Prenylnaringenin is an inhibitor of multidrug resistance-associated transporters, P-glycoprotein and MRP1. *European Journal of Pharmacology, 644*, 32–40.

Wilson, D. G. (1975). Plant remains from the Graveney boat and the early history of *Humulus lupulus* L. in W. Europe. *New Phytologist, 75*(3), 627–648.

Wohlfart, R., Hänsel, R., & Schmidt, H. (1983). The sedative-hypnotic principle of hops. 4. Pharmacology of the hop substance 2-methyl-3-buten-ol. *Planta Medica, 48*, 120–123.

Wu, C. N., Sun, L. C., Chu, Y. L., Yu, R. C., Hsieh, C. W., Hsu, H. Y., & Cheng, K. C. (2020). Bioactive compounds with anti-oxidative and anti-inflammatory activities of hop extracts. *Food Chemistry, 330*, 127244.

Yajima, H., Noguchi, T., Ikeshima, E., Shiraki, M., Kanaya, T., Tsuboyama-Kasaoka, N., Ezaki, O., Oikawa, S., & Kondo, K. (2005). Prevention of diet-induced obesity by dietary isomerized hop extract containing isohumulones, in rodents. *International Journal of Obesity, 29*, 991–997.

Yamazaki, T., Takahashi, C., Taniguchi, Y., Narukawa, M., Misaka, T., & Ano, Y. (2020). Bitter taste receptor activation by hop-derived bitter components induces gastrointestinal hormone production in enteroendocrine cells. *Biochemical and Biophysical Research Communications, 533*(4), 704–709.

Zanoli, P., & Zavatti, M. (2008). Pharmacognostic and pharmacological profile of *Humulus lupulus* L. Review. *Journal of Ethnopharmacology, 116*, 383–396.
Zanoli, P., Rivasi, M., Zavatti, M., Brusiani, F., & Baraldi, M. (2005). New insight in the neuropharmacological activity of *Humulus lupulus* L. *Journal of Ethnopharmacology, 102*, 102–106.
Zanoli, P., Zavatti, M., Rivasi, M., Brusiani, F., Losi, G., Puia, G., Avallone, R., & Baraldi, M. (2007). Evidence that the beta-acids fraction of hops reduces central GABAergic neurotransmission. *Journal of Ethnopharmacology, 109*(1), 87–92.
Zhou, D., Wang, C., Li, X., Zhao, Y., Jing, J., Ma, Y., et al. (2018). Dietary functional flavonoids as natural hepatoprotective agents against acute liver injury from hop (Humulus lupulus L.). *Journal of Functional Foods, 45*, 471–479.

Index

K. Carbone (ed.), *All About Hops: The Crop, its Cultivation, and its Uses in Brewing and Beyond*, https://doi.org/10.1007/978-3-031-96606-4

T

U

V

X

MIX
Papier aus verantwortungsvollen Quellen
Paper from responsible sources
FSC® C105338

If you have any concerns about our products,
you can contact us on
ProductSafety@springernature.com

In case Publisher is established outside the EU,
the EU authorized representative is:
Springer Nature Customer Service Center GmbH
Europaplatz 3, 69115 Heidelberg, Germany

Printed by Libri Plureos GmbH
in Hamburg, Germany